Stress, Workload, and Fatigue

HUMAN FACTORS IN TRANSPORTATION

A Series of Volumes Edited by
BARRY H. KANTOWITZ

Stress, Workload, and Fatigue

Edited by

PETER A. HANCOCK
University of Minnesota

PAULA A. DESMOND
Texas Tech University

2001

LAWRENCE ERLBAUM ASSOCIATES, PUBLISHERS
Mahwah, New Jersey   London

Lawrence Erlbaum Associates, Inc., Publishers
10 Industrial Ave.
Mahwah, NJ   07430

Cover design by Kathryn Houghtaling Lacey

Library of Congress Cataloging-in-Publication Data

Stress, workload, and fatigue / edited by Peter A. Hancock, Paula A.
    Desmond.
        p.   cm.
    Includes bibliographical references and index.
    ISBN 0–8058–3178–9
    1. Job stress. 2. Employees–Workload. 3. Fatigue.   I. Hancock,
    Peter A., 1953– . II. Desmond, Paula A.
HF5548.85.S765   2000
158.7–dc21   99–38851
                                                            CIP

Printed in the United States of America

10   9   8   7   6   5   4   3   2   1

For my mother
— P.A.H.

To the memory of Dr. Thomas Hoyes,
a dear friend and a wonderful scholar.
— P.A.D.

Contents

WORKLOAD

Theory

FATIGUE

Theory

Research

Practice

Commentary

Series Foreword

Barry H. Kantowitz
University of Michigan Transportation Research Institute

The domain of transportation is important for both practical and theoretical reasons. All of us are users of transportation systems as operators, passengers, and consumers. From a scientific viewpoint, the transportation domain offers an opportunity to create and test sophisticated models of human behavior and cognition. This series covers both practical and theoretical aspects of human factors in transportation, with an emphasis on their interaction.

The series is intended as a forum for researchers and engineers interested in how people function within transportation systems. All modes of transportation are relevant, and all human factors and ergonomic efforts that have explicit implications for transportation systems fall within the series purview. Analytic efforts are important to link theory and data. The level of analysis can be as small as one person, or international in scope. Empirical data can be from a broad range of methodologies, including laboratory research, simulator studies, test tracks, operational tests, fieldwork, design reviews, or surveys. This broad scope is intended to maximize the utility of the series for readers with diverse backgrounds.

I expect the series to be useful for professionals in the disciplines of human factors, ergonomics, transportation engineering, experimental psychology, cognitive science, sociology, and safety engineering. It is intended to appeal to the transportation specialist in industry, government, or academia, as well as the researcher in need of a testbed for new ideas about the interface between people and complex systems.

This volume achieves a major goal of the book series by demonstrating the interaction between practical and theoretical aspects of human factors. Each of the three sections is explicitly divided into four segments: theory, research, practice, and commentary. Chapters in the theory segment offer general models that explain stress, workload, or fatigue—the three main topics this book addresses. Chapters in the practice segment offer real-world examples of how theory might be applied or where additional theory might be useful. Chapters in the research segment form a transition from theory to practice. The commentaries attempt to integrate theory, research, and practice. Most of the domains examined refer to driving surface vehicles, but there are also chapters concerned with aviation and maritime systems; this demonstrates another goal of the series that transportation is a useful domain for testing models of human behavior. Forthcoming books in this series will continue this blend of practical and theoretical perspectives on transportation human factors.

Preface

The study of stress has a long and respected tradition in psychology, emanating from some of the earliest work on learning under states of compulsion. Chronic stress has also long been recognized as a precursor to illness and debilitation. Of similar vintage, fatigue has been a featured issue in industrial research since the turn of the century and the British Industrial Fatigue Research Board provided early insights which proved seminal with respect to disciplines such as occupational medicine, industrial hygiene and systems safety. In contrast, mental workload is a relatively new proposition growing out of the concerns of aviation psychologists faced with ever more loaded and taxed pilots in high-performance aircraft. A related cousin from aviation is the contemporary concern for situational awareness, a further face of performance capability now being applied in multiple contexts. Despite these divergent histories and backgrounds, it is our claim that each of these concepts, and indeed several others, share a commonalty as reflections of the energetic state of the individual involved. It is one of the major purposes of the present volume to seek out, describe, and explain these shared commonalties. To accomplish this aim, we have solicited contributions from many leading voices in these respective domains to provide insights concerning the present state-of-the-art in theory, research, and practice. We also wished to hear from outstanding theorists, researchers and practitioners whose depth of understanding allowed them to present overview commentaries on the respective areas of stress, workload and fatigue. Finally, we solicited a limited number of works which point explicitly to the potential and actual linkages across these three areas of energetic representation.

To understand and predict human performance response, we have to reach beyond the sterile, mathematical formulations of the traditional information-processing models to incorporate the emotive, affective, or more generally, the energetic aspects of cognition. These facets of response surface most readily and exert their greatest influence when the individual acts under stress, is faced by significant cognitive workload, or while in the grip of fatigue. However, energetic characteristics are pervasive and exert their vital and ubiquitous influence, even when they are not obviously in play as in the extreme circumstances noted. Indeed, one cannot hope to understand behavior without their inclusion and integration into models and theories. The human is a temporal and adaptive animal and so attempts to understand behavior which ignore or minimize such dynamic and energetic facets of behavior are doomed to failure. This text addresses such theoretical questions as one of its main thrusts. In addition to this drive for scientific understanding, there are many practical requirements in our progressively more utilitarian society which generate the need for a more fundamental understanding of this particular topic. The first is the growing necessity to operate systems around the clock. We truly do now have a "24-hour" society and its effects on behavior and performance are only now receiving the attention it deserves. A second is the growth of human-centered designs that now recognize explicitly the crucial contribution of the human operator to the efficiency of technologically redolent systems. Such systems demand an ever greater degree of prediction of operator response. At present, the sector of the scientific enterprise tasked with this need often cannot provide even an adequately acceptable approximation of response, much less accurate prediction.

Concerning stress, workload, and fatigue, fundamental and salutary questions have to be asked over our progress to date. Have we really made major, substantive steps in these respective areas in the last two decades? In respect of our state of knowledge, there are several standard works in the respective areas. In stress, we recall Broadbent's early text; and the edited volume by Hockey and the subsequent report of the influential International meeting edited by Hockey, Coles, and Gaillard have served as important landmarks. In mental workload, the book edited by Neville Moray and my own text with Najmedin Meshkati presented state-of-the-art developments to that time. In fatigue, Webb's book provided an important contribution and there have been others exploring the real-world impact of fatigue, such as Hartley's series on fatigue in driving. However, within the last decade, there are relatively few contributions with comprehensive coverage for the theorist, researcher, student, and practitioner to consult. One of our hopes for this work is that it serves this function at the turn of the decade, the century, and the millennium.

One problem that extends beyond the realm of stress, workload, and fatigue and indeed is a concern that affects all study of human behavior, centers on the contrast between information and understanding. On one hand stand committed empiricists who are convinced that we desperately need more experimental data to address the obvious holes in our present array of knowledge. On the other hand are ranged unrepentant theoreticians who are adamant that what is required are conceptual structures and frameworks to encapsulate and explain the data we already do have. Perhaps there will always be this fundamental tension. In respect of progress in behavioral research, these differing perspectives have recently been aired and Howard (1998) stated a significant concern about the focus on experimentation in commenting:

It sometimes seems that the main purpose of psychology is not to understand the mind and behavior by discovering principles and interrelating them into theories but to do experiments. Articles are seen as worthwhile only if they present new data and suggest new experiments to gather yet more data. As a result, psychologists are drowning in a sea of data that few try to pull together. I sometimes feel that an article that presents a complete, final, accurate theory of some domain will be rejected by every journal because "it suggests no avenue for further research." (p. 69)

Though not quite so vehement as Howard, we have sympathy with this position. It seems each time we turn around that there is yet another new journal or text, that we feel guilt about not having yet acquired, let alone read. And all the time the pile of experimental papers one promises to get to, grows; reprints and preprints from friends and colleagues jostle with manuscripts for obligatory review. We each have the chronic angst that major advances have been made about which we know nothing and are just about to express our profound and dismal ignorance. This is especially true in interdisciplinary areas like stress, workload, and fatigue where advances in fundamental neurophysiology (e.g., Robbins & Everitt, 1996) are ranged alongside practical applicational developments in organizational design (e.g., Theorell, 1997).

One (flight) reaction is simply to throw one's metaphorical hands up in horror and admit the inevitability of defeat. An alternative (fight) reaction is to recognize our personal limits and seek coherence and consensus through collegial collaboration. The present text seeks specifically to address this issue[1] through an integrative glance that allows for the incorporation of knowledge from the areas of stress, workload, and fatigue and also energetic constructs beyond even these broad domains. By doing so, we hope we have demonstrated that they are not distinct and separate phenomenon but, in actuality, are only different facets of the same phenomenon. We are not yet so buried in literature that we cannot acknowledge the comparable and laudable efforts of our many colleagues, several of whom kindly agreed to contribute to the present volume. It is our hope that the present work can provide illumination along the path to eventual integration and a benchmark of progress on a number of fronts at this auspicious moment in time.

—P. A. Hancock
—P. A. Desmond

---

[1] However, such an effort has the paradoxical effect of adding to the burden of information load on flooded researchers. In many ways, it is a signal to noise problem in which if both signal and noise grow proportionally with noise being larger, signal, then gets progressively harder to distinguish. One might hope that a true and useful signal will shine through, a shaky hope in a relativistic world.

PART I

Stress

THEORY

Levels of Transaction: A Cognitive Science Framework for Operator Stress

Gerald Matthews
University of Cincinnati

Perhaps the most remarkable accomplishment of research on stress and performance is the quantity of data accumulated. A recent review (Smith & Jones, 1992a) runs to three volumes, with 36 chapters on distinct stress factors. With the benefit of hindsight, it was never likely that this glut of data would submit to a single, general theory. The failure of arousal theory to do so is well documented (e.g., Neiss, 1988). In place of a general theory, contemporary research offers two key insights: first, that stress states require multidimensional description (Hockey, 1984), and second, that stress is a dynamic phenomenon associated with active, effortful attempts at adaptation (Hancock & Warm, 1989).

Hockey's (1984) cognitive state model is perhaps the most widely accepted framework for systematizing stress and performance data. Hockey's account began by demonstrating various deficiencies of traditional arousal theory as a general explanation. He advocated instead the detailed assessment of the performance consequences of individual stressors, by using a set of key indicators of cognitive function such as attentional selectivity and short-term memory. Different stressors induce different "cognitive patternings" of information-processing change. This descriptive scheme matches the contemporary zeitgeist, in which work on stressors such as noise, heat, and so on constitute rather separate research foci (apart from occasional "interaction studies"). However, the scheme does not in itself explain the data, and it provides only an outline description. Stressors such as noise may be fractionated almost indefinitely: White noise, speech, aircraft noise, traffic noise, office noise, and so forth may all have somewhat different cognitive patternings. Similarly, Hockey's performance criteria omit significant aspects of performance (vigilance, dual task, decision making), and also treat as unitary rather heterogenous constructs such as short-term memory (STM).

Broadbent (1971) proposed an influential adaptive model in which an "upper level" of control compensated for effects of task and environmental factors, so that performance change is often smaller than expected. Hancock and Warm (1989) discussed the dynamics of compensatory effortful control and suggested that both underload and overload tend to induce dynamic instability, so that performance becomes more vulnerable to environmental stressors. In such cases, there is a qualitative, and sometimes catastrophic, change in the mode of system functioning. In somewhat similar vein, Hockey (1986, 1997) distinguished several discrete control modes. In a "strain" mode, the system maintains performance under increasing demands by increasing effort, but at the cost of increased discomfort and physiological costs. Alternatively, in a passive or fatigued mode, the system lowers performance targets and reduces effort. These theories emphasize the role of the operator as an active agent, making decisions about how to *cope* with task demands. They link stress during performance to wider theories of stress and emotion as expressions of transactions between person and environment (Lazarus & Folkman, 1984).

Such explanatory frameworks are an important advance, but still leave questions open. As Hockey (1997) described, effortful control of performance is expressed in subjective experience, in cognitions of the task, in physiological response, and, under certain conditions, in objective behavior and performance. It is uncertain which of these factors indexes causal agents and which is an epiphenomenon. Sanders (1990), for example, made the biological bases for effortful control the principal causal factors, whereas Kluger and DeNisi (1996) emphasized the individual's beliefs about the task. Hancock and Warm (1989) referred both to "isomorphism between physiological action and psychological response" (p. 533) and to the "meaning sought by the individual perceiver . . . contingent upon previous experience with both task and stress" (p. 529).

In this chapter, I argue that stressors affect a variety of qualitatively different mechanisms, operating at a multiplicity of levels, ranging from single-cell response (Beatty, 1986) to complex decisions about coping with the task environment (Kluger & DeNisi, 1996). Explaining observations related to individual mechanisms provides a catalogue of minitheories, but an explicit cognitive science framework is required for a more systematic account of the different types of explanation required in stress research. A related question is whether a "cognitive patterning" is just an arbitrary collection of independent performance shifts, or whether it has some functional unity. I argue that in many cases patternings have a coherence that reflects the relational nature of "stress," as a representation of person–environment interaction. Patternings may relate to adaptation and strategy, the focus of transactional models of stress (Lazarus & Folkman, 1984). They may also relate to changes in the basic parameters controlling processing (the "cognitive architecture"), in which case the patterning may derive coherence from the underlying neural systems shaped by natural selection.

THE TRANSACTIONAL THEORY OF STRESS

Most contemporary stress researchers have accepted the view of Lazarus and Folkman (1984) that stress is a quality of transactions between person and environmental demands. These authors (p. 24) defined stress as "a relationship between the person and the environment that is appraised by the person as taxing or exceeding his or her resources and endangering his or her well-being." More recently, Lazarus (1991) suggested that stress, as a topic, should be accommodated in emotion research. Negative emotions, such as anxiety and sadness, are expressions of core relational themes, which describe the person–environment transaction. For example, anxiety represents facing an uncertain, existential threat.

The transactional model is also concerned with finer grained analysis of the cognitive processes that control the person–environment interaction. Stress reactions are influenced by evaluation or appraisal of environmental demands. Appraisals may be divided into primary (personal significance of events) and secondary (coping ability) appraisals. Some authors have distinguished different levels of appraisal differing in "automaticity" and accessibility to consciousness (Van Reekum & Scherer, 1997). Transactional theory emphasizes the person's active attempts to deal with external demands, over what may be extended periods. Lazarus and Folkman (1984) distinguished two fundamental categories of processing: *problem* or *task focused* and *emotion focused*. Task-focused coping is directed toward changing external reality, whereas emotion-focused coping aims to change the way the person feels or thinks about the source of stress. Many authors (e.g., Cox & Ferguson, 1991; Endler & Parker, 1990) have seen attempts at *avoidance* of demands as a third basic category. The transactional theory is supported by many studies demonstrating that stress outcomes relate to appraisal of environmental demands and the person's choice of coping strategy (see Zeidner & Endler, 1996).

Coping overlaps with the concept of self-regulation (Matthews, Schwean, Campbell, Saklofske, & Mohamed, 2000), which may be defined as "proximate motivational processes by which persons influence the direction, amount, and form of committed effort during task engagement" (Kanfer, 1990, p. 222). Effort may be internally directed (emotion-focus) as well as task directed. For example, progress toward personal goals may be effected by reappraisal of previous outcomes in a more positive light or by lowering

personal standards. Stress reactions are controlled by self-regulative processing constructs, including the stable knowledge structures that support self-
beliefs and motivations (Wells & Matthews, 1994), processing routines for
self-monitoring and self-evaluation (Kanfer, 1990), metacognitive beliefs
about the utility of emotion-focused coping (Cartwright-Hatton & Wells,
1997), and coping skills (Matthews & Wells, 1996). Wells and Matthews
(1994) suggested that self-regulation is organized at three levels: a lower,
automatic level that generates intrusive thoughts, an executive level that regulates coping, and schema-like self-knowledge in long-term memory.

PERFORMANCE AND THE TRANSACTIONAL MODEL

Traditional research on stress and performance largely assumes a *stimulus
definition* of stress as a set of *stressors:* noise, heat, and so forth. This approach
lends itself to experimental control and manipulation of stress, through
varying noise intensity, for example. However, in stress research generally,
stimulus definitions have been discredited for some time (Cox, 1978), because the stress reactions induced by a given stimulus vary from person to
person and occasion to occasion. For example, reactions to noise stimuli
vary with perceptions of control and utility (Jones & Davies, 1984). It is not
so much the stimulus that is important, but appraisals of the stimulus and
the success or failure of coping efforts (Lazarus & Folkman, 1984).

Transactional theory highlights the limitations of conceptualizing performance change as a fixed response to "stressful" stimuli, but it has neglected objective performance change as a stress outcome, and it is vague
about information processing. Transactional stress processes may generate
both *direct* and *indirect* effects on performance, mediated by coping and self-
regulative processes (Matthews & Wells, 1996). Direct effects follow from
task-related coping efforts, such as voluntary decisions to adopt a risky
speed–accuracy trade-off, to prioritize one of two tasks in a dual-task situation, or to monitor a particular source of stimuli. Of course, task-focused
coping does not necessarily function as intended. Indirect effects follow
from emotion-focused coping, which may initiate self-referent processing
that interferes with concurrent task processing. Such *cognitive interference*
is best known from test anxiety research (Sarason, 1988), in which performance deficits are typically characterized as a reallocation of attentional resources or working memory from the task at hand to processing internal
worries (Zeidner, 1998).

The transactional model also neglects biologically based stressor effects.
Some stressors, such as toxins, might conceiveably impair performance without any changes in the person's self-cognitions or subjective awareness and
so are beyond the scope of the transactional model. More typically, many

stressors affect both neural functioning and the person's cognitions. The transactional model then applies to the extent that the person's cognitions of the stressor and its effects on physiology drive performance change, which may or may not be the case. In practice, it is often difficult to ascertain the degree of interaction between neural functioning and cognitive stress processes and, hence, the applicability of the model.

In summary, the transactional model tells us that, in a particular experimental setting, we should ask how the individual appraises the stressor, what self-regulative challenges are posed, and how the person chooses to cope with the conjoint demands of task and stressor. The model also suggests some outline explanations for effects of the more "cognitive" stressors such as noise and anxiety on performance. Set against these theoretical strengths and its breadth of application are its lack of reference to objective performance and neglect of neurally controlled performance effects. Next, I discuss how these difficulties may stem from confusion of levels of explanation for performance effects.

A COGNITIVE SCIENCE FRAMEWORK FOR STRESS AND PERFORMANCE

According to the "classical theory of cognitive science" (Newell, 1980; Pylyshyn, 1984), any cognitive phenomenon has three complementary levels of explanation, relating to the biological hardware, the symbolic computational machinery, and the adaptive goals of the system. Usually, one or another level is most appropriate for a specific observation. Consider noise effects on performance. First (biological explanation), noise may affect neural functioning, via increased cortical arousal, for example. Second (symbol-processing explanation), noise may influence the formal characteristics of computation, by changing the system parameter of total resource availability. If noise simply diverts resources to processing the noise stimulus (distraction), there is no change in computational characteristics: All that has changed is the input to the "program." Third (adaptive explanation), noise may affect the personal meaning or significance of the task. Perhaps it is seen as a challenge to be overcome rather than an irrelevance, leading to a change in motivation or strategy.

Levels of Explanation in Stress Research

The classical theory has been applied most commonly to topics such as reasoning and language processing and so tends to neglect stress factors. Pylyshyn's (1984) explanatory framework may be modified for use in indi-

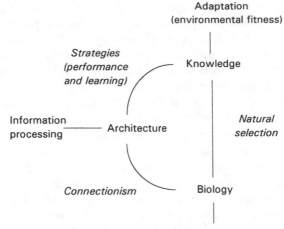

FIG. 1.1.1.   Levels of explanation for a cognitive science of
stress and personality. Adapted from "An Introduction to the
Cognitive Science of Personality and Emotion," by G. Matthews, 1997, in G. Matthews (Ed.), *Cognitive Science Perspectives on Personality and Emotion* (p. 19), Amsterdam: Elsevier.
Copyright 1997 by Elsevier. Adapted with permission.

vidual difference and stress research as shown in Fig. 1.1.1 (Matthews,
1997a). Pylyshyn's three levels of explanation are conceptualized not as a hierarchy but as a loop. Neural functioning is based on the adaptive successes
and failures of previous generations: People whose brains are equipped for
successfully handling stressful situations are more likely to pass on their
genes to their offspring.

Pylyshyn insisted that computations are necessarily symbolic. At the symbol level, he distinguished the algorithm as an abstract program description
from the functional architecture that actually implements the algorithm in
real time, by using basic resources, memory space, and so forth. In the stress
context, the functional architecture is of more interest; it is unlikely that, for
example, stress affects Chomskyian rules for grammatical transformation
(at all). In addition, the commitment to symbolic processing may be premature. The subsymbolic models provided by connectionism are becoming
an increasingly important tool in modeling relations between affect and
cognition (e.g., Matthews & Harley, 1996; Siegle & Ingram, 1997). There are
theoretical grounds too for associating emotion with nonpropositional
codes (Oatley & Johnson-Laird, 1987). I use the term *cognitive architecture* to
refer to the formal parameters of computation, without commitment to a
particular conception of processing operations.

The adaptive level is variously described as referring to meaning, knowledge (of the goal-relevance of actions), intentionality, and rationality, in the sense of optimizing behavior for goal attainment (Anderson, 1990). Two senses of "adaptation" should be distinguished. The Lazarus transactional model is concerned, for the most part, with adaptive processes that are accessible to consciousness and can be verbalized. Adaptation may also refer to environmental fitness caused by natural selection. Both kinds of explanation are covered by the "adaptive level," but there is no simple relation between them. I emphasize adaptation in the sense used by Lazarus, which I believe has made better progress in providing falsifiable hypotheses than has evolutionary psychology.

Explanatory Limitations of Processing Models

Within the stress literature, there are models at each of the three levels. For example, anxiety may be described neurologically (Gray, 1982), as biases in processing stages (Williams, Watts, MacLeod, & Matthews, 1988), or as an adaptive theme (Lazarus, 1991). Information-processing models predominate in contemporary research, because of their capacity to explain the task-dependence of stressor effects (cognitive patterning). However, they are of limited explanatory value, as I now discuss, using resource theories of stressor effects as an example.

Resources may be defined as a metaphorical pool of energy for processing, which can be allocated flexibly and in graded fashion to a variety of different computations. Interactions between stressors and task demands support models of resource availability and allocation. Various stressors, including noise (Fisher, 1986), heat (Hancock, 1986), anxiety (Zeidner, 1998), subjective tiredness (Matthews & Davies, 1998), and prolonged work (Warm, 1993) tend to impair performance most reliably when the task is attentionally demanding. Stressor effects generalize across a variety of qualitatively different tasks, implying that they influence a general resource, rather than a single critical process (Matthews & Davies, 1998).

Resource theories are scientifically valuable as a reasonably successful basis for predicting stressor effects on novel tasks, when other factors specific to the stressor are taken into account. However, they are ambiguous at an explanatory level. There are at least three different bases for changes in the "functional resources" available for processing:

1. *Change in parameters of the architecture.* The stressor may actually change the total quantity of resources available (Humphreys & Revelle, 1984). Architecture may be investigated without reference to biology. However, a change in the architecture is perhaps more credible when it can be linked to changes in biological and/or neural functioning. Hancock (1986), for example, showed that resource loss from thermal stress is contingent on breakdown of

thermoregulation and established a strong correspondence between biological and cognitive functioning.

2. *Change in task demands.* As Näätänen (1973) observed in an arousal theory context, stressors tend to impose distracting internal stimuli (discomfort) or external stimuli (e.g., noise). Processing distraction may interfere with task-related performance. In this case, it is not the architecture that changes, but the nature of the task. The issue is which components of the architecture tend to be overloaded by the additional information. Where the distracting stimuli are appraised as threatening or uncontrollable, stress-mediated effects, such as changes in perceptions of the task (Jones & Davies, 1984) should be distinguished from changes caused by processing of additional stimuli.

3. *Changes in strategy.* Methods for testing resource theories successfully discriminate loss of resources from changes in strategic allocation of resources across different task components (Matthews & Margetts, 1991). However, there may be a general withdrawal of resources from the task because the person chooses to focus attention on internal thoughts and worries (Sarason, 1988), as a consequence of coping through emotion-focus. Such strategic choices require explanation at the adaptive level.

Hence, we can have radically different explanations for similar stress × task demand interactions. Stressor effects on specific processes are often similarly ambiguous, given that most processing, even of an "automatic" nature, tends to be sensitive to strategy change (Cohen, Dunbar, & McClelland, 1990). Distinguishing explanations requires a stronger focus on the constructs that bridge the explanatory levels. Connectionist models may be seen as both "neural nets" and as subsymbolic information-processing models. Similarly, strategies bridge the architectural and adaptive levels. A strategy represents both a sequence or mode of organization of computations and the person's goals and intentions. In the sections that follow, I look, first, at the interface between neural function and the cognitive architecture and, second, at the interplay between the architecture and the person's efforts at adaptation.

BRIDGING BIOLOGICAL AND COGNITIVE-ARCHITECTURAL LEVELS OF EXPLANATION

Limitations of Neuroscience

The progress made by cognitive neuroscience in describing the neural bases for specific processing functions indicates the potential importance of neurological explanations for stressor effects (e.g., Derryberry & Reed, 1997).

Historically, biological approaches have tended to promise more than they have eventually delivered. For example, performance data have frequently been interpreted in terms of arousal theory and the Yerkes–Dodson Law (Broadbent, 1971), but the deficiencies of this approach are well known and need not be reiterated here (see Hockey, 1984; Matthews & Amelang, 1993). Nevertheless, biologically based research can show two solid accomplishments. First, a variety of biological agents influence performance, including drugs, viruses, and biological rhythms (see Smith & Jones, 1992a). Such studies implicate a variety of independent physiological systems, such as circuits controlled by a particular neurotransmitter, circadian oscillators, and hormones. Second, although the problems of psychophysiology overlap with those of arousal theory, psychophysiological measures are sometimes enlightening about the nature of performance change, as indexes of compensatory effort, for example (Hockey, 1997).

However, psychobiological accounts of stressor effects have limited explanatory power. Consider caffeine effects (see Lieberman, 1992, for a review). Through double-blind studies, we can obtain a cognitive patterning for caffeine, such as its facilitative effect on demanding attentional tasks. Quite a lot is known about the pharmacology of caffeine. Its blockade of adenosine receptors leads to central nervous system arousal, which may in turn influence processing constructs such as attentional resources. However, the explanation is not so simple. Despite its methodological strengths, caffeine research illustrates various problems of biologically based research:

We do not know which specific neural systems mediate biological stressors. According to Daly, Shi, Nikodijevic, and Jacobson (1994, p. 202) "[C]affeine, through blockade of adenosine receptors, would be expected to indirectly influence the function of most neuronal pathways in the brain." Caffeine affects the release of a wide variety of neurotransmitters, including noradrenaline, dopamine, serotonin, acetylcholine, glutamate, and γ aminobutyric acid (GABA). Even if we could isolate a particular neurotransmitter, it is likely to contribute to interacting multiple systems, with differing consequences for performance. For example, Robbins (1986) described dorsal and ventral noradrenergic systems supporting different behavioral functions: A noradrenergic agonist might influence performance through either or both systems. In practice, much psychobiological performance research is based on a narrow empiricism that focuses on tasks, rather than on the neural and/or cognitive mechanisms intervening between experimental manipulation and performance. Neurological methods manipulating performance, such as drug administration, often produce such complex and contingent effects that it is hard to infer mechanisms.

We do not know how neural response is moderated by real-life experiences. Caffeine is an "ecological" stressor to which people have differing experiences. Animal research shows that chronic caffeine intake influences the density of a variety of receptor types (Daly et al., 1994). In humans, chronic caffeine use may affect both performance and physiological responses (Smith, 1994).

We do not have good biological models of personality and individual difference factors. The effects of most stressors on psychophysiology and performance are moderated by personality factors (e.g., Humphreys & Revelle, 1984). Caffeine increases autonomic arousal in extraverts, but actually decreases tonic and phasic electrodermal activity in introverts (B. D. Smith, 1983). It shows complex interactions with extraversion/impulsivity in its effects on performance (Humphreys & Revelle, 1984), but biological explanations for personality effects on performance have numerous shortcomings (Matthews, 1997b).

We do not have good control over cognition. Experimental control is better for drugs than for most other biological agents, but the double-blind design may not control for cognitive reactions (Rohsenow & Marlatt, 1980). In the case of caffeine, Kirsch and Weixel (1988) showed that decaffeinated coffee may induce substantial changes in blood pressure and alertness. Even in a double-blind design, subjects' responses to the decaffeinated beverage were controlled by cues that influenced their beliefs, such as the amount of coffee taken from the jar. Kirsch and Weixel showed that beliefs about coffee effects predicted performance change; Fillmore and Vogel-Sprott (1992) produced performance change by manipulating expectancy alone.

There are also conceptual and methodological problems associated with trying to explain cognitive-architecture–level phenomena in terms of neural constructs. The neural implementation constrains processing, but architecture-level constructs cannot necessarily be reduced to neural-level constructs. Neurological methods, however sophisticated, do not reveal the nature of the abstracted codes that support processing any more than probing a computer with a voltmeter gives us a description of a symbolic programming language such as C++ (see Pylyshyn, 1984). Localization of the neural systems that mediate stressor effects is important but insufficient: We need an understanding of how those systems represent information-processing constructs.

Connectionism and the Biological-Cognitive Interface

The best way to connect cognitive and neural levels of explanation is through neural network or connectionist models. Network models may account for performance on a variety of ostensibly different tasks, and they are well suited to accommodating the complexities associated with distinct but interacting systems or modules. Some stressors, notably drugs, may act directly on network function, whereas others act via some other agent, such as amino acids in the case of foodstuffs. Modeling requires, first, a specification of the neural nets influencing performance, and, second, a specification of how stressors influence the neural nets. However, bridging the conceptual divide between neural and cognitive explanations remains problematic, because of the important differences between typical connectionist models and actual neural functioning (e.g., Smolensky, 1988). Anderson (1990) claimed that such models should be seen as descriptions in terms of cognitive architecture rather than of neural hardware. More generally, we should view connectionism as a modeling tool rather than a commitment to a particular level of description.

The "bridge" between the two levels of explanation thus is made up of two piers yet to meet. The pier built from biology includes simulations based on what is known about neuroanatomy. For example, Banquet et al.'s (1997) model of memory is comprised of an overall neural architecture integrating cortical and limbic circuits, together with a detailed connectionist simulation of hippocampal learning of spatial and temporal information. The network model is based on the neurophysiology of the dentate gyrus and brain area CA3 (see also Gray, 1982). It is primarily concerned with memory, but it also has the potential for modeling arousal and stress effects. The connectionist part of the model incorporates cholinergic modulation of learning, and the more general model describes top-down influences of emotion and motivation controlled by pathways from the prefrontal cortex to the hippocampus (entorhinal cortex). It is relatively straightforward to adapt such a model to predict the effects of, for example, emotions and cholinergic agonists on certain memory tasks. Simulation of the representation of memories in the prefrontal cortex might then succeed in attaching the neural descriptions that the model currently offers to the subsymbolic cognitive description provided by appropriate connectionist models, completing the bridge.

The cognitive architectural pier focuses on modeling performance data and working downward to neural descriptions. Cohen and Servan-Schreiber (1992) described connectionist models for attentional tasks including the continuous performance test and the Stroop test. In contrast to Banquet et al. (1997), the simulations were based on theories of cognitive architecture, not neuroanatomy. Cohen and Servan-Schreiber (1992) showed that schizophrenics' deficits in attention may be modeled by varying the gain parameter of the network. They suggested that this cognitive architectural parameter is controlled by dopaminergic modulation of the prefrontal cortex. Further simulation, based on the neuronatomy of the prefrontal cortex and other structures implicated in attention, might, in principle, complete the bridge to the neural level of description.

In summary, biological approaches to stress and performance have failed to develop their considerable potential, for both methodological and conceptual reasons. Cognitive science provides a framework for deciding on the level of explanation appropriate for a particular research problem. In some cases, such as drug effects on single neuron functioning (Beatty, 1986), a purely neurological explanation may suffice. More typically such accounts do not explain the variation of stressor effects on performance with task parameters. The optimistic view is that we can identify mutually compatible cognitive-architectural and neural descriptions through connectionism. Whether this goal is attainable in practice remains to be seen. It may turn out that the networks that best represent neuroanatomy and the networks that best represent cognitive models are simply too far apart for compatibility of description. However, at least for relatively simple tasks, current research provides grounds for optimism.

A final point is that an exclusive focus on the cognitive consequences of neural functioning leads to a neglect of the relational aspects of biological responses. Neural influences on architecture are presumably selected by evolution to prepare the organism for information processing that meets the needs of the transaction, so that neural activity may be seen as an index of a biological transaction between organism and environment. I return to the relational significance of the neural-architectural interface in a later section.

THE COGNITIVE ARCHITECTURE, ADAPTATION, AND STRATEGY

Standard information-processing models are often ambiguous over the role of strategy. The effect of a stressor on a particular process might follow from either a change in the architecture or from strategic control of the process. Most researchers probably have some working conception of processing that distinguishes automatic from controlled or executive processes. Such distinctions are important but insufficient. On the one hand, it is becoming increasingly clear that automatic processing is modified by strategic control, even in such seemingly strategy-free contexts as processing of subliminal stimuli (Dagenbach, Carr, & Wilhelmson, 1989) and the standard Stroop test (Besner, Stoltz, & Boutilier, 1997). Furthermore, automaticity may be a consequence of learning rather than of architecture (Matthews & Harley, 1996). If noise influences an automatic process, is the effect due to a transient change in architecture or to the person's previous learning in noisy environments?

On the other hand, variation in strategic processing does not necessarily reflect variation in strategy *choice*, because, with multiple levels of attentional control, strategy operates indirectly on behavior, through biasing lower level processing (e.g., Norman & Shallice, 1985). For example, Rabbitt's (e.g., 1989) studies of speed–accuracy trade-off indicated that strategic control is overlaid on involuntary variation in speed in the overall reaction time band, such that strategic intervention is initiated when involuntary increase in speed triggers an error. An effect of stress on speed–accuracy trade-off might reflect either involuntary variation in time-to-completion of process-ing (a consequence of architecture) or strategy for error handling. Strategy change may be a secondary consequence of architectural change. Broad-bent (1978) discussed how several apparently independent consequences of noise, such as increased selectivity in processing, may all reflect the person's strategic attempts to deal with loss of capacity, although, equally, selectivity change may sometimes be involuntary.

"Strategy" is useful as a bridging construct only if we can define and coor-dinate the differing perspectives on strategy afforded by the architectural and adaptational levels of explanation. At the cognitive-architectural level, a strategy is simply a description of (qualitatively or quantatively) different sequences of processing performed on a given set of inputs, supported by the same underlying architecture. For example, we can model linguistic and spatial strategies on the picture-sentence verification task from performance data (Hunt & McLeod, 1978), without reference to intentionality. The pro-cessing supporting strategy choice may be modeled through decomposi-tion of executive processing (Shallice, 1988) or through activation of "goal" or "task demand" units (Cohen et al., 1990). However, such models do not provide the rational principles for the person's strategy use guiding self-regulation and coping (Matthews & Wells, 1996). At the adaptive level of explanation, strategies are conceptualized as the choice of processing se-quences or organization to meet adaptive goals in a given environment. In principle, we should be able to integrate (and distinguish) the two levels of description by describing the same strategy in both processing and adaptive terms. Research on (1) anxiety and (2) environmental stressors indicates how this may be done.

Strategy and Architecture:
Levels of Explanation for Anxiety Effects

Anxiety is a stress factor defined by response rather than stimulus. Like other stress factors, though, it is associated with quite a complex cognitive patterning (Eysenck, 1992), which includes both a deficit in resources or working memory and a bias in selective attention that prioritizes threat stim-uli. Attempts at applying arousal theory to the data have been largely unsuc-cessful, because worry rather than emotional arousal is more predictive of performance impairment (Mueller, 1992). Cognitive neuroscience models are rather more successful. Derryberry and Reed (1997) have shown that anxiety relates to at least two neural attentional systems, controlling slower disengagement from threat stimuli and faster focusing on local threat infor-mation. At one level, anxiety can be investigated as a biocognitive response.

However, most recent research effort has been applied to development of conventional information-processing models for anxiety effects. Such

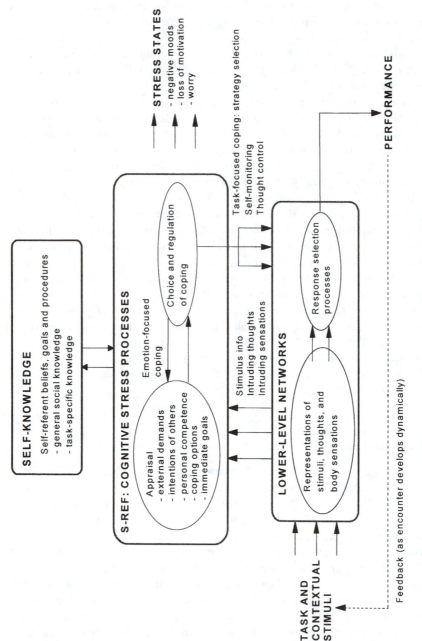

FIG. 1.1.2.   The S-REF architecture for stress and performance.

18

models may emphasize either inputs to the system or change in architecture. In Bower's (1981) network theory, anxiety may be seen (most simply) as activation of an emotion node. Performance in an anxiety state reflects the concurrent activation of threat- and task-related nodes. Alternatively, anxiety may influence not just activation but the underlying parameters that govern network response to input. Williams et al. (1988) proposed an architectural hypothesis of this general kind. They proposed that motivationally significant stimuli undergo a series of analyses controlled by discrete processing stages. Preattentive processing assigns a threat value to the stimulus and allocates resources to it. State and trait anxiety biases operation of these stages. Such ideas have been influential in developing a picture of anxiety as the outward sign of a bias in the cognitive architecture. Trait-anxious individuals are prone to stress and state anxiety because their preattentive processing signals that the world is a threatening place (MacLeod & Hagen, 1992).

In contrast, Matthews and Wells (1999) claimed that experimental and clinical studies of anxiety and attentional bias demonstrate the central role of strategic processes. Bias induced by a prime or cue is found only when the time lag between prime and "target" stimulus is sufficient for strategic processing (e.g., Calvo & Castillo, 1997). Bias is sensitive to contextual factors influencing expectancy, such as trial blocking (Richards, French, Johnson, Naparstek, & Williams, 1992). Clinical insights into anxiety show the importance of self-beliefs and their guidance of active self-regulation. Beck (1987) described the nature of schemas in anxiety, which shape people's awareness of threat and their attempts to cope with threat perceptions. Threat is attributed not just to external stimuli, but also to internal thoughts and bodily sensations that intrude into consciousness, seemingly autonomously. In other words, anxious and nonanxious individuals may use the same architecture to implement differing strategies.

A Cognitive Science of Anxiety: The S-REF Model

Wells and Matthews (1994) developed the Self-Regulative Executive Function (S-REF) model of anxiety. In cognitive science terms, the model attempts to specify some of the important aspects of both the cognitive architecture and adaptation (Matthews & Wells, 1999). A three-level architecture is outlined, including (a) a lower, "automatic" level, which generates intruding thoughts spontaneously; (b) a supervisory executive (the S-REF), whose functioning is resource limited; and (c) a library of items of self-knowledge in long-term memory (LTM). In response to threatening stimuli, the S-REF retrieves generic plans for coping from LTM and modifies them to handle the immediate situation. Figure 1.1.2 outlines the components of the architecture, although it is not intended to model executive control of lower level functioning in detail. Self-knowledge is both a component of the architecture, as representations of generic action plans or schemas, and an adaptive-level construct, referring to self-regulative beliefs about the use of coping options to support personal motivations.

The figure does not show in detail the various pathways through which the S-REF may influence lower level processing. Task-focused coping may be directed toward both earlier and later stages of lower level processing, including qualitative choices of processing routine and allocation of attention and effort (see Norman & Shallice, 1985). People may choose to monitor internal representations, for threat content, for example, or they may attempt to suppress thoughts and sensations.

Anxiety relates to the content of various plans held as self-knowledge, so that strategy choice, rather than individual differences in architecture, mediates anxiety effects on performance. Anxious individuals are more likely to recycle information in the S-REF, coping through the emotion-focused strategy of perseverative worry or rumination, which leads to cognitive interference and performance deficit. Likewise, anxiety is associated with monitoring for threat and selective attention bias, an instance of task-focused coping, in which the "task" reflects personal goals, rather than instructions given in an experimental setting. Anxiety may also sometimes affect another task-focused strategy of increasing effort directed to the experimental task, as an attempt to compensate for other, maladaptive strategies (Eysenck & Calvo, 1992). These strategy effects may be modeled at the level of the cognitive architecture, through processing models. For example, Matthews and Harley (1996) presented a connectionist model of anxiety, which shows how strategically controlled threat monitoring might be implemented.

At the adaptive level, choice of coping reflects people's self-regulative goals and their beliefs about how those goals may be achieved. Coping strategies may be seen as rational attempts at adaptation to the external and internal environment, although, in stress-vulnerable individuals, such attempts are often based on faulty appraisals. Anxiety relates to various biases in self-regulation, such as a general tendency toward larger discrepancy between actual and ideal status (Matthews, Schwean, et al., 2000), and to activation of specific discrepancies such as threats to self-preservation goals (Oatley & Johnson-Laird, 1987). Coping also reflects metacognitive goals, in that anxious individuals tend to place particular importance on controlling their internal thoughts, an influence on the use of emotion-focused coping and worry (Cartwright-Hatton & Wells, 1997). Explaining anxiety effects on performance requires an understanding of both the motivations of the anxious individual (adaptive level) and computational models of how those motivations are expressed in strategy choice (cognitive architecture level).

Environmental Stressors and Strategy

Coping strategy may also be an important component of response to environmental stressors. Noise has been characterized as strengthening the use of the dominant strategy (Smith & Jones, 1992b), which may be a form of task-focused coping. S. Cohen's (1980) classic studies of noise aftereffects demonstrated performance impairment associated with passivity and reluctance to exert effort, apparently a preference for avoidance coping rather than task-focused coping. Noise aftereffects are perhaps an instance of a more general tendency for fatigue to induce reluctance to exert effort (Holding, 1983). Jones (1984) suggested that noise tends to induce appraisals of loss of competence and various strategic attempts to maintain competence, which may be abandoned if the task is prolonged (as in aftereffect studies).

Fatigue effects demonstrate the importance of investigating mechanisms rather than tasks. In general, task-induced fatigue effects tend to be more detrimental when the task is attentionally demanding (e.g., Warm, 1993), implying a loss of functional resources. However, contrary to resource theory, Desmond and Matthews (1997) found that drivers are able to compensate behaviorally for task-induced fatigue when task demands are high (in simulated driving). Performance appears to break down only when the task is relatively easy (straight-road driving). The effect is attributed to failure to apply sufficient task-focused coping because (a) subjects report less coping, and (b) steering reversal data show that active control of the task is diminished. Drivers also report reduced active coping following long real-world drives (Desmond, 1997). Speculatively, loss of active coping may be attributed to subjects' beliefs that the goal of accuracy of lateral tracking is less important when the road is straight, and they can address the conflicting goal of reducing discomfort by reducing effort (see chap. 1.5, this volume). Other fatigue manipulations, such as sleep deprivation, might also influence architectural properties. At the least, however, demonstrating such effects requires control for changes in coping.

The Role of Knowledge and Contextualization

The self-knowledge component of the S-REF model is *contextualized,* in that it represents personal experience of specific types of context and situation. Test anxiety results not just from a general tendency toward negative self-appraisals, but also from past experience of the relevant context, such as failure to study effectively (Mueller, 1992). Modeling of longitudinal data shows that low achievement at one school grade predicts lower expectancy at the next school grade, which in turn predicts anxiety (Pekrun, 1992). Matthews (chap. 1.5, this volume) discusses how driver stress is not just a general expression of the individual's stress vulnerability, but instead reflects the driver's beliefs specifically about the driving context and how to cope with it.

Knowledge has been largely neglected in research on stress and perform-ance, perhaps because experimental control of stress generates the illusion that the person comes to the experiment as a tabula rasa. This is not the case, especially for the stressors of most practical significance: those operat-ing in the real world. As already indicated, people bring to the laboratory beliefs about agents such as caffeine (Kirsch & Weixel, 1988) and noise (Jones & Davies, 1984). Such beliefs may or may not influence performance on a given task. However, accommodating the role of prior knowledge re-quires an understanding of how beliefs influence coping in laboratory envi-ronments and how the coping strategies concerned influence information processing. Hence, laboratory studies call for conceptual clarity and careful design in distinguishing architectural change from the subject's concerns about the personal significance of the performance environment.

FITTING THE PIECES TOGETHER: LEVELS OF TRANSACTION

So far, this chapter has focused mainly on explanations for discrete re-search findings: the effect of a stressor on a particular task, for instance. We have seen that such findings may be interpreted in radically different ways. Some stressor effects should not be seen as relating to "stress" at all; they simply reflect changes in task demands handled routinely by the cognitive architecture. Changes in the way in which the person processes information may be attributed either to changes in parameters of the cognitive archi-tecture, perhaps attributable to changes in neural function, or to strategy changes reflecting transactional stress processes associated with appraisal and coping.

Next, I turn to the further problem of the piecemeal dissection of stressor effects to which contemporary stress research lends itself. This approach is descriptively essential, but it fails to provide a satisfactory basis for either inter-relating effects of different stressors (we can multiply stressors and tasks indefinitely) or for explaining the cognitive patterning of the stressor as a whole. Hockey's (1984) account of different modes of control of pro-cessing offers a partial solution. However, the different modes have not been fully operationalized empirically; it is not always clear whether given modes influence strategy or architecture or both, and it is unclear which of the at-tributes of processing modes are causal influences on cognition and per-formance. In this section, I argue that patternings of information-processing change under stress reflect two qualitatively different levels of transaction: the biocognitive and the cognitive adaptive. In both cases, the key point is that independent changes in processing function may be linked to a common transactional theme. At the biology–architecture interface, the transaction reflects the person's inbuilt neurological capacities for handling evolutionarily significant demands of different kinds. At the knowledge–architecture interface, the transaction reflects the key relational themes describing the person's orientation to task demands. Further questions arise about the alignment of the levels.

The Biocognitive Transaction

The relational basis for neural influences on architecture is most easily illustrated with reference to animal research. Gray's (1982) behavioral inhibition model of anxiety describes several outputs (attention, arousal, inhibition), which are neurologically distinct, but functionally coordinated. They subserve a common function of interrupting ongoing behavior and preparing the organism for dealing with environmental attributes such as threat and novelty. Artificial intelligence (AI) researchers have independently arrived at similar analyses of the utility of anxiety (e.g., Simon, 1967). Hence, "neural anxiety" represents a description of the relation between an organism and its representation of the environment. It filters environmental attributes for current relevance (punishment signals, innate fear stimuli, etc.) and biases the organism toward certain ways of acting on the environment (passive avoidance). Gray's (1982) theory has various difficulties in explaining the human anxiety data (Matthews & Gilliland, 1999), but it provides a template for conceptualizing stress-induced biocognitive changes as aspects of a functional system. The biocognitive description of preparedness for threat leads naturally into concerns with its adaptive value in the evolutionary sense, although, in humans, such an analysis probably requires reference to volition, personal meaning, and strategic control of behavior, that is, adaptation in the sense used by Lazarus.

The cognitive patternings of biologically based stressors such as drugs may show functional coherence, if we can distinguish effects on architecture from effects on motivation and strategy. Such an approach works only to the extent that the biological agent may be mapped onto a small number of underlying functional systems, and it is still an open question whether drug effects are "clean" enough for this to be done. Plausibly, the common effects of stimulants might be explained in terms of a dopaminergic reward system (Depue, 1995), for example. Perhaps the cognitive patterning of caffeine facilitates reward-related activity, and its diverse effects on performance reflect this common theme. On the other hand, if the agent affects many systems, a coherent account of performance change may not be attainable. The best that can be done in this case is a descriptive account of the influence on architecture, which may still be practically valuable.

Another application is provided by Thayer's (1989, 1996) concept of bioenergetic systems, expressed at the subjective level by energetic arousal and tense arousal. These systems support preparedness for vigorous motor activity and reacting to threat, respectively. Various effects of energy on performance have been demonstrated (see Matthews & Davies, 1998, for a review). Energy tends to enhance attention, but its effects are moderated by task parameters such as display size in visual search tasks and stimulus degradation and memory load in vigilance. Energy is consistently related to performance efficiency on demanding attentional tasks, but not to strategy (e.g., Matthews & Westerman, 1994), implying that energy correlates with some parameter of the cognitive architecture. Matthews and Davies (1998) developed the hypothesis that energy is related to availability of resources for visual attention. Energy may relate to other components of the architecture, such as those supporting verbal learning and reaction time (Thayer, 1978). Such correlates may contribute to the functionality associated with energy, over and above the contribution made by enhancement of attentional resources.

Energy probably reflects the coordinated operation of several neural systems (Thayer, 1989, 1996). The performance consequences of high energy identified by Matthews and Davies (1998) parallel those of arousal assessed psychophysiologically (Munro, Dawson, Schell, & Sakai, 1987) and manipulated pharmacologically (e.g., Rohrbaugh et al., 1988), supporting a biocognitive interpretation of the data. Matthews and Davies (1998) speculated that subjective energy may be a marker for a dopaminergic attentional system. It remains to be seen whether energy effects may be directly linked to neural systems by connectionist modeling. If not, energy may still be identified with an integrated biocognitive response, described at the cognitive architecture level, even if a detailed account of its neural underpinnings is difficult to achieve.

The Cognitive-Adaptive Transaction

The identification of the adaptive factors around which strategy changes cohere is relatively easy because, with some attention to methods, we can use self-report measures to investigate a person's goals and intentions. (The difficult part of the enterprise is mapping self-report onto processing.) The Wells and Matthews (1994) S-REF model describes how effects of anxiety and other negative emotions follow from the person's self-regulatory goals and associated self-beliefs. What is perhaps missing from current research is an account of the main adaptive challenges faced by individuals without clinical pathology in dealing with real-world and laboratory performance tasks. In this section, I describe some recent work that discriminates three different aspects of adaptation and relates it to "state" constructs.

TABLE 1.1.1

Three Cognitive-Adaptive Syndromes in Task-Induced Stress
and Their Associations With Appraisal and Coping Measures

	Task engagement	Distress	Worry
Principal scales	Energetic arousal	Tense arousal	Self-consciousness
	Motivation	Low hedonic tone	Low self-esteem
	Concentration	Low perceived control	Worries about the task
			Worries about personal concerns
Appraisals	High demands	High overall workload	Self-appraisals (?)
	High effort	Threat	
Coping	Task focus	Emotion focus	Emotion focus
			Avoidance

Note. Adapted from "Validation of a Comprehensive Stress Questionnaire," by G. Matthews et al., 1999, in I. Mervielde, I. J. Deary, F. De Fruyt, and F. Ostendorf (Eds.), *Personality Psychology in Europe* (Vol. 7), Tilburg: Tilburg University Press.

Matthews et al. (1999) identified 10 state dimensions through item factor analysis, in the traditional domains of affect (mood), conation (motivation), and cognition. A second-order factor analysis of the 10 factorial scales identified three broad stress syndromes, task engagement, distress, and worry, summarized in Table 1.1.1. These syndromes are systematically related to the cognitive process variables described by the Lazarus transactional model of stress. Studies of occupational stress and of test anxiety showed that subjects' states in a controlled performance context were predictable from their context-linked appraisals and coping strategies. Personnel who reported using emotion-focus and avoidance strategies to deal with high workload showed relatively high levels of distress and worry following performance of a demanding working memory task in the workplace. Thayer's (1996) analysis of basic moods suggests that they are output from the biocognitive transaction, but they also participate in wider cognitive-adaptive syndromes, in which affect is enmeshed with conscious cognitions and motivation.

Further evidence has been obtained by measuring state and stress processes concurrently. In several of the studies reviewed by Matthews et al. (1999), workload was measured by the NASA-TLX (Hart & Staveland, 1988). Although the NASA-TLX is used in human factors research to assess properties of tasks, it actually requires respondents to appraise the task and their reactions to it. As such, responses are open to the same biases as any other appraisals (see Wells & Matthews, 1994), as shown by their variation with subject mood (Matthews & Westerman, 1994). Across several studies ($N = 567$), Matthews, Joyner, et al. showed that distress correlated at .57 with overall workload. Task engagement was more weakly related to overall workload ($r = .21$) but more strongly related to workload patterning: high mental demands and effort, but good performance. Workload correlates of worry were of small magnitude; the syndrome probably relates more to self-appraisal than to task appraisal. Hence, stress state reflects appraisal of task demands.

In a recent, unpublished study (see Matthews, Schwean, et al., 2000, for a summary), Sian Campbell and I assessed subjects' appraisal of a demanding "rapid information-processing" task, by using the Lazarus dimensions. We also developed a coping questionnaire specifically geared to the performance context, assessing dimensions of task focus, emotion focus, and avoidance, as previously defined (Matthews & Campbell, 1998). Regression analyses showed that about 50% to 60% of the variance in state measures was predictable from appraisal and coping. Distress appeared to represent an integration of process measures: high threat appraisal and use of emotion focus in preference to task focus and avoidance. The other two stress state dimensions related to coping only: engagement to high task focus/low avoidance and worry to high emotion focus/high avoidance.

Matthews et al. (1999) inferred that stress states relate to the principal transactional themes associated with performance environments. Engagement relates to commitment of effort to the task. High engagement is a property of tasks that force effort out of the subject, such as speeded working memory tasks, and of individuals willing to exert effort, such as those with personality traits of conscientiousness and self-efficacy. Distress relates to overload and making the best of circumstances under which successful performance is perceived as unattainable. It is higher when task demands are high and when individuals are neurotic. Worry relates to self-regulation, when the nature of the transaction calls for people to pull back mentally from a task and consider their personal motives and attitudes.

These adaptive level state constructs may be expressed in strategy change demonstrated through objective performance measures. Matthews (chap. 1.5, this volume) discusses how changes in simulated driving consequent on stress and fatigue manipulations corresponded to subjective change. For example, fatigue was associated with subjective disengagement, reduced task-focused coping, and, on some measures, performance impairment (Desmond & Matthews, 1997). Recent studies (see Matthews & Campbell, 1998) suggested that performance tends to be enhanced by task focus and impaired by avoidance and emotion focus, although, as with all stress factors, associations between coping and performance vary across task components. Assessment of stress states allows stressors to be grouped together on the basis of commonality of state change. For example, driving fatigue, noise aftereffects, and sleep deprivation may all operate as "disengagement" stressors. Such stressors should induce common styles of coping and patterns of performance change.

Alignment of Levels of Transaction

The final element in the present analysis of stress is the inter-relation between the two levels of transaction. When straightforward opportunities or threats arise, the relational themes of the two levels are probably aligned, such that external circumstances elicit biocognitive preparedness congruent with the person's cognitions of the situation. In other circumstances, there may be conflict. Many laboratory studies contrive to set the two levels against each other. Pay a person a small fee to perform a task when sleep deprived, and performance may reflect the outcome of a conflict between the rest-seeking tendencies dominating the biocognitive transaction (e.g., microsleeps and attentional lapses) and the intention to perform well. Conversely, subjects performing on the driving simulator sometimes report finding themselves involuntarily drawn into involvement with the task, despite their awareness of the artificiality of the situation. Task stimuli appear to elicit biocognitive reactions more appropriate to the real-life environment. Speculatively, the function of moods such as energy and tension may be to communicate the content of the biocognitive transaction to the adaptive level, to assist "top-down" conflict resolution. Hockey's (1984) control modes may describe the configuration of the two types of transaction: "Strain mode" perhaps refers to voluntary compensation for an inadequate biocognitive reaction to the task. According to the present view, there may be separate effects on performance of the two levels of transaction.

Over longer time intervals, there may be synergistic relations between biocognitive dispositions and acquired skills for handling demanding environments. Personality traits such as extraversion-introversion may reflect stable adaptations to specialized environments (see Matthews, 1999, for a detailed account), as shown in Fig. 1.1.3. Extraversion is a partly inherited trait (Loehlin, 1992), so we can envisage heredity as providing extraverts with a distinctive cognitive architecture. Empirically, extraversion seems to be associated with a cognitive patterning including a variety of distinct processing strengths and weaknesses (Matthews, 1997b). Extraverts tend to perform better on verbally coded dual tasks, demanding STM tasks, and memory retrieval tasks, whereas introverts have an advantage in vigilance and in reflective problem solving. This cluster of correlates of extraversion may prepare the extravert for acquiring skills such as rapid verbal responding and handling demanding social interactions, which are adaptive in the environments favored by extraverts, such as parties and high-pressure occupations. Conversely, introverts are cognitively adapted to solitary, low-information environments. Effects of extraversion on strategy, interests and motivation (see Matthews, 1999) indicate that the trait is also associated with adaptive qualities that overlay and interact with its architectural properties. Other traits may be analyzed similarly.

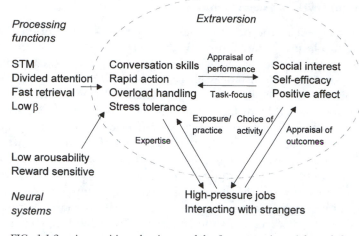

FIG. 1.1.3.   A cognitive-adaptive model of extroversion. Adapted from
"Extraversion, Emotion and Performance," by G. Matthews, 1997, in G. Mat-
thews (Ed.), *Cognitive Science Perspectives on Personality and Emotion* (p. 433),
Amsterdam: Elsevier. Copyright 1997 by Elsevier. Adapted with permission.

CONCLUSIONS

The starting point for this chapter was the diversity of "stressor" effects on
performance and the failure of general theories. The conceptual framework
(*not* a theory) described here differentiates two levels of transaction between the operator and the task environment. Hancock and Warm (1989)
were right to identify "physiological-psychological isomorphism" and "individual meaning" as critical to stress reactions, but these two aspects of stress
relate to different levels of transaction. Cognitive science requires differentiation of neural, cognitive-architectural, and adaptive levels of description
for phenomena. One or other is usually more appropriate, depending on
circumstances. A deeper understanding of stressor effects requires consideration of the interplay between levels, as expressed by the concept of levels of
transaction. The biocognitive transaction refers to changes in architecture
driven by patterned changes in neural net processing, in preparation for
handling evolutionarily significant challenges. Specific "biological" stressors
may induce either functionally coordinated changes of this kind or complex, arbitrary changes with no functional unity.

The cognitive-adaptive transaction is the level described by contemporary stress theory. Its application to performance studies requires a specification of how contextualized knowledge influences selection of coping
strategies and the consequences of coping for information processing, intended and unintended. Adaptation to performance environments seems

to be organized around the three transactional themes of commitment to effort, overload, and self-regulation. Measurement of the corresponding modes of adaptation provides a tool for relating adaptive-level stress processes to strategies specified in processing terms. Work on anxiety lends itself naturally to this approach, but it may also contribute to explaining effects of other stressors such as noise and fatigue.

Future research may complete the explanatory loop shown in Fig. 1.1.1 by exploring the evolutionary basis for the biocognitive transaction. The patterning of biologically driven changes in architectural parameters presumably reflects selection for changes in processing capability with adaptive or survival value. A better understanding is needed of the environmental challenges that have driven selection of integrated biocognitive responses such as the energetic arousal response. The context sensitivity and strategic nature of many stressor effects show the need to avoid a crude evolutionary determinism. Performance research is not well served by the evolutionary psychological concept of the "module," which is apt to be another name for a homunculus. Understanding stress responses as biological adaptations must be tempered by an appreciation of the limitations of the biocognitive machinery and the need for culture-bound and contextualized cognitive-adaptive control.

ACKNOWLEDGMENT

Some of this research was supported by a Medical Research Council Grant (G9510930).

REFERENCES

Anderson, J. R. (1990). *The adaptive character of thought.* Hillsdale, NJ: Lawrence Erlbaum Associates.

Banquet, J. P., Gaussier, P., Dreher, J. C., Joulain, C., Revel, A., & Günther, W. (1997). Space-time, order, and hierarchy in fronto-hippocampal system: A neural basis of personality. In G. Matthews (Ed.), *Cognitive science perspectives on personality and emotion* (pp. 123–189). Amsterdam: Elsevier Science.

Beatty, J. (1986). Computation, control and energetics: A biological perspective. In G. R. J. Hockey, A. W. K. Gaillard, & M. G. H. Coles (Eds.), *Energetics and human information processing* (pp. 43–52). Dordrecht, Netherlands: Martinus Nijhoff.

Beck, A. T. (1987). Cognitive models of depression. *Journal of Cognitive Psychotherapy, 1,* 5–37.

Besner, D., Stoltz, J. A., & Boutilier, C. (1997). The Stroop effect and the myth of automaticity. *Psychonomic Bulletin and Review, 4,* 221–225.

Bower, G. H. (1981). Mood and memory. *American Psychologist, 36,* 129–148.

Broadbent, D. E. (1971). *Decision and stress.* London: Academic Press.

Broadbent, D. E. (1978). The current state of noise research: Reply to Poulton. *Psychological Bulletin, 85,* 1052–1067.

Calvo, M. G., & Castillo, M. D. (1997). Mood-congruent bias in interpretation of ambiguity: Strategic processes and temporary activation. *Quarterly Journal of Experimental Psychology, 50A,* 163–182.

Cartwright-Hatton, S., & Wells, A. (1997). Beliefs about worry and intrusions: The Meta-Cognitions Questionnaire and its correlates. *Journal of Anxiety Disorders, 11,* 279–296.

Cohen, J. D., Dunbar, K., & McClelland, J. L. (1990). On the control of automatic processes: A parallel distributed processing account of the Stroop effect. *Psychological Review, 97,* 332–361.

Cohen, J. D., & Servan-Schreiber, D. (1992). Context, cortex and dopamine: A connectionist approach to behavior and biology in schizophrenia. *Psychological Review, 99,* 45–77.

Cohen, S. (1980). After effects of stress on human performance and social behavior: A review of research and theory. *Psychological Bulletin, 88,* 82–108.

Cox, T. (1978). *Stress.* London: Macmillan.

Cox, T., & Ferguson, E. (1991). Individual differences, stress and coping. In C. L. Cooper & R. Payne (Eds.), *Personality and stress: Individual differences in the coping process* (pp. 7–32). Chichester, England: Wiley.

Dagenbach, D., Carr, T. H., & Wilhelmson, A. (1989). Task-induced strategies and near-threshold priming: Conscious influences on unconscious perception. *Journal of Memory and Language, 28,* 412–443.

Daley, J. W., Shi, D., Nikodijevic, O., & Jacobson, K. A. (1994). The role of adenosine receptors in the central action of caffeine. *Pharmacopsychoecologia, 7,* 201–213.

Depue, R. A. (1995). Neurobiological factors in personality and depression. *European Journal of Personality, 9,* 413–439.

Derryberry, D., & Reed, M. A. (1997). Motivational and attentional components of personality. In G. Matthews (Ed.), *Cognitive science perspectives on personality and emotion* (pp. 443–473). Amsterdam: Elsevier Science.

Desmond, P. A. (1997). *Fatigue and stress in driving performance.* Unpublished doctoral dissertation, University of Dundee.

Desmond, P. A., & Matthews, G. (1997). Implications of task-induced fatigue effects for in-vehicle countermeasures to driver fatigue. *Accident Analysis and Prevention, 29,* 513–523.

Endler, N., & Parker, J. (1990). Multidimensional assessment of coping: A critical review. *Journal of Personality and Social Psychology, 58,* 844–854.

Eysenck, M. W. (1992). *Anxiety: The cognitive perspective.* Hillsdale, NJ: Lawrence Erlbaum Associates.

Eysenck, M. W., & Calvo, M. G. (1992). Anxiety and performance: The processing efficiency theory. *Cognition and Emotion, 6,* 409–434.

Fillmore, M. T., & Vogel-Sprott, M. (1992). Expected effect of caffeine on motor performance predicts the type of response to placebo. *Psychopharmacology, 106,* 209–214.

Fisher, S. (1986). *Stress and strategy.* Hillsdale, NJ: Lawrence Erlbaum Associates.

Gray, J. A. (1982). *The neuropsychology of anxiety: An enquiry into the functions of the septo-hippocampal system.* Oxford: Oxford University Press.

Hancock, P. A. (1986). Sustained attention under thermal stress. *Psychological Bulletin, 99,* 263–281.

Hancock, P. A., & Warm, J. S. (1989). A dynamic model of stress and sustained attention. *Human Factors, 31,* 519–537.

Hart, S. G., & Staveland, L. E. (1988). Development of a multidimensional workload rating scale: Results of empirical and theoretical research. In P. A. Hancock & N. Meshkati (Eds.), *Human mental workload* (pp. 139–183). Amsterdam: Elsevier.

Hockey, G. R. J. (1984). Varieties of attentional state: The effects of the environment. In R. Parasuraman & D. R. Davies (Eds.), *Varieties of attention* (pp. 449–483). New York: Academic Press.

Hockey, G. R. J. (1997). Compensatory control in the regulation of human performance under stress and high workload: A cognitive-energetical framework. *Biological Psychology, 45,* 73–93.

Holding, D. H. (1983). Fatigue. In G. R. J. Hockey (Ed.), *Stress and fatigue in human performance* (pp. 145–168). Chichester, England: Wiley.

Humphreys, M. S., & Revelle, W. (1984). Personality, motivation and performance: A theory of the relationship between individual differences and information processing. *Psychological Review, 91,* 153–184.

Hunt, E. B., & MacLeod, C. M. (1978). The sentence-verification paradigm: A case study of two conflicting approaches to individual differences. *Intelligence, 2,* 129–144.

Jones, D. M. (1984). Performance effects. In D. M. Jones & A. J. Chapman (Eds.), *Noise and society* (pp. 155–184). New York: Wiley.

Jones, D. M., & Davies, D. R. (1984). Individual and group differences in the response to noise. In D. M. Jones & A. J. Chapman (Eds.), *Noise and society* (pp. 125–153). New York: Wiley.

Kanfer, R. (1990). Motivation and individual differences in learning: An integration of developmental, differential and cognitive perspectives. *Learning and Individual Differences, 2,* 221–239.

Kirsch, I., & Weixel, L. J. (1988). Double-blind versus deceptive administration of a placebo. *Behavioral Neuroscience, 102,* 319–323.

Kluger, A. N., & DeNisi, A. (1996). The effects of feedback interventions on performance: A historical review, a meta-analysis, and a preliminary feedback intervention theory. *Psychological Bulletin, 119,* 254–284.

Lazarus, R. S. (1991). *Emotion and adaptation.* Oxford: Oxford University Press

Lazarus, R. S., & Folkman, S. (1984). *Stress, appraisal and coping.* New York: Springer.

Lieberman, H. R. (1992). Caffeine. In A. P. Smith & D. M. Jones (Eds.), *Handbook of human performance, Vol. 2: Health and performance* (pp. 49–72). London: Academic Press.

Loehlin, J. C. (1992). *Genes and environment in personality development.* Newbury Park, CA: Sage.

MacLeod, C., & Hagen, R. (1992). Individual differences in the selective processing of threatening information, and emotional responses to a stressful life event. *Behaviour Research and Therapy, 30,* 151–161.

Matthews, G. (1997a). An introduction to the cognitive science of personality and emotion. In G. Matthews (Ed.), *Cognitive science perspectives on personality and emotion,* (pp. 3–30.) Amsterdam: Elsevier.

Matthews, G. (1997b). Extraversion, emotion and performance: A cognitive-adaptive model. In G. Matthews (Ed.), *Cognitive science perspectives on personality and emotion* (pp. 399–442). Amsterdam: Elsevier.

Matthews, G. (1999). Personality and skill: A cognitive-adaptive framework. In P. L. Ackerman, P. C. Kyllonen, & R. D. Roberts (Eds.), *Learning and individual differences: Process, trait, and content determinants.* Washington, DC: American Psychological Association Press.

Matthews, G., & Amelang, M. (1993). Extraversion, arousal theory and performance: A study of individual differences in the EEG. *Personality and Individual Differences, 14,* 347–364.

Matthews, G., & Campbell, S. E. (1998). Task-induced stress and individual differences in coping. In *Proceedings of the 42nd Annual Meeting of the Human Factors and Ergonomics Society* (pp. 821–825). Santa Monica, CA: Human Factors and Ergonommics Society.

Matthews, G., & Davies, D. R. (1998). Arousal and vigilance: The role of task factors. In R. R. Hoffman, M. F. Sherrick, & J. S. Warm (Eds.), *Viewing psychology as a whole: The integrative science of William N. Dember* (pp. 113–144). Washington, DC: American Psychological Association Press.

Matthews, G., & Gilliland, K. (1999). The personality theories of H. J. Eysenck and J. A. Gray: A comparative review. *Personality and Individual Differences, 26,* 583–626.

Matthews, G., & Harley, T. A. (1996). Connectionist models of emotional distress and attentional bias. *Cognition and Emotion, 10,* 561–600.

Matthews, G., Joyner, L., Gilliland, K., Campbell, S. E., Huggins, J., & Falconer, S. (1999). Validation of a comprehensive stress state questionnaire: Towards a state "Big Three"? In I. Mervielde, I. J. Deary, F. De Fruyt, & F. Ostendorf (Eds.), *Personality psychology in Europe* (Vol. 7, pp. 335–350). Tilburg: Tilburg University Press.

Matthews, G., & Margetts, I. (1991). Self-report arousal and divided attention: A study of performance operating characteristics. *Human Performance, 4,* 107–125.

Matthews, G., Schwean, V. L., Campbell, S. E., Saklofske, D. H., & Mohamed A. A. R. (2000). Personality, self-regulation and adaptation: A cognitive-social framework. In M. Boekarts, P. R. Pintrich, & M. Zeidner (Eds.), *Handbook of self-regulation* (pp. 171–207). New York: Academic Press.

Matthews, G., & Wells, A. (1996). Attentional processes, coping strategies and clinical intervention. In M. Zeidner & N. S. Endler (Eds.), *Handbook of coping: Theory, research, applications* (pp. 573–601). New York: Wiley.

Matthews, G., & Wells, A. (1999). The cognitive science of attention and emotion. In T. Dalgleish & M. Power (Eds.), *Handbook of cognition and emotion* (pp. 171–191). New York: Wiley.

Matthews, G., & Westerman, S. J. (1994). Energy and tension as predictors of controlled visual and memory search. *Personality and Individual Differences, 17,* 617–626.

Mueller, J. H. (1992). Anxiety and performance. In A. P. Smith & D. M. Jones (Eds.), *Handbook of human performance. Vol. 3: State and trait* (pp. 127–160). London: Academic Press.

Munro, L. L., Dawson, M. E., Schell, A. M., & Sakai, L. M. (1987). Electrodermal lability and rapid vigilance decrement in a degraded stimulus continuous performance test. *Journal of Psychophysiology, 1,* 249–257.

Näätänen, R. (1973). The inverted-U relationship between activation and performance: A critical review. In S. Kornblum (Ed), *Attention and performance* (Vol. 4, pp. 110–117). New York: Academic Press.

Neiss, R. (1988). Reconceptualizing arousal: Psychobiological states in motor performance. *Psychological Bulletin, 103,* 345–366.

Newell, A. (1980). Physical symbol systems. *Cognitive Science, 4,* 135–183.

Norman, D. A., & Shallice, T. (1985). Attention to action: Willed and automatic control of behaviour. In R. J. Davidson, G. E. Schwartz, & D. Shapiro (Eds.), *Consciousness and self-regulation: Advances in research* (Vol. 4, pp. 1–17). New York: Plenum.

Oatley, K., & Johnson-Laird, P. (1987). Towards a cognitive theory of emotions. *Cognition and Emotion, 1,* 29–50.

Pekrun, R. (1992). Expectancy-value theory of anxiety: Overview and implications. In D. G. Forgays, T. Sosnowski, & K. Wrzesniewski (Eds.), *Anxiety: Recent developments in cognitive, psychophysiological and health research* (pp. 23–41). Washington, DC: Hemisphere.

Pylyshyn, Z. W. (1984). *Computation and cognition: Toward a foundation for cognitive science.* Cambridge, MA: MIT Press.

Rabbitt, P. M. A. (1989). Sequential reactions. In D. Holding (Ed.), *Human skills* (pp. 147–170). Chichester, England: Wiley.

Richards, A., French, C. C., Johnson, W., Naparstek, J., & Williams, J. (1992). Effects of mood manipulation and anxiety on performance of an emotional Stroop task. *British Journal of Psychology, 83,* 479–491.

Robbins, T. W. (1986). Psychopharmacological and neurobiological aspects of the energetics of information processing. In G. R. J. Hockey, A. W. K. Gaillard, & M. G. H. Coles (Eds.), *Energetics and human information processing* (pp. 71–90). Dordrecht, Netherlands: Martinus Nijhoff.

Rohrbaugh, J. W., Stapleton, J. M., Parasuraman, R., Zubovic, E. A., Frowein, H. W., Varner, J. L., Adinoff, B., Lane, E. A., Eckardt, M. J., & Linnoila, M. (1988). Dose-related effects of ethanol on visual sustained attention and event-related potentials. *Alcohol, 4,* 293–300.

Rohsenow, D. J., & Marlatt, G. A. (1980). The balanced placebo design: Methodological considerations. *Addictive Behaviors, 6,* 107–122.

Sanders, A. F. (1990). Issues and trends in the debate on discrete versus continuous processing of information. *Acta Psychologica, 74,* 123–167.

Sarason, I. G. (1988). Anxiety, self-preoccupation and attention. *Anxiety Research, 1,* 3–7.

Shallice, T. (1988). *From neuropsychology to mental structure.* Cambridge: Cambridge University Press.

Siegle, G. J., & Ingram, R. E. (1997). Modeling individual differences in negative information processing biases. In G. Matthews (Ed.), *Cognitive science perspectives on personality and emotion* (pp. 301–353). Amsterdam: Elsevier Science.

Simon, H. A. (1967). Motivational and emotional controls of cognition. *Psychological Review, 74,* 29–39.

Smith, A. P., & Jones, D. M. (Eds.). (1992a). *Handbook of human performance* (3 vols.). London: Academic Press.

Smith, A. P., & Jones, D. M. (1992b). Noise and performance. In A. P. Smith & D. M. Jones (Eds.), *Handbook of human performance. Vol. 1: The physical environment* (pp. 1–28). London: Academic Press.

Smith, B. D. (1983). Extraversion and electrodermal activity: Arousability and the inverted U. *Personality and Individual Differences, 4,* 411–419.

Smith, B. D. (1994). Effects of acute and habitual caffeine ingestion on physiology and behavior: Tests of a biobehavioral arousal theory. *Pharmacopsychoecologia, 7,* 151–167.

Smolensky, P. (1988). On the proper treatment of connectionism. *Behavioral and Brain Sciences, 11,* 1–74.

Thayer, R. E. (1978). Toward a psychological theory of multidimensional activation (arousal). *Motivation and Emotion, 2,* 1–34.

Thayer, R. E. (1989). *The biopsychology of mood and arousal.* Oxford: Oxford University Press.

Thayer, R. E. (1996). *The origin of everyday moods.* New York: Oxford University Press.

Van Reekum, C. M., & Scherer, K. R. (1997). Levels of processing in emotion-antecedent appraisal. In G. Matthews (Ed.), *Cognitive science perspectives on personality and emotion* (pp. 259–300). Amsterdam: Elsevier.

Warm, J. S. (1993). Vigilance and target detection. In B. M. Huey & C. D. Wickens (Eds.), *Workload transition: Implications for individual and team performance* (pp. 139–170). Washington, DC: National Academy Press.

Wells, A., & Matthews, G. (1994). *Attention and emotion: A clinical perspective.* Hove: Lawrence Erlbaum Associates.

Williams, J. M. G., Watts, F. N., MacLeod, C., & Mathews, A. (1988). *Cognitive psychology and emotional disorders.* Chichester, England: Wiley.

Zeidner, M. (1998). *Test anxiety: The state of the art.* New York: Plenum.

Zeidner, M., & Endler, N. S. (Eds.). (1996). *Handbook of coping: Theory, research, applications.* New York: Wiley.

Zeidner, M., & Saklofske, D. (1996). Adaptive and maladaptive coping. In M. Zeidner & N. S. Endler (Eds.), *Handbook of coping: Theory, research, applications* (pp. 505–531). New York: Wiley.

An Information-Processing Model of Operator Stress and Performance

Keith C. Hendy
Philip S. E. Farrell
Defence and Civil Institute of Environmental Medicine

Kim P. East
Atomic Energy of Canada Limited,
Mississauga, Ontario, Canada

> *Insofar as living organisms perform the functions of a communication system, they must obey the laws that govern all such systems.* —Miller, 1971

There are few quantitative models in cognitive psychology. This chapter details an attempt to devise such a model in the context of human information processing. The work was guided by the need to develop models of operator workload and performance that could be embedded in predictive methods for human engineering analysis—specifically task network simulation. For a general description of this procedure and of its origins in task time-line analysis, see Meister (1985).

It has been said that information-processing and energetic models hold the greatest promise for estimating operator loading (e.g., Gopher & Donchin, 1986; Hendy, Liao, & Milgram, 1997). The model described here combines a simple information-processing (IP) paradigm (Hendy et al., 1997) with the tenets of William T. Powers' perceptual control theory or PCT (Power, 1973). The combined IP/PCT model provides a unifying framework that integrates knowledge about operator workload, performance, and error production with concepts such as situation awareness. Most important, the IP/PCT model traces the dependencies that link these factors (Hendy, 1995).

From the IP/PCT model, it is argued that the underlying stressor that determines operator performance, error production, and judgments of workload is *time pressure*. More specifically, it is claimed that *all* factors that affect operator workload can be reduced to their effect on this variable. Hence, time pressure is claimed to be the principal stressor in the human information-processing context. However, the IP/PCT model also incorporates the more global effects of various psychological and physiological stressors that act diffusely on the arousal and activation mechanisms.

This chapter describes the IP/PCT model; presents evidence supporting the IP/PCT model from two experiments by using a task that simulates a complex air traffic control (ATC) environment; and discusses some practical applications of the model to the fields of human performance prediction and operator decision making.

In the hierarchical structure of the combined model presented here, PCT refers to the structure of human–human and human–machine interaction, whereas the IP model describes the nature of the active information-processing components in that structure. Therefore, it is appropriate to describe PCT first and then place the IP model, in context, in the PCT framework. For a detailed description of the IP/PCT model, and particularly its adaptation for implementation in task network simulation, refer to Hendy and Farrell (1997).

PERCEPTUAL CONTROL THEORY

The presence of feedback is essential to the control of goal-directed human activity, according to William T. Powers' perceptual control theory model (Powers, 1973). Powers' model is multilayered, with multiple goals providing the reference points for a hierarchical organization of control loops. These loops provide control at many levels—from the lowest levels of sensory processing, upward to the satisfaction of abstract goals such as the need for self-esteem and actualization. In PCT terms, an emitted action or behavior is in response to the presence of an error, or difference, signal. The emitted action is transmitted purposefully, with the intention of changing the state of the world so that the operator's perception can be made to match a desired state or goal, which reduces the error signal to zero. It is a fundamental thesis of PCT that it is the perception that is controlled, *not* the behavior.

The Perceptual Control Loop

The PCT model for a single level of abstraction in the control hierarchy is illustrated in Fig. 1.2.1. A matrix and vector formulation is used as an

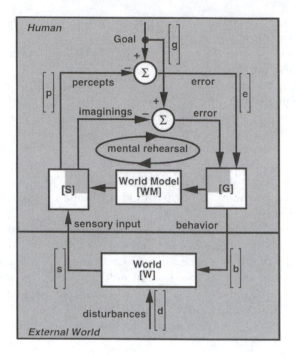

FIG. 1.2.1.   The perceptual control theory (PCT) model.

approximate representation of the multidimensional nature of the model. The *goals, errors, behaviors, disturbances, sensory inputs,* and *percepts* are shown in vector form (i.e., **g, e, b, d, s, p**), whereas the transfer functions **G, W,** and **S** are shown as matrices. An inner loop is also shown, involving the transformation matrix labeled *World Model* (**WM**). This loop represents the role of processes that are entirely internal, such as imagination, mental rehearsal, thinking, reasoning. This inner loop involves subsets of **S** and **G** (these are shown shaded in Fig. 1.2.1) as no overt behaviors are emitted, and sensory inputs are not necessarily involved—hence the transformation processes required by these activities are bypassed. The internal loop might be associated with, among other things, the processes of building Level 2 and 3 situation awareness—namely, *comprehension of the situation* and *projection of the current situation into the future*—or the higher levels of the three-stage process described by Thompson, Jamieson, and Hendy (1997) as *building awareness of the implications* of the current system state and the *formulation of plans* to cope with those implications. For simplicity, the vector and matrix notation is not developed for this internal loop.

The perceptual control model is a dynamic, adaptive model. All terms of **g, S, G,** and **WM** are capable of change because of interactions with the external world and with inner-loop activity. In general, the contents of **g**, **S**, **G**, and **WM** are available to all loops that are either under control or potentially might be under control. Hence, adaptations of the mental model owing to the control of one loop potentially flow to the control of another—now or in the future. In general, the transfer functions that form the terms of **S**, **G**, and **WM** have latencies or transport delays associated with them. These latencies are related to the resolution of uncertainty and flow directly from the decision times of the IP model.

From Fig. 1.2.1, it can be seen that perceptions and behaviors are shaped by the terms of the transfer functions **S** and **G** as follows:

$$\begin{bmatrix} e \end{bmatrix} = \begin{bmatrix} g \end{bmatrix} - \begin{bmatrix} p \end{bmatrix} \tag{1}$$

$$\begin{bmatrix} p \end{bmatrix} = \begin{bmatrix} S \end{bmatrix} \begin{bmatrix} s \end{bmatrix} \tag{2}$$

$$\begin{bmatrix} b \end{bmatrix} = \begin{bmatrix} G \end{bmatrix} \begin{bmatrix} e \end{bmatrix}, \text{ and} \tag{3}$$

$$\begin{bmatrix} s \end{bmatrix} = \begin{bmatrix} W_1 \end{bmatrix} \begin{bmatrix} b \end{bmatrix} + \begin{bmatrix} W_2 \end{bmatrix} \begin{bmatrix} d \end{bmatrix}. \tag{4}$$

The set of all $\mathbf{g} \in \{\mathbf{g}_1, \mathbf{g}_2, \ldots \mathbf{g}_n\}$ represents the active goal states at a single level of abstraction. It should be noted that a full multilevel representation of the PCT structure is beyond this simple formulation, as in PCT the output of one loop can become the goal for another loop. Hence, a simple orthogonal basis, necessary for the rigorous application of the formalism represented by Equations 1 through 4, does not exist.

In combination, **S**, **G**, and **WM** represent what might be called the operator's mental model. Both inner and outer loops are assumed to be adaptive; therefore, **S**, **G**, **WM**, and **g** change with time as learning takes place. The outer loop (**g**, **S**, and **G**) can also adapt owing to inner loop activity, through various deductive and inductive processes that effect the operator's goals and knowledge states.

Error Correction and Goal Achievement

In PCT, an action is initiated by an error signal that is equal to the difference between the desired state (the reference or goal) and the perceived state (see Equation 1). Both the perception that is formed and the action that is initiated by this error signal are shaped by the structure of the operator's understanding of how the world functions (**S** and **G**—see Equations 2 and 3) or, more correctly, by an understanding of what one must perceive from the world, and do to the world, to achieve desired outcomes. This understanding includes, among other things, a knowledge of the relations between objects in the external world, assumed control-display relations, and the dynamic and static properties of the objects one is interacting with. This structure is part of the operator's mental model and is resident in the transfer functions represented by the matrices **S** and **G**. The operator's goals and world model **WM** also affect the strategies that are chosen for implementing actions (affecting the terms of **G** that are chosen to operate on the error vector **e**).

It will be seen that the term *error* has different meanings in the context of the IP and PCT models. In the PCT sense, errors are associated with the output of a comparator, shown in Fig. 1.2.1 by the symbol Σ. The comparator merely performs the algebraic operation $x - y$. In this sense **e** could be considered a *difference* vector rather than an *error* vector. The term *error* often carries with it the connotation of an uncorrected deviation from the goal, whereas under PCT the difference vector should approach zero after some period (the settling time of the loop). Feedback is essential if this difference signal is to be reduced toward zero (this also requires, in conventional control theory terms, that loop gain is negative[1] and $\gg 1$). Differences are not reduced if the **G** and **S** components of the mental model remain inappropriate to the selected goal **g** (e.g., an erroneous internal representation of the localization of a control in space, an assumed incorrect control-display relation) — at least to the extent that negative loop gain *is* or *is not* achieved.

Feedback is also essential to the adaptation of the mental model with learning. PCT predicts that as long as feedback loops remain intact and the following conditions are satisfied — there is sufficient time; the operator can learn to form an appropriate mental model (has the enabling knowledge, perceptual capabilities, etc.); and the operator can make the required actions (manipulative skill, language requirements, etc.) — the system should self-correct or adapt such that the difference term is finally reduced to criterion levels. Hence, under the assumptions stated, PCT predicts that an operator eventually achieves the goal if the control loop remains unbroken. Note that the mental model does not have to be strictly *correct*, but **S** and **G** must result in negative loop gain if the error vector is to be reduced to acceptable limits. An appropriate mental model would result in a small error variance with a corresponding short settling time. An inappropriate mental model would result in either divergent behavior or in a large error variance with long settling times for goal achievement.

---

[1] The concept of negative loop gain implies that the feedback signal reduces the overall gain of the system. It is the total gain around the loop, *including* the sign of the comparator (both $x - y$ and $x + y$ conventions exist) that must be taken into account. Positive loop gain would augment the input signal and cause divergent behavior (Thomason, 1956).

Mental Models and Situation Awareness

Although one might associate the union of **S**, **WM**, and **G** with the *mental model* in a global sense, not all the possible hierarchies of control loops are directly relevant to the decisions that are made and the actions that are produced in performing a particular job of work. For example, control of the autonomic system (e.g., core temperature, heart rate, and breathing rate) has relatively little effect on the behaviors of interest to many systems designers or analysts. The interesting part of the mental model—particularly when concerned with the concept of situation awareness (SA)—is that part of **S**, **WM**, and **G** with direct relevance to the task at hand. Collectively, the notion of situation awareness (SA) represents the terms of **S**, **WM**, and **G** relevant to the loops that are either under active control, or might come under active control, in the performance of a task—particularly, or perhaps specifically, those terms of **S**, **WM**, and **G** that are unique to the current situation. Owing to the transient nature of this knowledge, SA can be seen as conceptually distinct from the solid base of declarative knowledge that is stable or changes slowly over time. This distinction, however, is of little relevance in the context of PCT discussions. Any distinction one might want to make would be restricted to the loops that are served by one type of knowledge over the other, and not by the nature of the way in which the knowledge operates to shape perceptions and form actions. In general, loops may be under either conscious or unconscious control, particularly those loops associated with autonomous behaviors (Endsley, 1993).

The following definitions of the mental model and situation awareness are consistent with the position stated earlier (from Hendy, 1995). For the purposes of PCT, these definitions place situation awareness in the domain of the mental model rather than separate from it, just as memory might be considered a continuum, ranging from long-term memory to short-term and working memory:

> The *mental model* is that part of the operator's internal state that contains the knowledge and structure, including what might be termed situation awareness, necessary to perform a task. As such, the operator's mental model directly shapes the operator's actions and determines the potential to perform in accordance with system demands.

> The term *situation awareness* particularly relates to that dynamic and transient state of the mental model produced by an ongoing process of information gathering and interpretation during the performance of some job of work. Although the concept can be generalized to all tasks, no matter what their complexity, the term *SA* is usually used when considering tasks that have strategic and tactical components such as flying an aircraft, controlling or monitoring a plant, or making tactical and strategic decisions.

It is recognized that memory structures almost certainly involve different storage mechanisms (e.g., synaptic versus chemical), but this does not change the essential nature of the perceptual control model. The transformations that are applied to the error vector **e**, to produce a behavioral output **b**, draw on a hierarchy of loops that use long-term declarative structures as well as transient or situation-specific knowledge. Some of the transformations may be at the level of pattern recognition; others may involve rules or algorithmic problem-solving strategies. The actual transformations that apply to the control of a specific loop, on a specific occasion, depend on the strategy that is chosen.

The transfer functions **S** and **G**, in general, involve transport delays. In tracking tasks, these delays have been attributed to latencies in the visual and neuromuscular systems (Allen, McRuer, & Thompson, 1989). In more complex decision-making tasks, these latencies are most likely dominated by the processing time for higher level decision making (i.e., involving **G**). The introduction of transport delays is approximately equivalent to the addition of a further lag term in the transfer function. The net result is a reduction in the bandwidth of the closed loop response (Van de Vegte, 1986) and a degraded ability to deal with rapidly varying information-processing demands. Transport delay also reduces the phase margins for stable closed-loop performance. We later argue that these transport delays are directly attributable to the processes described by the IP model.

Finally, Endsley's Level 2 and Level 3 SA (Endsley, 1993) can be interpreted in PCT terms. Level 2 SA goes beyond simple awareness to a knowledge of the implications of the situation. In PCT terms, lower level control (controlling direction of gaze, auditory attention, etc.) passes state information (stored in short-term and working memory) up to higher level processing loops. These loops insert that state knowledge into memory traces, possibly at Rasmussen's rule-based level (Rasmussen, 1983), to arrive at new knowledge that addresses the implications of this state (Level 2 SA). This involves the internal control loop of Fig. 1.2.1. In this sense, SA might be considered the human's adaptation to the changing world state (Hendy, 1995). New loops are brought into active control until some action is emitted which will change the world state such that the current goal is perceived to be met. Level 3 SA (projection into the future) is essentially a predictive or lead term. Applying lead to a heavily lagged system can result in more rapid settling times, but phase margin and stability can be sacrificed, particularly if the variables being operated on, in the external world, are nondeterministic.

Evidence for Perceptual Control Theory

The abstract nature of mental models and situational awareness make them generally difficult to observe. PCT, in contrast, seems to be evident in everyday activities such as driving a car, shopping, or building a house. Indeed, many expressions used to describe behavior have strong control connotations, for example: "locus of control," "a person's need for control," "they show controlling behavior," "they must be in control," "they have lost or are out of control." Most likely, a person does not randomly decide to do any activity, but has some goal in mind. Generally, we must also monitor the external world (feedback) to determine its current state with respect to our goals. The perceived current state of the world, and our goals, affect the decision-making process. The resultant behaviors exert influence in the external world, which in turn provides sensory information back to the human in a closed loop fashion. This is the essence of perceptual control theory.

Human behavior is commonly described with action verbs, rather than perceptual states. For instance, going from the first floor of a building to the top floor, one might *walk* to the elevator, *press* the door button, *enter* the elevator, *press* the numbered button, *wait* for the doors to close then open, and *exit* the elevator. This sequence of events might lead readers to believe that human beings are directly controlling their actions, perhaps in response to some stimulus. However, implicit in this elevator scenario is the goal of reaching the top floor, the initial perception of being on the first floor, and a series of actions that move the current perception closer to the goal state.

In 1913, the father of behaviorism, John B. Watson, stated that individuals were made, not born (cited in Myers, 1989). This philosophy led to a stimulus–response (S–R) model of human behavior (John, Rosenbloom, & Newell, 1985; Myers, 1989; Weiten, 1989) that has been propagated throughout this century. The S–R model remains the fundamental principle in psychology that PCT challenges. Psychologists in the mid-1950s began to dissect the S–R model's *black box* into perceptual, cognitive, decision-making, and psychomotor processes (see Card, Moran, & Newell, 1983; Knight, 1987; Wickens, 1992, for further discussion). This new paradigm was sometimes referred to as the stimulus–cognition–response (S–C–R) model. Yet, even with the inclusion of a central-processing component, human intentions and goals were rarely explicitly considered.

In the 1960s and 1970s, engineers explored control theory models to describe human behavior where reference signals and feedback became explicit (see Hess, 1987, for a detailed discussion). This volume of literature accurately describes human behavior during tracking tasks and comes closest to resembling PCT. A structural and philosophical difference between optimal control models (OCMs) and PCT is that the reference state is not represented as part of the human transfer function in OCM. This implies that the goal is generated somewhere in the world other than in the human. PCT takes an opposite view and depicts goals as being initiated by the human operator.

In his book *Engineering Psychology and Human Performance*, Wickens (1992) devoted a chapter to representing the human as a system transfer function for manual tracking tasks. The chapter begins with a single-input, single-output description of the transfer function (essentially a S–R representation). As the external conditions become less predictable, changes to the basic model, which would account for disturbances acting on the system, are discussed. At the end of the chapter, an optimal control model that incorporates a *goal* input to optimize control is presented. Such positions are supportive of the PCT case.

Direct experimental evidence for PCT comes from Bourbon (Bourbon, 1990, 1996), who tested PCT experimentally by using a manual tracking task under two conditions. The first condition yielded a value for the only free parameter K (the loop gain function resembled an integral controller). The second condition introduced a random unknown disturbance into the loop. The PCT model, with K from Condition 1, accounted for over 99% of the variance when stimulated with the Condition 2 disturbance. The test was repeated 5 years later and yielded a correlation of $r = 0.99$. The PCT model accounts, at least in part, for human information processing and human behavior in manual tracking tasks.

PCT is said to be evident in systems other than human tracking, such as in genetic reorganization, self-realization, and social systems (Robertson & Powers, 1990). Yet, few experimental studies attempted to validate PCT in these more complex environments. Two ongoing studies explore PCT concepts in such environments and also depart from the more traditional situation of compensatory or pursuit tracking. One study deals with the comparison of two aircraft instrument interfaces: the first designed with traditional engineering techniques and the other designed with PCT concepts (Farrell & Semprie, 1997). A second study examines the cognitive compatibility (defined as "the facilitation of goal achievement through the display of information in a manner which is consistent with internal mental processes and knowledge, in the widest sense, including sensation, perception, thinking, conceiving and reasoning"—UTP-7, 1994, p. 53) of interfaces in a tactical simulator environment. This study uses PCT to establish hypotheses related to the strategies participants might employ.

People well versed in this theory see evidence for PCT all around them. Preliminary experimental evidence makes PCT hard to ignore as a description of human dynamics. At the very least, PCT makes goal-driven behavior and feedback explicit and has been shown to model human information processing and behavior with considerable accuracy in some tasks.

Summary

PCT is a model for all goal-directed human activity. Indeed, PCT is said to describe the goal-directed activity of all organisms, from simple single-cell identities, to complex collections of cells (such as a human), to groups and societies of these complex organisms. Perceptual control theory makes explicit the existence of a goal, the shaping of perceptions and behaviors by an internal knowledge state, and the essential requirement of feedback for goal achievement.

PCT provides a framework in which an information-theoretic model, such as the IP model, can be readily absorbed. Perceptual control theory brings all the rigor of engineering control theory into the human domain. From PCT, we can see that the dynamic behavior of the PCT model is bandwidth limited, with this limitation coming from lags and delays in the terms of the transfer functions contained in the matrices **G** and **S**. The IP model provides a mechanism for explaining these delays and shows how strategy selection provides a trade-off between speed of response and absolute accuracy of performance. By going to a less accurate, less computationally intensive strategy, transport delays are less, and the dynamic response increases because of the increased bandwidth. Transport delays are a product of the time required to process the information associated with selecting and forming an action at a finite processing rate.

THE INFORMATION-PROCESSING MODEL

The information-processing (IP) model (Hendy & Farrell, 1997; Hendy et al., 1997) acts in the PCT loop at all points where information is actively processed, for example, where the error signal is transformed into an action; in the internal processing loop (thinking, reasoning, deducing, etc.); and where incoming sensory information is transformed into a perception of the state of the world.

Following Miller's lead (Miller, 1971), as outlined at the beginning of this chapter, one can argue that it is possible to characterize the information-processing load of any human activity in terms of the amount of uncertainty to be resolved. Conceptually, it should be possible to actually quantify the amount of uncertainty in terms of bits of information. Although in practice this usually turns out to be impossibly difficult (Sanders, 1979), it is the concept, rather than the practice, that is important to this argument.

Although *task load* is quantified here in terms of bits of uncertainty, the actual amount of information to be processed is a function of the *strategies* chosen by the human in formulating a solution, as well as of the statistical characteristics of the task to be performed. The choice of strategy determines the number and complexity of decisions to be made and depends on many things, including the level of training, confidence, experience, accuracy requirements, consequences of failure. The strategy selected may also vary adaptively as the human attempts to satisfy demands with limited processing capacity.

The IP model is based on the notion that information is processed by many interacting structures. In this context, the term *processing structure* refers to the neural pathway(s) involved in processing the information for a particular task (e.g., see Detweiler & Schneider, 1991). Interference in multiple concurrent task performance is assumed to depend on the amount of physical overlap between the structures involved in processing each task, with information processed serially in any given structure. The selection of a specific strategy is assumed to involve a specific set of processing structures; different strategies, in general, involve different structures. If there is a constant strategy, task interference is seen by an increase in the processing or decision time for one or more of the tasks (Hawkins et al., 1979), as a direct consequence of the competition for serial resources. Note that the concept of *overlap* is assumed to exist at the neural level. This is consistent with the architecture described by Detweiler and Schneider (1991) for a connectionist model of skill acquisition.

The Human Information Processor

The IP model starts with the assumption that human cognition can be attributed to a capacity-limited information processor or processors (see also Hendy & Farrell, 1997; Hendy et al., 1997). What one associates with *task performance*, and what ultimately results in the experience we call *workload*, is considered to be the direct result of the action of this processor. Figure 1.2.2 brings together many of the concepts associated with the workload construct and attempts to show the relations that exist between them. The essential claims of the IP model stem from the following two equations:

$$\text{Task Load} \div \text{Processing Rate} = \text{Decision Time}$$
$$\text{Decision Time} \div \text{Time Available} = \text{Time Pressure}.$$

The equations trace the conversion of *task loading* into a *time pressure* variable. To assist in interpreting the semantic model of Fig. 1.2.2, consider the following statement for one of several pathways in the figure, starting with *Time Pressure* and ending at *Task Load:* "*Time Pressure* drives an *Effort*ful process that causes the information-processing system to *adapt* by *decreasing* the *Task Difficulty,* which is *associated* with the *Task Load.*" In the IP model, it is assumed that the effective rate at which information is processed by a specific processing structure is limited. Furthermore, a more binding assumption is made that processing rate is not only limited, but is fixed (e.g., see Kinchla, 1980) for a given structure (say, C bits per second). *Task difficulty* is unequivocally associated with the amount of information to be processed (bits). *Capacity* is equated to the processing rate (in bits/s).

The selected strategy also characterizes the *depth of processing* and sets the total amount of information to be processed. Under a constant processing

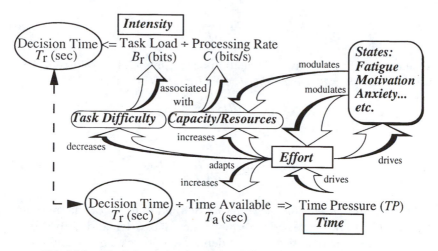

FIG. 1.2.2.   The information-processing (IP) model for the human operator showing the relations between some of the principal constructs associated with the cognitive aspects of task loading.

rate, this also determines the time required to arrive at a decision (T_r). Time
and the amount of information B_r to be processed are related linearly by the
fixed processing rate, as follows:

$$T_r = \left(\frac{1}{C}\right) B_r \text{ seconds.} \tag{5}$$

Although the processing rate is assumed to be constant for a specific
structure, it may vary from structure to structure. From PCT, any action or
perception is likely to involve many hierarchically organized processing
structures. Hence, B_r and C represent the collective action of all loops that
are involved.

The term *depth of processing* is associated with all of the following aspects:
the selection of a strategy, the involvement of specific processing structures,
and the amount of information to be processed in carrying out this strategy
to criterion performance (with deeper processing involving more information processed, which, in turn, leads to more accurate performance). Processing structures can be, in the most general case, mutually orthogonal and
thus compatible with theories such as Wickens' multiple resource (MRT)
model (Wickens, 1992).

The time it takes to make a decision, when considered in isolation, is not
likely to be a source of stress. Suppose, however, that T_r is compared with the
time available for reaching a decision T_a. Then it can be shown that *time pressure (TP)*, as defined by the ratio:

$$TP = \frac{T_r}{T_a},\tag{6}$$

is a measure of the relative information-processing load as follows. If the average *rate of information-processing demand* (i.e., *RID* bits/s), is

$$RID = \frac{B_r}{T_a},\tag{7}$$

then:

$$TP = \frac{T_r}{T_a} = \left(\frac{1}{C}\right)\frac{B_r}{T_a} = \frac{RID}{C}.\tag{8}$$

Furthermore, if *processing resources* are defined in terms of channel capacity, then *TP* can be expressed in terms of the relative demand for processing resources (i.e., the ratio between the demanded processing rate and the limiting rate of *C* bits/s). It will be assumed that *all* human problem solving has a defined window of opportunity, whether it derives directly from the dynamics of the system or indirectly from the need, for example, to retain the outcomes of intermediate problem-solving steps in memory.

Obviously, channel overload occurs when *TP* is greater than 1, and therefore there is an a priori justification for claiming *TP* as a measure of the information-processing load. When demand (*RID*) exceeds capacity (*C*), rate of output can no longer meet demand. As *RID* increases, rate of information processed (*RIP*) eventually reaches, and then maintains, the limiting rate of *C* bits/s, as shown in Fig. 1.2.3. The shaded areas in Fig. 1.2.3 indicate regions where demand exceeds the dynamic response capabilities of the human information-processing (HIP) system.

However humans are adaptive goal-driven beings, and as $TP \to 1$ it is expected that the HIP adapts, by investing *effort*, consciously and/or unconsciously, to try to match demand with capacity. In the IP model, *effort* is equated with the metacontroller's activity (see also Jex, 1988) in reducing the mismatch between demand and capacity, in other words, by attempting to keep the *TP* ratio below 1. In PCT terms, metacontroller activity represents high-level control that is expected to flow outward and downward to other, more diffuse, adaptive mechanisms such as *arousal* and *activation*. These are seen as general priming mechanisms, adapting more slowly than task-level loops and showing sensitivity to the long-term effects of various psychological and physiological states (including the long-term effects of the *effortful* adaptation to processing demand). Generally, their effects are expected to be seen in all processing structures, rather than focused on specific structures.

From Fig. 1.2.2, it can seen that there are three possibilities by which the HIP can reduce an information-processing load mismatch. According to the IP model, they are:

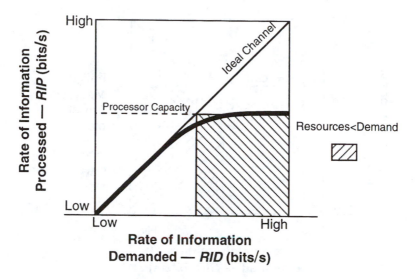

FIG. 1.2.3. Hypothesized relations for the limited capacity human information processor, compared with an ideal unlimited channel, in terms of rate of information demanded, and rate of information processed.

1. By reducing the task load or the amount of information (B_r) to be processed.
2. By increasing the time available (T_a) for solving the problem.
3. By increasing the channel capacity C.

The amount of information to be processed can be reduced in a variety of ways, for example:

- By eliminating high-level processing loops (e.g., perhaps related to monitoring and checking intermediate results).
- By changing the strategy or depth of processing, say by changing the type of processing behavior from knowledge-based to rule-based to skill-based activity, or from algorithmic processing to direct memory retrieval (see also Rasmussen, 1983).

Similarly, strategies also exist for prolonging the time available for decision making, for example, by allowing error to accumulate before corrective action is taken or by slowing a vehicle and extending the decision time line.

The first two methods—reducing the amount of information to be processed or increasing the time available—are assumed to be the primary mechanisms for adaptation to the information-processing load. It has been assumed that the rate of processing (C) remains relatively constant (Kinchla, 1980), although channel capacity might vary in the presence of changing

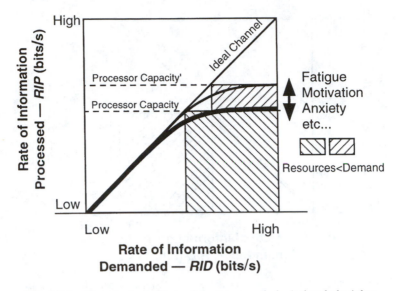

FIG. 1.2.4. Hypothesized effects of changing psychological and physiological states on human information-processing rates.

physiological and psychological states that eventually affect the state of activation. Hence, the time to process task-related information should remain
essentially constant under a constant strategy and a constant state of activation, regardless of the imposed time pressure. In the IP model, variations
in channel capacity are associated with diffuse stress responses caused by
changes in the human's physiological and psychological states. This is shown
in Fig. 1.2.4 by two input–output curves, corresponding to different channel
capacities in the same processing structure (the thinner line representing
the higher level of activation).

It is accepted that if the human chooses a less computationally intensive
(less information) decision-making strategy, the actual number of decisions
per second might increase (this might be termed the *apparent processing rate*)
even though the internal rate of information processing (in terms of the actual bits/s) remains constant (this is the *actual processing rate*). In this case,
each decision involves less information processed, with consequences for
the accuracy of performance (the relation between strategies and the level
of performance is expanded in the following discussion).

Time Pressure, Performance, and Error Production

For the IP model, the relations between rate of information demanded, rate
of information processed (which will be less than or equal to *RID* as deter-

mined by the channel capacity), performance, and error are assumed to be
as shown in Fig. 1.2.5. The input–output relationship, seen previously in
Figs. 1.2.3 and 1.2.4, is shown in Fig. 1.2.5 to illustrate the hypothesized relations between processing capacity and the outcome measures of *performance*
and *error* production. The effect of different activation states on the outcome measures is also shown in this figure.

In Fig. 1.2.5, *performance* is defined explicitly as the ratio of information
processed to information demanded. If there is a constant processing rate,
this is proportional to

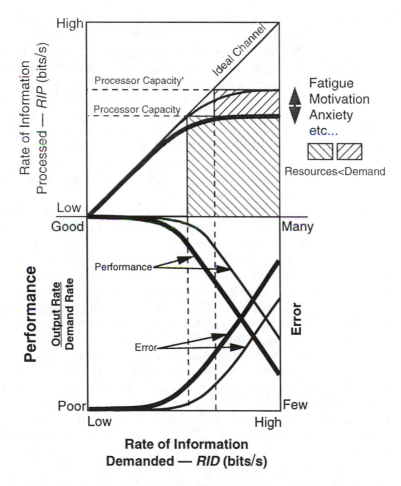

FIG. 1.2.5. Hypothesized relations between rate of information demanded (*RID*), rate of information processed (*RIP*), performance, error,
channel capacity, and the effects of various psychological and physiological
stressors that change the state of activation and arousal.

$$\frac{RIP}{RID}. \tag{9}$$

It should be noted that *response time* is not associated with *performance* in Fig. 1.2.5. Response time has a different, and quite explicit, place in the IP model. As indicated on the right-hand side of Fig. 1.2.5, *error* is hypothesized to be related to the amount of information that remains unprocessed. As the rate of information demanded increases and resources become insufficient (assuming a constant C), the performance ratio decreases with a corresponding rise in the number of errors.

The absolute level of performance—what might be called *accuracy* or *precision*—and the degree of error resulting are contingent on the strategy chosen for processing the information. Remember that a particular strategy involves specific processing structures. Performance using simple heuristics may not be as good as more structured rule-based problem solving or the decisions resulting from a detailed algorithmic approach. However, if it is accurate enough for the requirements of the task, it establishes a standard of performance. Therefore, the chosen strategy establishes the benchmark against which *zero error* is judged. The relative increase in errors shown in Fig. 1.2.5, as a result of information shed, must be compared with the relevant standard. Therefore, Fig. 1.2.5 shows the time-constrained performance of a given information-processing structure.

Whereas Fig. 1.2.5 describes time-constrained problem solving under a constant strategy (therefore a constant information-processing structure), Fig. 1.2.6 presents the problem in terms of the strategy chosen, or *depth of processing* (hence invoking different IP loops in the PCT sense). The dependent variables in Fig. 1.2.6 are *decision time, information processed, accuracy,* and *error.* In the top half of the figure, both decision time and the amount of information processed increase with the depth of processing. As it is assumed that the information-processing rate is constant, the relation between decision time and the amount of information processed is always one to one. However, as previously noted, it is assumed that channel capacity (bits/s) may vary under the effects of various activating factors that act on neural transmission times. Therefore, although a given strategy always involves the same amount of uncertainty to be resolved, the calibration between *bits* and *seconds* depends on the maximum processing rate. The representation of a moving scale with a set point in Fig. 1.2.6, is intended to convey this variable calibration.

Note that to maintain a distinction with performance, as defined in relative terms in Fig. 1.2.5 for time-constrained problem solving, the term *accuracy* is used in the bottom half of Fig. 1.2.6 to denote the outcome of the *absolute* amount of uncertainty resolved. Error in this case is assumed to depend on the difference between the amount of information actually processed

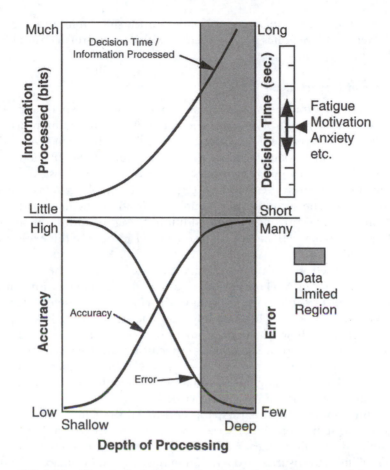

FIG. 1.2.6.   Hypothesized relations for the absolute performance of the human information processor in terms of depth of processing, decision time, information processed, accuracy, and error. Because different levels of psychological and physiological stressors might affect channel capacity, the calibration between information processed and decision time is shown as being variable.

and the amount of information processing required for truly *error-free* performance in an absolute sense. This might be considered to be the gold standard, or the standard of all standards, and represents the most precise performance the human is capable of, with all externally imposed time pressures removed and with the best hierarchy of processing structures. Once the data-limited region is entered, neither error nor accuracy measures show further improvement. The data-limited region reflects fundamental limitations in the ability to discriminate, typically owing to the capability of human senses, rather than to information-processing limits at a cognitive level. Hence, this region is not represented with the time-related variables of Fig. 1.2.5.

Evidence in Support of the IP Model

To date, evidence for the IP model comes from two related experiments, both using an air traffic control simulation to manipulate information-processing load. The first experiment, investigating the relation between time and intensity effects in workload, led to the formulation of the IP model as a framework for interpreting the results that were obtained. This experiment has been described in the literature (Hendy et al., 1997) and is only summarized here. The second experiment arose from predictions made from Experiment 1 and provided useful additional insights into the theoretical underpinnings of the IP model. A more detailed account of this experiment is presented.

The task used in both experiments simulates many of the features of radar control in an ATC environment, although the fidelity with which it represents *real* operations is irrelevant in the context of these experiments. The ATC simulation, which is arguably a complex cognitive task, was specifically designed to allow for the manipulation of certain parameters thought to drive information-processing load—specifically the *time* load and the *intensity* load. In this context, *intensity* load refers to the assumed intensity of attentional demands made by the task and was thought to be related to the total amount of information to be processed.

For these experiments, participants controlled simulated aircraft targets on a representation of a radar screen, to satisfy certain goals related to the scheduled destinations of each aircraft. Participants were briefed as to their responsibilities to ensure safe operation in the airspace (avoid loss of separation, etc.); to route aircraft to their nominated destinations (compass point, airfield, altitude, etc.); and to provide expeditious control (keep path lengths appropriately small).

The variables that can be manipulated in this simulation include:

- The number, type (heavy fixed wing, light fixed wing, and helicopter), and proportion of aircraft by type.
- Target speed (in two increments relative to the cruising speed of the type).
- The schedule (random arrivals, a constant number of aircraft under control, and a variable airspace load that increases and decreases linearly for a preset number of cycles).
- The number of aircraft (e.g., either the number on the screen, the total number presented over the complete session time, or the peak value, depending on the type of schedule).

- The update interval (the time interval after which the screen updates and all targets move on to their next point in space—this determines the time during which all controlling actions must be made).
- The number, location, and runway orientation of airfields.

Control is exercised by mouse commands to heading, altitude, and speed buttons on the screen. All aircraft are cooperative and comply with the commanded speeds, headings, and altitudes at the first opportunity (at the next screen update).

Experiment I

Task.   For Experiment 1, a constant schedule was used. In this type of schedule, the number of targets increases during the first 20% of the session time, from zero at the start of the session to a pre-set value (N). After this time, as soon as a target departures the airspace, it is immediately replaced with a new target, randomly generated in terms of point of arrival, altitude, and point of departure. All aircraft were of the same type (hence speed and rate of climb/descent). Speed control of targets was disabled. Only heading and altitude control were available to participants.

Variables.   Two independent variables were manipulated, namely, the number of aircraft under control in the constant schedule (N) and the update interval for the simulation (ΔT). The following values were used in a balanced five by five repeated-measures design. For reasons described in Hendy et al. (1997), a completely counterbalanced design was not employed.

Number of aircraft (N)	2	3	4	5	6
Update interval (ΔT seconds)	3	6	9	12	15

The dependent variables were:

- *OWL* from subjective workload ratings obtained from the NASA Task Load Index (Hart & Staveland, 1988).
- *errors,* or the total number of missed routings, missed time slots, loss of separation, or unsafe altitude commands (cleared down to zero feet, but not in a position to land at an airport).
- *correct,* or the number of aircraft successfully handled (handed off at their correct destination in the allowed time—note that these aircraft could have been involved previously in a loss of separation incident).
- *success,* or the ratio of the number of aircraft successfully handled (*correct*) to the total number of aircraft available for control.
- *path,* or the actual length of the flight path, compared with the shortest or most expeditious path.

Participants.   Four men and two women participated in this experiment. Ages ranged from 20 to 26 years, and none had previous experience of air traffic control. Participants were remunerated for their involvement, receiving a bonus if they completed the experiment. All completed the experiment.

Hypotheses.   It was hypothesized that the number of aircraft N directly manipulated the amount of information that each participant had to deal with. The update interval ΔT, on the other hand, determined the time during which all decisions had to be actioned. Hence, ΔT manipulated the time available (T_a) in the IP model sense. If the total amount of information to be processed in a given update interval is a function of N, then according to the IP model, the predicted time pressure would be (see the Appendix to this chapter for details of the following derivations):

$$TP = \left(\frac{1}{C}\right)\frac{f(N)}{\Delta T}. \tag{8}$$

If there is a constant rate of processing, the following predictions were made:

Hypothesis 1: $OWL = f_1\{TP\}$

Hypothesis 2: $errors = f_4\{TP - 1\}$

Hypothesis 3: $correct = f_2\{TP\}$

Hypothesis 4: $success = f_3\left\{\frac{1}{TP}\right\}$

Hypothesis 5: $path = f_5\{TP\}.$

Modeling.   A variety of checks were made on the data to detect any obvious biases introduced by the lack of counterbalancing in the experiment. No biases were seen. Two-way repeated measures analyses of variance (ANOVA) were applied to each of the dependent variables. The results of these analyses can be found in Hendy et al. (1997).

Data for the Task Load Index (*tlx*), and the performance variables, were fitted by using various linear and nonlinear models in the two independent variables N and ΔT and the derived variable TP. This analysis assumes that both the chosen strategies and the level of activation remain constant. Both the $\mathbf{b}_0(1 + \mathbf{b}_1 N)$ and the $\mathbf{b}_0(1 + \mathbf{b}_1 N + \mathbf{b}_2 N^2)$ forms for B_r were tried (see the Appendix for details of these derivations). Although many models were in-

TABLE 1.2.1

Models for the Variables *tlx, correct, success,* and *errors* in Terms of the Independent Variables N, ΔT, and the Derived Variable *TP.* For Each Variable, the First Model Is Based on the Underlying Regression Equation of the ANOVA. The Second Model Derives From the IP Model Predictions

Model	Origin	Proportion of Variance
$tlx = \mathbf{b}_0 + \mathbf{b}_1 N + \mathbf{b}_2 \Delta T$	ANOVA	0.46
$tlx = \mathbf{c}_0(1 + \exp(\mathbf{a}_0 TP))$	IP model	0.45
$correct = \mathbf{b}_0 + \mathbf{b}_1 N + \mathbf{b}_2 \Delta T + \mathbf{b}_3 N \Delta T$	ANOVA	0.87
$correct = \mathbf{a}_0 + \mathbf{a}_1 TP + \mathbf{a}_2 TP^2 + \mathbf{a}_4 TP^4$	IP model	0.88
$success = \mathbf{b}_0 + \mathbf{b}_1 N + \mathbf{b}_2 \Delta T + \mathbf{b}_3 N \Delta T$	ANOVA	0.58
$success = 1/[1 + (\mathbf{a}_0 TP^{\mathbf{p}})]^{0.5}$	IP model	0.67
$errors = \mathbf{b}_0 + \mathbf{b}_1 N + \mathbf{b}_2 \Delta T + \mathbf{b}_3 N \Delta T$	ANOVA	0.64
$errors = \exp[\mathbf{a}_0 + (\mathbf{a}_1 TP - 1)]$	IP model	0.69

Note.   See the Appendix for details of these derivations.

vestigated, an internally consistent set was chosen as being the most representative. The selected models for the primary variables, and the proportions of the variance they accounted for, are shown in Table 1.2.1. The results for *path* were weak and are not shown.

In judging the *reasonableness* of the results from the modeling exercise, it was assumed that $\mathbf{b}_0$, $\mathbf{b}_1$, and $\mathbf{b}_2$ should all be positive. This assumption was based on the belief that as each new aircraft appeared on the screen, it *added* linearly to the amount of information to be processed because of its presence alone ($\mathbf{b}_1 N$), and *added* nonlinearly owing to its relations with other targets ($\mathbf{b}_2 N^2$). Therefore, in Experiment 1, models with negative coefficients were rejected on these a priori grounds.

Under these assumptions, the linear form of $B_r = \mathbf{b}_0(1 + \mathbf{b}_1 N)$ was preferred over the second order relation as it gave a consistently reasonable fit over the range of workload and performance measures. This results in an expression for time pressure as follows:

$$TP = \frac{\mathbf{t}_0(1 + \mathbf{b}_1 N)}{\Delta T}. \tag{9}$$

For consistency the same value of $\mathbf{b}_1 = 1.55$ was used in all models. This value came from the best fit for the *tlx* data. A value of $\mathbf{t}_0 = 0.75$ seconds was estimated from the best fit for *correct.*

Results.   Plots of the main independent variables from Experiment 1, for all subjects, under all conditions, are shown in Fig. 1.2.7. These data are plotted as a function of the derived variable *time pressure.* Taking the model-

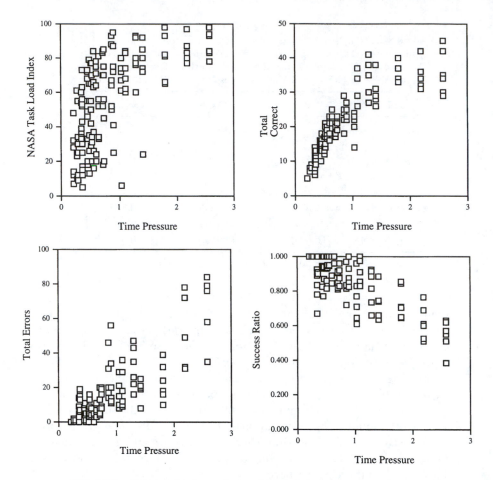

FIG. 1.2.7.   Plots of the dependent variables *tlx, correct, errors,* and *success*
from Experiment 1, against the derived variable time pressure.

ing exercise (see Table 1.2.1) overall, the results for the variables *correct, success,* and *errors* were interpreted as being consistent with the hypothesis that
time pressure was a factor in determining the performance of the human
information processor in this environment. For *tlx,* the fit derived from the
IP model performed no better than the relation derived from the ANOVA.
This result, however, is not interpreted as being inconsistent with IP model
predictions. For a more detailed discussion of the results of this experiment,
see Hendy et al. (1997).

For comparison, particularly with Fig. 1.2.5, the best-fitting models for *tlx,
correct, errors,* and *success* are shown in Fig. 1.2.8. The following equivalencies
were assumed (Hendy et al., 1997):

Rate of information demanded (*RID*) $\propto$ *TP.*
Performance $\equiv$ *success.*
Errors $\equiv$ *errors.*
Rate of information processed (*RIP*) $\propto$ *correct.*

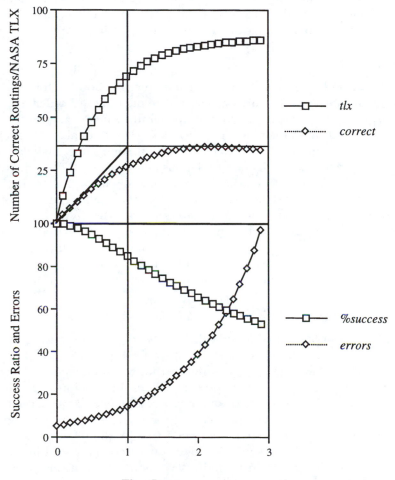

Time Pressure

FIG. 1.2.8.   *Top panel:* Plot of the dependent variables NASA Task Load Index (*tlx*), and the total number of aircraft correctly routed to their destinations (*correct*), versus the derived variable time pressure (*TP*).

  Bottom panel: Plot of the dependent variables percent success ratio (*% success*), and the total number of errors made (*errors*), versus the derived variable time pressure (*TP*).

The relations for *correct, success,* and *error* mimic the forms hypothesized in Fig. 1.2.5 quite reasonably. Note that the vertical line in Fig. 1.2.8 (at *TP*= 1) marks the separation between the region where resources exceed demand (i.e., *TP*< 1) and that where resources do not meet the demand (*TP*> 1). The horizontal line in the upper panel of Fig. 1.2.8, at the peak value of *correct*, is associated with the limited capacity of the processing structures involved in servicing the task. The line joining the origin and the point of intersection between the vertical and horizontal lines can be equated to the input–output relation of an ideal information-processing channel.

Experiment 2

The failure, in Experiment 1, of the second-order relationship $\mathbf{b}_0(1 + \mathbf{b}_1 N + \mathbf{b}_2 N^2)$ to reliably explain more of the variance than the first-order relationship $\mathbf{b}_0(1 + \mathbf{b}_1 N)$ was attributed to incompletely developed expertise among the participants in this complex cognitive task (Hendy et al., 1997). It was hypothesized that participants concentrated on controlling individual aircraft targets while largely ignoring relations between targets. This interpretation was consistent with results that had been obtained previously (Burbank, 1994). Hendy et al. (1997) suggested that the relative contribution of the N^2 term might be an indicator of gathering expertise in the task, as participants went beyond a simple target-by-target strategy and started to be sensitive to patterns of targets. Experiment 2 set out to examine this hypothesis (see East, 1996a, 1996b; East, Hendy, & Matthews, 1996).

One might also expect that the values of the parameters $\mathbf{t}_0$, $\mathbf{b}_1$, and $\mathbf{b}_2$ in

$$TP = \frac{\mathbf{t}_0(1 + \mathbf{b}_1 N + \mathbf{b}_2 N^2)}{\Delta T}, \tag{10}$$

would vary from individual to individual, depending on the strategies that each employed in interacting with the task. In Experiment 1, it was not possible to individually calibrate the models because of insufficient data points over the complete range of time pressure values. In addition, the requirement that $\mathbf{b}_1$ and $\mathbf{b}_2$ always be positive was questionable, as further analysis showed that this condition is not necessary for *TP* to monotonically increase with *N*. Indeed, as will be seen, the notion that *TP* should always increase with *N* may also be inappropriate. These issues were addressed in Experiment 2, along with the sensitivity of the N^2 term to increasing proficiency at the task.

Task.   The simulation environment for this experiment was the same as in Experiment 1. A constant schedule was again used for the main trials. Values of *TP* were preselected in nominal 0.25 steps, over the range 0.5 to 2.75, using the best estimate for *TP* derived from the pooled data of Experiment 1, namely:

$$TP = \frac{0.75\,(1 + 1.55N)}{\Delta T}. \tag{11}$$

Variables.   Values of the variables N and ΔT were chosen to satisfy the pre-selected targets for *TP*. Because N is obviously an integer value, and ΔT must be chosen from the values available in the simulation software—these were {3, 6, 9, 12, 15} seconds—the target values of *TP* were approximated only, as shown in Table 1.2.2.

Design.   The experiment consisted of 14 daily sessions with each session lasting approximately 1.5 to 2 hours. Day 1 was a training and familiarization session. On Day 14, a series of trials related to measuring the effects of over-load on situation awareness were performed. These conditions are irrele-vant in the current context. The main experiment was conducted on Days 2 to 13. On even-numbered days, participants performed Conditions 1, 3 , 5, 7, and 9 from Table 1.2.2. On odd-numbered days, the performed Condi-tions 2, 4, 6, 8, and 10. The order of presentation of conditions was random-ized to prevent any systematic bias in workload ratings.

Over the 12 days of the main experiment, the complete set of experimen-tal conditions was experienced 6 times (these are labeled Blocks 1 to 6; see Table 1.2.3).

Participants.   Potential participants were screened by using a version of the computer game Tetris as an indicator of some of the spatiotemporal skills that were thought to be associated with performance in the ATC task

TABLE 1.2.2
Values of the Independent Variables N and ΔT Used in Experiment 2

Condition	Estimated Time Pressure	ΔT	N
1	0.47	9	3
2	0.73	9	5
3	0.99	9	7
4	1.25	9	9
5	1.48	6	7
6	1.68	6	8
7	2.07	6	10
8	2.19	3	5
9	2.58	3	6
10	2.97	3	7

Note.   The estimated time pressure (*TP*) was calculated from Equation (13).

TABLE 1.2.3
Ordering of Conditions for Experiment 2 Into 6 Blocks of Trials

Day	Block	Conditions (Random Order)
1	—	Training sessions
2	1	1, 3, 5, 7, 9
3	1	2, 4, 6, 8, 10
4	2	1, 3, 5, 7, 9
5	2	2, 4, 6, 8, 10
6	3	1, 3, 5, 7, 9
7	3	2, 4, 6, 8, 10
8	4	1, 3, 5, 7, 9
9	4	2, 4, 6, 8, 10
10	5	1, 3, 5, 7, 9
11	5	2, 4, 6, 8, 10
12	6	1, 3, 5, 7, 9
13	6	2, 4, 6, 8, 10
14	—	SA conditions

Note. Days 2 to 13 cover the main experiment.

(East, 1996b). From those who completed the screening process, seven males and two females were selected to participate in the main experiment. Their ages ranged from 30 to 50 years. This group represented, according to availability, the three highest scorers, the three lowest, and three who scored in the middle of the range. A double blind approach was taken, and neither participants nor the experimenter knew who fell into which group. All participants received remuneration for their involvement.

Hypotheses. It was hypothesized that over the six blocks of trials, participants would gain expertise in performing the ATC task. This expertise should be indicated, in general, by improvements in all the objective performance indicators (*correct, errors, success*) and reductions in perceived workload under identical external loading conditions. Through developing expertise, it was expected that participants would start to attend to the relations between aircraft targets. This is captured in the following quotation:

> It might be hypothesized that the ability to deal with spatial relationships between aircraft targets characterizes expert rather than novice performance. Hence, models for experts might be expected to be more sensitive to the N^2 term than was the case in this study. If this hypothesis is substantiated, the relative loading of the second order term could be an index of expertise for this task. Generally, one would expect the value of b_1 to get smaller, and the ratio $b_2:b_1$ to get larger, with practice. (Hendy et al., 1997, p. 45)

In other words, it was hypothesized that the relative contribution of the quadratic term in

$$TP = \frac{t_0(1 + \mathbf{b}_1 N + \mathbf{b}_2 N^2)}{\Delta T}, \tag{14}$$

would increase over blocks.

Modeling.   With the design of this experiment, there were sufficient data points (10 conditions) to develop models for each participant, for each block of the experiment. Because the data from Experiment 1 showed a tendency for a delayed onset of increased workload ratings with time pressure (Hendy et al., 1997), a PROBIT model of the form

$$OWL = \mathbf{c}_0 ZCF\left[\frac{(-\mathbf{c}_1 + TP)}{\mathbf{c}_2}\right], \tag{12}$$

was used in preference to the exponential chosen for Experiment 1 (ZCF represents the cumulative normal probability function—it should be noted that although PROBIT models were tried in Experiment 1, they did not improve the fit for the pooled data relative to the exponential relation). The data for each subject, for each block of trials, were fitted with the PROBIT model of Equation 15. This gave 54 separate models (9×6) in all.

Results.   Corrected r^2 values ranged from 0.000 to 0.991 with the majority (40 out of 54) exceeding 0.800. The second-order model accounted for a greater proportion of the variance, compared with the linear model, in all but 2 of the 54 cases. For 1 of these 2 cases, the difference was only 0.001. For the other, the second-order model failed to converge, whereas the first-order model did, but with a low $r^2 = 0.107$. Overall, as reported by East (1996b), the r^2 for the linear models was high and left little headroom for improvement.

Although the original hypotheses anticipated predictable changes in the $\mathbf{b}_2 : \mathbf{b}_1$ ratio, it can be shown that owing to the nature of the model used and the acceptable ranges of its parameters (see East, 1996b), this ratio is essentially meaningless with respect to the purpose for which it was proposed—specifically, tracking the development of expertise via the relative contribution of the quadratic (N^2) term as compared directly with that of the linear (N) term. A much better demonstration of the relative changes in processing strategy as expertise is gained, and the effect of these strategy changes on information processing load, is shown in Fig. 1.2.9.

If the original hypotheses are to be borne out, it is expected that as expertise is gained the predicted time pressure for processing a single aircraft would decrease over time, and subjects would provide evidence that they were processing information over and above a simple linear increase, as each additional aircraft target came under control.

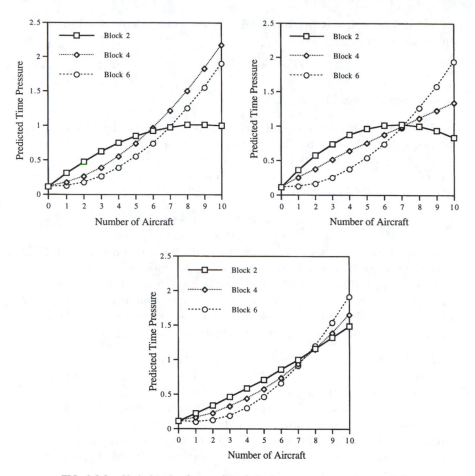

FIG. 1.2.9. Variation in the predicted time pressure versus external demand for three subjects in the medium spatiotemporal group (as selected
from their Tetris scores) early during the experiment (block 2), midway
(block 4), and at the end of the experiment (block 6).

In Fig. 1.2.9, the predicted time pressure is shown for each of the three
subjects from the medium spatiotemporal group, as functions of the number of aircraft under potential control. It was found that participants from
the medium group showed greater changes that might be attributed to improving expertise (East, 1996b) than did those from the low or high groups.
The update interval was set at 9 seconds for these calculations, and $t_0 = 1s$
was used to set the y-axis intercept. Hence,

$$TP = \frac{(1 + \mathbf{b}_1 N + \mathbf{b}_2 N^2)}{9}. \tag{16}$$

The values of b_1 and b_2 were derived from the best-fitting second-order models for *OWL,* for each of the three subjects, for each block of the experiment. $t_0 = 1s$ is common to all models, and hence it is not individually calibrated as are b_1 and b_2. $t_0 = 1s$ gave the most consistent conditions for convergence.

The slope of the line from $N = 0$ to $N = 1$ represents the incremental increase in time pressure, because of the introduction of a single aircraft target. If participants were processing just this amount of new information for each new target, say ignoring the need for aircraft separation, then *TP* should continue to increase linearly with N. On the other hand, if the *TP* versus N relation is concave downward, then participants were not even keeping up with the basic demands as each new target arrived. This suggests that information was being shed and, therefore, the actual time pressure the person experienced climbed sublinearly and may have plateaued or even decreased in a U-shaped fashion if he or she gave up on the task (this would actually represent a change of strategy—targets not attended to do not add directly to the information-processing load, but might add indirectly to the load because of increasing clutter on the screen). This type of relation was associated with novice performance (East, 1996b). On the other hand, if the relation is concave upward, *TP* was increasing supralinearly with N. In this case, subjects were processing more than just the information associated with the presence of a target in isolation.

From Fig. 1.2.9, it can be seen that, for all participants in the medium spatiotemporal group:

- The *TP* associated with controlling one target decreased from Block 2 to Block 4 to Block 6.

- The *TP* versus N relation went from concave downward or essentially linear to concave upward over blocks.

Therefore, increasing upward curvature is associated with increasing levels of expertise in this task even if the measure for representing developing curvature originally proposed, the ratio of $b_2 : b_1$, fell short of the mark owing to the erroneous assumption that these coefficients must always be positive.

In Fig. 1.2.10, the mean *OWL* and mean total number of *errors* data are shown plotted against the mean predicted time pressure values calculated from Equation (16), for each of the 10 conditions of Experiment 2 (see Table 1.2.2). These means were calculated by averaging over blocks (6) and subjects (3), in each of the high, medium, and low spatiotemporal groups. For comparison, *OWL* and *errors* are also shown plotted against time pressure values from the linear model of Equation (13). Whereas Equation (13) represents the best-fitting model for the pooled data of Experiment 1, the predicted time pressure values from the second-order model are individually calibrated. The effect of the individual calibration appears to be to col-

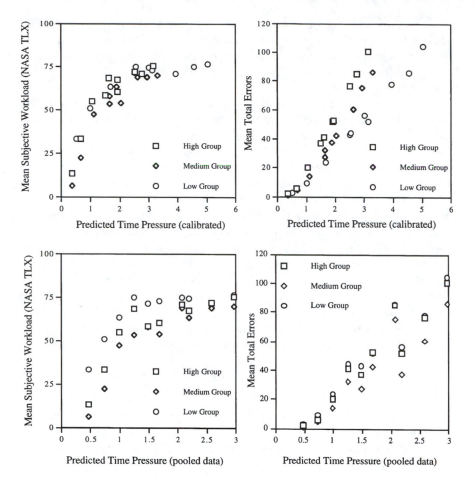

FIG. 1.2.10.   *Top panel:* Plot of the mean values of *OWL* and total errors against the individually calibrated predicted time pressure values from Equation 16.

Bottom panel: Plot of the mean values of *OWL* and total errors against the predicted time pressure values from the best-fitting linear model of Experiment 1 (see Equation 13).

lapse the subjective workload values into a single relation, while splitting out the error data into three distinct curves that, it might be argued, represent three different processing strategies used by the low, medium, and high spatiotemporal groups (see also East, 1996b).

If the concept of time pressure is robust, and of course if the measurement of workload is reliable, then this is what might be expected. All subjects should report the same experience of workload under the same internally experienced values of time pressure. However, the absolute number of errors they commit depends on the strategy chosen. The strategy in turn provides the calibration between external load factors (in this case N and ΔT) and the perceived time pressure.

Summary

The IP model provides a consistent framework that encompasses many aspects of human information processing, for example: automatic versus controlled processing; capacity-limited performance; multiple task performance; effects of operator strategies; serial and parallel processing behaviors; continuous tasks; speed–accuracy trade-offs; effects of training and skilled performance; and awareness for self-report. Furthermore, the IP model captures each of the three elements of workload that have provided the direction for much of the workload research over recent years: a factor due to time; a factor due to the intensity of effort; and a factor due to psychological and or physiological stresses.

Most important, it does so in an integrated framework that not only includes these factors in isolation, but establishes the relations between them. The IP model can be distinguished from classical resource theory through its explicit representation of *time*. Time is at the essence of the IP model; it is inextricably bound up in its various elements, through processing rates and decision times. The temporal aspects of the IP model establish the dynamic behavior of the information-processing components in a perceptual control loop. This directly links the IP model with perceptual control theory.

The IP model assumes that incoming tasks compete for processing structures that, once allocated to a particular task element, process information in a serial fashion. While a particular structure is engaged in processing some element of a given task, it is unavailable to process elements of other tasks. It is assumed that multiple structures exist for processing, and, to the extent that tasks that overlap in the time domain share part or all of a common structure, multiple task interference is seen as increased decision times because of serial processing of some or all the task elements. Furthermore, it is assumed that those components of overlapping tasks that involve a common structure are actually processed serially by time multiplexing.

From the IP model, if the strategy chosen for loop control is one of low precision, then a loss of accuracy results from unprocessed information or imprecise rules—despite the possibility that loop closure in the PCT sense is achievable so that **p** matches the set goal **g**. The overload condition (*time pressure* > 1) occurs when the dynamics of the demand function exceed the bandwidth of the closed loop system. Information is shed, which results in *errors*. Hence, errors are associated both with the transient behavior of the adaptive control loop (i.e., in terms of the remaining distance from the goal in PCT terms) and with unprocessed information resulting from a low-precision strategy or channel bandwidth limitations (from the capacity limitation of the IP model). In the former case, errors disappear over time, given sufficient loop gain. In the latter case, errors remain until the strategy changes or the overload is removed.

APPLICATIONS OF THE IP/PCT MODEL

The work that led to the IP/PCT model was motivated by a very practical problem: how does one inject human factors knowledge into design at the very earliest opportunity—during the concept and feasibility phases of system development? Workload and performance prediction is an integral part of a structured approach to human engineering front-end analysis (MIL-STD-46855, 1994). Because this early involvement suggests an approach based on modeling or simulation, the IP/PCT model had to take quantitative form, as a purely descriptive model would have fallen short of this goal. Yet the IP/PCT model is more than just a series of equations; it also provides a comprehensive descriptive framework that can be applied in a variety of circumstances.

Following are some examples of how the IP/PCT model has been put to use. The first example demonstrates the use of the IP/PCT model in a manner consistent with its heritage, that is, in a direct quantitative sense. Other examples show the use of the IP/PCT model in a wider sense, providing a descriptive framework for investigating human performance under time stress and prescribing solutions to real world problems.

Task Network Simulation

The IP/PCT model can be given mathematical form for use in predictive tools for human factors engineering. Indeed, the IP/PCT model has been coded into a software tool for task network analysis. Called the integrated performance modelling environment (IPME), this software is commercially available. IPME contains both the IP/PCT model and the prediction of operator performance (POP) model (Jordan et al., 1996) from the United Kingdom's Centre for Human Sciences at Farnborough. An activity is underway, within the nations of The Technical Cooperation Program (TTCP),[2] to competitively validate models for this type of application. This validation will include the IP/PCT model, POP, W/INDEX (North & Riley, 1989), Win-Crew (Archer & Lockett, 1997), and the VACP model of Aldrich, Craddock, and McCracken (1984).

---

[2]TTCP is an agreement between the governments of Australia, Canada, New Zealand, United Kingdom, and the United States to support collaborative research.

In the IPME implementation, time-multiplexed serial processing, as assumed for the IP model, is used to develop interference effects during multitask concurrent processing. Task interference is manifested directly as an increase in the completion time of the tasks. This interference is graded according to the extent of the assumed competition for common processing structures. For details of this implementation see Hendy and Farrell (1997).

Issues in Team Performance

PCT describes a world in which humans and machines interact through the medium of their external environment. During the performance of a team task, members act on certain variables in their external world until the goals of the team members are perceived to be met or until the team stops attending to the task. If this process is to converge, then all members of the team, including the nonhuman members, must be working to the same goals or at least to a consistent set of subgoals. This requirement is predictable from the literature on the stability of multicontroller systems (Van de Vegte, 1986). Team performance breaks down if multiple controllers try to act on the same or coupled variables in the environment while pursuing different goals or incompatible subgoals. Also, because the actions that team members initiate are formed by their internal state of knowledge, these mental models should be identical or at least compatible. If this is not the case, then one controller perceives the actions of the other controller(s) as disturbances in the environment that must be acted on. Under these conditions, controllers are fighting for control, with each controller trying to undo the actions of the other. Because of latencies in the output side of the control loop, divergent behavior may result (e.g., the avoidance behavior of two people meeting head on in a narrow corridor).

Interestingly, this view of human–human interaction applies equally to human–machine interaction. A nonhuman team member (e.g., a Flight Management System or FMS) has certain objectives (goals) programmed into it by the system designers. It also takes commands from the human team members. These become the top-level "goals" of the machine, which initiates a series of lower level control loops in response to the various algorithms built into the box. It is often the human's lack of understanding of these lower level goals that causes problems in highly automated systems. Hence, the PCT framework is a completely general framework for the analysis of human–machine–human systems in any combination.

The Role of Communication
in Human–Machine–Human Interaction

Building on the concepts presented here and starting with the notion of an all capable individual—the *Superoperator* who is able to accomplish the team task alone—one can trace the emergence of communication between team members and the requirement for leadership by using the PCT framework. As one moves from (e.g., see Hendy, 1998) the *Superoperator* to the *Dream Team* (not quite as capable as *Superoperator* but having identical life experiences that ensure identical knowledge structures up to the time when they commence the task) to the real world (consisting of a collection of individuals with some common training and life experience, but from many diverse backgrounds), a structure emerges that suggests that communication is required to fulfill a number of functions. These are:

- Allocate responsibilities (to prevent duplication of task activity and manage individual task load).
- Pass the results of internal processes (i.e., thinking without observable action) so that the internal mental models of team members remain the same.
- Provide feedback on system states that are not in the common locus of attention (includes both asking for and offering information).
- Direct attention (i.e., inviting direct observation rather than using verbal descriptions to convey information).
- Establish common goals.
- Build common mental models of system states (this pertains only to the mental models that each member of the team might call on to control the loops he or she is responsible for—not everyone needs to know everything, and in a time-pressured environment this is a luxury one cannot afford; however, the alignment of mental models should be a goal if time permits—this allows for appropriate decisions in the presence of unexpected or novel events).

In addition, when we are dealing with human–human interaction rather than strictly human–machine interaction, a secondary communication role emerges related to establishing and maintaining effective external communications, which includes:

- Establishing leadership.
- Establishing the level of trust.
- Establishing the authority gradient.
- Establishing receptiveness, attentiveness, cooperativeness, assertiveness, and so on.

This structure provides a taxonomy for verbal protocol analysis. The same structure could be applied to nonverbal communication.

Crew Resource Management (CRM)

Crew resource management, or more generally team resource management (TRM), has evolved over the years from an early emphasis on individual management styles and interpersonal skills to the more recent focus on error management and decision making (Helmreich, 1996; Hendy & Ho, 1997). Yet CRM/TRM has previously lacked a theoretical basis to guide the development of training and assessment programs. The IP/PCT model fills this gap.

Resource management can be seen to be the management of *time* (from the IP model), *knowledge* (goals and mental models from PCT) and *attention* (what you are controlling, again from PCT). A proposed training program in aviation decision making has been designed, based directly on the framework provided by the IP/PCT model (Hendy & Ho, 1997). Classroom training is intended to cover the following 11 modules:

- Introduction to the decision-making loop.
- Leadership and followership.
- The emergence of communication and leadership functions.
- Sensation/perception.
- Goal setting.
- Action selection.
- Resource management.
- Management of control and attention.
- Management of time pressure.
- Management of knowledge.
- Communication.

These modules encompass each of the static components of the PCT loop, along with the emergent properties identified for multicontroller systems. The IP model provides the dynamic aspect of information processing in the PCT loop. A proposed assessment instrument has also been developed, based on the degree to which individuals and the team manage the three elements of *time, knowledge,* and *attention.*

CONCLUSIONS

This chapter describes a model of human information processing that provides a concise theoretical framework for integrating a great deal of what we know about operator performance, error production, and perceptions of workload under stress. This model, the information processing/perceptual control theory (IP/PCT) model is primarily concerned with the stress imposed by time, which is said to directly influence primary decision-making loops. However, the model also encompasses the effects of more diffusely acting physiological and psychological stressors resulting from drugs, alcohol, fatigue, anxiety, and so on. The IP/PCT model is both an analytical model and a descriptive model. Unlike arbitrary models that are empirically derived to fit available data, the parameters of the IP/PCT model, because it is rooted in theory, have meaning in a human information-processing sense. Also because of its analytic nature, the IP/PCT model can be instantiated into software for quantitative workload and performance prediction.

As a descriptive framework, the model provides a powerful structure for thinking about problems of human information processing. To date, the IP/PCT model has provided insight into issues related to team performance and the emergence of communication and leadership in teams and groups. The model has also provided clear guidance in formulating a syllabus for aviation decision-making training and assessment.

REFERENCES

Aldrich, T. B., Craddock, W., & McCracken, J. H. (1984). *A computer analysis to predict crew workload during LHX scout-attack missions* (Vol. 1: MDA903-81-C-0504-1-84[B]). Fort Rucker, AL: United States Army Research Institute Field Unit.

Allen, R. W., McRuer, D. T., & Thompson, P. M. (1989). Dynamic systems analysis programs with classical and optimal control applications of human performance models. In G. R.McMillan, D. Beevis, E. Salas, M. H. Strub, R. Sutton & L. van Breda (Eds.), *Applications of human performance models to systems design* (pp. 169–484). New York: Plenum Press.

Archer, R. D., & Lockett, J. F. (1997). WinCrew—A tool for analyzing performance, mental workload and function allocation among operators. In E. F. Fallen, M. Hegan, L. Bannon, & T. McCarthy (Eds.), *Proceedings of the First International Conference on Allocation of Functions, Volume II: Short Papers* (pp. 157–160). Louisville: IEA Press.

Bourbon, W. T. (1990). On the accuracy and reliability of predictions by control-system theory. *Perceptual and Motor Skills, 71,* 1331–1338.

Bourbon, W. T. (1996). On the accuracy and reliability of predictions by Perceptual Control Theory: Five years later. *The Psychological Record, 46,* 39–47.

Burbank, N. (1994). *The Development of a task network model of operator performance in a simulated air traffic control task* (DCIEM-94-05). New York, Ontario, Canada: Defence and Civil Institute of Environmental Medicine.

Card, S. K., Moran, T. P., & Newell, A. (1983). *The psychology of human computer interaction.* Hillsdale: Lawrence Erlbaum Associates.

Detweiler, M., & Schneider, W. (1991). Modeling the acquisition of dual-task skill in a connectionist/control architecture. In D. L. Damos (Ed.), *Multiple-task performance* (pp. 69–99). London: Taylor & Francis.

East, K. P. (1996a). *Validation of the IP model.* (PW&CS Contract No. W7711-5-7258/001-TOS for DCIEM, North York, Canada). Toronto, Ontario, Canada: Kim East Consulting.

East, K. P. (1996b). *Workload, performance and the development of situational awareness: An information processing approach.* Master's Thesis, University of Guelph, Guelph, Ontario, Canada.

East, K. P., Hendy, K. C., & Matthews, M. (1996). Validation of an information processing-based model for workload and performance prediction. *Proceedings of the Human Factors Society of Canada's 29th Annual Conference* (pp. 155–160). Mississauga, Ontario, Canada: Human Factors Association of Canada.

Endsley, M. R. (1993). Situation awareness in dynamic decision making: Theory. In *Proceedings of the First International Conference on Situational Awareness in Complex Systems.*

Farrell, P. S. E., & Semprie, M. A. H. (1997). *Layered protocol analysis of a control display unit* (97-R-70). North York, Ontario, Canada: Defence and Civil Institute of Environmental Medicine.

Gopher, D., & Donchin, E. (1986). Workload—An examination of the concept. In K. R. Boff, L. Kaufman, & J. P. Thomas (Eds.), *Handbook of perception and performance, Volume II: Cognitive processes performance* (pp. 41-1–41-49). New York: Wiley.

Hart, S. G., & Staveland, L. E. (1988). Development of a NASA-TLX (Task Load Index): Results of empirical and theoretical research. In P. S. Hancock & N. Meshkati (Eds.), *Human mental workload* (pp. 139–183). Amsterdam: Elsevier Science.

Hawkins, H. L., Olbrich-Rodriguez, E., Halloran, T. O., Ketchum, R. D., Bachmann, D.B., & Reicher, G. M. (1979). *Preparation cost and dual task performance: Further evidence against a general time sharing factor* (Contract No. N14-77-C-643). Arlington, VA: Personnel and Training Research Program, Office of Naval Research.

Hehmreich, R. L. (1996). *The evolution of crew resource management.* Paper presented to the IATA Human Factors Seminar, Warsaw, Poland. Available: http://www.psy.utexas.edu/psy/helmreich/main.htmm#new.

Hendy, K. C. (1995). Situation awareness and workload: Birds of a feather? In *Proceedings (CP-575) of the AGARD AMP Symposium on Situation Awareness: Limitations and Enhancement of the Aviation Environment* (pp. 21-1–21-7). Neuilly-sur-Seine, France: Advisory Group for Aerospace research and Development.

Hendy, K. C. (1998). CRM: More than just talk, talk, talk . . . *Flight Safety Spotlight, 98*(1), 12–15.

Hendy, K. C., & Farrell, P. S. (1997). *Implementing a model of human information processing in a task network simulation environment* (97-R-71). North York, Ontario, Canada: Defence and Civil Institute of Environmental Medicine.

Hendy, K. C., & Ho, G. (1997). *Human factors of CCC-130 operations—Volume 5: Human factors in decision making* (DCIEM 98-R-18). North York, Ontario, Canada: Defence and Civil Institute of Environmental Medicine.

Hendy, K. C., Liao, J., & Milgram, P. (1997). Combining time and intensity effects in assessing operator information processing load. *Human Factors, 39,* 30–47.

Hess, R. A. (1987). Feedback control models. In G. Salvendy (Ed.), *Handbook of human factors* (pp. 1212–1242). New York: Wiley.

Jex, H. R. (1988). Measuring mental workload: Problems, progress, and promises. In P. A. Hancock & N. Meshkati (Eds.), *Human mental workload* (pp. 5–38). Amsterdam: Elsevier Science.

John, B., Rosenbloom, P., & Newell, A. (1985). A theory of stimulus-response compatibility applied to human–computer interaction. In M. Helander (Ed.), *Proceedings of the ACM CHI'85 Conference on Human Factors in Computing Systems* (pp. 213–219). New York: Association for Computing Machinery.

Jordan, C. S., Farmer, E. W., Belyavin, A. J., Selcon, S. J., Bunting, A. J., Shanks, C. R., & Newman, P. (1996). Empirical validation of the prediction of operator performance (POP) model. In *Proceedings of the Human Factors and Ergonomics Society's 40th Annual Meeting* (pp. 39–43). Santa Monica, CA: Human Factors and Ergonomics Society.

Kinchla, R. A. (1980). The measurement of attention. In R. S. Nickerson (Ed.), *Attention and performance VIII* (pp. 213–238). Hillsdale, NJ: Lawrence Erlbum Associates.

Knight, J. L., Jr. (1987). Manual control and tracking. In G. Salvendy (Ed.), *Handbook of human factors* (pp. 182–218). New York: Wiley.

Meister, D. (1985). *Behavioral analysis and measurement methods.* New York: Wiley.

MIL-STD-46855 (1994). *Human engineering requirements for military systems, equipment and facilities.* Washington, DC: Department of Defence.

Miller, G. A. (1971). What is information measurement? In W. C. Howell & I. L. Goldstein (Eds.), *Engineering psychology—Current perspectives in research* (pp. 132–142). New York: Appleton-Century-Crofts.

Myers, D. (1989). Chapter 9: Learning and thinking. In A. Vinnicombe & D. Posner (Eds.), *Psychology* (2nd ed., pp. 229–253). New York: Worth.

North, R. A., & Riley, V. A. (1989). W/INDEX: A predictive model of operator workload. In G. R. McMillan, D. Beevis, E. Salas, M. H. Strub, R. Sutton, & L. van Breda (Eds.), *Applications of human performance models to systems design* (pp. 81–89). New York: Plenum Press.

Pew, R. W., Baron, S., Feeher, C. E., & Miller, D. C. (1977). *Critical review and analysis of performance models applicable to man–machine systems evaluation* (AFOSR-TR-77-0520). Bolling Air Force Base, Washington, DC: Air Force Office of Scientific Research.

Powers, W. T. (1973). Feedback: Beyond behaviorism. *Science, 17,* 351–356.

Rasmussen, J. (1983). Skills, rules and knowledge: Signals, signs and symbols and other distinctions in human performance models. *IEEE Transactions on Systems, Man, and Cybernetics, 13,* 257–266.

Robertson, R. J., & Powers, E. T. (Eds.). (1990). *Introduction to modern psychology: The control-theory view.* Gravel Switch, KY: The Control Systems Group.

Sanders, A. F. (1979). Some remarks on mental load. In N. Moray (Ed.), *Mental workload: Its theory and measurement* (pp. 41–77). New York: Plenum Press.

Storer, J. E. (1957). *Passive network synthesis.* New York: McGraw-Hill.

Thomason, J. G. (1956). *Linear feedback analysis.* London: Pergamon Press.

Thompson, M. M., Jamieson, D. W., & Hendy, K. C. (1997). Assessing mental model content and function: A metric of effective crew performance. In R. S. Jensen, & L. A. Rakovan (Eds.), *Proceedings of the 9th international symposium on aviation psychology* (pp. 911–976). Columbus, OH: Aviation Psychology Laboratory, Ohio State University.

UTP-7. (1994). *Minutes of the annual meeting, 21–24 June 1994, Melbourne, Australia.* (minutes of Technical Panel UTP-7). Washington, DC: The Technical Cooperation Program.

Van de Vegte, J. (1986). *Feedback control systems.* Englewood Cliffs, NJ: Prentice-Hall.

Weiten, W. (1989). *Psychology: Themes and variations.* Pacific Grove, CA: Brooks-Cole.

Wickens, C. D. (1992). *Engineering psychology and human performance* (2nd ed.). New York: HarperCollins.

Appendix

Derivation of Equations
for the Prediction of Dependent Variables
OWL, Success, Correct, Errors, and *Path*

OWL	Operator Workload.
correct	Number of aircraft correctly handled.
success	Ratio of the number of aircraft successfully handled to the total number presented.
errors	Total number of operational errors such as near misses, collisions, missed destinations.
path	Ratio of the actual path length to the shortest path from entry to exit.
T_a	Time available for processing.
B	Amount of information actually processed in time T_a.
B_r	Amount of information required to be processed (this is a function of the task and the strategy chosen for solution).
T_r	Time required to process B_r bits of information.
C	Channel capacity (bits s^{-1})
TP	Time pressure (T_r/T_a).
N	Number of aircraft.
ΔT	Update interval.
T_{sess}	Total session time.
n_{sess}	Number of update intervals per session.
I_{sess}	Total amount of information processed in one session.
ΔI	Information processed per update interval.
ΔI_{shed}	Information unprocessed or shed in an update interval.
I_{shed}	Total amount of information shed.

The information processing model presented here provides a strong framework for the following quantitative derivations. However, in all cases certain assumptions have had to be made about the specific form of various equations not covered by the model predictions. Although various methods are used to establish the nature of these underlying relations (e.g., choosing functions that satisfy the problem constraints), the final choice of function is somewhat arbitrary. Therefore, in several cases, alternative forms are offered, even if constraints are potentially violated in the interests of parsimony. In the realm of empirical model building, one must learn to live with such conflicts and rely on one's judgment. The more theoretically derived and constrained relations provide some possibility of extrapolation, but those that are determined by purely empirical methods do not.

INFORMATION TO BE PROCESSED

Assume

$$B_r = f\{N\}$$
$$= \mathbf{b}_0(1 + \mathbf{b}_1 N + \mathbf{b}_2 N^2),$$

The first-order term in N represents the amount of information due to each aircraft (the routine load—see Arad's ATC model; Pew, Baron, Feeher, & Miller, 1977), while the second-order term is introduced as a result of the amount of information due to the number of pairwise relations (Arad's airspace load). Strictly, the number of pairwise relations is given by the combinatorial form:

$$\binom{N}{2} = \frac{N(N-1)(N-2)!}{2(N-2)!}$$

$$= \frac{N^2 - N}{2},$$

hence, the first-order term N in fact contains contributions from both the routine load and the airspace load.

Whether the second-order term is necessary is likely to depend on a subject's strategy. If the subject is not actively working to avoid conflicts between aircraft (i.e., attending to intertarget relations only whenever a collision or near miss is imminent), a first-order form is sufficient. That is:

$$B_r = \mathbf{b}_0(1 + \mathbf{b}_1 N).$$

In this case, solving for the intercept $\mathbf{b}_0$, that is when $N = 0$,

$$TP = \frac{T_r}{T_a} = \frac{B_r}{C\Delta T} = \frac{\mathbf{b}_r}{C\Delta T}$$

where $T_a = \Delta T$. Assume that the minimal decision time (equivalent to setting a value of $TP = 1$) for this task involves making a single glance at the screen, to confirm the absence of targets ($N = 0$). Assuming that such a glance takes $\mathbf{t}_0$ seconds, then

$$1 = \frac{\mathbf{b}_0}{\mathbf{t}_0 C},$$

or

$$\frac{\mathbf{b_0}}{C} = \mathbf{t_0}.$$

Therefore, in general, for this task

$$TP = \frac{\mathbf{t_0}(1 + \mathbf{b_1}N + \mathbf{b_2}N^2)}{\Delta T}.$$

MODEL FOR OPERATOR WORKLOAD (OWL)

In general,

$$TP = \frac{T_r}{T_a}$$

$$T_r = \frac{B_r}{C}.$$

Then as,

$$T_a = \Delta T,$$

$$TP = \frac{B_r}{C\Delta T}$$

$$= \frac{1}{C}\frac{f\{N\}}{\Delta T}.$$

Assuming that channel capacity is constant and absorbing its value into $f\{N\}$, then

$$TP = \frac{f\{N\}}{\Delta T}.$$

Assume

$$OWL = f_1\{TP\}.$$

Here it is assumed that OWL is the result of an effortful adaptation to the imposed task demand as represented by TP. It is the perceived effort that is reported as subjective workload.

Constraints

1. For any N, $OWL \rightarrow 0$ as $\Delta T \rightarrow \infty$, and
   $OWL \rightarrow 1$ as $\Delta T \rightarrow 0$.
2. For any ΔT, $OWL \rightarrow 0$ as $N \rightarrow 0$, and
   $OWL \rightarrow 1$ as $N \rightarrow \infty$.

One class of functions that satisfies these constraints is the class of asymptotic growth functions. Alternatively, a simple polynomial in TP could be tried if strict adherence to the constraints can be waived. For example,

$$OWL = c_0[1 - \exp(TP)]$$

$$= c_0 \left[1 - \exp\left(t_0 \frac{(1 + b_1 N + b_2 N^2)}{\Delta T}\right)\right]$$

or, alternatively,

$$OWL = a_0 + a_1 TP + a_2 TP^2 + \ldots$$

$$= a_0 + a_1 \left[\frac{(1 + b_1 N + b_2 N^2)}{\Delta T}\right] +$$

$$a_2 \left[\frac{(1 + b_1 N + b_2 N^2)}{\Delta T}\right]^2 +$$

$$a_3 \left[\frac{(1 + b_1 N + b_2 N^2)}{\Delta T}\right]^3 +$$

$$\ldots$$

Note that the simple form of Arad's ATC model (see Pew et al., 1977), considering routine and airspace loads only, reduces to

$$OWL = a_0 + a_1 \left[\frac{N + b_2 N^2}{\Delta T}\right].$$

This is equivalent to the first-order polynomial form with no constant term (i.e., the 1), in the expression for B_r. It should be noted that even if the simple polynomial form provides a good statistical fit to experimental data, it can be considered to be an empirical fit only, as it does not satisfy the assumed constraints at $OWL = 1$. This derivation provides a linkage between the framework of the IP model and Arad's ATC model and points to potential problems with Arad's quantification of the problem.

MODEL FOR THE NUMBER OF CORRECT ROUTINGS

Assume the number of aircraft successfully routed to their destinations is

$$correct = h\{\text{total information transmitted}\}.$$

Information processed per update interval (ΔI) is given by

$$\Delta I = B_r = (CT_a) TP \text{ for } TP < 1 \text{ and,}$$
$$\Delta I = CT_a = \text{constant, for } TP \geq 1.$$

Because there is evidence to suggest that the change to resource-limited performance is not abrupt, however (see Wickens, 1992), some sort of non-linear form (either exponential or polynomial) is favored over this piece-wise linear relation (this evidence perhaps reflects slight moment-to-moment variations in channel capacity or perhaps minor changes in strategy, which change the apparent throughput). Whatever the cause, it appears that this transition is associated with a gradual rather than an abrupt change.)

A function that approximates this relation is

$$\Delta I = C\Delta T[1 - \exp(-TP)].$$

Number of update intervals per session:

$$n_{\text{sess}} = \frac{T_{\text{sess}}}{\Delta T}, \text{ then}$$

the total amount of information processed in the session is

$$I_{\text{tot}} = n_{\text{sess}}\Delta I$$

$$= \frac{T_{\text{sess}}}{\Delta T} C\Delta T[1 - \exp(-TP)]$$

$$= T_{\text{sess}}C[1 - \exp(-TP)].$$

Finally, assume

$$correct = h(I_{\text{tot}}),$$

that is, that the number of aircraft correctly routed varies directly with the number of correct decisions that have been made. Consequently, it is reasonable to assume that this relation is a linear one and thus is of the form

$$correct = \mathbf{c}_0[1 - \exp(-TP)].$$

As mentioned earlier, if we accept some constraint violation, a monotonic polynomial of the form

$$correct = \mathbf{a}_0 + \mathbf{a}_1 TP + \mathbf{a}_2 TP^2 + \ldots$$

could also be used as an alternative formulation.

MODEL FOR SUCCESS RATIO

The success ratio is defined as:

$$success = \frac{\text{aircraft successfully handled}}{\text{total number of aircraft}}$$

Assume

$$success = g\left\{\frac{B}{B_r}\right\}.$$

If we assume, as a limit, error-free performance for $TP < 1$, then as TP increases and eventually exceeds unity (overload), B, the amount of information actually processed in time T_a, should approach the available time multiplied by the processing rate. In other words, for $T_a < T_r$ (i.e., $TP \geq 1$),

$$B = C\Delta T.$$

Therefore, substituting for B and $B_r = TP.C.\Delta T$ above

$$success = f_3\left\{\frac{1}{TP}\right\}.$$

Hence, *success* is an inverse relation in TP.

Constraints

1. For any N, $success \rightarrow 1$ as $\Delta T \rightarrow \infty$ and
   $success \rightarrow 0$ as $\Delta T \rightarrow 0$.
2. For any ΔT, $success \rightarrow 1$ as $N \rightarrow 0$, and
   $success \rightarrow 0$ as $N \rightarrow \infty$.

A fitting function with these characteristics is the Butterworth function (Storer, 1957) of the form

$$success = \frac{1}{[1 + (TP)^p]^{-0.5}}.$$

Alternatively, as with earlier models, if we accept constraint violation, a simple polynomial form could also be used.

MODEL FOR NUMBER OF ERRORS

Assume that errors are a function of the amount of information shed. For each update interval, the amount of information shed (ΔI_{shed}), or not processed, is given by:

$$\Delta I_{shed} = k\{B_r - B\}$$

$$= k\left\{B\left[\frac{B_r}{B} - 1\right]\right\}.$$

For $T_a < T_r$ (i.e., $TP \geq 1$)

$$B = C\Delta T,$$

hence,

$$\Delta I_{shed} = kC\Delta T\left[\frac{f(N)}{C\Delta T} - 1\right].$$

The total amount of information shed during the session time is therefore

$$I_{shed} = n_{sess}\Delta I_{shed}$$

$$= kT_{sess}C\left[\frac{f(N)}{C\Delta T} - 1\right].$$

Assuming that errors are a function of the total amount of information shed,

$$errors = f_4\{(TP - 1)\},$$

where the constants k and $T_{sess}C$ have been absorbed into the coefficients of $f_4\{(TP - 1)\}$.

Constraints

1. For any N, $errors \to 0$ as $\Delta T \to \infty$, and
   $errors \to \infty$ as $\Delta T \to 0$.
2. For any ΔT, $errors \to 0$ as $N \to 0$, and
   $errors \to \infty$ as $N \to \infty$.

These constraints are satisfied by a simple exponential of the form

$$errors = \exp[\mathbf{a}_0 + \mathbf{a}_1(TP - 1)].$$

Alternatively, a polynomial fit could be used if constraint violation can be tolerated.

MODEL FOR PATH LENGTH

One might assume that the choice of an optimal strategy might require additional processing resources, over those required to perform the essential aspects of this task, to service such a high-level activity. The IP model does not provide a clear indication of the nature of the underlying association, if indeed there is one. The simplest approach is to assume *path* is related to TP, that is,

$$path = f_5\{TP\}.$$

In the absence of more specific guidance, a simple polynomial fit will be used, of the form

$$path = \mathbf{a}_0 + \mathbf{a}_1 TP + \mathbf{a}_2 TP^2 + \ldots$$

RESEARCH

Stress and Teams: Performance Effects and Interventions

Jeanne L. Weaver
Clint A. Bowers
Eduardo Salas
University of Central Florida

The investigation of stress and its effects on human performance is a critical area of psychological research. Few would debate the importance of this field of inquiry, but our understanding in this area is far from complete. Similarly, although the importance of teams in the workplace has been established (Sundstrom, De Meuse, & Futrell, 1990), little is known about the impact of stress on the performance of work teams. Because teams are frequently called on to perform tasks that are complex, highly interdependent, and dangerous (Salas & Cannon-Bowers, 1997), it is imperative that we understand how teams function in stressful environments. Thus, optimizing the performance of teams in aviation, military command and control, emergency operations, and other complex environments is mandated.

In light of this, the current chapter first reviews current literature about the performance effects of stress in teams. Next, we discuss literature on approaches for the amelioration of stress effects in teams. These approaches include both theoretical and empirical attempts to develop techniques that help minimize any deleterious effects of stress. Finally, a charter for future research in this area is outlined.

STRESS AND HUMAN PERFORMANCE

Although researchers have studied stress phenomena at the individual level for many years, it has been frequently noted that the area is rife with definitional and methodological problems. One of the currently best-accepted definitions of stress is put forth by Lazarus and Folkman (1984) among others. This definition of stress has already been adopted for use in the human performance and stress literature (see Driskell & Salas, 1996; Prince, Bowers, & Salas, 1994). Lazarus and Folkman defined stress as "a particular relationship between the person and the environment that is appraised by the person as taxing or exceeding his or her resources and endangering his or her well-being" (Lazarus & Folkman, 1984, p. 19). Clearly, the interaction of stressful agents and the human system of evaluation and appraisal are integral to this theory of stress (Lazarus, 1966). The role of interpretation of stressors in the stress response is emphasized, with the "resources" component of the definition referring to what one draws on to cope.

Given that stress is defined as this "interaction" between event and interpretation, it is necessary to separately define the stimulus event. A *stressor* is typically defined as the precipitating event or agent that potentially threatens an organism's well-being (Baum, Singer, & Baum, 1981). One differentiation that has been made between types of stressors is that of Driskell and Salas (1991), who proposed a distinction between ambient and performance-contingent stressors (Driskell & Salas, 1991). Ambient stressors are factors that are part of the environment or background, whereas performance-contingent stressors are stressors that are directly tied to task performance. The implication of this distinction is that, unlike ambient stressors, performance-contingent stressors can be alleviated by successful task performance. According to Driskell and Salas (1991), it is important to understand the latter stressors because of their presence in operational settings. For example, aircrews faced with life-threatening emergencies are able to eliminate these stressors through effective performance.

It is particularly critical in these times for researchers, trainers, applied scientists, and even the public to consider the need for extending knowledge in this research area (Driskell & Salas, 1996). These authors provided several reasons for this statement. First, the complexity and capabilities of the systems that exist in the modern world create task-performance situations in which there is little, or more frequently, no margin for error. Second, the pervasiveness of stress, in both everyday and high-demand performance situations, likewise increases the urgency of the need for the accumulation of knowledge in this area. Finally, past research has indicated that the potential effects of stress are varied in nature. Specifically, stressor effects have been found with regard to physiological changes, emotional reactions, cognitive effects, and changes in social behavior, among others (see Driskell & Salas, 1996, for additional discussion). However, given the popular present view that stress is an interactive phenomenon, it is essential to mention the mediating events between stimulus exposure and outcome as well.

We have already discussed the utility of Lazarus' definition of stress, along with mention of his model of stress. This model of stress emphasizes appraisal, resources, and coping behaviors. Coping is defined as "cognitive and behavioral efforts to manage specific external and/or internal demands that are appraised as taxing or exceeding the resources of the person" (Lazarus & Folkman, 1984, p. 141). Lazarus and Folkman (1984) also explained that coping serves two functions: (a) *problem-focused* coping manages or alters the problem in the environment causing distress, and (b) *emotion-focused* coping regulates emotional responses to problems. Furthermore, coping in and of itself is not typically inherently good or bad but is either appropriate or inappropriate for particular types of stressor situations. The concept of coping has been heavily discussed in clinical psychology literature and consequently has often been considered to have questionable use for applications to literature with an emphasis on stress and human performance (Hogan & Hogan, 1982). However, as the previous definition suggests, these "behavioral efforts" (i.e., coping) can be considered simply as the process/behaviors used in the confrontation of a stressor and therefore might possess some utility in describing stress phenomena with performing work teams. In fact, the idea of considering the *appropriateness* of coping in certain task-performance situations has the direct implication that *patterns* of coping that are associated with positive performance outcomes might be identified. These patterns might include team member interactions and/or individual methods for managing affective responses in the face of stressors. A similar line of research has already been conducted (see Alkov, Borowsky, & Gaynor, 1982; Picano, 1990).

Because this line of research is a critical one and stress has been investigated from so many perspectives, it likely behooves researchers from different disciplines to capitalize on the advances in others' knowledge to the degree practicable. Specifically, although it is crucial that we consider the distinctions between team and individual stress, it is also reasonable to glean whatever might be useful from the existing individual stress literature (e.g., useful models, moderator variables, assessment methods) to improve our cognizance of the manner that stress affects teams. However, the next section describes several relevant distinctions that have been made between individual and team stress and that provide a point of departure for our review of stressor effects in teams.

STRESS AND TEAM PERFORMANCE

As a first distinction between the team and individual stress literatures, it has been suggested that a "new" class of stressors be considered as appropriate for investigations of stress in teams. First, it is necessary to clarify what the term *team* means. In the confines of this chapter, *team* refers to two or more individuals working toward a common goal in an interdependent fashion (Salas, Dickinson, Converse, & Tannenbaum, 1992). Furthermore, the teams we consider are primarily those with a high degree of interdependence and an almost constant need for coordination among the team's members. These teams must often gather their information from multiple sources and engage in frequent communication with their teammates (Salas & Cannon-Bowers, 1997). In view of the nature of such teams, a new class of stressors labeled teamwork stressors is warranted. Such stressors are defined as "stimuli or conditions that (a) directly impact the team's ability to interact interdependently or (b) alter the team's interactive capacity for obtaining its desired objectives" (Morgan & Bowers, 1995, p. 267). These are variables that are postulated to have a direct impact on the team's interaction and coordination. These teamwork stressors might include such variables as team workload, team size, and time pressure. This chapter on "teamwork stress" highlights the importance of considering critical distinctions between stress and individuals' performances versus the influence of stress on the performance of teams.

The proposition of such a class of stressors captures a core issue in advancing our understanding of the stress-team performance relationship. Team performance appears to be particularly susceptible to the effects of stress because there is a requirement for teams to maintain acceptable performance, by interacting effectively with their team members, *in addition* to the need for individuals to maintain their own performances. For teams, then, there are both intra- and interindividual processes to consider. In fact, team stress researchers have proposed that being in a team increases demand under conditions of stress (Kanki, 1996; Morgan & Bowers, 1995).

It is suggested that teams, like individuals, possess, to a greater or lesser extent, resources that contribute to their ability to cope under stress. It is likely, therefore, that this additional level of demand adds to the degree of "resource strain" on a team and its members with consequent outcomes (e.g., team and individual performance) being impacted by this coping ability. All these factors must be considered in attempts to improve our grasp of the issues that are related to stress and teams. Specifically, the interactive nature of this research area must be addressed. Consequently, our definition of team stress is a direct outgrowth of Lazarus' definition of individual stress. *Team stress* is a particular relationship between the team and its environment, including other team members, that is appraised by the team members as taxing or exceeding their resources and/or endangering their well-being.

PERFORMANCE EFFECTS OF STRESS IN TEAMS

This section reviews literature on the performance effects of stressors on team performance. The review includes studies from military psychology literature, as well as studies on the performance of "emergency" or critical incident teams (e.g., Flin, Slaven, & Stewart, 1996). Consequently, the stressors described here range from laboratory studies of time pressure and threat to the investigation of factors influencing the decision-making performance of emergency workers. For the sake of brevity, the studies in this review include only relevant research conducted in the last decade.

Team Research in the 1980s and 1990s

Laboratory Research

One line of research undertaken by Driskell and Salas (1991) has investigated team decision making under stress, in one of relatively few laboratory investigations of stress and team performance. These researchers sought to investigate the centralization of authority hypothesis (i.e., the degree to which stress is related to group status and decision making) by using 78 male navy recruits as subjects. First, a check of the stress manipulation found that stress condition subjects were more likely to report that they were excited and felt panicky in comparison to the no-stress condition subjects. Results also indicated that status was a significant determinant of group interaction in that low-status members deferred to high-status members, while high-status members failed to defer to low-status members. However, the introduction of stress caused both low- and high-status members to become more willing to defer. The authors concluded that training for optimal performance under stress conditions should address team member interactions through interventions targeted toward the behavior of both low- *and* high-status team members.

In an effort to determine the influence of individual differences on anxiety and team performance, Weaver, Bowers, and Morgan (1996) used a shock threat as stressor. Specifically, this research sought to investigate the effects on team performance, individual anxiety, and coping, of a stressor presented as either related to the task being performed or as unrelated to the task being performed. These two test conditions were also considered in comparison to a baseline–no stress condition. As expected, teams composed of low self-control members reported higher anxiety than teams composed of high self-control members, and there was greater anxiety reported for all teams, in the two stressor conditions, in comparison with the no-stress condition. However, the stressor manipulation was not associated with any performance differences on the team decision-making task, on any of the four performance measures of interest, across the three conditions. Although it was hypothesized that high self-control teams would outperform low self-control teams on all of the four team performance measures (i.e., team score, penalty points, query time, and number of items queried), there was a significant effect of self-control only for penalty points. Specifically, high self-control teams scored fewer penalty points than low self-control teams. In summary, with regard to performance effects, the stressor manipulation had no effect on any measure of performance, and self-control influenced only one performance measure out of four.

With regard to the one significant finding about performance, it was found that high self-control teams were more attentive to more critical targets. This finding indicates that high self-control teams were more proficient at recognizing the threat of closer targets and at performing in such a way as to minimize the threat. It is possible that one's self-control has implications for one's ability to perform, particularly with regard to time pressure situations. The performance findings, or lack thereof, of this study are somewhat illustrative of the unpredictability of stressor effects in team performance research as the next several studies reviewed reveal as well.

Stress and Naturalistic Decision Making

Serfaty, Entin, and Volpe (1993) investigated the effects of ambiguity, uncertainty, and time pressure on team decision making performance. The task used simulated operations in a naval combat information center (CIC). Similar to earlier findings (see Lanzetta), Serfaty and his colleagues found that the relation between level of stressor and team performance was not necessarily linear. Specifically, increased uncertainty was not directly associated with team error rate. However, increased ambiguity was associated with increased errors. Finally, the relation between time pressure and error rate was curvilinear, with errors increasing from low to moderate time pressure and no difference between low and high time pressure. The authors commented that these results showed "complex patterns of the way different stressors combined to generate stress and affect the team decision and coordination strategies" (Serfaty et al., 1993, p. 1230). Specifically, the authors observed changes in the teams' communication processes that might account for such performance findings. These researchers concluded that adaptability is the key to superior team performance.

Weaver, Bowers, and Morgan (1994) also conducted a study to assess the effects of time pressure and ambiguity on team performance. This study likewise used a simulated CIC decision-making task. The simulated command and control task was the tactical naval decision-making system (TANDEM). In contrast to the findings of Serfaty and his colleagues, ambiguity had no significant performance effects in this study, although increased time pressure was associated with degraded team performance. Unfortunately, there were no communication data collected in this study that might shed light on these contrary findings.

Urban, Weaver, Bowers, and Rhodenizer (1995) likewise sought to study the effects of time pressure, as well as resource demand and structure, on team performance and communication over time. Time pressure significantly degraded team performance relative to the baseline and resource demand conditions. In contrast, teams in the resource demand condition did not exhibit performance inferior to that of teams in the baseline condition. However, the high resource demand teams did exhibit fewer statements about the availability of resources than teams in the baseline or time pressure conditions. Time pressure and resource demand did appear to exert differential effects on team performance. Although the communication results of this study offered some explanation for the teams' strategies to cope with resource demand, the authors recommended that further research attempt to determine strategies that might help offset the deleterious effects of time pressure on team performance.

Implications of stress have also been considered with regard to the performance of aircrews (Kanki, 1996). Kanki considered the influence of stressors originating from three sources: stress originating among individuals of a team; stress generated by the relationship of the team to the larger organization, including other teams in the organization; and stress generated by the team's relation to its environment (i.e., natural and operational). We briefly describe several studies reviewed by Kanki that are illustrative of the former and latter stressor categories. Kanki described one study that tested the influence of leader personality on crew effectiveness. The most consistently effective aircrews were those led by persons who were high in levels of instrumentality, expressivity, and achievement striving. This leadership style was in contrast to two others: captains characterized by high levels of negative expressivity and low levels of instrumentality, or captains with high verbal aggressiveness, negative instrumentality, and competitiveness.

Kanki also reviewed a study conducted to determine the influence of new technology (i.e., automation) on aircrew performance. The study indicated that increased automation was associated with slightly poorer performance and higher reported workload. However, neither frequency nor types of errors could be identified that would explain the differences in workload ratings and performance of the two conditions of automation. Kanki pointed out that it is important for the aviation industry to test new technologies before implementation to prevent aircrews from being forced into "on the job" training. In summary, these studies exemplify that stressors can function to influence aircrew performance from at least several levels.

Field Research.   Efforts to understand the influence of stressors on team performance have also been explored in hospital settings. Xiao, Hunter, Mackenzie, Jefferies, and Horst (1996) investigated the influence of task complexity for emergency medical care workers. Xiao and colleagues stated that such a study was needed to determine the implications of task complexity for team coordination. They conducted the study specifically to compare one emergency medical procedure (i.e., tracheal intubation) under two conditions of task urgency. This procedure was chosen in particular because it is considered by medical personnel to be stressful and high in workload. Data on 48 intubation procedures were collected over a 3-year period. Patients were divided into two conditions by using the urgency of the procedure as the means for categorization. By comparing the task characteristics of the two conditions created, the authors were able to derive four task complexity components that are inherent in the tasks performed by emergency medical team personnel. These four components are concurrent, multiple tasks, changing plans, uncertainty, and compressed work procedures and high workload. The authors concluded that such task complexity components challenge team coordination and consequently increase the possibility of team coordination errors. Furthermore, the authors recommended that these errors might be minimized through the design of task procedures and through training in explicit communications. Other approaches to improving the performance of medical teams are described later in the section on stress interventions for medical care workers.

An additional type of emergency team has also received the attention of researchers. The final types of teams to be considered here are emergency response, offshore oil and gas teams. Flin (1996) has investigated the decision-making performance of such teams to stressors such as fires, explosions, and blowouts. These teams are geographically distant on oil platforms and thus, unlike personnel of similar installations onshore, are unable to depend on outside emergency services to aid them in such emergencies. Consequently, these teams are required to respond to any emergencies that occur and to take appropriate actions. These actions might include evacuation of personnel, cessation of operations, and appropriate interactions with others (e.g., coast guard employees, onshore management). Flin stated that although such intense conditions can have some performance-enhancing effects for incident commanders and their teams, a number of negative effects have been reported for personnel working in such acute stress conditions. These effects can encompass behavioral, cognitive, physiological, and emotional domains. Furthermore, such effects can be present in the immediate task performance situation as well as in the longer term for some oil worker personnel. Some of the short-term effects can range from aggressiveness, irritation, and apathy to tunnel vision, reduced concentration, and distorted time perception. Her review listed other outcomes as well. These include acting more rapidly than necessary, over-reliance on familiar responses, and problems with thinking ahead. Past analyses of disasters on these rigs (e.g., *Piper Alpha, Ocean Odyssey*) have indicated that the performance effects associated with the stress of encountering disasters on these rigs can have disastrous consequences. Statements of observers clearly indicate that procedures broke down and leadership was lacking. Consequently, subsequent attempts have been made to improve the selection and training of offshore installation managers and their emergency response teams. These procedures are discussed in the section on interventions.

Summary

This section has reviewed literature on the direct and indirect effects of stressors on team performance. This review reflects that the research in this area, while finding some evidence for consistent stressor effects, is often characterized by equivocality as well. For example, time pressure often acts to influence performance in a deleterious manner. However, there is also evidence that the effects of time pressure are not necessarily linear (e.g., Serfaty et al., 1993). There may be something particularly unique about the way that time pressure influences performance, or these findings may be indicative of the manner in which most stressors act on performance. Because much of the literature in this area has used time pressure as the stressor of choice, the complex nature of its relation to performance has been revealed sooner perhaps than other stressors that might also have critical implications for successful task performance in real world situations. Only by conducting more research on the effects of other stressors will the answer to this question be revealed. Furthermore, findings about the effects of time pressure indicate that research should be designed in such a way as to reveal these complex relations if they do exist.

The literature on the effects of stressors other than time pressure is relatively minimal. The effects exerted by such stressors as ambiguity, resource demand, uncertainty, and threat are even less understood. The literature that does exist seems to indicate that more research on these stressors is warranted. Of the studies reviewed here, it was seen for ambiguity, for example, that it can negatively affect performance in some situations and fail to influence performance in others. Clearly, the operational definition of a stressor such as ambiguity has direct implications for its resultant effects on performance. Creating an analogue of ambiguity in an experimental situation is particularly challenging because by definition it often refers to the extent that information is available for some task performance situation. This creates a difficult situation in the laboratory when attempting to determine whether ambiguity negatively affects performance because it is challenging to determine whether it does so through action as a stressor or simply because it is more difficult to arrive at a correct conclusion with less available information.

The studies reviewed about the influence of threat were also equivocal. Although one study indicated that the teams' process changed under stress, it was also seen from other evidence that threat can be present with no influence on performance. Thus, more research is necessary to improve our knowledge of the effects of threat because threat is clearly a component of

TABLE 1.3.1
Performance Effects

Stressor	Study Type	Findings	Source(s)
Threat	Laboratory	Both low and high status members were more willing to defer under stress	Driskell & Salas, 1991
Threat	Laboratory	No effects of threat on team performance	Weaver, Bowers, & Morgan, 1996
Ambiguity, uncertainty, and time pressure	Laboratory	Uncertainty unrelated to errors, increased ambiguity was associated with increased errors, time pressure and error rate were related in a curvilinear manner to errors	Serfaty, Entin, & Volpe, 1993
Time pressure and ambiguity	Laboratory	No effects of ambiguity on performance, time pressure degraded performance	Weaver, Bowers, & Morgan, 1994
Time pressure and resource demand	Laboratory	Time pressure degraded performance relative to baseline condition, while resource demand did not	Urban, Weaver, Bowers, & Rhodenizer, 1995
Leadership style and automation	Laboratory	Leadership style was related to performance, and automation was associated with slight performance decrements	Kanki, 1996
Task complexity	Field	Recommendations provided about training and design based on observations with emergency medical care workers	Xiao, Hunter, Mackenzie, & Jefferies, 1996
Emergency offshore oil and gas accidents	Analytical	Reviewed short- and long-term effects of stress on leaders and their teams	Flin, 1996

many team task performance situations. These studies have revealed, however, that although the link between stress and performance is far from direct, the effects of stress on the teams' processes are typically far more visible. It is clear that to improve the state of knowledge in this area, it is beneficial, whenever possible, to include measures of team process to better explicate the adaptation of the team to the stress situation. Finally, we should also focus attention on the most appropriate measures of process that allow us to reveal these relations. Table 1.3.1 provides a summary of the research reviewed in this section. The table summarizes each reference cited in this section with regard to the stressor(s) investigated, the type of study undertaken, and the major findings of each.

In general, this review has revealed the following:

1. Time pressure is associated with degraded team performance. However, there is evidence that the relation of time pressure to performance degradation is likely curvilinear in nature.
2. The effects of time pressure on a team's performance are influenced
   by the characteristics of its members and their interactions with one
   another.
3. Although there is some equivocality about the effects of threat, it
   also frequently acts to degrade team performance.
4. The extent of performance deterioration under threat is likely
   strongly influenced by the salience of the threat.
5. Although it is probable that ambiguity/uncertainty in increasing
   amounts is associated with the deterioration of team performance,
   to date the data are equivocal.
6. Data about the influence of such variables as automation and leader
   characteristics acting as stressors have shown that the team's performance can indeed be degraded by negative leader characteristics
   and increased workload.
7. The degree of team performance decline as a function of such variables as leader characteristics and automation is likely determined
   by team process factors.

HOW DO WE PREVENT
TEAM PERFORMANCE DECREMENTS?

Although the literature about the performance effects of stress in teams is
far from complete with regard to contributing a comprehensive understanding of the various effects of stressors, researchers have begun to develop approaches that act to "manage" stress in teams. These approaches are varied
in nature. While some have focused on training interventions, others have
attempted to contribute to our grasp of this problem by identifying factors
that might prove critical in offsetting the effects of stressors in teams. Consequently, this section describes theoretical and empirical attempts to increase
our ability to alleviate deleterious effects on performance.

As early as 1949, Janis reported that when soldiers were asked to identify
the training they felt they were lacking, they expressed a desire for training
under realistic battle conditions characterized by threat, noise, heat, and
high workload. It is clear that these soldiers were stating the need for provision of techniques that would enable them to tolerate conditions of stress in
a manner that would compromise neither their performance nor their personal safety. The development of such techniques is beginning to reach a
point where guidance can be offered about training and/or personnel selection that provide some prophylactic action against stressor effects.

Literature Reviews and Laboratory Research

As our first example, research conducted by Weaver and her colleagues (Bowers, Weaver, & Morgan, 1996; Weaver, Morgan, Adkins-Holmes, & Hall, 1992) identified a number of variables that might serve to moderate the effects of stress on team performance. The main premise of the review was that a clear comprehension of stress and its effects can be gained only by identifying and investigating factors that influence the relation between stressors and outcomes. By explicating the role of these so-called moderator variables, researchers might be better equipped to predict the effects of stress, to conduct research that allows an unobscured view of stress effects, and to design effective interventions for the control of stress.

The variables identified in this review were social support, locus of control, perceived control, trait anxiety, self-efficacy, self-control, and experience. It was concluded from the review that control perceptions might be a unifying construct that could explain the relation among these variables. In other words, these variables might serve as moderators by minimizing cognitive, affective, and behavioral reactions to stressors. The review also discussed the possible implications and predictions associated with incorporating knowledge about these variables. For example, it was hypothesized, based on the findings of this review, that teams with members low in trait anxiety would be expected to outperform teams composed of high-anxiety members under high-stress conditions. It was concluded that increased control perceptions might function to minimize the "detrimental effects of stress on decision making by preserving problem solving and decision making functions, regulating attention and effort, and allaying emotional reactivity" (Weaver et al., 1992, p. 48). However, a statement of caution was provided as well. Most of these variables are "individual difference" variables. Better predictions about their influence on team level interactions require further empirical investigations into their potential role as stress moderators for teams.

Several empirical studies have been conducted in an attempt to determine the moderating influence of other variables as well. For example, Zaccaro, Gualtieri, and Minionis (1995) undertook an empirical investigation of the extent that task cohesion can act as a facilitator of team decision making under temporal urgency. These researchers manipulated two levels of task cohesion and temporal urgency to study the effects on performance and communications of three-person teams of undergraduates. Their results indicated that high task cohesion was able to mitigate the effects of temporal urgency on team performance. Specifically, high task-cohesive, high temporal urgency teams, were able to perform as well as teams with low temporal urgency in both of the conditions. However, performance was significantly degraded for low task-cohesion teams exposed to high time pressure. There was also an overall finding that high task-cohesive teams spent more time planning and exchanging information in the planning period and communicated more task-relevant information during the performance period than low task-cohesive teams. The authors hypothesized that under conditions without time pressure, low task-cohesive teams were able to compensate for their lack of planning. In contrast, lack of planning in high time pressure conditions is associated with the deterioration of performance. The manipulation of task cohesion consisted of an incentive given for the outperformance of other teams and of emphasizing the importance of the team performing effectively. Such a manipulation appeared to create at least a fair analogue of task cohesion in "real" task performance situations because there are typically incentives, sometimes critical, for good performance as well as for the realization of the importance of the task being performed. However, such a manipulation does not account for the possible influence of social cohesion that is an issue in real-world team performance situations. The authors concluded that the results of this study imply that training should emphasize the goals of the teams' tasks and building the collective efficacy (i.e., the team's ability to meet the demands of the situation) of the team.

Volpe, Cannon-Bowers, Salas, and Spector (1996) also conducted an experiment to test the extent that another variable might mitigate a stressor's effects. The researchers manipulated two levels each of workload and cross-training to assess the effects on task performance, communication, and teamwork. Specifically, teams who received cross-training had better teamwork than teams without cross-training, and cross-trained teams also used more efficient communication than non-cross-trained teams. Furthermore, cross-trained teams were more effective in terms of performance. However, cross-training did not interact with the workload in the manner expected. Workload did significantly affect teamwork and communication in the expected direction. Although this study does indicate the general utility of cross-training for team performance, the findings are less clear with regard to its effectiveness in mitigating the effects of workload.

The research reviewed in this section has focused first on efforts that have sought to identify variables that might mitigate the effects of stressors through the conduct of literature reviews for this purpose, and second on empirical tests of such variables. The remainder of this section addresses approaches that have specifically attempted to derive training approaches for the management of stress in teams.

Training Approaches

Much of the literature on team stress management exists in the aviation domain. The training methods used there are likewise being considered with regard to their use for teams in other areas (e.g., emergency workers, anes-thesiology). Before describing these approaches in more detail, we first briefly consider Johnston and Cannon-Bowers' (1996) work on stress expo-sure training (SET). The emphasis of SET is on extending earlier work con-ducted within the clinical domain with the goal of determining how "train-ing for stress exposure can be designed to enhance performance in various task environments" (Johnston & Cannon-Bowers, 1996, p. 224). These re-searchers also predicted that SET would be appropriate for adaptation to team-level applications.

According to Johnston and Cannon-Bowers, there are three primary ob-jectives of SET. These three objectives are to enhance trainees' familiarity with the stress environment, to build trainee confidence about perform-ance, and to build the skills that promote effective performance under stress. Stress exposure training attempts to meet these goals by providing two components: training in stress coping and instructional design. There are three phases required to provide both of these components. The first phase provides information about the common reactions to encountered stressors. The second phase provides training that focuses on the acquisition of stress-coping skills through practice and feedback. In particular, this phase attempts to develop the trainees' ability to maintain their awareness of their own reactions to stress so that they are cued to invoke the appropriate stress-reduction skill. In the third phase of training, trainees are required to practice their coping skills while being exposed to stressors. The primary outcome associated with successful resolution of this training phase is im-proved cognitive and psychomotor performance under stress.

The effectiveness of this three-phase training model was assessed via a lit-erature review that considered 37 articles reported since the 1970s and through consideration of a SET meta-analysis, conducted by Saunders, Driskell, Johnston, and Salas (1996). Johnston and Cannon-Bowers con-cluded, based on the studies considered, that the majority of the research conducted indicates support for the effectiveness of SET. More specifically, SET has been shown to have a positive influence on a myriad of perform-ance behaviors, to improve perceptions of one's self-efficacy, and to reduce perceived anxiety. As mentioned previously, this article does not specifically address the utility of SET for team applications. However, past research on its effects has indicated effectiveness for a variety of subject populations regardless of occupation, anxiety level, or age. Consequently, it is apparent, as these authors have suggested, that it would be useful to determine the extent that SET could serve as an effective method of training teams to per-form effectively under stress.

The aviation domain has received a considerable amount of researchers' attention, with regard to the development of approaches for the manage-ment of stress in aircrews. In particular, Prince et al. (1994) sought to contribute to this area by considering conceptual approaches to the area of stress and human performance, research on stressors in the cockpit, studies of the effects of stress on crew processes, and investigations of the effects of stress on aeronautical decision making (ADM). A primary goal of the effort was to integrate these literatures to create a training program for the reduction of stress effects on ADM. The approach for the development of this training involved the combination of existing knowledge about stress management/exposure training with existing knowledge about aircrew coordination training (ACT).

In essence, Prince and her colleagues recommended that the three-phase training model described by Johnston and Cannon-Bowers (1996) be applied to training seven dimensions of behaviors that are believed to represent the key behavioral elements of an effective program in ACT. (See Prince et al., 1994, for a thorough discussion of these dimensions and their classification into two categories of behaviors: *prevention-focused* behaviors and *problem-focused* behaviors.) The ultimate goal of this training is for aircrews to maintain flight performance under increased levels of stress while practicing these newly acquired behaviors corresponding to all seven dimensions of behaviors. The authors proposed that establishing the effectiveness of such a program necessitates the measurement of attitudes, stress perceptions, and performance.

Many of the ideas first used in crew resource management (CRM) with aircrews have subsequently been adopted for use in the medical domain. In particular, Howard, Gaba, Fish, Yang, and Sarnquist (1992) have noted that there are striking similarities between flight-deck operations and the behavioral requirements of the operating room. One similarity is the high proportion of errors attributed to human error in these task performance situations (i.e., 65–70%). Consequently, these researchers have adopted the training techniques of the aviation industry to serve as a model for training development in anesthesia crisis management (ACRM). In testing the application of CRM to this environment, results indicated that pre- and post-test measures revealed that although residents' scores improved significantly following training, experienced anesthesiologists' scores did not. The authors noted that behavioral responses during the scenarios are likely more indicative of effective or ineffective crisis resource management than are the written exam scores. Those scenarios revealed inadequate team coordination in several instances (e.g., failed communications, poor task delegation).

Based on the findings of their training program, the authors proposed a number of essential components for successful crisis resource management training. They proposed the necessity for an "emergency procedures" manual listing critical incidents with appropriate guidelines for their resolution, realistic hands-on simulation experiences, debriefings of simulation performance, and information dissemination about human error and decision making. In addition to the inclusion of these components, the authors noted that such training, like CRM in aviation, should be ongoing to maximize the acquisition and maintenance of individual and operating-team member skills.

Armour (1995) has also noted the need for such training in team member interactions to contribute to stress reduction. Armour studied teams employed by the Minnesota Department of Transportation. The three goals of the intervention with these teams were to correct negative interactions among team members, to identify and formalize positive modes of interaction among members, and to reinforce/establish norms of respect for employees via exercises involved in the creation of the code of conduct. As a part of a team-building exercise, the code of conduct exercise was undertaken to create an agreed-on way of team members interacting with one another. To prevent employees from perceiving the code being thrust on them, each work unit worked together to create its own agreed-on code. The process had three phases. First, the code was created through the use of confidential individual interviews, a meeting to reveal thematic data gathered through the interviews, and exercises in which ideal changes were generated and then incorporated into the consensus code. Second, regular follow-up unit meetings were held to discuss concerns that arose with regard to behaviors that went against the code. Finally, in the third phase, the facilitator returned to gauge whether the code was being implemented appropriately and whether any changes were indicated. Armour noted that of the 10 units surveyed who had undergone the intervention, all responded that their work unit atmosphere had improved at least somewhat, with one half indicating it had improved "quite a bit." The author noted that a strength of this technique is that it provides a mechanism for team members to voluntarily reduce their own levels of stress.

Flin's research about team critical incident management represents the final area to be discussed. In the section on the effects of stressors, we briefly discussed the research of Rhona Flin and her colleagues in the area of the offshore oil and gas industry (Flin et al., 1996). Flin and Slaven were commissioned to assess the selection and training of the OIMs following inquiries about the fatal emergency incidents in the North Sea. Consequently, interviews were conducted to gather data on current selection and training procedures for OIMs in crisis management. In addition, interviews were also conducted with other emergency services and military bases to study their selection and training procedures, as well as with OIMS to obtain their descriptions of factors that influence decision making on the platforms. Their research revealed that the selection for the OIM positions has changed over time. Although historically OIMs had been mariners with significant experience at sea, with the accompanying training in emergency management, the new generation of OIMs was often without experience in this area. Consequently, at present, training procedures have been established that incorporate simulated emergency command training for OIMs and their teams. Flin and her colleagues (1996) noted that since the *Piper Alpha* disaster, oil companies have been made responsible for assuring that their offshore employees are competent in emergency management. (See Flin et al. [1996] for discussion of these performance criteria provided by the Unit of Competence Controlling Emergencies.) According to the standards, performance about these criteria is to be assessed on at least three accident scenarios. Teams and their leaders can then be judged for their performance in these simulated activities. Unfortunately, little research has yet to be conducted to determine the effectiveness of such programs. However, Flin and her colleagues advised that such training is inadequate without attention to training of team skills. Consequently, efforts are underway to implement such training in existing simulated exercises (see Flin, 1995). Efforts such as these represent a positive model for the development of knowledge about team stress and interventions, because of the systematic approach used in tackling this problem area.

Summary

It is clear from this section that a number of potentially useful interventions for the alleviation of stress in teams have been, or are being, developed. The work of Flin and her colleagues particularly appears to hold promise as a model for developing interventions that might optimize the performance of teams in situations characterized by the presence of stressors. Likewise, the work of Johnston and Cannon-Bowers, although primarily addressing the issue of stress in individuals, makes a strong contribution to the development of interventions that are effective for minimizing the effects of stress. Such a foundation is necessary to begin to develop stress exposure training interventions appropriate for use with task-performing teams. Finally, the work of other researchers discussed here (e.g., Helmreich; Armour) indicates the importance of considering team member attitudes and interactions to minimize negative stress outcomes. Thus, interventions targeting such behaviors appear to be warranted and have previously been used successfully in aviation environments. However, although the aviation area has already made enough progress in its training efforts to serve as a model for the training development efforts of others, the ideas for additional training of Prince and her colleagues described here indicate that even in this domain there is still progress to be made. Table 1.3.2 summarizes the research reviewed in this section with regard to the implications of the work and its primary contribution.

This section has shown support for several issues related to the prevention of team performance decrements:

1. Team cohesion has been shown to mitigate the effects of at least one stressor (i.e., time pressure) with regard to team performance.
2. Stress inoculation techniques (i.e., SET) have been shown to have valuable utility in the prevention of degraded performance, improvement in self-efficacy, and the reduction of perceived anxiety.
3. In spite of compelling data with regard to SET and the performance of individuals, only preliminary efforts have been attempted to test the utility of SET with operational teams.
4. Critical incident techniques appear to hold strong potential for improving the ability of practitioners to provide adequate training to teams of varying types (e.g., anesthesiology, offshore oil and gas industry) by using appropriate simulations.

TABLE 1.3.2
Research with Implications for Team Stress Interventions

Applications	Description of Intervention	Findings	Source(s)
Training	Cohesion	Demonstrated that high task cohesion may act to mitigate the effects of time pressure	Zaccaro, Gualtieri, & Minionis, 1995
Training	Cross-training	Tested effects of cross-training as a moderator of workload and team performance. No interaction of cross-training and workload	Volpe, Cannon-Bowers, Salas, & Spector, 1996
Training	SET	Proposed that SET be adapted for use in task performance situations	Johnston & Cannon-Bowers, 1996
Training	SET	Demonstrated the general effectiveness of SET for a variety of subjects and behaviors	Saunders, Driskell, Johnston, & Salas, 1996
Training	SET merged with ACT	Proposed the integration of SET and ACT to develop a more effective stress-training program for the aviation domain	Prince, Bowers, & Salas, 1994
Training	ACRM	Sought to use CRM in aviation as a model for the development of training in anesthesia crisis management	Howard, Gaba, Fish, Yang, & Sarnquist, 1992
Training	Team skills	Demonstrated improved work environment via improved team member interactions	Armour, 1995
Training and selection	Critical incident mgmt.	Demonstrated use of simulation for establishing of cooperative relationships with target group	Flin, Slaven, & Stewart, 1996

5. Crisis resource management and SET development efforts for teams must also consider training about team skills.

A CHARTER FOR THE FUTURE

The preceding discussion of the effects of stress in teams and of possible alternatives for the "management" of such stressors has indicated the importance of this area of psychological research. In addition, this review also serves to identify areas in need of further work as well as challenges that researchers might yet face in their attempts to generate meaningful conclusions in this area. Thus, this final section synthesizes the literature described earlier to provide prescriptive guidance for increasing our ability to limit the effects of stressors on team performance. It is our premise that research attention to these issues significantly contributes to our ability to predict stressor effects and to design interventions to ameliorate their effects. Specifically, this section outlines a research agenda that considers a number of propositions. These propositions describe steps that could be taken to significantly increase our ability to intervene with regard to performance effects of stress for teams. In addition, issues that should be considered with regard to each proposition are mentioned as well.

Proposition I:
There Is a Need to Develop Effective Simulations

One of the most critical steps toward making progress in the team stress area is the development of task-appropriate simulations. Rhona Flin's research stands as one example of the use of critical incident techniques to develop beneficial simulations. Such efforts appear to hold strong potential for improving the ability of practitioners to provide adequate training for a variety of operational teams. Although researchers have noted the importance of the use of simulation in the past, the requirement for simulations in this area is particularly critical if significant progress is to be made to ameliorate the effects of "realistic" stressors. Furthermore, there is a need for the development of methodologies that allow useful data to be acquired in a way that minimizes cost and maximizes research findings.

In doing so, however, several issues must be considered. First, the simulations developed must adequately represent the complexity of the team task performance situation. Past research has considered the interdependent nature of team tasks as a critical variable related to team performance (Saavedra, Earley, & Van Dyne, 1993). It is likely that the role of interdependence as an influential factor for teams performing under stress is no less crucial. The level of interdependence varies depending on the task being performed and the goal and feedback structure of the task performance environment. It is desirable to simulate, to the degree possible, realistic levels of task complexity. By doing so, researchers are better prepared to obtain accurate data about stressor effects and to design and test interventions for coping with stressors in team task performance situations. Thus, this appears to be a crucial issue to be considered with regard to simulation development.

Additional issues that need to be considered here are the ability of researchers to gain access to appropriate populations of subjects and the cost of conducting such research. It is often difficult for researchers to gain access to operational teams to perform research of the type recommended here. In addition, development of simulations with appropriate levels of fidelity is often cost prohibitive. In answer to these issues, it likely behooves researchers to form partnerships in the manner of Flin and her colleagues to provide information that is beneficial to all parties concerned.

Proposition 2: Increased Accumulation of Data About Stressor Effects Is Necessary

As the literature review of stressor effects revealed, there is still a great deal of information to be gleaned in this area. Although time pressure has been shown to be associated with degraded team performance, there is also evidence that the relation of time pressure to performance degradation might actually be curvilinear in nature. Furthermore, it is likely that the effects of time pressure on a team's performance are influenced by the characteristics of its members and possibly their interactions with one another. Consequently, although the effects of time pressure have received relatively extensive study by researchers, there is still much to be resolved in this area.

One highly operationally relevant stressor has yet to receive adequate attention by team stress researchers. Although some data indicate that team performance is degraded under conditions of threat, we are far from knowledgeable about the effects of this critical stressor on teams. Because the extent of performance deterioration under threat is likely to be strongly influenced by the salience of the threat, methods for studying this stressor under realistic simulated task conditions must be developed. These simulated task conditions would also provide increased opportunities to learn more about such stressors as workload, uncertainty, and resource demand. As we have seen, the data are equivocal about the influence of these stressors.

The equivocality of past research is likely due at least in part to the existence of moderator variables as described earlier. It is possible that resources contributed by these moderator variables are differentially effective depending on the stressor source. To test this proposition, future research might consider whether some resources are more useful for some stressors than others with regard to different types of outcomes. For example, perhaps cohesion would buffer stressor effects for a "team level" (e.g., workload) stressor but provide few resources to offset the effects of a stressor that has its primary impact at the individual level (e.g., personal threat). All these issues are related to the complexity of the team task performance situation. This complexity creates an inordinate number of factors that must be considered to significantly expand our expertise in this area.

An additional issue in attempting to classify stressor effects is manipulating stressors in such a way as to remain within the guidelines of ethics and yet to create stressors that are salient enough to be realistic. Although "stress" research is difficult to justify, a component of threat is necessary to provide a realistic approximation of a stressor situation. For stress researchers to make meaningful contributions to knowledge in this area, research must address this issue. Simulations provide a reasonable response to this problem. By using participants that understand the nature and importance of the experimental situation and thus approach their participation seriously (e.g., as aviators most often do), the data obtained would probably be that much more useful in terms of their generalization to real-world situations.

Proposition 3: The Development and Testing of SET Is Needed for Team Applications

An additional step that appears strongly indicated at this time is the programmatic development of SET programs for work team applications, particularly for naturalistic, high-stress occupations (e.g., firefighters, emergency response teams, military teams, etc.). To date, we have yet to fully explore the design and utility of such interventions for teams in high-stress situations. It is also reasonable that these stress interventions need to be designed to consider and be effective for tasks of varying levels of interdependence.

It is hypothesized that stress interventions in highly interdependent task situations would likely require more attention to stress associated with interactions of members than those in low-interdependence tasks. For such work teams, it is important that their training include training targeted specifically toward the mitigation of stressor effects owing to or related to the interactions of the team's members. At this point, the most obvious response to this need would be to target training toward the acquisition of team skills and cohesion in addition to the goals more typically associated with SET. As our review noted, preliminary efforts have been begun in this area; however little data have been acquired to test the effectiveness of such programs.

Proposition 4: There Is a Need to Study Team-Coping Processes

The final proposition to be considered here is one that was mentioned early on in this chapter with regard to team-coping patterns. The idea proposed here is that previously developed simulations be used with actual team task operators as test subjects who are then exposed to stressors, with the goal of identifying the most appropriate patterns of coping in those situations. In essence, what is proposed is that the *appropriateness* of coping in certain task performance situations can be ascertained, by determining which *patterns* of coping are associated with positive performance outcomes. These patterns might include team member interactions and/or individual methods for managing effective responses in the face of stressors. By taking such an approach, "expert copers" might be identified and used as "models" of characteristics to be trained for less proficient copers.

It is also recommended that as much field research data as possible be accumulated about the coping patterns of teams under stress. Specifically, it is suggested that in addition to gathering data in simulated task situations, there is also a need to study teams in their natural environments. For example, one might observe firefighters as they interact and perform in response to actual emergencies. These behaviors could then be analyzed in relation to measures of performance obtained from supervisors, team leaders, and so on to reach conclusions about the utility of particular coping patterns. Depending on the theoretical model selected to guide such research, other measures (e.g., appraisal, performance expectations) might also provide insight into the stress–team performance relation. The theoretical model of stress emphasized in this chapter has been that of Lazarus and Folkman (1984). Other similar models that have been proposed (Driskell & Salas, 1996) also hold potential for guiding programs of research in this area. However, to date, relatively few efforts have been made to develop a comprehensive model for the explanation of stress in *teams*.

CONCLUDING REMARKS

Although this review has demonstrated the existence of an accumulation of knowledge, there still remain many unanswered questions. We hope that the research charter described here will serve as a point of departure for the development of more programmatic research efforts in this area. The tasks to be pursued are the development of good simulations, increased knowledge about stressor effects on team performance, the test and development of SET for application to operational team environments, and attempts to identify patterns of coping most often associated with positive performance outcomes with stressor exposure.

These tasks are believed to represent major milestones that must be achieved if significant advances are be made in this research area. It is likely that frequent communication among the members of this research community would facilitate the development of programmatic research efforts that would, in turn yield substantial data of merit. Given the criticality of this area of psychological research, the state of knowledge has clearly not kept pace. By attending to the tasks delineated here, psychological researchers can make a significant contribution to the training and performance of teams in a variety of operational environments.

REFERENCES

Alkov, R. A., Borowsky, M. S., & Gaynor, J. A. (1982). Stress coping and the U.S. Navy aircrew factor mishap. *Aviation, Space, and Environmental Medicine, 53*(11), 1112–1115.

Armour, N. L. (1995). The beginning of stress reduction: Creating a code of conduct for how team members treat each other. *Public Personnel Management, 24*(2), 127–132.

Baum, A., Singer, J. E., & Baum, C. S. (1981). Stress and the environment. *Journal of Social Issues, 37,* 5–35.

Bowers, C. A., Weaver, J. L., & Morgan, B. B., Jr. (1996). Moderating the performance effects of stress. In J. E. Driskell & E. Salas (Eds.), *Stress and human performance* (pp. 163–192). New York: Lawrence Erlbaum Associates.

Driskell, J. E., & Salas, E. (1991). Group decision making under stress. *Journal of Applied Psychology, 76,* 473–478.

Driskell, J. E. & Salas, E. (1996). *Stress and human performance.* New York: Lawrence Erlbaum Associates.

Flin, R. H., (1995). Crew resource management: Training teams in the offshore oil industry. *Journal of European Industrial Training, 19*(9), 33–37.

Flin, R. H. (1996). *Sitting in the hot seat.* New York: Wiley.

Flin, R., Slaven, G., & Stewart, K. (1996). Emergency decision making in the offshore oil and gas industry. *Human Factors, 38*(2), 262–277.

Hogan, R. & Hogan, J. C. (1982). Subjective correlates of stress and human performance. In E. A. Alluisi & E. A. Fleishman (Eds.), *Human performance and productivity: Stress and performance effectiveness* (pp. 141–163). Hillsdale, NJ: Lawrence Erlbaum Associates.

Howard, S. K., Gaba, D. M., Fish, K. J., Yang, G, & Sarnquist, F. H. (1992). Anesthesia crisis resource management training: Teaching anesthesiologists to handle critical incidents. *Aviation, Space, and Environmental Medicine, 63*(9), 763–770.

Johnston, J. H., & Cannon-Bowers, J. A. (1996). Training for stress exposure. In J. E. Driskell & E. Salas (Eds.), *Stress and human performance* (pp. 223–256). New York: Lawrence Erlbaum Associates.

Kanki, B. G. (1996). Stress and aircrew performance: A team-level perspective. In J. E. Driskell & E. Salas (Eds.), *Stress and human performance* (pp. 127–162). New York: Lawrence Erlbaum Associates.

Lazarus, R. S. (1966). *Psychological stress and the coping process.* New York: McGraw-Hill.

Lazarus, R. S., & Folkman, S. (1984). *Stress appraisal and coping.* New York: Springer.

Morgan, B. B., Jr., & Bowers, C. A. (1995). Teamwork stress: Implications for team decision making. In R. A. Guzzo & E. Salas (Eds.), *Team effectiveness and decision making in organizations* (pp. 262–290). San Francisco: Jossey-Bass.

Picano, J. J. (1990). An empirical assessment of stress-coping styles in military pilots. *Aviation, Space, and Environmental Medicine, 61,* 356–360.

Prince, C., Bowers, C. A., & Salas, E. (1994). Stress and crew performance: Challenges for aeronautical decision making training. In N. Johnston, N. McDonald, & R. Fuller (Eds.), *Aviation psychology in practice* (pp. 286–305). Brookfield, VT: Ashgate.

Saavedra, R., Earley, P. C., & Van Dyne, L. (1993). Complex interdependence in task-performing groups. *Journal of Applied Psychology, 78*(1), 61–72.

Salas, E., & Cannon-Bowers, J. A. (1997). Methods, tools, and strategies for team training. In M. Quinones & A. Ehrenstein (Eds.), *Training for a rapidly changing workplace: Applications of psychological research* (pp. 249–279). Washington, DC: American Psychological Association Press.

Salas, E., Dickinson, T. L., Converse, S. A. & Tannenbaum, S. L. (1992). Toward an understanding of team performance and training. In R. Swezey & E. Salas (Eds.), *Teams: Their training and performance* (pp. 3–29). Norwood, NJ: Ablex.

Saunders, T., Driskell, J. E., Johnston, J. H., & Salas, E. (1996). The effect of stress inoculation training on anxiety and performance. *Journal of Occupational Health Psychology, 1*(2), 170–186.

Serfaty, D., Entin, E. E., & Volpe, C. (1993). Adaptation to stress in team decision making and coordination, In *Proceedings of the 37th annual meeting of the Human Factors and Ergonomics Society* (pp. 1228–1232). Santa Monica, CA: Human Factors and Ergonomics Society.

Sundstrom, E., De Meuse, K. P., & Futrell, D. (1990). Work teams: Applications and effectiveness. *American Psychologist, 45*, 120–133.

Urban, J. M., Weaver, J. L., Bowers, C. A., & Rhodenizer, L. (1995). Effects of workload and structure on team processes and performance: Implications for complex team decision making. *Human Factors, 38*(2), 300–310.

Volpe, C. E., Cannon-Bowers, J. A., Salas, E. & Spector, P. E. (1996). The impact of cross-training on team functioning: An empirical investigation. *Human Factors, 38*(1), 87–100.

Weaver, J. L., Bowers, C. A., & Morgan, B. B., Jr. (1994). TANDEM: An empirical test of ambiguity and time-pressure as task parameters. In *Proceedings of the Applied Behavioral Sciences Symposium*. Colorado Springs, CO: U.S. Air Force Academy.

Weaver, J. L., Bowers, C. A., & Morgan, B. B., Jr. (1996). The effect of individual differences on anxiety and team performance. *Journal of the Washington Academy of Sciences, 84*(2), 68–93.

Weaver, J. L., Morgan, B. B., Jr., Adkins-Holmes, C., & Hall, J. (1992). *A review of potential moderating factors in the stress-performance relationship.* (Tech. Rep. No. NTSC TR-92-012). Orlando, FL: Naval Training Systems Center.

Xiao, Y., Hunter, W. A., Mackenzie, C. F., Jeffries, N. J., & Horst, R. L. (1996). Task complexity in emergency medical care and its implications for team coordination. *Human Factors, 38*(4), 636–645.

Zaccaro, S. J., Gualtieri, J. & Minionis, D. (1995). Task cohesion as a facilitator of team decision making under temporal urgency. Special Issue: Team processes, training, and performance. *Military Psychology, 7*(2), 77–93.

On Grasping a Nettle and Becoming Emotional

Alan F. Stokes
Rensselaer Polytechnic Institute

Kirsten Kite
Human–Machine Solutions, Ltd.

> *"When I make a word do a lot of work . . . I always pay it extra."*
> —H. Dumpty, cited in Carroll, 1871/1951, p. 231

By this measure, the word *stress* must be a Fortune 500 word, right up there with the top earners in the English language, for it certainly is required to do a very great deal of work. Look up *stress* in the *Oxford English Dictionary*, and you find no less than three full pages of tiny print giving different meanings and usages for the term. We begin this chapter by arguing that this very richness, which makes the term so subtle, complex, and useful in everyday speech, harms its utility as a scientific term or concept. The term *stress* is too polysemous, numinous, and inclusive to readily support the burden of empirical research that it has been made to carry. This has tended to undermine the coherence and interpretability of much of the corpus of work intended to elucidate the performance effects of psychological stress.

We then describe two traditional models or approaches to psychological stress, the stimulus-based (exogenous) and response-based (endogenous) approaches. Although these approaches no longer represent the consensus in theoretical stress research, they remain influential among applied psychologists, human factors practitioners, and decision researchers—often in the form of implicit assumptions that undergo little in the way of analytical scrutiny. Simpler formulations do lend themselves more easily to controlled experiments, statistical analyses, and the like, and it is understandable that applied researchers gravitate toward them for that reason alone. However,

neither stimulus- nor response-based models provide an adequate framework for a deeper understanding of the nature of stress—in applied contexts least of all. Among other things, they tend to gloss over the operation
of powerful, subtle, and complex *psychological* and affective processes. However, the primary point is not merely that exogenous and endogenous models of stress are theoretically impoverished (although they are), but that they
fail the test of practicality. That is, they provide no coherent foundation on
which to build a usable body of empirical findings; as a result, too little of
the existing research database can be generalized beyond individual studies
or applied to real-world situations.

A third model of stress, which in some ways integrates and builds on the
first two, is the transactional approach. This approach is the one more
favored by contemporary theorists and has certainly provided the basis for
a considerably more sophisticated understanding of psychological stress.
Nevertheless, we argue that even transactional models of stress remain inadequate and preserve a number of tempting oversimplifications for the researcher. Among other things, stress can still be treated as though it were a
variable, if only a mediating or intervening variable. One point developed in
this chapter is that the concept of stress (like that of arousal) as an irreducible, unidimensional state has no referent in the world of psychological
phenomena and that it is not usefully considered to be any kind of variable,
intervening or otherwise. The *word* "stress," we maintain, functions as a useful collective term in everyday speech; as such it covers a Venn diagram, as it
were, of meanings that only partially overlap with one another. Put another
way, there is no overarching concept of "stress" robust enough to make the
word into a scientifically useful term. As with Wittgenstein's famous "games"
example, the various ways in which the word is used are related more by
"family resemblance" than by any (nontrivial) common core meaning.

We suggest, therefore, that the transactional approach to stress is best
subsumed under the auspices of emotion research, in which the specific
emotions associated with "stress" (e.g., fear, sadness, anger) are addressed
directly. In particular, we wish to highlight the significance of the notions of
action readiness and *control precedence* in emotion and consider their implications for the study of human performance under stress. Emotion, we maintain, needs to be recognized for what it is—a primary, central, *integral* (and
sometimes autonomous) component of human cognition, not an optional
(or intervening) extra, tacked on as an afterthought to an essentially "cold"
information-processing system.

Finally, we argue that emotion research itself needs to be viewed in (or at
least be informed by) a broader context of evolutionary psychology and Darwinian anthropology. It is, of course, important and often illuminating to
consider the context in which emotions evolved—what they are *for*—but
this is not merely because of the need for intellectual roots and cogency in

modeling. There are also compelling applied reasons, not least the requirement that human performance research be clear about its aims and claims. Although these should include a consideration of whether any "stress" emotions putatively linked to performance changes are causal, pancultural, and monomorphic (implying universal and generalizable), we focus primarily on the explanatory potential of the action-oriented components of "affect programs."

MEANINGS OF STRESS

Everyday Usages

St. Augustine once noted that the concept of time is something we all understand—that is, until we stop to really think about it. Then the concept seems to dissolve into many imponderables. In some ways, the concept of stress is similar. One only has to observe reactions to such questions as "Is stress an emotion?" "Is stress, by definition, always bad for you?" or "Is anxiety a cause or a result of stress?" In any group of people, such questions generally result in a barrage of arguments replete with personal definitions, overlapping concepts, prescriptions, and anecdotes. As already noted, the *Oxford English Dictionary* (OED) devotes three pages of tiny print to the word, and much of this remains even after eliminating no longer used archaisms, arcane legal usages, and grammatical, prosodic, and phonetic meanings (for example, stress as emphasis, timing, or loudness in an utterance). Among the usages the OED records can be found stress as strain, hardship, adversity, suffering, affliction, injury, force, pressure, compulsion, overwork, harassment, and fatigue. There is also engineering stress (e.g., stress cracks in a wing spar), physiological stress, and financial stress, as well as psychological stress.

Amid this profusion of meanings, an interesting and important dichotomy can be identified. On the one hand, there is stress viewed as an agent, circumstance, situation, or variable that disturbs the "normal" functioning of the individual; on the other hand, there is stress seen as an effect—that is, the disturbed state itself. Obviously, even without the optimistic notion of a stress-free normality (both reflected and perpetuated in standard "cold" information-processing models of cognition), this bifurcation of meaning is arguably the most fundamental source of the confusion surrounding the stress concept. Twentieth-century science has done little to reconcile these meanings: On the contrary, the dichotomy has led to divergent schools of thought, which, in turn, have had an important effect on the ways in which stress is considered in applied psychological research, including human factors and industrial/organizational psychology.

Scientific Usages

As noted already, the word *stress* as it used in everyday language is rich, poly-semous, and subtly dependent on context. However, the same qualities that make it a powerful and useful term for common conversation make it a problem for researchers, who promptly narrow and redefine the term for the purpose of scientific inquiry. This in turn makes it that much more difficult to gauge the relevance of many research results to specific opera-tional contexts.

This in no way means, however, that science has converged on a precise, generally agreed-on definition of stress; on the contrary, the word has been associated with important conceptual confusions (Elliott & Eisdorfer, 1982). Stress has been used as a catchall term for actual or presumed anxiety-elicit-ing events, for psychological and emotional states, and for behavioral and physiological responses to particular events or circumstances. In fact, many stress researchers have come to regard stress not as a variable itself, but rather as an organizing concept, "a rubric consisting of many variables and processes" (Lazarus & Folkman, 1984, p. 22). This is an important and often misunderstood idea. It does not, of course, mean that there are no models or theories of stress: There are plenty from which to choose. However, many applied studies, not least those in human factors, are couched in no coher-ent model of stress whatsoever, and fewer still explicitly present any kind of model. This chapter goes on to describe three such approaches or philoso-phies in stress research: stimulus-based models, response-based models, and transactional models (Cox, 1978). The three approaches are not necessarily mutually exclusive. The three approaches emphasize, respectively, situ-ational variables, generalized responses (especially biochemical responses), and intervening psychological variables—that is, those processes involved in individual assessment of threat (Fisher, 1983).

MODELS OF STRESS

Stimulus-Based Approaches

In applied areas of psychology, many workers have for the last several decades proceeded on the basis of a simple and often tacit exogenous con-ception of stress—one that focuses attention almost exclusively on external events or conditions rather than on the subjective experience itself. In such a view, there is a tendency for situational variables that are assumed to be aversive to be simply labeled a priori as *stressors*. Typical examples of such variables include workload, time pressure, noise, and also "life events" such as promotion, moving house, and marriage.

In keeping with this philosophy, a good deal of applied research has consisted of selecting a given variable, manipulating it experimentally, and
styling the manipulation as "stress." Thus, time restrictions or increases in
workload (for example) are sometimes treated as though they were inherently stressful and may be labeled as stressors irrespective of whether the individuals studied actually experience or report any distress or discomfort
whatsoever. This approach (and also the physiological response-based approach, discussed subsequently) has been criticized on the grounds that
stress becomes "merely a convenient label and collective noun indicating
certain environmental and organismic conditions" (Sanders, 1983, p. 62).

Certainly the list of factors that have been named as stressors is impres-
sively long: It features almost every imaginable physical, environmental, and
social condition and includes army food, the weather, poverty, and even so-
called future shock — the "shattering stress and disorientation that we induce
in individuals by subjecting them to too much change in too short a time"
(Toffler, 1970, p. 2). The point is not that these factors cannot be stressful,
but rather that almost anything that can be named (from fame to unchanging weather) can, under some imaginable circumstance or other, induce
psychological stress. The only thing that these factors have in common is
that they can be appraised as stressful. (By the same token, it is possible for
one environmental stimulus or factor to provoke either a stressed or an
unstressed reaction, depending on circumstances.) Simply labeling examinations, emergencies, noise, and the like as stressors, therefore, makes no
special claim about these factors and, in the absence of some theoretical
underpinning, leaves us none the wiser. Indeed, we are worse off, as a con-
found may have been introduced into experimentation: the performance
effects from Factor X being rolled in with the performance effects of psychological stress occasioned by the individual's *interpretation* of Factor X.

It could be argued, of course, that humans (and other species) do appear
to have a repertoire of more or less innate fears and "stress" responses presumably handed down by evolution (fear of falling from heights, distress at
the sound of a baby crying, and so on). The fear-of-falling response, for example, is reliably observable even in infants less than 1 year old who have
had no experience of falling. This suggests that the fear is innate; moreover,
a Darwinian rationale is hardly implausible, given our arboreal heritage.
The environmental factors that trigger such "hardwired" responses[1] are the
best and arguably the *only* real candidates for consideration as stressors in
the orthodox, deterministic use of that term. In this sense, one might say

---

[1] Of course, what may be innate may sometimes be better described as a preparedness to
learn to fear. Children learn to fear snakes very readily, despite their absence from modern
urban life, but they do not readily develop analogous fears of doors or drawers, even after
trapping their fingers in them.

that there actually is such a thing as a stimulus that is "inherently" stressful (although important questions remain about eliciting conditions, the role of learning, and individual variability—bungee jumpers, for example, actively seek falling experiences). Ironically, however, such factors have attracted little or no attention from applied stress researchers.

Stress Versus Strain. One variable that has often been suggested as an external stressor is anxiety (Hamilton, 1980). A common alternative view is that anxiety is not an external factor causing stress, but is an internal factor (a result, not a cause of stress). This external/internal dichotomy is often expressed in a distinctive way in the stimulus-based approach to stress, that is, by using an engineering analogy: the "stress and strain" model. This model holds that just as stress is an external force—for example, mechanical loading applied to a wing spar resulting in strain in the spar—so human stress is best viewed as an external factor that produces "strain" in the person. Regrettably, there are a number of serious problems with this otherwise neat and tempting formulation. Wing spars do not, of course, interpret their circumstances or evaluate the stress they are under. Thus, "strain" has no real psychological meaning, that is, no cognitive or emotional component to it—the analogy is empty. Moreover, all wing spars react similarly to similar stresses, and an (uncompromised) spar stressed this week responds identically to the way it responded when it was stressed last week. This is not, of course, an adequate description of human stress response. It is a striking fact, often observed in times of war, that similarly qualified and trained persons may react very differently in identical circumstances (Glass & Singer, 1972; Grinker & Spiegel, 1945). Indeed, individuals do not react in the same way even at the most basic physiological level, as research on endocrine responses has shown (Rose, 1980; Strelau, 1989). What is more, not only is it true that what stresses Mr. Smith may not stress Ms. Brown, it is equally true that what Smith interpreted as stressful last week he may not interpret as stressful this week. In short, two major shortcomings of the stimulus-based approach in general and the engineering analogy in particular are that they ignore individual differences and simply omit the emotional component of the experience that is the very hallmark of the everyday concept of human stress. "Men and their organizations are not machines, even if they have machine-like aspects," noted Cox (1978, p. 14), "and the [stress–strain] analogy breaks down rather too easily."

Response-Based Approaches

In contrast to stimulus-based views of stress, response-based approaches focus not on the external circumstances assumed to induce stress reactions, but on the reactions themselves. In fact, in this model, the responses (or pat-

terns of responses) displayed in a given situation are considered to be the
defining parameter of stress. In theory, this conception of stress could incorporate many different categories of responses—behavioral, affective, cognitive, and possibly others. In actuality, this has not been the case to any great
extent. Historically, the most extensively studied type of stress response has
been the physiological. In part, this emphasis has its roots in investigations
conducted early in the 20th century by Yerkes and Dodson (1908). This research has had an extensive impact on subsequent thinking about stress and
is the origin of the oft-repeated claim that the relation between performance and stress can be graphed as a normal distribution or inverted U-curve
(see Stokes & Kite, 1994, for a critical review of this idea). However, perhaps
the most often cited body of work is that of Selye (1956), whose work has exerted a profound influence on response-based approaches to stress, and, indeed, on popular conceptions of stress in general.

Selye, who was studying physiological responses to injury, emotion, and
other intense stimuli, observed that while some of these reactions were
clearly linked to particular stimuli, others seemed to be less specific and
tended to appear in a wide variety of aversive or demanding situations. This
latter group included increases in heart rate, respiration, adrenaline output,
and a number of other metabolic and endocrine functions associated with
autonomic nervous system activity. This generalized, "nonspecific" physiological reaction is typically considered to represent an increase in "arousal,"
a hypothetical construct usually taken to mean the basic energetic state of
an organism. There are important problems with this construct, but for the
time being it is sufficient to note that various biochemical and psychophysiological measures have been proposed as indexes of arousal. Selye's importance does not stem from this, however. (He was by no means the first to
observe and describe the phenomenon of physiological arousal, which can
be traced back to Cannon, 1915.) Rather, Selye's influence comes from the
way in which he associated arousal with the idea of stress, using a construct
that became popularized with the term *general adaptation syndrome,* or GAS:
general, because it seemed to represent a systemic, nonspecific reaction
common to many different categories of stimulus; *adaptation,* insofar as its
function was to prepare the individual to respond to an external threat,
presumably either by warding off the attack or escaping—the so-called
fight-or-flight reaction.

However, Selye's description of the process as a *syndrome* (a term more
commonly associated with pathology) hints at the fact that he did not regard
it as universally adaptive. First, it is not without metabolic cost to the organism; in other words, it consumes energy. Second, it has been Selye's contention that repeated or long-term activation of the arousal response can
lead to depletion of the very neurochemicals and other physical resources
that give rise to or define it in the first place. This consequently leaves the

individual ill-equipped to cope with further threats that arise. He or she may also become subject to feelings of exhaustion and, possibly, a compromised immune system with increased vulnerability to opportunistic diseases. Although Selye's model of stress has been superseded by others (as discussed later), these latter insights do seem to represent an important and enduring contribution to stress theory.

Nevertheless, after Selye's model, a good deal of empirical research on stress responses has proceeded as though physiological arousal is essentially a measure of psychological distress. Measurements of heart rate, respiration, and skin conductance, not to mention plasma phospholipids, urinary catecholamines, and the like, have the cachet of objective, quantitative, "hard" science. The difficulty, however, lies in deciding what these various results actually tell us about stress. After all, similar symptoms of adrenergic arousal (the adrenaline "rush") are associated with exhilaration, illness, effort, keen anticipation, and sexual activity. (We would expect to find "stresslike" physiological changes in, for example, thrilled teenagers enjoying hectic fairground rides—but we have not seen this, or, indeed, any equivalent control condition used in most applied research on stress and performance!) Certainly, among stress theoreticians the (physiological) response-based approach to stress has come under a good deal of criticism, insofar as it suffers from shortcomings similar to the stimulus-based approach. Both tend to consider stress effects purely in terms of a simple stimulus–response or direct cause-and-effect relation, essentially bypassing the role of the individual as a thinking, reflective, purposive, emotionally engaged participant in the process (Lazarus, DeLongis, Folkman, & Gruen, 1985).

In this sense, both stimulus- and response-based approaches are curiously nonpsychological. They acknowledge little or no role for cognitive appraisal or mediation between stimulus and response. Neither view of stress has anything to say about the *perception* of threat or challenge, a perception presumably influenced by the individual's (or crew's) purposes, goals, and hypotheses about (and interpretations of) events in particular situational contexts. As a consequence, these approaches, although very widely used, in fact have limited usefulness as conceptual frameworks for stress research in many military and civilian contexts. Cox (1985) has pointed out that "from reading the popular literature on stress, the researcher believes that [a simple physiological index unarguably related to stress] exists." He adds, "Alas, it is a myth borne out of hope rather than understanding" (p. 1155).

Transactional Approaches

A third view of stress began to take shape in the late 1970s and 1980s, with the development of what has been called the transactional approach (Cox &

Mackay, 1976; Fisher, 1983; McGrath, 1976; Welford, 1973). This approach was described as a radical redirection in stress research (Coyne & Lazarus, 1980), and by the 1990s was to become the consensus among stress theorists (see, e.g., Cox, 1987). Transactional models conceptualize stress as inhering neither in the person nor in the environment as such. Rather, stress has come to be seen as a function of the interaction *between* the individual and his or her environment. That is, it inheres in the nature of the encounter, as Lazarus, an important theorist, has often put it (see, e.g., Lazarus & Folkman, 1984). These encounters have affective meaning because the individual has beliefs, goals, and intentions, and the environment imports threats, challenges, opportunities, and risks. The individual's process of evaluating such factors in the light of personal motivations or agendas lies at the core of both the stress experience and the nature of coping with that stress. Thus, there are no psychological "stressors" in any absolute, objective sense (other than those relating to Darwinian processes discussed earlier), and neither is stress a variable in the normal scientific sense of the word.

Transactional approaches, then, emphasize the role of individual appraisal in the human stress response. Rather than focusing exclusively on precipitating factors (stimuli) or on responses to these factors, the concern is with the interpretation or appraisal of situations in terms of their demand and the individual's perception of his or her coping resources. Transactional models must, of course, recognize the importance of stimulus events and of physiological and behavioral responses, but in an important sense they are more "psychological" than either the stimulus- or response-based approaches to stress, in that they acknowledge the subjective nature of stress and emphasize the mental processes that mediate the individual's reactions.

Transactional models have been widely used in studies of workplace stress, as well as in a range of "high-stress" contexts, from civil disasters and terrorism to surgery and skydiving (Ayalon, 1983; Cohen & Ahearn, 1980; Epstein, 1983; Janis, 1983). However, the influence of this approach is only just beginning to be felt among human factors specialists, many of whom were schooled in the earlier stimulus- or response-based models of stress. As Hammond (1990) pointed out in a comprehensive review of the literature on stress and judgment, human factors researchers have tended not to cite the stress literature from personality research, clinical studies, or social psychology, even though this is where most theoretical progress in the transactional approach has been made.

Perhaps one reason that applied human factors research has lagged behind theoretical advances lies in the fact that transactional models of stress are more complex than their stimulus- and response-based predecessors and are therefore more difficult to operationalize experimentally. If stress is no longer conceptualized as a variable inhering either in the person or in the environment, it passes a fortiori beyond the grasp of straightforward metrics such as heart rate or ambient noise levels. Human factors researchers have tended to fight shy of approaches that might bring with them greater methodological challenges and more experimental headaches than stimulus- or response-based models, and this transactional models undoubtedly do. It can be done, however, as a number of studies have demonstrated. These include a series of simulation-based experiments on pilot decision making conducted at the University of Illinois (Stokes, Barnett, & Wickens, 1987; Stokes, Belger, & Zhang, 1990; Stokes, Kemper, & Marsh, 1992; Stokes & Raby, 1989), as well as a number of applied studies in the field of military operations. One of these, a joint Australian and Swedish study, examined cognitive appraisal and coping processes in the brief time before ejection from fast-jet fighters (Larsson & Hayward, 1990). Another, a German Army study of stress in actual low-altitude night helicopter operations, systematically evaluated (along multiple dimensions) all three elements: the pilot, the situation, and the interaction between the two (Harss, Kastner, & Beerman, 1991).

These and other studies have demonstrated that the transactional approach can be implemented in fields where the assumptions underpinning stimulus- and response-based models have gone largely unchallenged for several decades. We now go on to consider in greater detail the concept of cognitive appraisal, which has been an integral aspect of most transactional approaches to stress, and to suggest that this approach, too, suffers from certain limitations.

Cognitive Appraisal in the Stress Literature. As one author has put it, "The analysis of perceptions of . . . stressful settings is potentially an extremely powerful one" (Jones, 1991, p. 72). The key word here is "perceptions." Although one view of stress might be to consider it as resulting from a mismatch between the demands of a situation and the individual's ability to meet those demands, the notion of cognitive appraisal introduces the element of subjectivity. In this framework, stress is viewed as the result of a mismatch between individuals' *perceptions* of the demands of the task or situation and their *perceptions* of the resources for coping with them. What this means is that individuals who, for example, *over*estimate their resources or *under*estimate the demands of the task continue unstressed until something—some feature of a deteriorating situation, say—prompts a reevaluation of demand or resources. Likewise, individuals who either overestimate demand or underestimate their coping ability may respond negatively, irrespective of the "objective" circumstances. Such an appraisal can result in feelings of helplessness and resignation in situations that could, in fact, be mastered. There is certainly evidence that such misperceptions can lead directly to stress and error in decision making (Stokes & Kite, 1994).

Cox and Mackay (1976) proposed a model of stress in which the situation's actual demands, as well as the individual's actual capabilities, are each evaluated through the filter of perception and then compared against one another in the process of cognitive appraisal. An imbalance between the two perceptions results in stress, which itself can be analyzed in terms of various manifestations (e.g., psychological, physiological, cognitive, behavioral). Although this model is more comprehensive than many, there are additional variables that researchers have identified as being relevant to the appraisal process. For example, McGrath (1976) defined stress in terms of *three* elements: perceived demand, perceived ability to cope, and perception of the importance of coping—that is, the extent to which the demands of the situation threaten the goals or aspirations of the individual. This makes intuitive sense: It seems unlikely that stress is proportional to the perceived mismatch between demand and capability irrespective of how critical that mismatch is. Perhaps even a modest imbalance may be very stressful if life is at stake. Conversely, a profound skill deficit may have little implication for psychological stress if the situation is one in which the individual has no need or expectation of excelling.

Another variable that has sometimes been related to stress is uncertainty (Warburton, 1979). This may well constitute a significant fourth element in cognitive appraisal models because small demand/resource imbalances may also create stress through the increased uncertainty of the outcome where success is important. Indeed, it seems likely that stress may also occur where ability to cope is perceived as positive but marginal, risks are perceived to be high, and success is therefore deemed to be vital.

The concept of cognitive appraisal, then, is a cornerstone of transactional approaches to stress and represents an important leap in sophistication over traditional, "black box" models of stress. Even this approach, however, has suffered from some important limitations. Consider the following encapsulation of the cognitivist position: "An event does not become a stressor until a cognitive processing system has identified it as such on the basis of existing long-term memory data" (Hamilton, 1982, p. 117). At the very least, this statement leads to some rather odd-sounding inferences: For example, are infants less capable, or indeed, incapable of experiencing stress, to the extent that they lack the pertinent long-term memory data? What should we assume about amnesiacs, Alzheimer's patients, and so forth? The more important point is this: Without denying the importance of purely cognitive or intellectual processes, it seems rather arbitrary and gratuitous to simply stop there, merely to avoid grappling with the Pandora's box of affective states that the stress rubric glosses over. We are not here advocating the opening of the box—it has, of course, always been open (in folk psychology and for as long as academic psychology has existed). From the standpoint of the applied stress researcher, it is more a matter of looking inside.

STRESS AND EMOTION

Consider the following scenario. You are riding in a tour bus visiting sites of archeological interest when suddenly the bus is hijacked by a group of masked individuals. They inspect passports, and upon seeing yours they bundle you off the bus unceremoniously and lead you at gunpoint to a jeep. There you are blindfolded, handcuffed, and pushed into the back seat. You feel the jeep move off as several gunshots ring out. For the next 30 minutes, you are able to consider your situation without visual distractions as you bounce along a dusty trail.

The hijackers hold almost all the resources worth having—surprise, a plan, weapons, strength of numbers, local knowledge, transport, teamwork, unhindered sight, and so on. You have a watch you cannot see, a pen you cannot reach, and, in your back pocket, a paper party hat from the previous night's hotel reception, but you do have intelligence and fortitude. What are the demands of the situation? First, you must stay cool, clearheaded, and alert. You feel you may have had the internal resources to manage that, but those shots! Where is your spouse, still in the bus? Why did they pick you? Are you to be shot because of your nationality? If ever an encounter made demands on you, this is one—staying cool is not enough. You absolutely must break free, escape, get help, and rescue the others. Success is imperative.

Analysis of the situation certainly lends itself to a transactional perspective. There is a large perceived (and actual) resource/demand mismatch, and overcoming this to succeed is vital, unlikely, and fraught with peril. The model certainly predicts the emergence of considerable psychological stress. The question arises, however, why stop there, at this very crude macroscopic level of analysis, as though stress were an irreducible, unidimensional state? If stress is, as we cited earlier, "a rubric consisting of many variables and processes," why terminate the analysis at the level of a rubric? In so doing, we blanket and bundle together many potentially interesting, important, and dynamically changing processes that we should be teasing apart, not least powerful emotional and emotional-cognitive responses that in some sense "make up the stress." These emotional processes can be expected to exert a forceful influence on observable behavior (and indeed, they do, arguably via mechanisms such as "control precedence," discussed later).

With this perspective in mind, let us return to your experience on the bus and your plight as a hostage. Immediately before the attack, you were feeling *happy*. The bus screeched to a halt. You were *startled* and then *curious*. The assault began, and you were *surprised* or *shocked*. This quickly turned to *fear*, which itself subsided slightly, turning into moderate *relief*, as it became clear that the raiders were taking control of the bus and were not bent on destroying it and its passengers. Perhaps you felt a flicker of *excitement* (if less than the young boy opposite you who, having been *bored* on the trip, welcomed the drama as a great romp). As passports were examined, you became *anxious* and *apprehensive,* and, as the boy was knocked back into his seat with a rifle butt, *angry* and *outraged.* You stifled your urge to intervene on the boy's behalf—a fact that was later to create feelings of *guilt* in you.

We could continue this scenario at length, deploying further emotion terms such as "despair," "apathy," "terror," "hate," and "sadness." The point, we trust, however, has been made: The word "stress," well-remunerated and tireless worker although it may be, cannot be paid enough to make it cover one tenth of the psychological richness of such a scenario. It could be objected that it is unreasonable to suggest that a plain technical term from the researcher's toolbox should in any way be able to capture or encompass a diverse set of colorfully descriptive terms, essentially literary devices, such as those woven into the scenario here, but "stress" is neither a plain nor a technical term (as already discussed). More important for the present discussion, the psychological richness of the passage is not a mere matter of lexical choice in an essentially literary endeavor. Emotion terms do purport to have referents in the world of psychological phenomena and arguably denote often complex but coherent psychophysiological states or processes with quite distinct behavioral concomitants (action preferences and readiness). As such they cannot, without loss, be lumped together or subsumed under the rubric "stress" (save for the shorthand of everyday discourse) or be adequately explicated by reference to the literature on stressed performance. As Lazarus (1991) put it, "Although stress and coping are still important, social scientists have begun to realize that these concepts are part of a larger rubric—the emotions. Appraisal must now be made to account for the differences among distinctive negative emotions such as anger, fright, anxiety, guilt, shame, sadness, envy, jealousy, and disgust, and positive emotions such as happiness, pride, love, and relief" (p. vii).

Many applied psychologists interested in "stressed" performance have been reluctant to grasp the nettle, however, and we (the authors) do not exclude ourselves from this category. This reluctance may be, in part, because there are several distinct theoretical approaches to emotion, each with its own voluminous literature. Moreover, the most promising approaches for applied psychology may not be the easiest to understand, and those that are easier to understand are less useful to the empirical researcher. For example, one of the more accessible frameworks is "feelings theory," which begins with the proposition that emotions are a matter of introspection—a combination of subjective quality and intensity of sensation. For many, this kind of thinking is hopelessly "woolly," even sentimentalist. An understandable response might be, better to go with stimulus- or response-based stress models, which at least support objective definition and quantification, than to invoke intangible "feelings." Actually, however, in the emotions research community itself, the feelings approach has been roundly criticized for its numinous-ness and lack of conceptual rigor, and several other major and competing literatures exist.

One of these is conceptual analysis, which, with its roots in Wittgenstein-ian philosophy, rests heavily on a belief/desire psychology and a rigorous ex-ploration of the ways in which emotion terms are deployed in everyday speech and writing. This again is not a literature in which many applied psy-chologists feel at home. It can be a daunting task to wade through such analyses, and it is quite possible to re-emerge without much sense of pro-gress or enlightenment. To give a glimpse of this, recall that early in the hi-jack passage we used the term *startle*. Startle is treated as a bona fide emotion by some researchers, but is excluded by others (e.g., Ekman, Friesen, & Simons, 1985; Lazarus, 1982). Even *sleepiness* has been taken as an emotion by one set of workers (Watson & Tellegen, 1985)! Thus, the status of several of the psychological descriptors in the hijack passage is debatable, if concep-tual analysis is where one starts. Among the arguments for such an approach appears the following eminently reasonable one: Insofar as what counts and what does not count as an emotion presumably influences the set of phe-nomena that a theory of emotion must explain, the nonemotion-state weeds must first be separated from the emotion-state flowers. Otherwise, any the-ory of emotion (perhaps a cognitive-appraisal-based model, for example) might have to cover *sleepiness* (or *exhaustion*, perhaps *sickness*). In fact, much of this preliminary work has already been accomplished. (For a good exam-ple of some of the conceptual groundwork done in this field, see Ortony, Clore, & Collins, 1988.)

There is, however, an important criticism of any approach that con-fines itself to conceptual analysis and that is necessarily informed by folk-psychological concepts and usages. Popular usage cannot be the final ar-biter of scientific categories, because folk categories do not necessarily map onto categories identified by science. A number of such categories spring readily to mind: racial groups, bugs, fish, constellations, dinosaurs, and ver-min. In the United States, a vernacular derivative of vermin, "varmints," in-cludes both rabbits (lagomorphs) and their predator, coyotes (canids), as well as several other unrelated taxa. Conceptual analysis would classify mushrooms as vegetables, but the fact that they are widely referred to as veg-etables in everyday life (even by competent botanists) does not make them botanically so. (The same is true of tomatoes, technically a fruit although never referred to as such in everyday parlance.) Moreover, terms and lin-guistic categories also differ from one language to another, and there is no reason to ascribe special status to English or to any other single language. Everyday terminology, including that of folk psychology, has a range of objec-tives and functions that are simply different from those underpinning science and research. Folk psychology does more than merely describe emotions;

it also prescribes emotional behavior and promotes collective myths about it, as social construction theories of emotion emphasize (Griffiths, 1997). Thus, although the study of the everyday application of terms in the realm of emotional behavior may illuminate public beliefs about that realm and naming conventions in it, if we "want to know about emotion, rather than what is currently believed about emotion, [conceptual] analysis must proceed hand in hand with the relevant empirical sciences" (Griffiths, 1997, p. 7).

In our view, one of the most empirically well-founded approaches to emotion, and also a promising one from the point of view of the "applied stress researcher," is the affect program theory (see, e.g., Ekman, 1977). This approach posits the existence of prepared, coordinated sets of adaptations, stored as a pattern or template and enabled by features of the "encounter" (to use Lazarus' transactional stress term). When an affect program is triggered in an individual, there is said to follow a stereotypical pattern of short-term psychophysiological changes. These include, for example, neurochemical, cardiopulmonary, and other somatic changes, as well as facial expressions, bodily postures, voice qualities, and (crucially for our purposes) a particular action readiness, potential, or predisposition. In short, each emotion comes as a "package deal." "Strong" versions of the theory imply actual genetic preprogramming; "weaker" versions take the program notion a little less literally and treat the patterning of the psychophysiological changes more as emotion-specific "syndromes" (Lazarus, 1991). Such syndromes may represent the operation of a "virtual program" (Griffiths, 1997), being the output of a number of interlinking, parallel control mechanisms. These mechanisms may range from the hardwired end of the spectrum (as we have suggested) through to very plastic, culturally transmitted modes of responding. Certainly, there is compelling evidence from the social construction theories of emotion (noted earlier) that at least some features of the package may be shaped by culture. That is, among other things, non-self-conscious adherence to social norms may be involved. (Doubters are referred to Ronald Simons' remarkable 1996 book on startle, with its description of *latah*, an often elaborate hyperstartle response found in Malaysia.) Affect program theory has its origin in Darwin's (1872) work on facial expressions, and its background in evolutionary theory could not be more impeccable. However, there is no need to confine the theory to genetic influences alone, for it is well able to absorb cultural influences in the development and elaboration of affect programs. In Ekman's hands, affect program theory has developed considerable evidence that particular emotions do indeed unfold in unique patterns. Although the theory is limited to the so-called "lower emotions" of disgust, fear, contempt, joy, sadness, anger, and surprise, these categories do include some key emotions that an applied stress researcher might wish to investigate. Moreover, the concept of "action readiness" gives affect program theory a powerful functionalist framework for considering the behavioral effects of "stress," and this is considered next.

FROM FEELING TO ACTING:
WHAT EMOTIONS ARE FOR

Frijda (1986, 1994, 1996) is one of a number of workers (going back to Darwin) who have advocated a functionalist approach to emotion. Noting the emotions' power "to guide attention, to distract and interrupt, and, first and foremost, their character of impulse" (1996, p. 5), he suggested that the ultimate function of emotions is to potentiate or stimulate behavioral responses that, in turn, influence the person–environment relation in the service of some adaptive goal. In the case of some emotions, such as fear and anger, the link with adaptive behavior is direct and obvious. Fear motivates individuals to protect themselves from the fear-inducing stimulus. Anger's concomitant, aggression, may influence the behavior of others in ways advantageous to oneself. Frijda argued that this *action readiness,* as he termed it, is no mere spin-off of emotion, but rather the whole point—its raison d'être. This is apparent with infants, for example, who enter the world already able to mold their environments in adaptive ways (i.e., by signaling their needs to adults). As children grow in cognitive sophistication, they increasingly learn to suborn their reactions to proximate stimuli in the service of other, more distal goals. Emotions may therefore not be outwardly expressed, nor are their behavioral concomitants always carried out. Thus, action readiness refers not to action itself but to the control exerted by the emotional system over behavioral *priorities,* as Panksepp (1996) put it.

Allied with action readiness is the concept of *control precedence,* which refers to the manner in which emotional states not only motivate certain behaviors, but also preempt other behaviors—and cognitions—associated with different, competing goals and issues. A soldier takes fright and flees from enemy fire, disregarding other priorities of duty and comradeship (to say nothing of military penalties for desertion). A jealous husband who discovers his wife in a compromising position responds with murderous aggression, not considering in the heat of the moment whether this might provoke a dangerous retaliation. (The wife herself has also quite likely been acting on certain emotions that have crowded out competing claims.) The person overcome by grief may disregard an entire range of adaptive goals and activities.

The constructs of action readiness and control precedence bring with them certain implications for the likely evolutionary antecedents of emotion in humans and other higher vertebrates. Some workers have proposed that emotions may have arisen from sensorimotor reflexes and other "hardwired" response patterns, which in many species constitute the entire behavioral repertoire. Evidence for this view is certainly not contradicted by the enormous body of research on the role of neurochemicals in controlling emotions (Panksepp, 1993). However, emotions are more sophisticated than mere reflexes, in that they are not rigidly and automatically coupled with specific actions (Lazarus, 1991).

The relation between emotion and adaptive behavior is not necessarily a simple one. First, not all manifestations of an emotion need be adaptive for the emotion to be so, any more than all manifestations of blood clotting are adaptive. The problem does, however, take on a new dimension when an emotion seems routinely negative and nonadaptive, such as grief, which, as noted earlier, may cause the sufferer to neglect more useful activities. Frijda has even suggested that control precedence may sometimes lead to action *un*readiness and that this, too, may be adaptive. (A putative example is the tonic immobility or "freezing" occasionally observed in frightened animals and sometimes in humans, discussed at length in Stokes & Kite, 1994.) It has sometimes been suggested that the adaptive function of grief is primarily preemptive—to motivate the individual to prevent situations where grief would occur (Averill, 1968). This idea, for what it is worth, could also apply to some other emotions such as shame and guilt. Another more promising view has been suggested by Tooby and Cosmides (1990), who presented a theory of grief (and some other emotions) in terms of a "recalibrating function." They suggested that the function of these emotions is to readjust or recalibrate variables in schemas—reevaluating the weightings given to factors in circumstances always clouded with partial information and uncertainty. An analogy might be made with what occurs in times of financial crises and stock market crashes: Such crises are unpleasant and sometimes traumatic, but in their aftermath stock values are readjusted in a more realistic direction, and stability is restored.

Which Comes First: Thinking or Feeling?

Cognitive appraisal models of stress have traditionally been grounded in "cold" information-processing models of cognition. That is, they assume (perhaps only implicitly) that human mental processes are characterized primarily by a nonemotional, computationally based assessment of environmental stimuli. Emotions or "stress" may follow in the wake of this assessment, but they cannot, in this model, precede it. There is in fact a debate going back some years on this question. Zajonc (1984) argued for the "primacy of affect," noting that emotional reactions could be evoked by stimuli that had not been processed at the cognitive level. Lazarus' (1984) rejoinder, published in the same journal, defended the importance of cognitive appraisal. The problem may partly be a matter of definitional slippage: Is "cognition" understood to refer to a conscious, deliberate, intellectual process, or does it also include the most basic processes of physical perception? We have already noted that Hamilton's (1982) purely cognitivist stance ("an event does not become a stressor until a cognitive processing system has identified it as such on the basis of existing long-term memory data") falls short from an ontogenetic standpoint if nothing else. In some respects, the dichotomy implied by the Zajonc/Lazarus interchange is a false one, for the "primacy of affect" and the "primacy of cognition" scenarios both reflect certain aspects of human experience. Griffiths (1997) has observed that in some situations (particularly, perhaps, novel ones), cognitive-level analysis may lead to a perception of impending harm (or advantage, or whatever), and an emotional response then follows. However, it is entirely possible for emotional responses and cognitive appraisals to be in conflict with one another: We may feel upset about something that we "know" we should not feel upset about, or we may think that we ought to feel some emotion that we do not in fact feel, or, as Zajonc noted, we may experience emotional responses before any cognitive input has occurred at all. (Griffiths referred to the latter phenomenon as "reflex emotion.")

Can We Keep Cognition and Emotion Apart?

Lazarus, who in 1984 had argued for the primacy of cognition, was himself to eventually view cognition and emotion in terms of a feedback loop—albeit one that remains *initially* cognitive: "Cognitive activity causally precedes an emotion in the flow of psychological events, and subsequent cognitive activity is also later affected by that emotion" (1991, p. 127). New evidence does more than merely challenge this "new orthodoxy," however. Recent findings reported by LeDoux (1996) and by Damasio (1994) have further muddied the traditional distinction between affect and cognition, suggesting that emotional functioning may play an important and primary role in processes once regarded as purely "cognitive."

LeDoux (1996) presented evidence for a rapid "low road" process in which salient or warning information is detected and processed through the thalamus and amygdala for emotional assessment *prior to* the completion of normal "high road" recognitional processes of the cerebral cortex. This track contrasts with the slower default condition in which nonurgent material is recognized by the cortex and only subsequently passed down to the amygdala for emotional evaluation. The "low road" process, although vague on the identification of the threat stimulus, is swift in its identification *as* a threat stimulus and can initiate appropriate action (or more precisely, the somatic precursors to the action, e.g., escape). LeDoux suggested that evolutionary processes have selected for the preservation of the "low road" pathway, despite increasing cortical development, because the time savings, although they may be minute, nevertheless convey a significant adaptive advantage over evolutionary time. The mechanism of this rapid response to a "stressor," of course, relies not at all on appraisal processes in higher cognition or indeed on any cognitive appraisal as the term is normally used in stress or emotions literature.

In his 1994 book *Descartes' Error,* Damasio made the claim that emotional processing, far from interfering with reasoning (as in the "unruly passions" stereotype), is in fact "central to rationality" (p. 115). Damasio outlined a view of emotion as a powerful signal system that provides (roughly as pain does for the body) an early warning to the cognitive system as to whether things are going well or badly and indicates which things need attention. He proposed a scenario wherein emotions help to streamline the process of reasoning and decision making by providing associations to previously experienced situations. This, he argued, serves to highlight the desirability of some decision options while eliminating undesirable choices from further analysis.

The significance of Damasio's perspective is that it emerged from research on individuals with frontal lobe damage, many of whom displayed profound impairment both in their experience of emotion and in their ability to make decisions. He described, for example, a patient whose frontal lobes had been damaged by a tumor. The patient's intellectual faculties were seemingly intact: He could discourse with ease on politics, business, and the events of the day, and he performed normally on a variety of standard tests of cognitive function. Yet this previously successful professional could no longer hold a job. He was unable to manage a schedule or prioritize even the simplest activities, and he made one disastrous decision after another. Nor did he learn from his mistakes, even when the consequences of his poor planning were pointed out to him. Concomitant with this were an extreme emotional flatness and detachment at odds with the personality he had had before the tumor. Nothing seemed to bother or upset him, not even the loss of his job, life savings, and marriage. In evaluating this individual, Damasio began to suspect that the patient's absence of emotion and his inability to make decisions were intertwined. It was as if he could not choose among competing options because he had no reason to value one more than another, and he could not avert personal disaster because the prospect of it did not alarm him.

This hypothesis led to a series of studies of the relation between emotional functioning and decision making in frontal-lobe-damaged patients. For example, when shown upsetting pictures (e.g., of homicides), these individuals could recognize the material as disturbing and describe why it was (or ought to be) disturbing, and yet, as some reported, they did not actually experience any particular emotion. These self-reports were consistent with measures of skin conductance responses: Normal individuals reacted sharply to the pictures, but the clinical patients displayed no response at all. Another study addressed decision making in a specially designed gambling game, which required participants to repeatedly choose cards from any of four decks. Each card, when selected, indicated a dollar amount that the participant was immediately given; a random few, however, also carried a penalty that the participant had to pay before continuing. No additional rules were announced, but with successive choices it became clear that two of the decks yielded moderate rewards and occasional, also moderate penalties, whereas the other two yielded somewhat higher rewards combined with catastrophically high penalties. It was clear, that is, to normal participants, who increasingly favored the first two decks as the game progressed. Frontally damaged patients, in contrast, made such poor choices that they often became "bankrupt" before the game was over. Interestingly, they were in fact able to figure out which decks were "unlucky," but this knowledge did not lead to a rational course of action. Further investigations revealed that normal participants actually started adopting an advantageous strategy *before* they had consciously figured out the rules, leading to the suggestion that emotional responses to prior mistakes, processed through subcortical structures loosely termed the limbic system, were guiding their assessment of the game before the higher cortex of the brain had perceived the nature of the threat (Bechara, Damasio, Tranel, & Damasio, 1997).

As with LeDoux, these findings run counter to the more traditional conception of emotions as occurring only in response to an initial, high-level cognitive appraisal and also to the view that emotional states are incompatible with rational analysis and decision making. Damasio (1994) has acknowledged that situations exist where emotional responses (particularly to perceived threat) are likely to hinder rather than help optimal functioning, but suggested that even here emotions are necessary to such functioning:

> Ultimately, the question raised here concerns the type and amount of somatic marking applied to different frames of the problem being solved. The airline pilot in charge of landing his aircraft in bad weather at a busy airport must not allow feelings to perturb attention to the details on which his decisions depend. And yet he must have feelings to hold in place the large goals of his behavior in that particular situation, feelings connected with the sense of responsibility for the life of his passengers and crew, and for his own life and that of his family. (p. 195)

In other words, the question is one of distal versus proximal goals, and of the sometimes conflicting emotions that these separate objectives may evoke. Panksepp (1996) made this point nicely in discussing the relation between emotion and cognition in the developing child: Young infants respond to their environment in immediate and stereotyped ways, but as cognitive sophistication increases these more "instinctive" emotional reactions become progressively modulated (or "decoupled"; Lazarus, 1991). Human emotions, moreover, cannot be adequately understood without reference to their neurological basis, which, in turn, should be placed in evolutionary context:

Existing evidence overwhelmingly supports the premise that the basic emotions arise from genetically ingrained neuronal operating systems that were
laid down in subcortical areas fairly early in mammalian brain evolution. This
knowledge has, I believe, profound implications for the future development
of psychology as a coherent science as well as for a more general understanding of certain key aspects of "human nature." (Panksepp, 1996, p. 31)

Some emotion researchers have indeed been drawing on an explicitly Darwinian perspective. Transactional models of stress and emotion are clearly
concerned with adaptation, in the sense that they focus on the relation between the individual and his environment and the individual's success or failure in this relation. Concurrent with this development has come an increasing
interest in evolutionary perspectives on human psychology, although individual researchers have differed somewhat as to emphasis. It has become almost a commonplace to point out that our modern, urbanized environment
represents only a tiny fraction of the history of our species and that the remaining 99% or so can be characterized almost entirely by the foraging lifestyle of our Pleistocene ancestors. (If one considers the several million years
occupied by the hominid lineage as a whole, the fraction of time represented
by "modern" life becomes infinitesimal indeed.) The evolved features of the
human mind should presumably reflect this, and indeed it is not difficult to
generate plausible-sounding hypotheses along these lines. (For example,
many typical phobias fall into this category, as with the hypothetical child
cited earlier who was afraid of snakes but not of trapping his fingers in doors.)

In many respects, evolutionary psychology provides a natural partner
both to stress research and to work on the emotions. Traditionally, aversive
emotions have tended to be placed in the purview of clinical psychology,
which brings with it a particular agenda and set of assumptions (i.e., that
such feelings are a sign of "pathology" that needs to be "cured"). From an
evolutionary perspective, of course, even unpleasant emotions may play an
important role not only in the adaptation of the individual to the environment, but in actual survival and inclusive fitness. (Two obvious—perhaps
hackneyed—examples are fear of predators and anxiety for one's offspring.) If, likewise, the absence of a given emotion has a negative effect on
fitness and if there is any differential genetic component in the tendency to
experience the emotion, then its representation in the population increases
with successive generations. Evolutionary psychology differs from its controversial predecessor, sociobiology (Wilson, 1975), in that it focuses not on the
evolution or the adaptiveness of behaviors per se, but on the psychological
mechanisms presumed to underpin behavior (see, e.g., Barkow, Cosmides,
& Tooby, 1992). This level of analysis provides a far more fertile ground for
investigation; also, contrary to some perceptions, it in no way excludes the
role of individual learning, experience, culture, or other environmental
components of human psychology and behavior.

Acceptance of Darwinian theory in the study of human psychology has unfortunately been hindered by a number of popular misconceptions about the nature of evolution by natural selection. To address just a few: First, it is not necessary or even accurate to assume that if a trait exists it must have arisen exclusively via natural selection. Second, a trait need not be adaptive 100% of the time to be selected for, as long as its presence confers even a slight statistical advantage overall—recall LeDoux's thalamus–amygdala pathway advantage. (This is an important point to consider when analyzing stress reactions that otherwise seem mystifyingly counterproductive.) Third, a trait may have been selected for in the past without being adaptive in present-day conditions.

One more important criticism that has been leveled at the field of evolutionary psychology itself concerns the propensity to generate entertaining but ultimately untestable "just-so stories" (Gould, 1978). These scenarios seek to explain some current human characteristic in terms of what is assumed to have been adaptive in the ancestral, Pleistocene environment (as the storyteller imagines it to have been). Griffiths (1997), for example, has warned against "taking the mere plausibility of an evolutionary explanation as adequate confirmation of that explanation" (p. 91). There is no disputing that verificationist reasoning has been endemic in this area, but mainstream psychology has paid too little attention to the converse argument. That is, models and explanations need to put to the test of evolutionary plausibility, and those that *fail* the test need to be reassessed. We include in this category any theory of the emotions or of stress that is limited primarily to conscious, deliberative intellectual processes carried out by modern, adult *Homo sapiens*. Such theories, not least those built around the precepts of formal logic, tend to proceed as though the human mind in some sense dropped down from heaven, ready formed, into a phylogenetic vacuum. For underpinning research, a functional, evolution-based approach to emotion is particularly useful as a source of hypotheses (as opposed to conclusions). For example, the argument that the emotion "grief" has, in part, the function of "retooling" schemas for changed circumstances is capable of expansion and empirical testing. Parents who have lost a child to drunken drivers, cults, drugs, or firearms sometimes channel considerable resources into mounting public campaigns against the perceived threat. That is, under some conditions the affect program includes some very active public behaviors of social consequence. In the case of the bereaved parents, it is a testable question whether this kind of retroactive "locking the barn door" behavior is correlated with other factors affecting the parents' inclusive fitness. These might include the number of remaining reproductive years, the presence of other surviving offspring, or other variables impinging on the parents' genetic survival. Wright (1994) has even hypothesized that the perceived reproductive value of the lost child may play an important role.

In the framework outlined here, affect program theory and the concepts of action readiness and control precedence may also hold promise for *applied* stress researchers. As an example, Stokes and Kite (1994) pointed out that existing stress literature has been unable to explain why stressed performance can be associated either with functional paralysis (undercontrolling a system) or with exaggerated psychomotor output (overcontrolling a system); neither has any model been established for predicting when each of these occurs and in which individuals (if at all). However, there may be value in asking what effects could be expected if "the stress" is experienced primarily as fear, as anger, or as other emotions or emotion combinations. Put another way, how much does it depend on the affect program "run" by the individual? The prioritization effects of an emotion's control precedence may be quite significant here (popular accounts of "road rage" come to mind, for example).

CONCLUSION

What we have attempted in this chapter is to set out the limitations of *stress* as the primary organizing concept underpinning work on the energetics of human performance. This is not because the numinous character of the term itself implies that there is no phenomenon to explore or that research in stimulus-based, response-based, or transactional frameworks has made no contribution to understanding. Rather, we suspect that there may be a point in such research programs at which the limitations of the stress concept give diminishing returns. At that point, we suggest, it is worth "unpacking" the stress concept into its emotional constituents—physiological, cognitive, and affective—to give better resolution to the emerging picture of energetic effects on human performance. It is true that this enterprise is fraught with complications, not least when the "evolutionary plausibility" yardstick must also be applied. Nevertheless, emotion research has provided some powerful tools that applied psychologists can and should make use of—tools oriented as much to action, behavior, or performance as to inchoate "feelings," or even more so. There are, we believe, worthwhile returns for those who are prepared to grasp the nettle—and "become emotional."

REFERENCES

Averill, J. R. (1968). Grief: Its nature and significance. *Psychological Bulletin, 70,* 721–748.
Ayalon, O. (1983). Coping with terrorism: The Israeli case. In D. Meichenbaum & M. E. Jaremko (Eds.), *Stress reduction and prevention* (pp. 293–339). New York: Plenum.
Barkow, J. H., Cosmides, L., & Tooby, J. (Eds.) (1992). *The adapted mind.* New York: Oxford University Press.

Bechara, A., Damasio, H., Tranel, D., & Damasio, A. (1997). Deciding advantageously before knowing the correct strategy. *Science, 275* (5304), 1293–1294.

Cannon, W. B. (1915). *Bodily changes in pain, hunger, fear, and rage.* New York: Appleton.

Carroll, L. (1951). *Through the looking glass.* Garden City, NY: Nelson Doubleday. (Original work published 1871)

Cohen, R. E., & Ahearn, F. L., Jr. (1980). *Handbook for mental health care of disaster victims.* Baltimore: Johns Hopkins University Press.

Cox, T. (1978). *Stress.* London: Macmillan.

Cox, T. (1985). The nature and measurement of stress. *Ergonomics, 28,* 1155–1163.

Cox, T. (1987). Stress, coping and problem solving. *Work and Stress, 1,* 5–14.

Cox, T., & Mackay, C. (1976, November). *A psychological model of occupational stress.* Paper presented at the meeting of the Medical Research Council, Mental Health in Industry, London.

Coyne, J. C., & Lazarus, R. S. (1980). Cognitive style, stress perception, and coping. In I. L. Kutash, L. B. Schlesinger, & Associates (Eds.), *Handbook on stress and anxiety* (pp. 144–158). San Francisco: Jossey-Bass.

Damasio, A. (1994). *Descartes' error: Emotion, reason, and the human brain.* New York: Grosset/Putnam.

Darwin, C. (1872). *The expression of the emotions in man and animals.* New York: New York Philosophical Library.

Ekman, P. (1977). Biological and cultural contributions to body and facial movement. In J. Blacking (Ed.), *The anthropology of the body* (pp. 39–84). ASA Monograph 15. London: Academic Press.

Ekman, P., Friesen, W. V., & Simons, R. C. (1985). Is the startle reaction an emotion? *Journal of Personality and Social Psychology, 49,* 1416–1426.

Elliott, G. R., & Eisdorfer, C. (1982). Conceptual issues in stress research. In G. R. Elliott & C. Eisdorfer (Eds.), *Stress and human health: Analysis and implications of research* (pp. 11–24). New York: Springer.

Epstein, S. (1983). Natural healing processes of the mind: Graded stress inoculation as an inherent coping mechanism. In D. Meichenbaum & M. E. Jaremko (Eds.), *Stress reduction and prevention* (pp. 39–66). New York: Plenum.

Fisher, S. (1983). Memory and search in loud noise. *Canadian Journal of Psychology, 37,* 439–449.

Frijda, N. H. (1986). *The emotions.* Cambridge: Cambridge University Press.

Frijda, N. H. (1994). Emotions are functional, most of the time. In P. Ekman & R. J. Davidson (Eds.), *The nature of emotion: Fundamental questions* (pp. 112–122). New York: Oxford University Press.

Frijda, N. H. (1996). Passions: Emotion and socially consequent behavior. In R. D. Kavanaugh, B. Zimmerberg, & S. Fein (Eds.), *Emotion: Interdisciplinary perspectives* (pp. 1–27). Mahwah, NJ: Lawrence Erlbaum Associates.

Glass, D. C., & Singer, J. E. (1972). *Urban stress: Experiments on noise and social stressors.* New York: Academic Press.

Gould, S. J. (1978). Sociobiology: The art of storytelling. *New Scientist, 80,* 530–533.

Griffiths, P. (1997). *What emotions really are: The problem of psychological categories.* Chicago: University of Chicago Press.

Grinker, R. R., & Spiegel, J. P. (1945). *Men under stress.* New York: McGraw-Hill.

Hamilton, V. (1980). An information processing analysis of environmental stress and life crises. In I. G. Sarason & C. D. Spielberger (Eds.), *Stress and anxiety* (Vol. 7, pp. 13–30). New York: Hemisphere.

Hamilton, V. (1982). Cognition and stress: An information processing model. In L. Goldberger & S. Breznitz (Eds.), *Handbook of stress: Theoretical and clinical aspects* (pp. 105–120). New York: Free Press.

Hammond, K. (1990). *The effects of stress on judgment and decision making: An overview and arguments for a new approach.* Unpublished manuscript, University of Colorado at Boulder.

Harss, C., Kastner, M., & Beerman, L. (1991). Personality, task characteristics, and helicopter pilot stress. In E. Farmer (Ed.), *Stress and error in aviation* (pp. 3–14). Aldershot, England: Avebury Technical.

Janis, I. (1983). Stress inoculation in health care: Theory and research. In D. Meichenbaum & M. E. Jaremko (Eds.), *Stress reduction and prevention* (pp. 67–99). New York: Plenum.

Jones, D. M. (1991). Stress and workload: Models, methodologies and remedies. In E. Farmer (Ed.), *Stress and error in aviation* (pp. 71–77). Aldershot, England: Avebury Technical.

Larsson, G., & Hayward, B. (1990). Appraisal and coping processes immediately before ejection: A study of Australian and Swedish pilots. *Military Psychology, 2,* 63–78.

Lazarus, R. S. (1982). Thoughts on the relations between emotion and cognition. *American Psychologist, 37,* 1019–1024.

Lazarus, R. S. (1984). On the primacy of cognition. *American Psychologist, 39,* 124–129.

Lazarus, R. S. (1991). *Emotion and adaptation.* New York: Oxford University Press.

Lazarus, R. S., DeLongis, A., Folkman, S., & Gruen, R. (1985). Stress and adaptational outcomes: The problem of confounded measures. *American Psychologist, 40,* 770–779.

Lazarus, R. S., & Folkman, S. (1984). *Stress, appraisal, and coping.* New York: Springer.

LeDoux, J. (1996). *The emotional brain: The mysterious underpinnings of emotional life.* New York: Simon & Schuster.

McGrath, J. E. (1976). Stress and behavior in organizations. In M. D. Dunnette (Ed.), *Handbook of industrial and organizational psychology* (pp. 1351–1395). Chicago: Rand-McNally.

Ortony, A., Clore, G., & Collins, A. (1988). *The cognitive structure of emotions.* Cambridge: Cambridge University Press.

Panksepp, J. (1993). Neurochemical control of moods and emotions: Amino acids to neuropeptides. In M. Lewis & J. Haviland (Eds.), *The handbook of emotions* (pp. 87–107). New York: Guilford Press.

Panksepp, J. (1996). Affective neuroscience: A paradigm to study the animate circuits for human emotions. In R. D. Kavanaugh, B. Zimmerberg, & S. Fein (Eds.), *Emotion: Interdisciplinary perspectives* (pp. 29–60). Mahwah, NJ: Lawrence Erlbaum Associates.

Rose, R. M. (1980). Endocrine responses to stressful psychological events. *Advances in Psychoneuroendocrinology, 3,* 251–276.

Sanders, A. F. (1983). Towards a model of stress and human performance. *Acta Psychologica, 53,* 61–97.

Selye, H. (1956). *The stress of life.* New York: McGraw-Hill.

Simons, R. C. (1996). *Boo! Culture, experience, and the startle reflex.* New York: Oxford University Press.

Stokes, A. F., Barnett, B., & Wickens, C. D. (1987). Modeling stress and bias in pilot decision making. *Proceedings of the Human Factors Association of Canada, 20,* 45–48.

Stokes, A. F., Belger, A., & Zhang, K. (1990). *Investigation of factors comprising a model of pilot decision making, Part II: Anxiety and cognitive strategies in expert and novice aviators.* Urbana-Champaign: University of Illinois Aviation Research Laboratory.

Stokes, A. F., Kemper, K. L., & Marsh, R. (1992). *Time-stressed flight decision making: A study of expert and novice aviators.* Urbana-Champaign: University of Illinois Aviation Research Laboratory.

Stokes, A. F., & Kite, K. (1994). *Flight stress: Stress, fatigue, and performance in aviation.* Aldershot, England: Avebury Technical.

Stokes, A. F., & Raby, M. (1989). Stress and cognitive performance in trainee pilots. *Proceedings of the Human Factors Society, 33,* 883–887.

Strelau, J. (1989). Individual differences in tolerance to stress: The role of reactivity. In C. D. Spielberger, I. G. Sarason, I. G., & J. Strelau (Eds.), *Stress and anxiety* (Vol. 12, pp. 155–166). New York: Hemisphere.

Tooby, J., & Cosmides, L. (1990). The past explains the present: Emotional adaptations and the structure of ancestral environments. *Ethology and Sociobiology, 11,* 375–424.

Toffler, A. W. (1970). *Future shock.* New York: Random House.

Warburton, D. (1979). Physiological aspects of information processing and stress. In V. Hamilton & D. Warburton (Eds.), *Human stress and cognition* (pp. 33–65). New York: Wiley.

Watson, D., & Tellegen, A. (1985). Toward a consensual structure of mood. *Psychological Bulletin, 98,* 219–235.

Welford, A. T. (1973). Stress and performance. *Ergonomics, 15,* 567–80.

Wilson, E. O (1975). *Sociobiology: The new synthesis.* Cambridge, MA: Harvard University Press.

Wright, R. (1994). *The moral animal.* New York: Vintage Books.

Yerkes, R. M., & Dodson, J. D. (1908). The relation of strength of stimulus to rapidity of habit-formation. *Journal of Comparative and Neurological Psychology, 18,* 459–482.

Zajonc, R. B. (1984). On the primacy of affect. *American Psychologist, 39,* 117–123.

1.5 A Transactional Model of Driver Stress

Gerald Matthews
University of Cincinnati

Diary studies of driving (e.g., Gulian, Glendon, Matthews, Davies, & Debney, 1990) suggest that mild stress symptoms are frequently experienced during driving. Such symptoms include unpleasant emotion, worry, and minor health problems. Even the familiar experience of commuting may be a significant source of stress (Novaco, Stokols, & Milanesi, 1990). Stress reactions to driving may impair performance and compromise safety (see chap. 1.8, this volume). Exposure to severe life events such as bereavement or divorce is associated with heightened risk of a motor vehicle accident (McDonald & Davey, 1996; Selzer & Vinokur, 1974). Hence, driver stress research contributes to understanding of the way that real-world stressors influence cognition and performance, life stress, and transportation human factors and road safety.

There are two initial difficulties in constructing a model of driver stress and performance. First, there is considerable scope for confusion between stress-related constructs (Koslowsky, 1997). It is important to distinguish symptoms versus underlying causal processes; state versus trait expressions of stress; person versus situational influences on stress; and general stress-related factors (e.g., negative affectivity, life events) versus factors specific to driving. Establishing such distinctions requires a coherent theoretical framework. Second, much of the research in the area has neglected objective measures of performance. Various cognitive mechanisms may contribute to performance impairment (Sivak, 1981), and neither subjective data nor accident data identify the information-processing functions that mediate stress effects on impairment.

This chapter reviews the development of a model of driver stress that aims to integrate both subjective and objective data. It is organized as follows. First, I describe the theoretical framework for the research provided by

Lazarus and Folkman's (1984) transactional model of stress. Next, I discriminate multiple dimensions of vulnerability to driver stress and describe their associations with self-reported behavioral and affective stress outcome measures. Further studies showed that individual differences in driver stress relate to the appraisal and coping constructs described by the transactional model and suggest an overall model of driver stress traits. The next part of the chapter reviews behavioral studies of driver stress by using a driving simulator to assess performance. The transactional model provides a basis for predicting associations between driver stress factors and efficiency and style of performance. Finally, I discuss two studies investigating information-processing mechanisms that may mediate an association between sensitivity to emotional distress and attentional impairment. I conclude by summarizing the cognitive-adaptive basis for various aspects of driver stress.

THE TRANSACTIONAL MODEL OF DRIVER STRESS

According to transactional models (Cox & Ferguson, 1991; Lazarus & Folkman, 1984; chap. 1.1, this volume), stress arises out of dynamic transactions or encounters between person and environment. Development of the transaction depends on *cognitive stress processes:* appraisal of the encounter and choice and regulation of coping strategies. Gulian, Matthews, Glendon, Davies, and Debney (1989) proposed that driver stress may be generated by cognitive appraisals that the demands of the task tax or exceed the driver's capabilities and coping resources. Appraisals depend on the interaction of situation and person factors. Situational influences on task demands include physical factors such as visibility and traffic density, social factors as threats to self-esteem posed by other drivers, and factors extrinsic to the driving task such as time urgency imposed by work or personal preference. Effects of objective demands are mediated by appraisals: For example, subjective impedance seems more predictive of commuter stress than does objective impedance (Koslowsky, 1997). Appraisals are influenced by personality factors (Van Reekum & Scherer, 1997): Drivers probably differ in their predisposition to evaluate as threatening attributes of the driving environment. Coping too depends on both the opportunities afforded by the situation and personality.

Cognitive stress processes generate the various outcomes or symptoms of stress. Stress outcomes include transient *states* such as negative moods, lack of motivation, and worry, and, in more severe cases, longer lasting chronic symptoms. From the transactional perspective, performance change may be quite a complex outcome, dependent on the direct and indirect consequences of coping (discussed in chap. 1.1). Drivers may choose to adopt risky strategies, for example, or they may become distracted by their own negative cognitions.

In this chapter, I adopt a trait conception of personality, which assumes long-term stability of individual differences in behavior. Cognitive models of personality (Matthews, Schwean, Campbell, Saklofske, & Mohamed, 2000) propose that stability of behavior derives from stability of "self-knowledge" in long-term memory. Self-knowledge refers to declarative beliefs about personal goals and competence and to proceduralized processing routines for handling personally significant events and motivations (see Wells & Matthews, 1994, for a formal model). Personality traits may be both general in nature (e.g., neuroticism) and linked to specific contexts (e.g., test anxiety). Neuroticism is a marker for stress vulnerability in a variety of different contexts (Matthews & Deary, 1998) and so may contribute to driver stress. There may also be traits specific to driving: Even the most mild-mannered individual may turn into a demon behind the wheel. Self-knowledge biases processing in stressful encounters, generating correlations between personality and subjective stress and overt behavior. Figure 1.5.1 illustrates the overall framework adopted here. Its two key features are that cognitive stress processes mediate personality and environmental effects on stress outcome; and personality represents individual differences in self-knowledge, such as beliefs about personal competence and generic plans for coping. The figure does not represent the dynamic nature of the model: Processing changes over time in response to changing environmental conditions, performance feedback, and reappraisal. (The relation between cognitive stress processes and performance is discussed in detail in chapter 1.1, this volume; see Fig. 1.1.2.)

This chapter focuses on driver stress as a cognitive-adaptive phenomenon (chap. 1.1, this volume), although other levels of description are also potentially relevant. Loud music appears to enhance attention during driving because of its arousing effects rather than through influencing coping, for example (Matthews, Quinn, & Mitchell, 1998). There is also quite an extensive literature on psychophysiological reactions to driving (e.g., Fairclough, 1993), but it is difficult to link them directly to cognitive mechanisms and to performance (Gulian, 1987).

DIMENSIONS OF DRIVER STRESS VULNERABILITY

Development of the DBI and DSI

The Driving Behavior Inventory (DBI) was developed by Gulian et al. (1989) and Glendon et al. (1993). Gulian et al. had samples of company car drivers complete a questionnaire on their emotional reactions to driving and behavior in demanding driving conditions. Factor analysis of the item responses identified three principal dimensions of Aggression, Dislike of Driving, and

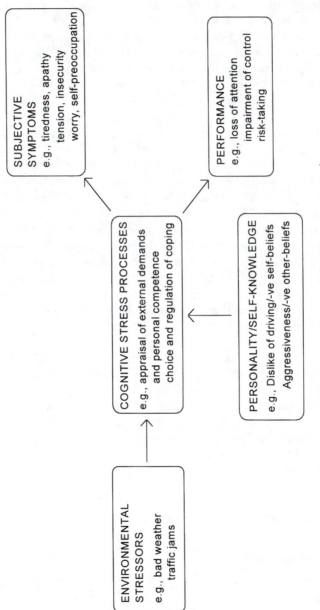

FIG. 1.5.1. A transactional framework for driver stress research.

136

Alertness, together with two minor dimensions related to overtaking. Aggression and Dislike corresponded to anger and anxiety responses to driving (e.g., Parry, 1968) and, at the extreme, to "road rage" and driving phobia. Alertness lacked emotional content and related to attempts to forestall stress through vigilance for potential dangers. Glendon et al. (1993) showed that DBI scores were stable across a 5-month period. Hence, the dimensions may be characterized as personality traits, linked to the driving context and representing vulnerabilities to qualitatively different types of stress outcome.

Development of the Driver Stress Inventory (DSI: Matthews, Desmond, Joyner, & Carcary, 1997) had two principal aims. The first was to add assessment of fatigue reactions and enjoyment of driving to the principal dimensions of the DBI. The second was to use the transactional framework shown in Fig. 1.5.1 as the basis for sampling. DSI scales are defined only by appraisal and subjective stress symptom items: Coping and behavioral outcomes require separate assessment. Five factors were extracted, including two factors similar to DBI Aggression and Dislike. Alertness was reconceptualized as Hazard Monitoring, relating specifically to active search for hazards. New factors of Thrill Seeking and Fatigue Proneness were also obtained. Sample items are shown in Table 1.5.1.

TABLE 1.5.1
Sample Items for Five Dimensions of Driver Stress Vulnerability

Dimension	Sample item
Aggression	I really dislike other drivers who cause me problems
	It annoys me to drive behind a slow moving vehicle
	Other drivers are to blame for difficulties I have on the road
Dislike of driving	I feel tense or nervous when overtaking another vehicle
	I find myself worrying about my mistakes when driving
	I am disturbed by thoughts of an accident or the car breaking down
Hazard monitoring	I make an effort to look for potential hazards when driving
	I try very hard to look out for hazards even when it's not strictly necessary
	I make an effort to see what's happening on the road a long way in front of me
Thrill seeking	I get a real thrill out of driving fast
	I like to raise my adrenaline levels while driving I sometimes like to frighten myself a little while driving
Fatigue proneness	I become inattentive to road signs when I have to drive for several hours
	My reactions to other traffic become increasingly slow when I have to drive for several hours
	I become sleepy when I have to drive for several hours

Some cross-cultural work has also been carried out on the driver stress questionnaires. Matthews, Tsuda, Xin, and Ozeki (1999) obtained factors resembling the three main DBI dimensions, but found that items relating to confidence and perceived control formed a separate factor, rather than attaching themselves to the Dislike dimension. Lajunen and Summala (1995) extracted Dislike, Aggression, and Alertness factors from Finnish data; and Matthews et al. (1997) have shown that the five DSI scales show similar validity coefficients in British and U.S. samples. The DSI also has good predictive validity in Australian drivers (Desmond, 1997).

Driver Stress Vulnerability and Other Stress Factors

Several studies have investigated associations between the DBI/DSI scales and measures of general personality and life stress factors. The broad trait of neuroticism or general affectivity operates as a general predisposition to stress reactions in a variety of real-world contexts (Matthews & Deary, 1998). Both three-factor (Eysenck & Eysenck, 1985) and five-factor (Costa & McCrae, 1992) models of traits concur that neuroticism is a fundamental aspect of personality. Dislike of Driving, as a predictor of emotional distress, might just be neuroticism in a different guise. Similarly, Aggression might largely reflect traits related to antisocial behavior: psychoticism in the case of the Eysenck personality model, and low conscientiousness and low agreeableness in the case of the five-factor model. Matthews, Dorn, and Glendon (1991) correlated the DBI traits with the scales of the EPQ-R (Eysenck, Eysenck, & Barrett, 1985). Modest correlations were found between Neuroticism (N) and Dislike and Aggression and between Psychoticism (P) and Aggression. Similar DBI/DSI correlates of N have been obtained by Lajunen and Summala (1995) in Finnish data, and by Fairclough (1997).

Matthews et al. (1997) gave drivers personality scales for the Five-Factor Model (Goldberg, 1992). N related not only to Dislike and Aggression, but also to Thrill Seeking and Fatigue Proneness. The next most predictive "Big Five" scale was Conscientiousness, which correlated at about .2 or so with higher Hazard Monitoring and lower Fatigue Proneness, Dislike of Driving, and Thrill Seeking. Several studies (Fairclough, 1997; Lajunen & Summala, 1995; Matthews et al., 1991; Matthews, Tsuda, et al., 1999) have investigated other personality correlates of the driver stress scales. Correlations, typically of .2–.3, have been found with cognitive failures (Dislike and Aggression), low self-esteem (Dislike and Aggression), interpersonal hostility (Aggression), Type A behavior (Aggression), everyday concentration (Alertness), sensation seeking (Aggression, Thrill Seeking), daytime sleepiness (Fatigue Proneness, Thrill Seeking), driver internal locus of control (Hazard Monitoring), and external locus (Dislike).

It has also been found that Dislike of Driving relates to life events (Gulian et al., 1989), minor hassles (Matthews et al., 1991; Matthews, Tsuda, et al., 1999), and overall level of stress responses in everyday life (Matthews, Tsuda, et al., 1999). As with broad personality measures, however, correlations rarely exceed .3–.4: Driver stress vulnerability traits are distinct from general personality and life stress measures. These data are limited by their cross-sectional nature, but they suggest that both life events and certain personality characteristics (especially neuroticism) may predispose the individual to driver stress. Presumably, the predisposition also depends on experiences and attitudes specific to driving. It is possible too that causality works in the reverse direction. Chronic stress caused by unpleasant driving experiences may feed back into personality change and susceptibility to life events.

Validation Against Stress Outcome Measures

Affect and Stress State.   Measures of driver stress vulnerability must predict state measures of stress reactions to individual drives. Reseach on mood disturbance has used the UWIST Mood Adjective Checklist (UMACL: Matthews, Jones, & Chamberlain, 1990), which assesses three fundamental dimensions of mood: energetic arousal, tense arousal, and hedonic tone (contentment vs. depression). Some studies have used an additional anger/ frustration scale, although this measure tends to be closely aligned with hedonic tone. Table 1.5.2 shows data from six studies in which the DBI or DSI was used to predict scores on the UMACL. In three field studies, the UMACL was completed in the driver's car immediately after a drive; the other studies were laboratory driving simulator studies. Correlates of stress vulnerability were similar in field and laboratory studies. Dislike of Driving was consistently related to postdrive tension and depression (low hedonic tone), and Aggression predicted postdrive anger. In field studies only, there was a tendency for Aggression to correlate with tension and Hazard Monitoring with energy. Similar correlates of Dislike and Aggression were found in a retrospective study that asked how drivers typically felt during various types of driving situations (Matthews, 1993). Ward, Waterman, and Joint (1998) found that DSI Aggression was substantially correlated with a scale assessing anger reactions to a variety of driving events.

Dislike of Driving tends also to correlate with predrive mood, presumably as a consequence of anticipation. Dorn and Matthews (1995) investigated the DBI as a predictor of *change* in mood as a consequence of simulated driving, controlling for individual differences in pretask mood. They showed that Dislike of Driving was more strongly related to mood deterioration when the drive required active interaction with other traffic (overtaking), as opposed to passive vehicle following. Aggression also related to increased unhappiness as a result of active traffic interaction. Hence, mood correlates

TABLE 1.5.2
Correlations Between DBI/DSI Scales and Mood
in Three Field Studies and Three Laboratory Studies

DBI scale	Study	N	Energy	Tension	Depression	Anger
Aggression	Field-1	50	−19	11	06	—
	Field-2	86	−06	25*	18	32**
	Field-3	104	−17	24*	23*	31**
	Lab-1	73	−04	21	19	—
	Lab-2	93	−22*	05	32**	—
	Lab-3	96	06	03	16	34**
Dislike of driving	Field-1	50	−21*	28*	33*	—
	Field-2	86	−20	22*	25*	00
	Field-3	104	−17	22*	23*	13
	Lab-1	73	−18	31**	31*	—
	Lab-2	93	−16	33**	23*	—
	Lab-3	96	−22	25*	22*	28**
Alertness/Hazard	Field-1	50	29*	−24	−22	—
Monitoring	Field-2	86	06	05	−04	−11
	Field-3	104	22*	−13	−11	−12
	Lab-1	73	10	11	−10	—
	Lab-2	93	09	−10	−09	—
	Lab-3	96	07	01	−17	−31**

Note. Field-1: Matthews et al. (1991; Study 4); Field-2: Matthews (1993); Field-3: Desmond (1997); Lab-1: Dorn and Matthews (1995; Sample 1); Lab-2: Dorn and Matthews (1995; Sample 2); Lab-3: Desmond (1997; fatigue condition). Field-3 and Lab-3 used the DSI; other studies used the DBI.
*P< .05.
**P< .01.

of the DBI and DSI are moderated by traffic conditions, as the transactional model of stress predicts. The third field study listed in Table 1.5.2 (Desmond, 1997) used a sample of Australian car drivers performing moderately lengthy drives (mean duration 78 minutes). Controlling for pretask mood, Desmond (1997) showed that Aggression related to increased anger and Dislike to increased tension. Studies of the DSI showed that Fatigue Proneness was the most consistent predictor of task-induced fatigued symptoms, and a cluster of stress reactions related to boredom and task disengagement in the laboratory (Matthews & Desmond, 1998), and in field studies (Desmond, 1997). Fatigue Proneness relates not just to tired mood but also to loss of task interest and motivation. Thrill seeking should predict more positive mood in traffic environments conducive to deliberate risk taking, but this hypothesis has not been tested.

States of stress are associated not just with affective and motivational outcomes, but also with awareness of cognitive state (Matthews, Joyner, Gilliland, Campbell, Huggins, & Falconer, 1999; chap. 1.1). Perhaps the most important cognitive stress symptom in the driving context is *cognitive interference:* the subjective expression of the diversion of attention from the task at hand onto internal worries (Sarason, Sarason, Keefe, Hayes, & Shearin, 1986). It is conceptualized here as a state rather than a process variable, that is, the person's awareness of intruding throughts, rather than the appraisal and coping processes that generate the thoughts concerned. Two field studies (Desmond, 1997; Matthews, 1993) tested for DBI/DSI correlates of cognitive interference. Matthews (1993) failed to show any associations between the DBI and interference, but Desmond (1997) obtained a significant correlation of .31 between Dislike and postdrive interference related to the driving task. Aggression and Fatigue Proneness also correlated with interference, to a lesser degree. The drive in the Desmond (1997) study was more demanding, as evidenced by state change data, which may explain why interference was more sensitive to stress vulnerability factors.

Self-Reported Behavior.   Item content (see Table 1.5.1) implies Aggression, Thrill Seeking and perhaps Fatigue Proneness might be associated with accident involvement. The first two scales have risk-taking connotations and so might relate to convictions for speeding and careless or dangerous driving. Alertness/Hazard Monitoring should presumably relate to greater safety.

The status of Dislike of Driving is more equivocal. On the one hand, life events likely to cause unhappiness and distress are associated with increased accident risk (Selzer & Vinokur, 1974). On the other hand, negative moods seem to be generally associated with caution and restraint in decision making (Forgas, 1995), which are desirable qualities in the driving context.

Consistent with expectation, Aggression, Thrill Seeking, and, less reliably, low Hazard Monitoring all relate to self-reported accident involvement (Matthews et al., 1991, 1997; Matthews, Tsuda, et al., 1999). Figure 1.5.2 shows how drivers reporting an accident in the last 3 years differed from accident-free drivers in the Matthews et al. (1997) study of British and North American drivers. For example, accident-involved drivers were, on average, .3 standard deviations higher in Thrill Seeking. Dislike and Fatigue Proneness do not seem to be associated with accidents. However, both Matthews et al. (1991) and Matthews, Tsuda, et al. (1999) found that high Dislike related to a lower incidence of speeding convictions, suggesting more behavioral caution in high Dislike individuals. This finding demonstrates that driver stress may not be entirely harmful; enjoyment of driving may lead to recklessness.

Driving behaviors may also be assessed by self-report. Reason, Manstead, Stradling, Baxter, and Campbell's (1990) Driving Behavior Questionnaire (DBQ) has been widely used. It distinguishes deliberate violations, such as choosing to break the speed limit, from unintended errors, such as selecting the wrong gear or failing to encode a highway sign. Violations appear to be

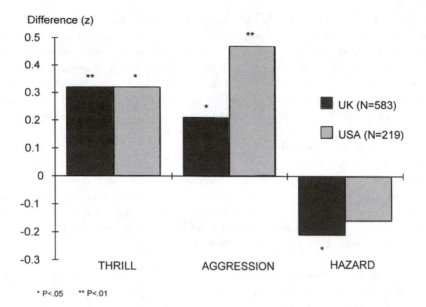

FIG. 1.5.2.   Differences in DSI scale scores between accident-involved and non-accident-involved drivers in British and North American samples. Data are from "A Comprehensive Questionnaire Measure of Driver Stress and Affect," by G. Matthews, P. A. Desmond, L. A. Joyner, and B. Carcary, 1997, in E. Carbonell Vaya and J. A. Rothengatter (Eds.), *Traffic and Transport Psychology* (p. 321), Amsterdam: Pergamon Press.

more predictive of future and past accident involvement than are errors (Stradling & Parker, 1997). Consistent with the DSI accident data reviewed in the previous section, Matthews et al. (1997) found that Thrill Seeking and Aggression were independently associated with greater DBQ violations, in both British and North American samples. Smaller, but still significant, negative correlations between Hazard Monitoring and violations were also found. Ward et al. (1998) obtained comparable results by using a self-report measure of violent reactions, such that Aggression was the strongest DSI predictor of violence. There were smaller, but significant, associations between violence and high Thrill Seeking and low Hazard Monitoring. Matthews et al. (1997) also used a scale for preferred speed, validated against onroad speed by West, French, Kemp, and Elander (1993). DSI correlates of this scale were similar to those of violations, but Dislike of Driving also predicted lower speed, consistent with the speeding conviction data (Matthews et al., 1991). Dislike of Driving, Fatigue Proneness, and (low) Hazard Monitoring were all associated with higher error rates: Thrill Seeking and Aggression were more strongly related to dangerous errors than to minor lapses. Dislike of Driving had a two-edged nature here: High scorers were safer with respect to lower speed, but potentially more dangerous with respect to commission of errors.

INDIVIDUAL DIFFERENCES IN COGNITIVE STRESS PROCESSES

The criterion validity studies confirmed that the DBI/DSI measures correlate with stress outcomes. The next stage of the research was to investigate the cognitive stress processes that may mediate relations between stress vulnerability factors and emotional and behavioral outcomes. The transactional approach accommodates the irrationality of drivers' cognitions. Everyday experience suggests that people often have distorted views of the threats and opportunities provided by driving and inflated conceptions of their driving skills. Perhaps driver stress vulnerability reflects the driver's characteristic cognitive distortions. The present research has investigated both the cognitive stress processes identified by Lazarus and other stress researchers: first, the appraisal of demands and one's personal competence to deal with those demands and second, choice of a coping strategy.

Appraisal

Sampling of items for the DSI distinguished appraisal items from affective outcome items. Factor analysis of the DSI items showed that the Dislike of Driving and Aggression factors were defined by both types of items. Dislike of Driving was associated with both emotional symptoms such as anxiety and cognitions about driving such as doubts about personal competence. Similarly, Aggression related to both feelings of anger and irritation and negative cognitions of other drivers' competence and intentions. Emotions and appraisals interlock at the trait level in defining stress vulnerability factors as cognitive-affective syndromes. On the DBI, the scales were defined by symptoms alone, so that they could be correlated with cognitive appraisal measures. Matthews (1993) collected post-task appraisal data in the study listed as Field-2 in Table 1.5.2. Dislike was associated with appraising the drive as uncontrollable and lacking in challenge, whereas Aggression related to threat appraisals. Unpublished analyses from this study showed that appraisals also correlated with ratings of environmental attributes such as poor visibility and road surface (threat), impedance (loss), and trip duration (challenge).

Appraisals of personal competence were investigated further by Dorn and Matthews (1995). Most people rated themselves as above average in safety and competence (e.g., Glendon, Dorn, Davies, Matthews, & Taylor, 1996). Drivers tended to have a false sense of safety, which relates to overestimation of personal control rather than to generalized optimism (McKenna,

Stanier, & Lewis, 1991). This cognitive bias was assessed by having respondents rate both themselves and a peer driver, someone of similar age and
gender, for safety, driving skills, and judgment. Dorn and Matthews (1995)
found that the Dislike of Driving scale was the strongest DBI predictor of
bias, but it was the *low* scorers who showed distorted appraisal. These individuals considerably overestimated personal judgment and skill and underestimated accident risk, compared with peer ratings. However, the high scorers, the drivers prone to emotional stress, gave similar ratings to self and
peers. Distress-prone drivers showed a kind of depressive realism (Alloy &
Abramson, 1979). Their appraisals appeared more accurate (at the group
level), but the cost of realism may be increased vulnerability to unhappiness
and anxiety. Conversely, aggressive drivers showed no bias in self-ratings but
appraised other drivers as lacking in skill (Dorn & Matthews, 1995). Lajunen
and Summala (1995, 1997) have also found that Dislike is negatively associated with drivers' self-concepts of being skillful.

The extent to which bias in appraisal is at least partially veridical remains
unclear. The behaviors of aggressive drivers, such as forcing other vehicles
to take evasive action, may well tend to elicit hostile reactions from other
drivers. In other words, negative appraisals of others may operate as a self-
fulfilling prophecy. It is possible too that drivers high in Dislike are genu-
inely deficient in driving skills, but, in this case, the objective performance
data discussed next suggest that there is no general deficit in competence associated with the stress vulnerability.

Coping

Evaluation of the traffic environment is dynamically related to choice of coping strategy. Coping reflects both people's appraisals of the coping options
afforded by the environment and their coping skills. The outcomes of coping feed back into appraisal. Coping may be expressed through overt behavioral responses or through internal processing only. According to Lazarus
and Folkman (1984), *problem-* or *task-focused* coping refers active attempts to
change external reality through behavioral response, such as reducing
speed following appraisals of lack of safety. By contrast, *emotion-focused* coping refers to attempts to deal with the stressor by reappraising one's emotional and cognitive reactions, through self-criticism or looking on the
bright side, for example. Cox and Ferguson (1991) distinguished a third
fundamental strategy of *avoidance,* which refers to attempts at ignoring the
stressor, often through self-distraction. A driver might listen to the radio or
talk with a passenger, for example. Lazarus and Folkman (1984) described
various coping strategies that extend the basic taxonomy of coping. These
include *confrontive coping,* a form of task focus that involves mastery of the
external challenge through self-assertion or conflict.

TABLE 1.5.3
Sample Items for Five Dimensions of Coping With Driver Stress

Dimension	Sample item
Confrontive coping	Showed other drivers what I thought of them
	Flashed the car lights or used the horn in anger
	Relieved my feelings by taking risks or driving fast
Task focus	Made sure I avoided reckless or impulsive actions
	Made sure I kept a safe distance from the car in front
	Tried to watch my speed carefully
Emotion focus	Blamed myself for getting too emotional or upset
	Wished I was a more confident and forceful driver
	Criticised myself for not driving better
Reappraisal	Tried to gain something worthwhile from the drive
	Felt I was becoming a more experienced driver
	Thought about the benefits I would get from the journey
Avoidance	Thought about good times I'd had
	Stayed detached or distanced from the situation
	Told myself there wasn't really any problem

In a retrospective study, Matthews (1993) had 166 drivers rate their use of six of the strategies described by Lazarus and Folkman (1984), across eight potentially stressful situations. Aggression related especially to confrontive coping ($r = .51$), whereas significant correlates of Dislike of Driving included self-criticism ($r = .23$) and escape avoidance ($r = .21$). Use of the rather general Lazarus and Folkman dimensions may not adequately capture coping strategies specific to driving, so that Matthews et al. (1997) developed a Driving Coping Inventory (DCI). Items were sampled to represent the main dimensions of coping identified in stress research, as they might be applied to driving. Table 1.5.3 summarizes the five factors extracted.

Confrontive coping strategies are clearly dangerous, because they involve antagonizing other drivers or risktaking. Task-focused strategies, on the other hand, are safety enhancing. Matthews, Quinn, and Mitchell (1998) measured situational coping in a simulator study and found that task focus was correlated with better vehicle control. Emotion focus here represents strategies of self-criticism and worry, which may be indirectly dangerous because attention is diverted from the driving task onto internal cognitions, causing cognitive interference (Matthews & Wells, 1996). Avoidance may also be associated with reduced attention to the task. Matthews, Quinn, and Mitchell (1998) found that both emotion focus and avoidance were associated with lower perceptual sensitivity in detecting "pedestrian hazard" stimuli. Koutsosimou, McDonald, and Davey (1996) found elevated levels of emotion-focused coping in traffic accident patients, compared with other surgery

patients. They suggested that use of emotion focus may be a direct vulnerability factor for accidents. Reappraisal is associated with more positive cognitions of the driving experience, which probably do not have the same potential for self-distraction.

The first three coping factors were substantially related to DSI scales, in
the sample of 533 British drivers. Confrontive coping was associated with
Aggression ($r = .58$) and Thrill Seeking ($r = .46$), but was negatively related
to Hazard Monitoring ($r = -.32$). Task focus was positively correlated with
Hazard Monitoring ($r = .43$). Emotion focus was most strongly associated
with Dislike of Driving ($r = .56$), and it also correlated with Fatigue Proneness ($r = .32$). Similar correlations have been found in a sample of drivers
from the United States (Matthews et al., 1997), and in a recent study of male
British drivers (Fairclough, 1997). These associations provide clues to the
behavioral consequences of the different stress dimensions. Aggressive drivers may be prone to violations and accidents because they use a rather maladaptive, confrontational form of coping under stress: Ward et al. (1998)
have confirmed the link between DCQ confrontational coping and violent
reactions. Similarly, Dislike of Driving may relate to error proneness because
of the distracting effects of emotion-focused coping.

Testing the Model: The Mediating Role of Cognitive Variables

Multivariate analysis of the Field-2 data collected by Matthews (1993) allows
a test of the role of appraisal and coping variables in generating subjective
stress symptoms. In the study concerned, 86 drivers completed state measures before and after a single drive and also postdrive questionnaires for
appraisal of task demands and coping. The coping measure here was a precursor of the DCQ (Matthews et al., 1997), in which emotion focus and reappraisal were not distinguished, so that four rather than five dimensions were
assessed. As shown in Table 1.5.2, Aggression related to anger and Dislike to
tension in this study. Table 1.5.4 summarizes data for three stress state measures: anger, tension, and task-related interference. The uncorrected Pearson correlations show a variety of significant predictors of mood and cognitive interference. Threat and loss/harm appraisal and confrontive coping
appeared to be generally associated with subjective stress.

To test for the mediating role of the cognitive stress process variables, a
multiple regression was run with each of the stress state measures as the dependent variable. Predictors were entered in four steps. First, pre-task state
was entered to control for initial level of stress; for instance, pretask tension
was the first predictor of post-task tension. Then, on successive steps, the four
appraisal variables, the four coping variables, and the three DBI scales were
entered. In each case, the appraisal and coping variables added significantly
to the variance explained, but the DBI scales did not, either collectively or
individually. DBI scales were unrelated to stress outcomes with cognitive

TABLE 1.5.4

Summary Statistics for Regression of Three Postdrive Stress State Variables
Onto Predrive State, Appraisal, and Coping

		Anger		Tension		Cognitive Int.	
		r	Partial r	r	Partial r	r	Partial r
Predrive state		51**	42**	56**	45**	42**	33**
Appraisal	Threat	56**	34**	58**	40**	47**	12
	Loss/harm	39**	12	41**	18	30**	20
	Challenge	00	−34**	−05	−29**	34**	07
	Control	−22*	00	−37**	−24*	−11	02
Coping	Confrontive	53**	36**	40**	00	36**	39**
	Task focus	16	06	07	−05	18	34**
	Emotion focus	09	14	23*	32**	49**	60**
	Avoidance	−19	03	−19	03	00	22

Note. r = uncorrected Pearson correlation.
Partial r = correlation corrected for all other predictors, in the final regression equation.
Cognitive Int. = task-related cognitive interference.
*P < .05.
**P < .01.
N = 86. Adapted from "Cognitive Processes in Driver Stress," by G. Matthews, 1993, in *Proceedings of the 1993 International Congress of Health Psychology.* Tokyo: ICHP. Copyright 1993 by ICHP.

stress process factors controlled, implying that the latter are the more proximal influence on stress state, consistent with the transactional model. Table 1.5.4 summarizes the regression statistics. The partial rs indicate the unique contribution made by each predictor in the final regression equation (forced entry of predictors, excluding DBI scales). Percentages of variance explained were 63% (anger), 57% (tension), and 58% (cognitive interference).

In each case, pretask state made a substantial contribution to the variance explained, indicating that individual differences post-task simply carry over from individual differences pretask. For the Table 1.5.4 regressions, appraisal and coping variables added from 26% to 44% extra variance; and, for each equation, several variables made independent contributions. High threat and low challenge were associated with both tension and anger, but confrontive coping predicted anger whereas lack of control and emotion focus predicted tension. Cognitive interference appeared to relate to coping rather than to appraisal, with emotion focus making the strongest contribution.

A MODEL OF DRIVER STRESS

The data reviewed so far suggest that Aggression and Dislike relate to broad cognitive-affective syndromes to which styles of appraisal and coping are

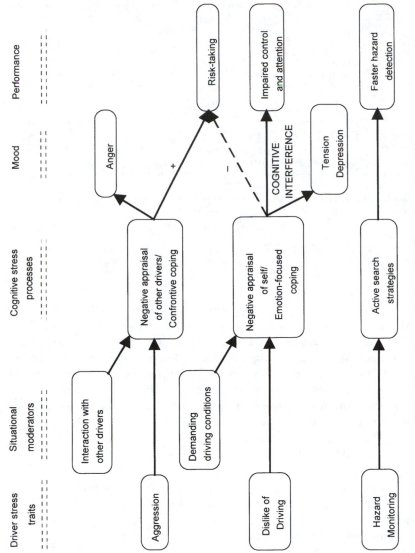

FIG. 1.5.3. A transactional model of three driver stress traits.

central. Alertness/Hazard Monitoring, Thrill Seeking, and Fatigue Prone-ness are somewhat narrower dimensions that currently are related more clearly to coping than to appraisal. Matthews, Dorn, et al. (1998) proposed a tentative causal model of the inter-relations between the three DBI stress vulnerability dimensions, information processing, and driver behavior, based on the evidence reviewed, and transactional principles, shown in Fig. 1.5.3. A more elaborate figure might also include life events and general personality traits as antecedents of driver stress states.

Aggression and Dislike relate to knowledge-based styles of cognition, which independently influence both affect and behavior. Specifically, Ag-gression is associated with the appraisal of other drivers as threatening and hostile and with the selection of confrontive coping strategies. Situations requiring active interaction with other drivers, such as driving in city traffic, provoke such cognitions to a greater degree in the more aggressive driver. Negative other-appraisal and confrontational coping generate feelings of frustration and anger, together with dangerous behaviors such as close fol-lowing, speeding, and risky overtaking.

Dislike of Driving relates to negative appraisals of driving competence and control and to use of emotion-focused coping strategies such as self-criticism, especially when driving conditions are demanding (e.g., poor visibility). Negative self-appraisals generate tension and depression and cognitive in-terference, which may impair attention and safety. However, such appraisals may compensate for these effects through biasing strategy choice towards behavioral caution, tending to increase safety. The compensation hypothesis explains the patterning of behavioral consequences evident in the self-reports of high Dislike drivers: more errors, slower speed, and no net affect on overall accident risk. Hazard Monitoring is perhaps a narrower dimen-sion, relating more strongly to cognition than to affective reactions. The high scorer on the scale actively monitors for hazards, but perhaps adopts a somewhat detached approach to driving, such that appraisals of driving do not impinge on appraisal of personal well-being.

STRESS AND DRIVER PERFORMANCE

The Matthews, Dorn, et al. (1998) transactional model describes both the relation between stress vulnerability and styles of cognition and the conse-quences of individual differences in cognition for performance. The next phase of the research aimed to relate driver stress to objective measures of performance in simulated driving. Broadly, the transactional model of driver stress shown in Fig. 1.5.3 predicts that Aggression should relate to risk-taking behavior, Dislike should relate to attentional impairment, and Hazard Monitoring should predict vigilance.

Testing For Moderator Effects

The transactional model specifies two types of moderator variables. The first type includes *situational stressors:* The appraisals and coping strategies typical of each dimension may be elicited only when the traffic environment makes certain kinds of demands (Dorn & Matthews, 1995). Hence, Aggression should be most predictive of performance when active interaction with other traffic is required, and Dislike should relate to performance mainly when the environment threatens personal safety or competence. The second type of moderator is the *information-processing demands* of the driving task, which may be particularly important for the cognitive interference effects linked by the model to Dislike of Driving. Anxiety research broadly suggests that cognitive interference operates through reducing the quantity of attentional resources available for performance (Sarason et al., 1986). However, resource theory provides only a partial account of interference phenomena and, in practice, is more successful for some tasks than for others (Matthews & Davies, 1998). If Dislike of Driving is associated with performance decrement, a resource hypothesis for the decrement requires careful testing by using task versions varying in attentional demands.

Predicted relations between driver stress traits and performance were tested by using a fixed-base computer-controlled driving simulator (Taylor, Dorn, Glendon, Davies, & Matthews, 1991). The driver views a scaled three dimensional display of the road ahead and other traffic on a 22-inch monitor. The car is controlled by using a steering wheel and brake and accelerator pedals. There are response buttons set into the steering wheel for use in attentional tasks. The main shortcomings are the lack of kinesthetic feedback when cornering, the compression of the field of view, and restricted screen resolution. However, the simulator shows what Sanders (1991) called functional fidelity: Behavior on the simulator is qualitatively similar to real-life behavior. For example, Dorn et al. (1992) showed that age and sex differences in performance observed in the field are reproduced on the simulator. In the stress context, as Table 1.5.2 shows, it seems as effective as real driving in eliciting emotional stress in vulnerable drivers. I discuss two sets of studies here. The first series of four studies simply tested for correlations between DBI variables and performance, across different types of driving situations. The second set of studies focused on attentional deficits associated with high Dislike of Driving. They investigated the moderating roles of situational stress and information-processing demands in more detail.

Correlational Studies of Driver Performance

Matthews, Dorn, et al. (1998) reported associations between DBI driver stress scales and performance in several different contexts: open-road driving, with no other vehicles present and free choice of speed; following another "lead" vehicle without overtaking (passing); driving among slow-moving cars, with limited opportunities to overtake.

Efficiency of vehicle control, indexed by variability of lateral position, was measured on open-road and following tasks. The cognitive interference hypothesis predicts that Dislike of Driving should be associated with greater positional variability (i.e., poorer lateral tracking). A subset of the subjects performing on the open road was required to respond to stimuli representing pedestrians, by using the steering wheel buttons, as a test of vigilance. It was predicted that Hazard Monitoring would relate to speed of response. Several measures relating to risk taking were available from the studies, including mean speed (open road, overtaking), mean distance from lead vehicle (following), and frequency of overtaking. High-risk overtakes (temporal distance to oncoming vehicle < 4s) were separately assessed. It was expected that Aggression would relate positively to these measures, especially in the overtaking task, because of the interaction with other traffic involved. Negative associations between Dislike and risk taking were predicted, on the basis that Dislike is associated with lack of confidence in driving competence. Gross control errors, such as driving off the roadway, were also assessed in some studies.

Table 1.5.5 summarizes the main relations between the DBI variables and performance obtained in these studies. Most of the predictions were confirmed. Dislike related to both impaired control and behavioral caution

TABLE 1.5.5
A Summary of Performance Correlates of the DBI

Task	N	Aggression	Dislike of Driving	Alertness
Open-road driving	140	—	Poorer control of lateral position	Lower speed Faster detection of hazards[1]
Vehicle following	140	—	Poorer control of lateral position Errors	—
Overtaking	65	Higher speed High-risk overtakes Errors	Fewer overtakes	—
Overtaking (young drivers)	93	Higher speed Overtakes High-risk overtakes Errors	Lower speed Fewer overtakes	—

Note.   [1]65 subjects performed hazard detection task. From "Driver Stress and Performance on a Driving Simulator," by G. Matthews et al., 1998. *Human Factors, 40.* Copyright 1998 by HFES. Reprinted with permission.

(reduced speed and overtaking). Tentatively, these correlates of Dislike may be related to cognitive interference and to coping through self-criciticism, respectively. Alertness (Hazard Monitoring) was associated with earlier discrimination of pedestrian hazards. Aggression related to risk taking, but only on the overtaking task, as expected, in two independent samples. Aggressive drivers not only drove faster and overtook more frequently, but they also committed more errors and engaged in more high-risk overtakes. However, more aggressive drivers did not seem to "tailgate" during vehicle following, as also expected, perhaps because they were prohibited from overtaking. Generally, the risk-taking behaviors of aggressive drivers may be a confrontive coping strategy adopted in response to appraisals of other traffic as threatening or frustrating.

Two other hypotheses for the correlation between aggression and risk taking can be eliminated. Aggression is not just a variant of sensation seeking, because aggressive drivers show no preference for higher speed in open-road driving. The situational moderator of other traffic is required to evoke confrontive coping. In addition, Aggression is not simply a consequence of lack of awareness of risk or risk acceptance: Aggression does not relate to biases in appraisal of safety and competence (Dorn & Matthews, 1995). Aggressive drivers' concerns with competing with other drivers may reduce their motivations toward safety, however, as shown in the questionnaire studies of Lajunen and Summala (1995, 1997).

INFORMATION-PROCESSING MECHANISMS FOR STRESS-INDUCED IMPAIRMENT

Theories of Attentional Impairment

Matthews, Dorn, et al. (1998) demonstrated performance correlates of the driver stress scales in line with those expected from the transactional model. Two further studies (Matthews, 1996; Matthews, Sparkes, & Bygrave, 1996) focused on the cognitive interference hypothesis for the association between Dislike of Driving and performance impairment. They tested two contrasting hypotheses for stress-induced impairment. First, impairment may reflect loss of attentional resources available for processing task stimuli. Resource explanations are prone to various methodological and theoretical difficulties (Navon, 1984; chap. 1.1), but they are attractive when impairments generalize across a wide range of qualitatively different tasks (Matthews & Wells, 1999). Attentional resource theory is, fundamentally, an account of performance variation under changing task loads (Townsend & Ashby, 1980). Hence, the basic prediction is that detrimental effects of Dislike of Driving should increase with the overall attentional demands of the driving task. The harder the task, the more vulnerable performance should be to diversion of resources to internal worries. Resource theory accounts are most attractive when load sensitivity is demonstrated across qualitatively different sources of load, that is, both task and stress factors. Dislike of Driving should be most detrimental to performance when an environmental stressor congruent with the concerns of high Dislike drivers is present.

Second, performance impairment may indicate the breakdown of dynamic effort regulation, in both underload and overload conditions (Hancock & Warm, 1989). Impairment reflects not so much a lack of capacity as a failure of adaptive efforts as compensation for task and environmental demands becomes progressively more difficult. The Hancock–Warm perspective has been applied to driving in fatigue studies conducted by Desmond and myself. Desmond and Matthews (1997) and Matthews and Desmond (1998) induced fatigue by using a demanding dual-task paradigm, combining driving with signal detection. Following fatigue induction, driving performance during "normal" single-task driving was assessed. Task demands were manipulated by comparing performance on straight and curved road sections. During fatigue induction, vehicle control and detection performance deteriorated more on straights than on curves. Aftereffects of the induction on subsequent driving were also more evident on straights than on curves. In other words, fatigue impaired driving when it was less rather than more demanding. This result makes no sense in terms of resource theory and contrasts sharply with results from sustained signal detection paradigms (See, Howe, Warm, & Dember, 1995). Instead, fatigue seems to make compensatory effort more dependent on immediate feedback from the task and less dependent on voluntary application of effort. This interpretation is supported by several additional features of the data. At a subjective level, induced fatigue was associated with loss of task engagement (as described in chap. 1.1, this volume), distress, and reduced active coping. Lower frequency of corrective steering responses demonstrated reduction in active control of performance. Performance was also sensitive to a motivational manipulation, but only on straight road sections (Desmond & Matthews, 1997).

Tests of Attentional Resource Theory

Matthews et al. (1996) tested the hypothesis that the lateral tracking impairment associated with Dislike of Driving is a consequence of reduced resource availability. During simulated driving, subjects were required to respond vocally to concurrent "grammatical reasoning" items, a task used by Brown, Tickner, and Simmonds (1969) to demonstrate dual-task interference in real-world driving. It aimed to counter methodological difficulties in testing resource theory (Navon, 1984) by setting up three independent tests for resource limitation, derived from Wickens' (1984) analysis:

1. *Difficulty sensitivity of dual-task interference:* The magnitude of interference should increase with difficulty of the driving task. Difficulty was manipulated by comparing performance on straights and curves.

2. *Modality sensitivity of interference:* Tasks in the same modality should show greater interference. Grammatical reasoning items were presented in auditory or visual modalities; visual presentation was expected to interfere more with driving.

3. *Performance tradeoff:* Resources released from one task should improve performance of the other. Priority afforded to reasoning and driving tasks was manipulated systematically, and performance operating characteristics (POCs) were constructed to assess tradeoff.

In fact, only the first of these hypotheses was supported, partially. Interference between reasoning and lateral control was found only when reasoning stimuli were auditory, on curved road sections. POCs showed that reasoning performance varied independently of driving performance. Resource theory did not explain dual-task interference in this paradigm. However, there was evidence for variation in effortful compensation, consistent with the Hancock–Warm model. On straight road sections, driving performance tended to improve during encoding of concurrent task stimuli and deciding on a response: Dual-task performance was actually superior to single-task performance.

Subjects were assigned to high and low Dislike groups on the basis of a median split. As in the earlier studies (Matthews, Dorn, et al., 1998), the high Dislike subjects showed significantly poorer control, assessed as heading error, in single-task driving (see Fig. 1.5.4). However, during dual-task performance, the difference between subject groups high and low in Dislike shrank to nonsignificance. Contrary to prediction from resource theory, the stress-related impairment was stronger when the driving task was relatively undemanding. Dislike also tended to be more detrimental on straight rather than curved road sections. The study also confirmed that Dislike of Driving is associated with cognitive interference, measured with a modification of Sarason et al.'s (1986) scale. It seems that, consistent with the Hancock and Warm (1989) model, drivers adapt quite successfully to the increased demands of dual task, perhaps through increased effort. The adaptive response may include suppression of cognitive interference, even in stress-vulnerable drivers. During single-task driving, the task may be appraised as less demanding of effort, and drivers fail to mobilize sufficient effort to maintain optimal performance. Under these circumstances, the stressed driver may divert attention to processing associated with worry.

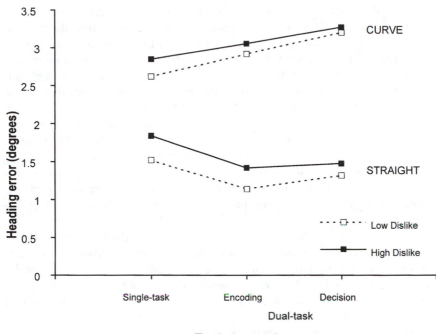

FIG. 1.5.4.   Heading error in simulated driving as a function of task de-
mands, road curvature, and dislike of driving. From "Stress, Attentional
Overload and Simulated Driving Performance," by G. Matthews, T. J.
Sparks, and H. M. Bygrave, 1996, *Human Performance, 9.*

Dislike of Driving, Induced Stress, and Compensatory Effort

The next step was to run a further test of the compensatory effort hypothesis for the Dislike of Driving effect (Matthews, 1996). This study aimed to extend previous work in three respects. First, the transactional model proposes
that high Dislike subjects are especially sensitive to traffic environments that
threaten their confidence in their personal competence. Thus the next
study included a stress manipulation of this kind, which exposed drivers to a
series of loss of control experiences. Second, the study also aimed to investigate subjective state change in more detail than in previous studies. Third,
the data so far have mainly concerned vehicle control, rather than specifically attentional tasks. This time, the study focused on attention to secondary task stimuli presented on road signs.

    Forty subjects performed a "winter drive," in which the car skidded un-
predictably and uncontrollably. After about 10 minutes, normal control was
restored, and the subject performed the attentional task, which involved

discriminating letter stimuli presented on road signs. By analogy with Posner and Cohen's (1984) studies, a spatial cue was used to indicate the side of the road at which the letter would appear, which was valid on 80% of trials. The main dependent measure was response time. Forty subjects performed in a control condition with normal vehicle control throughout. Subjects completed a battery of subjective state measures similar to those of Matthews et al. (1999; see also chap. 1.1, this volume) before and after performance. The general prediction was that the stress manipulation should have more pronounced effects on high Dislike than on low Dislike drivers. More specifically, the resource theory hypothesis predicts that high Dislike subjects should show impaired vehicle control and slower response to secondary task stimuli on curves rather than straights, whereas the effort-regulation hypothesis predicts that impairments should be stronger on straights.

The subjective data demonstrated that the stress manipulation was successful. Compared with the normal control drive, the winter drive caused large increases in tension, depression, and cognitive interference and a large decrease in perceived control. Furthermore, the stressor effect on cognitive interference was amplified in high Dislike drivers. They found that the loss of control experience was particularly worry inducing, consistent with the transactional model previously described.

Analyses of objective data showed stress effects on both heading error and response to secondary task stimuli. In general, the stress manipulation was detrimental to performance only in high Dislike subjects, a person × situation interaction consistent with the transactional model. The combination of high Dislike and imposed stress increased heading error mainly in single-task, straight-road driving: the easiest task condition. For the secondary task, there was a significant interactive effect of Dislike, level of stress, and curvature on average response time, illustrated in Fig. 1.5.5. As expected, Dislike was detrimental only in the stress condition, but high Dislike subjects were slower to respond only on straights. On curves, Dislike had no effect, although the stress manipulation tended to slow response. Once again, the data support an effort-regulation perspective. The loss of control experience induces especially high levels of cognitive interference in high Dislike subjects, but interference is only distracting when the task is relatively undemanding. In addition, the high Dislike–stress combination reduced use of the spatial cue on curves but not on straights, implying that "overload" conditions may induce strategy change, even if overall performance efficiency is unaffected. Hence, the "underload" Dislike deficit generalized to an explicitly attentional task, and it was contingent upon a theoretically derived stress manipulation. There was a convergence of subjective and objective consequences of high Dislike, that is, cognitive interference and impairment of attention.

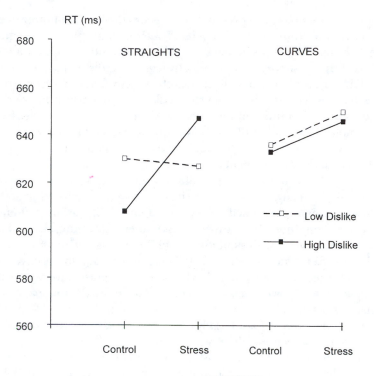

FIG. 1.5.5.   Mean response time to secondary task stimuli as a function of task demands, road curvature, and dislike of driving. From "Individual Differences in Driver Stress and Performance," by G. Matthews, 1996, in *Proceedings of the 40th Annual Meeting of the Human Factors and Ergonomics Society.*

CONCLUSIONS

The transactional model relates both subjective and behavioral aspects of driver stress to the information processing elicited by task demands. Appraisal and coping are the key cognitive processes that generate stress symptoms. Driver stress has various aspects: The nature of the processing elicited by the demands of driving determines the symptoms experienced. From the cognitive science perspective outlined by Matthews (1996; chap. 1.1, this volume), appraisal and coping support cognitive-adaptive transactions between driver and traffic environment. From the adaptive, self-regulatory perspective, the driver is seen as pursuing personal goals such as safety, comfort, and self-esteem in a sometimes demanding or hostile traffic environment (see chap. 3.3, this volume). Appraisal and coping reflect the driver's motives and beliefs about how to pursue personal goals. From the information-processing perspective, the transaction is associated with changing task strategies, including choice of maneuver (e.g., overtaking), attentional focus, and level of effort. Coping strategies and task strategies are distinct constructs (see chap. 1.1, this volume), but the research reviewed shows several convergences: between confrontive coping and risk-taking behaviors, between emotion focus and disruption of attention, and between reduced task-focused coping and reduced effortful control of vehicle trajectory. The transactional model, adapted to include information-processing constructs, provides the basis for explaining convergence.

The model contributes to explaining both sources and consequences of driver stress. As in cognitive-adaptive stress research generally, the appraisal and coping data implicate self-beliefs or self-knowledge as an influence on the stress process. Indeed, the driver stress vulnerability traits assessed by the DSI may be identified with stabilities of self-knowledge. The bias generated by individual differences in self-knowledge feeds through into individual differences in subjective state and behavior, although more empirical evidence on the role of self-knowledge is required. Furthermore, the low magnitude of correlations between DSI and general traits suggests that the self-knowledge that underpins driver stress is largely contextualized; that is, it reflects beliefs and experiences specific to driving.

Table 1.5.6 summarizes the most important correlates of the three original DBI scales. There is too little evidence on the two additional DSI scales to provide a transactional account of them, although Fatigue Proneness may be especially associated with motivational factors and reduced task focus (Desmond & Matthews, 1997; Matthews & Desmond, 1998). The transactional model attributes the various process and outcome correlates of Aggression primarily to the person's social beliefs about driving, consistent with the success of social psychological models in predicting intentions to violate (e.g., Parker, Manstead, & Stradling, 1995). The aggressive driver is disposed to interpret other drivers' actions as hostile and to react with potentially dangerous confrontive coping behaviors, especially when traffic is heavy or the driver is impeded. Aggression may be a consequence of both personality factors such as general hostility and learned social beliefs and norms. At an adaptive level, the aggressive driver appears to trade off reduced safety for greater awareness of potential threat from others. This tradeoff setting is maladaptive because it rests on an overestimation of the social and physical threats offered by other drivers, although aggressive drivers may to some extent elicit hostile reactions from other drivers through their own behaviors.

Central to Dislike of Driving are negative self-beliefs expressed most directly as appraisals of lack of competence and control. The high Dislike subject is also disposed to use rather negative forms of coping such as emotion

TABLE 1.5.6
Correlates of Three Driver Stress Traits

	Aggression	Dislike of Driving	Hazard Monitoring (Alertness)
General personality	Neuroticism Interpersonal hostility Type A	Neuroticism	Everyday concentration
Life stress factors		Life events/hassles Stress responses	
Situational moderators	Interaction with other traffic Impedance/ frustration	Task demands	?
Stress processes Appraisal	Negative appraisals of other drivers	Negative self- appraisals Low perceived control	?
Coping	Confrontive	Emotion focused	Task focus
Stress outcomes Subjective	Anger	Tension Unhappiness Cognitive interference	Energy ?
Performance (real-world)	Violations Errors Higher speed Accidents	Errors Lower speed	Fewer violations Fewer errors Lower speed Fewer accidents ?
Performance (simulator)	Risky overtaking Higher speed Errors	Less overtaking Lower speed Errors Impaired control Impaired attention	Faster hazard detection

focus. This pattern of cognition seems simultaneously to generate the subjective symptoms associated with Dislike, performance impairment, and behavioral caution. Task demands are the main factor that moderate the influence of Dislike on cognition, but the effect is subtle. Uncontrollable demands provoke the strongest subjective reactions, but low levels of task demands induce the strongest performance effects. The simulator studies (Matthews, 1996; Matthews et al., 1996) suggest that Dislike may be most damaging to performance when a driving experience that threatens personal competence is followed by a period of less demanding driving, providing the opportunity to reflect on the earlier event. Individual differences in Dislike seem to relate to an adaptive trade-off between contentment with driving and awareness of personal limitations as a driver. Self-awareness brings greater caution, but also greater cognitive interference, so that there seems to be no net effect on overall safety.

Hazard Monitoring indexes active attempts to anticipate danger, which presumably reflect beliefs about the importance of checking for hazards. It appears to be a largely beneficial aspect of "stress," although the actual impact of high Hazard Monitoring on accident involvement seems relatively small (Matthews et al., 1997). Further work on the situational moderators and performance correlates of this dimension is needed to develop the cognitive-adaptive perspective. Several questions defining the scale refer to hazard monitoring even when the road is familiar or "it's not really necessary." Possibly, high Hazard Monitoring is beneficial to safety only in relatively nonhazardous driving environments. Performance studies are required to investigate whether the focus of coping on monitoring for hazards trades off against other components of driving such as decision making. Search for hazards does not necessarily imply effective action in response to hazards, following detection.

In general, the transactional approach emphasizes the interplay between cognition and adaptation. The adaptive element is important because driver stress reflects the obstacles to personal goals encountered in the highway arena. Maladaptive responses to task demands are rational in terms of the driver's underlying beliefs and priorities. However, solely adaptive accounts are insufficient for detailed prediction of performance (as are social-psychological models), because of their neglect of information-processing factors. A key finding of the current research is that stress-related performance deficit is highly contingent on environmental and task factors. For the most part, stress-vulnerable drivers are not "accidents waiting to happen": They do not display a general impairment. Instead, individuals are vulnerable to rather specific sets of circumstances in which coping efforts translate into dangerous task strategies or impairment of information-processing routines. As Matthews and Desmond (chap. 1.8, this volume) describe, enhancement of driver safety thus requires interventions carefully targeted toward the maladaptive elements of the driver–traffic environment transaction.

ACKNOWLEDGMENT

I am grateful to Stephen Fairclough for comments on a previous draft of this chapter.

REFERENCES

Alloy, L. B., & Abramson, L. Y. (1979). Judgement of contingency in depressed and non-depressed students: Sadder but wiser? *Journal of Experimental Psychology, 108,* 441–485.

Brown, I. D., Tickner, A. H., & Simmonds, D. C. V. (1969). Interference between concurrent tasks of driving and telephoning. *Journal of Applied Psychology, 53,* 419–424.

Costa, P. T., Jr., & McCrae, R. R. (1992). Individual differences, stress and coping. In C. L. Cooper & R. Payne (Eds.), *Personality and stress: Individual differences in the coping process* (pp. 7–32). Chichester: Wiley.

Cox, T., & Ferguson, E. (1991). Individual differences, stress and coping. In C. L. Cooper & R. Payne (Eds.), *Personality and stress: Individual differences in the coping process* (pp. 7–32). Chichester, England: Wiley.

Desmond, P. A. (1997). *Fatigue and stress in driving performance.* Unpublished doctoral dissertation, University of Dundee.

Desmond, P. A., & Matthews, G. (1997). Implications of task-induced fatigue effects for in-vehicle countermeasures to driver fatigue. *Accident Analysis and Prevention, 29,* 513–523.

Dorn, L., Glendon, A. I., Hoyes, T. W., Matthews, G., Davies, D. R., & Taylor, R. G. (1992). Group differences in driving performance. In G. B. Grayson (Ed.), *Behavioural research in road safety* (Vol. 2, pp. 68–78). Crowthorne, England: TRRL.

Dorn, L., & Matthews, G. (1995). Prediction of mood and risk appraisals from trait measures: Two studies of simulated driving. *European Journal of Personality, 9,* 25–42.

Eysenck, H. J., & Eysenck, M. W. (1985). *Personality and individual differences: A natural science approach.* New York: Plenum.

Eysenck, S. B. G., Eysenck, H. J., & Barrett, P. (1985). A revised version of the Psychoticism scale. *Personality and Individual Differences, 6,* 21–29.

Fairclough, S. H. (1993). Psychophysiological measures of workload and stress. In A. M. Parkes & S. Franzen (Eds.), *Driving future vehicles* (pp. 377–390). London: Taylor & Francis.

Fairclough, S. H. (1997). *Driver impairment and individual differences.* (HUSAT Memo 1097R). HUSAT Research Institute, Loughborough University, England.

Forgas, J. P. (1995). Mood and judgement: The affect infusion model (AIM). *Psychological Bulletin, 117,* 39–66.

Glendon, A. I., Dorn, L., Davies, D. R., Matthews, G., & Taylor, R. G. (1996). Age and gender differences in perceived accident likelihood and driver competences. *Risk Analysis, 16,* 755–762.

Glendon, A. I., Dorn, L., Matthews, G., Gulian, E., Davies, D. R., & Debney, L. M. (1993). Reliability of the Driver Behaviour Inventory. *Ergonomics, 36,* 719–726.

Goldberg, L. R. (1992). The development of markers for the Big-Five factor structure. *Psychological Assessment, 4,* 26–42.

Gulian, E. (1987). *Driver stress: A literature review.* (CPERU Rep. No. 1.) Birmingham, England: Aston University.

Gulian, E., Glendon, A. I., Matthews, G., Davies, D. R., & Debney, L. M. (1990). The stress of driving: A diary study. *Work and Stress, 4,* 7–16.

Gulian, E., Matthews, G., Glendon, A. I., Davies, D. R., & Debney, L. M. (1989). Dimensions of driver stress. *Ergonomics, 32,* 585–602.

Hancock, P. A., & Warm, J. S. (1989). A dynamic model of stress and sustained attention. *Human Factors, 31,* 519–537.

Koslowsky, M. (1997). Commuting stress: Problems of definition and variable identification. *Applied Psychology: An International Review, 46,* 153–173.

Koutsosimou, M., McDonald, A. S., & Davey, G. C. L. (1996). Coping and psychopathology in surgery patients: A comparison of accident patients with other surgery patients. *British Journal of Health Psychology, 1,* 357–364.

Lajunen, T., & Summala, H. (1995). Driving experience, personality and skill and safety-motive dimensions in drivers' self-assessments. *Personality and Individual Differences, 19*, 307–318.

Lajunen, T., & Summala, H. (1997). Effects of driving experience, personality and driver's skill and safety orientation on speed regulation and accidents. In E. Carbonell Vaya & J. A. Rothengatter (Eds.), *Traffic and transport psychology: Theory and application* (pp. 283–296). Amsterdam: Elsevier.

Lazarus, R. S., & Folkman, S. (1984). *Stress, appraisal, and coping.* New York: Springer.

Matthews, G. (1993). Cognitive processes in driver stress. In *Proceedings of the 1993 International Congress of Health Psychology* (pp. 90–93). Tokyo: ICHP.

Matthews, G. (1996). Individual differences in driver stress and performance. In *Proceedings of the Human Factors and Ergonomics Society 40th annual meeting* (pp. 579–583). Santa Monica, CA: Human Factors and Ergonomics Society.

Matthews, G., & Davies, D. R. (1998). Arousal and vigilance: The role of task factors. In R. R. Hoffman, M. F. Sherrick, & J. S. Warm (Eds.), *Viewing psychology as a whole: The integrative science of William N. Dember* (pp. 113–144). Washington, DC: American Psychological Association.

Matthews, G., & Deary I. (1998). *Personality traits.* Cambridge: Cambridge University Press.

Matthews, G., & Desmond, P. A. (1998). Personality and multiple dimensions of task-induced fatigue: A study of simulated driving. *Personality and Individual Differences, 25*, 443–458.

Matthews, G., Desmond, P. A., Joyner, L. A., & Carcary, B. (1997). A comprehensive questionnaire measure of driver stress and affect. In E. Carbonell Vaya & J. A. Rothengatter (Eds.), *Traffic and transport psychology: Theory and application* (pp. 317–326). Amsterdam: Pergamon Press.

Matthews, G., Dorn, L., & Glendon, A. I. (1991). Personality correlates of driver stress. *Personality and Individual Differences, 12*, 535–549.

Matthews, G., Dorn, L., Hoyes, T. W., Davies, D. R., Glendon, A. I., & Taylor, R. G. (1998). Driver stress and performance on a driving simulator. *Human Factors, 40*, 136–149.

Matthews, G., Jones, D. M., & Chamberlain, A. G. (1990). Refining the measurement of mood: The UWIST Mood Adjective Checklist. *British Journal of Psychology, 81*, 17–42.

Matthews, G., Joyner, L., Gilliland, K., Campbell, S. E., Huggins, J., & Falconer, S. (1999). Validation of a comprehensive stress state questionnaire: Towards a state "Big Three"? In I. Mervielde, I. J. Deary, F. De Fruyt, & F. Ostendorf (Eds.), *Personality psychology in Europe* (Vol. 7, pp. 335–350). Tilburg: Tilburg University Press.

Matthews, G., Quinn, C. E. J., & Mitchell, K. J. (1998). Rock music, task-induced stress and simulated driving performance. In G. B. Grayson (Ed.), *Behavioural Research in Road Safety* (Vol. 8, pp. 20–32). Crowthorne, England: TRL.

Matthews, G., Schwean, V. L., Campbell, S. E., Saklofske, D. H., & Mohamed A. A. R. (2000). Personality, self-regulation and adaptation: A cognitive-social framework. In M. Boekarts, P. R. Pintrich, & M. Zeidner (Eds.), *Handbook of self-regulation* (pp. 171–207). New York: Academic Press.

Matthews, G., Sparkes, T. J., & Bygrave, H. M. (1996). Stress, attentional overload and simulated driving performance. *Human Performance, 9*, 77–101.

Matthews, G., Tsuda, A., Xin, G., & Ozeki, Y. (1999). Individual differences in driver stress vulnerability in a Japanese sample. *Ergonomics, 42*, 401–415.

Matthews, G., & Wells, A. (1996). Attentional processes, coping strategies and clinical intervention. In M. Zeidner & N. S. Endler (Eds.), *Handbook of coping: Theory, research, applications* (pp. 573–601). New York: Wiley.

Matthews, G., & Wells, A. (1999). The cognitive science of attention and emotion. In T. Dalgleish & M. Power (Eds.), *Handbook of cognition and emotion* (pp. 171–191). New York: Wiley.

McDonald, A. S., & Davey, G. C. L. (1996). Psychiatric disorders and accidental injury. *Clinical Psychology Review, 16*, 105–127.

McKenna, F. P., Stanier, R., & Lewis, C. (1991). Factors underlying illusory self-assessment of driving skill in males and females. *Accident Analysis and Prevention, 23,* 45–52.

Navon, D. (1984). Resources: A theoretical soupstone. *Psychological Review, 86,* 254–255.

Novaco, R. W., Stokols, D., & Milanesi, L. (1990). Objective and subjective dimensions of travel impedance as determinants of commuting stress. *American Journal of Community Psychology, 7,* 361–380.

Parker, D., Manstead, A. S. R., & Stradling, S. G. (1995). Extending the Theory of Planned Behaviour: The role of personal norm. *British Journal of Social Psychology, 34,* 127–137.

Parry, M. H. (1968). *Aggression on the road.* London: Tavistock.

Posner, M. I., & Cohen, Y. A. (1984). Components of visual orienting. In H. Bouma & D. G. Bouwhuis (Eds.), *Attention and Performance* (Vol. 10, pp. 531–556). Hillsdale, NJ: Lawrence Erlbaum Associates.

Reason, J., Manstead, A., Stradling, S., Baxter, J., & Campbell, K. (1990). Errors and violations on the roads: A real distinction? *Ergonomics, 33,* 1315–1332.

Sanders, A. F. (1991). Simulation as a tool in the measurement of human performance. *Ergonomics, 34,* 995–1025.

Sarason, I. G., Sarason, B. R., Keefe, D. E., Hayes, B. E., & Shearin, E. N. (1986). Cognitive interference: Situational determinants and traitlike characteristics. *Journal of Personality and Social Psychology, 51,* 215–226.

See, J. E., Howe, S., Warm, J. S., & Dember, W. N. (1995). Meta-analysis of the sensitivity decrement in vigilance. *Psychological Bulletin, 117,* 230–249.

Selzer, M. L., & Vinokur, A. (1974). Life events, subjective stress and traffic accidents. *American Journal of Psychiatry, 131,* 903–906.

Sivak, M. (1981). Human factors and highway-accident causation: Some theoretical considerations. *Accident Analysis and Prevention, 13,* 61–64.

Stradling, S. G., & Parker, D. (1997). Extending the theory of planned behaviour: The role of personal norm, instrumental beliefs and affective beliefs in predicting driving violations. In E. Carbonell Vaya & J. A. Rothengatter (Eds.), *Traffic and transport psychology: Theory and application* (pp. 367–376). Amsterdam: Elsevier.

Taylor, R. G., Dorn, L., Glendon, A. I., Davies, D. R., & Matthews, G. (1991). Age and sex differences in driving performance: Some preliminary findings from the Aston Driving Simulator. In G. B. Grayson & J. F. Lester (Eds.), *Behavioural research in road safety* (pp. 30–38). Crowthorne, England: Transport and Road Research Laboratory.

Townsend, J. T., & Ashby, F. G. (1980). *The stochastic modelling of elementary psychological processes.* Cambridge: Cambridge University Press.

Van Reekum, C. M., & Scherer, K. R. (1997). Levels of processing in emotion-antecedent appraisal. In G. Matthews (Ed.), *Cognitive science perspectives on personality and emotion* (pp. 259–300). Amsterdam: Elsevier Science.

Wells, A., & Matthews, G. (1994). *Attention and emotion: A clinical perspective.* Hove: Lawrence Erlbaum Associates.

West, R., French, D., Kemp, R., & Elander, J. (1993). Direct observation of driving, self reports of driver behaviour, and accident involvement. *Ergonomics, 36,* 557–567.

Ward, N. J., Waterman, M., & Joint, M. (1998). Rage and violence of driver aggression. In G. B. Grayson (Ed.), *Behavioural research in road safety* (Vol. 8, pp. 155–167). Crowthorne, England: TRL.

Wickens, C. D. (1984). Processing resources in attention. In R. Parasuraman & D. R. Davies (Eds.), *Varieties of attention* (pp. 63–101). New York: Academic Press.

PRACTICE

1.6

Stress in Ambulance Staff

Ian Glendon
Fiona Coles
Griffith University

A U.K. Association of Chief Ambulance Officers' report, *Ambulance 2000: Proposals for the Future of the Ambulance Service* (1990), maintained that "Ambulance service records show far too many instances of premature retirement through ill health and premature death due to stress related illnesses" (p. 19). The report recorded that the "intense pressure" results in few ambulance staff reaching retirement age without significant sickness or premature retirement on medical grounds. It drew attention to increased incidence of burnout among ambulance staff, brought on by service demands, rationalizing staff rosters, higher training levels, and increased responsibilities. As well as urging greater attention to improved recruitment and selection and increased awareness by management and staff of ways of identifying stress and developing coping strategies, it called for enhanced recognition of the role of accident and emergency ambulance teams. This chapter explores a number of the issues raised in this report by reviewing available literature on topics related to stress and coping among ambulance service personnel.

STRESS IN EMERGENCY SERVICES PERSONNEL

There is a growing body of literature on stress and coping strategies adopted by emergency service workers—fire, police, and ambulance personnel. In an annotated bibliography of 732 publications, Miletich (1990) found that two thirds of the material surveyed on emergency worker stress related exclusively to the police (see Table 1.6.1). Just under 7% of Miletich's sources related to ambulance/paramedic staff. The term *paramedic* is more commonly

TABLE 1.6.1
Distribution of Emergency Service Personnel References

Category/Service	Psychological Aspects	Physiological Factors	Family	Substance Abuse	Accidents	Suicide	Totals
Police Service	337	49	70	17	3	15	491 (67%)
Fire Service	23	125	—	3	13	—	164 (22%)
Ambulance/ Paramedic	33	5	—	3	10	—	51 (7%)
Police & Fire	2	3	—	—	—	—	5
Police & Ambulance	—	—	—	—	1	—	1
General/All Categories	10	6	—	—	4	—	20
Totals	405 (55%)	188 (26%)	70	23	31	15	732

Note. From *Police, Firefighter and Paramedic Stress: An Annotated Bibliography,* by J. J. Miletich, 1990, New York: Greenwood.

used in the United States, where paramedic trained staff may be associated with fire or police units, rather than being a separately identified service. Although arrangements differ between countries, the nature of the role is deemed to be sufficiently similar for research findings from different countries to be broadly comparable.

Miletich's (1990) bibliography is a useful compendium of topics or issues relating to stress among ambulance or paramedic staff. Over 80% of the studies considered either psychological or physiological aspects of stress in emergency services personnel. Table 1.6.2 summarizes topics from 46 sources reviewed by Miletich. These are from the psychological and physiological factors headings and cover ambulance/paramedic categories as well as those general publications in which paramedics or ambulance personnel are included. Several sources deal with more than one aspect of stress.

Individual Differences

Mitchell and Bray (1990) pointed out that emergency service workers accept challenges and voluntarily expose themselves to situations that few people who work in nonemergency fields would even consider. Similarities exist in respect of some of the stressors to which different categories of emergency service workers are exposed, for example, occasional or even frequent exposure to extreme danger, destruction, and human misery. Boyle (1997) notes that "emergency service workers are exposed to more emotionally distressing situations in the course of one shift than . . . many people would experience in a year" (p. 173). However, distinct variations in the type of stressors encountered and the reactions to them mean that there are also likely to be differences between the emergency services. For example, one study found that, compared with firefighters, ambulance service paramedics reported higher levels of job stress on questionnaire measures (Dutton, Smolensky, Leach, Lorimor, & Hsi, 1978). However, Young and Cooper (1995a, 1995b) found that compared with a normative group their sample of ambulance staff reported less Type A behavior.

It has been suggested that the personalities of emergency services personnel differ in key respects from those with less risky or demanding jobs. Mitchell (1986) suggested that ambulance personnel are idealistic, devoted,

TABLE 1.6.2
Summary of Topics Covered in 46
Publications on Paramedic/Ambulance
Personnel Reviewed by Miletich (1990)

Topic	Number of References
Job aspects (21)	
Job stress (various)	7
Burnout	6
Job satisfaction	2
Job turnover	2
Sickness absence	2
Organizational stressors/deficiencies	2
Coping (20)	
Coping with stressful aspects of job	13
(includes training, counseling, therapy, etc.)	
Personal health care/fitness	3
Managing anxiety/depression	2
Coping with patients' relatives/friends	1
Wellness programs (EAPs)	1
Trauma (7)	
Disaster case studies	2
Critical incident stress debriefing	2
"Trauma junkies" (thrive on challenge and	2
excitement associated with work)	
Coping with trauma	1
Death (6)	
Coping with death	4
Breaking news of death	1
Coping with grief	1
Occupational Health and Safety (4)	
Health hazards	2
Violence/vandalism	2

Note. From *Police, Firefighter and Paramedic Stress: An Annotated Bibliography,* by J. J. Miletich, 1990, New York: Greenwood.

goal oriented, histrionic, and dynamic and have high energy levels. Mitchell and Bray (1990) added to this list by suggesting that such workers are characterized by needing control; being obsessive/compulsive, highly motivated by internal factors, and action oriented; having a high need for stimulation; needing immediate gratification; being easily bored; being risk takers. Among other traits that have been identified as characteristic of ambulance workers is a strong affiliation need—wanting to help others. Glendon (1991) found that, compared with younger, less experienced ambulance staff, older, longer serving respondents in her sample had lower affiliation needs. Affiliation score was positively associated with perceived health, self-rated fitness, positive feelings about life, and problem-focused coping. Everly (1988) suggested that it may be an emotionally hardy type of person who self-selects into the emergency services. Studies on health professionals suggested that peer support, an internal locus of control, and work group support directly affect role ambiguity, which in turn influences work related stress, depression, and satisfaction with work (Revicki, Whitley, & Gallery, 1993).

In a combined study of over 2,000 firefighters and paramedics, Beaton, Murphy, Pike, and Jarrett (1995) carried out an analysis to reveal three clusters in their sample. On the basis of association with a range of reported somatic, behavioral, and psychological symptoms, Cluster 1 respondents were considered to be at high risk as reporting high levels of symptoms on most scales. Cluster 3 respondents were also considered to be at higher risk than the much larger Cluster 2 group. Together, the two higher risk groups represented less than 20% of those participating in the survey. However, given that their survey had a response rate of 51%, it remains possible that this figure does not represent the full extent of high-risk groups in the population of firefighters and paramedics.

Using the Occupational Stress Indicator, Young and Cooper (1995a, 1995b) compared stress outcomes for ambulance personnel and fire service staff. They found low job satisfaction to be a major stress symptom in the ambulance service but not in the fire service sample. Compared with working population norms, both samples showed significantly poorer mental and physical health and more job pressure. Compared with the fire service sample, the ambulance sample reported significantly more pressure from factors intrinsic to the job, career and achievement, and organizational structure and climate. Overall, compared with fire service personnel, ambulance staff in Young and Cooper's study reported higher levels of stress.

Like police and firefighters, the ambulance service is male dominated; typically, 90% of operational staff are male (Boyle, 1997). However, the ambulance service is unlike the other emergency services insofar as caring work is undertaken by men in public and in the company of other men. As Boyle (1997) noted, few male-dominated occupations have emotional labor as part of their skills repertoire. Of emotional management in ambulance staff, Boyle observed that "on the one hand, officers are expected to display the 'softer' emotions of compassion, empathy and cheerfulness in public regions whilst on the other hand refraining from the expression of grief, remorse or sadness in the company of other officers" (p. 1).

In a study focusing on locus of control (LoC) among ambulance staff, James and Wright (1993) found that external LoC was associated with higher levels of reported stress. They also found high internal LoC scores among their sample, which they accounted for by ambulance staff regarding themselves as highly trained and competent, thereby being capable of coping effectively with problems encountered in their work. However, James and Wright pointed out that high internality is a mixed blessing for ambulance staff. On the one hand, a belief in one's own competence and control over one's destiny could lead to an enhanced belief that one can cope successfully with difficult aspects of the job and could reduce stress. However, believing that one is responsible for adverse effects on patients could make the job more stressful. Thus, these authors did not find a negative relation between LoC and stress. They noted that ambulance staff who perceive strong influence from others and chance (i.e., externals) perceive their world more accurately than do their internal LoC colleagues. It seems that high externality and high internality among ambulance staff are both liable to lead to high levels of stress (Marmar, Weiss, Metzler, & Delucchi, 1996) and that the best option is a moderate degree of internality. Young and Cooper (1995a, 1995b) found that their sample of ambulance staff showed significantly more externality than did a norm group. Compared with a sample of fire service staff, ambulance staff reported higher externality on the "management processes" subscale, reflecting their relative perceived inability to influence those processes.

Organizational Studies

Identifying distinguishing personality factors among emergency services personnel has not secured as much research attention as the search for contextual features of their role, for example, organizational stressors. Cox, Leather, and Cox (1988) considered that, as well as coping with problems directly related to their jobs, those involved in health care may also experience stress arising from their lack of influence over organizational, political, and social issues. Young and Cooper (1995a, 1995b) considered that the most important predictor of both the mental and physical ill health of ambulance workers is pressure experienced from relationships with others. These authors argued that, for their sample, management was not developing relationships that were strong enough to initiate and encourage change.

An approach to the study of ambulance staff stress taken by a number of researchers is that of the survey—either or both of interviews and self-completion questionnaires (Glendon, 1991; James, 1988a). Glendon (1991) used a combination of in-depth individual and group interviews and participant observation to gather data for a larger scale questionnaire study in a large U.K. health authority. James (1988a) also used interviews to gather basic data from ambulance staff to develop alternative form questionnaires for wider distribution in another U.K. ambulance service. Both these studies found a considerable number of potential stressors among respondents to their initial data-gathering exercises.

Stress survey questionnaires to ambulance staff usually contain many items, and these need to be factor analyzed to make sense of a reduced field of categories. Typically, certain individual items stand out as being particularly stressful for respondents. Glendon (1991) found the five items that her respondents reported causing most job-related stress to be:

- Being with a bad driver at high speeds.
- Promotion of incompetent people.
- Insufficiently low retirement age.
- Lack of information about the organization's future.
- Dealing with the death of children.

The issue of road traffic accidents involving ambulances was addressed by Auerbach, Morris, Phillips, Redlinger and Vaughn (1987), who analyzed 102 such accidents. However, their conclusions were restricted to recommending that passenger restraints should be mandatory and that traffic signals and speed limits in urban settings be obeyed. Auerbach et al. (1987) also pointed out that "critical medical situations and prehospital anxiety are frequently translated into aggressive driving behavior" (p. 1490) and that ambulances are typically not designed for high speed driving.

James (1988b) categorized the variety and extent of demands made on ambulance service personnel in relation to emergency work and identified trauma—that is, accidents; medical emergencies; surgical emergencies; "special cases"—bomb alerts or major incidents; maternity cases; psychiatric cases; false alarms and hoax calls. In addition to these categories are the much more numerous nonemergency calls. Sparrius (1992) found that emergency work gave her respondents their greatest sense of achievement, while nonemergency patients were perceived to produce greater stress. James and Wright (1993) noted that the highest levels of stress among ambulance staff were less to do with the accidents, illness, incidents, and associated experiences and more to do with the way in which they were treated by other people, including superiors, colleagues, doctors, and hospital staff.

Some authors noted that certain aspects of the job can at one and the same time be potential stressors but can also be strong attractions of the job, for example, the unpredictability of demands from the amount and type of work (Dawson, 1990; James, 1988a; Robinson, 1984). Job features that are normally considered to be positive, such as variety, challenge, responsibility, and community service provision, may also induce stress. In a study of the relation between stress levels and the number of runs that a paramedic completed in a year, Mitchell (1984a) found stress levels to be highest when this figure was either greater than 600 or less than 200, suggesting a U-shaped stress function. Sparrius' (1992) respondents reported that a busy shift full of emergencies was less stressful than a slow shift. Boyle (1997) reported ambulance staff in her study expressing a preference for emergency calls over routine calls, because of the greater use of their skills in the former context. The downside of this is that involvement in too many emergency calls can result in emotional exhaustion.

Potential stressors for ambulance staff range from intrinsic components of the job to organizational and career structural issues. An open-ended question in Glendon's (1991) survey asking about the three current main causes of stress in their lives produced the highest responses for shiftwork/rosters/lack of leisure (39% of respondents mentioned this); finance/money (26%); critical of management/lack of information (22%). Robinson (1986) found that ambulance worker stress was predicted by high levels of shiftwork and job rotation. Table 1.6.3 summarizes findings from a number of independent studies that identify factors making up ambulance staff stress from analyzing survey data.

Research is generally lacking in the area of the interface between individual and organizational factors. For example, of the typically high affiliation needs of ambulance staff, Glendon (1991) noted that, if such workers are to perform to the best of their ability, then the support of a well-managed organization is particularly important. When this is not forthcoming, workers may be more susceptible to adverse experience of stress than are workers whose affiliation needs are of less intrinsic importance to their job. James and Wright (1991) pointed out that the major sources of stress for ambulance staff are extrinsic to the job—as reflected in the factor analyses (James, 1988b). However, they also noted that questionnaire surveys are likely to overestimate the extent to which job-extrinsic factors are identified. This is because admitting to experiencing stress in relation to handling patients could be construed as an admission that ambulance officers had doubts about their ability to cope effectively with this key aspect of the job—with possible adverse effects on their self-esteem and anxiety about future job performance. Stating that other people are the major source of stress is not threatening to one's self-esteem.

Developing earlier work, James and Wright (1991) interviewed 30 respondents from James' original questionnaire sample and found areas of stress that had not emerged from the earlier survey. These may be compared with the summary data shown in Table 1.6.3. Interview respondents consid-

TABLE 1.6.3

Summary of Factor Analytic and Other Studies of Ambulance Staff Stress

Study/Year	Management/ Admin/ Organizational Problems	Unfamiliar/ Difficult Duties	Work Overload/ Underload*	Interpersonal Relations**	Emergency Operational Aspects of Job***	Shiftwork/ Rosters	Communication Problems	Job Conditions/ Public Relations****	Employment Conditions	Family/ Social Aspects	Driving Stresses
James, 1988a, 1998b	✓	✓	✓	✓							
Robinson, 1988							✓	✓	✓	✓	
James & Wright, 1991			✓	✓	✓		✓	✓		✓	
Glendon, 1991	✓			✓	✓	✓				✓	✓
Hamil, 1991	✓				✓	✓		✓			
Winkler, 1980	✓				✓	✓	✓			✓	
Young, 1980	✓				✓	✓	✓			✓	

*For example, underuse of expertise/ability.
**For example, dealing with medical, nursing, and other hospital staff.
***Includes having responsibility for lives of others, colleagues, etc.
****Includes misuse/abuse of ambulance service.

174

TABLE 1.6.4
Major Emotional Stressors in Ambulance Work

- Death/serious injury of a colleague in the line of duty
- Suicide of a fellow officer
- Multiple casualty incidents
- Death/serious injury to children
- Attending scenes where victim is known/reminds staff of known/loved one
- Situations that threaten life/safety of staff
- Situations entailing prolonged rescue work
- Situations attracting undue/critical media attention
- Situations placing heavy/immediate responsibility of staff for saving lives
- Dealing with body parts
- Responding to many difficult situations in a short time
- Any incident in which circumstances are so unusual or sights/sounds so distressing as to produce an immediate/delayed emotional reaction

Note.   From *Parliamentary Select Committee of Inquiry into Ambulance Services—First Report,* by Queensland Parliament, 1990, Brisbane: Government Printing Office.

ered that "stress" was an inappropriate concept when applied to dealing with patients, their emotions more accurately being described in terms of sadness, concern, and sometimes frustration or impatience—particularly when time constraints affected the way in which ambulance staff wished to care for their patients.

From a small sample of interviewees in a South African ambulance service, Sparrius (1992) found a high level of negativity accorded by her respondents to organization-based stressors. Nineteen of twenty-two identified stressors she classified as organizational in origin. Noteworthy among these were paramilitary structure, management style, the disciplinary and control system, and limits to decision making. An important distinction is between what may broadly be termed organizational stressors and those that arise from emotional aspects of ambulance staff work. A Queensland Parliament (1990) report identified a dozen critical emotional stressors in ambulance work, which are outlined in Table 1.6.4.

Ethnographic Studies

Studies that examine particular stress components of ambulance staff work, either experimentally or via surveys, can provide valuable data in the form of pieces of the jigsaw. Ethnographic studies of ambulance staff (e.g., Boyle, 1997; Mannon, 1991; Metz, 1981; Palmer, 1983), not only complement more quantitative approaches, but can also provide a context in which results from such studies can be interpreted and understood. For example, using theoretical frameworks derived from emotionality in organizations, Boyle (1997) highlighted "how an organisation reconciles the emotional ambiguity that occurs when men perform public caring work within a traditionally . . . masculine organisational framework" (p. 245).

Public and Media Perceptions

Sparrius (1992) reported that extraorganizational stressors experienced by her South African sample of ambulance staff included low occupational status accorded by the public, health authorities, and even their own families. Lau (1988) identified a public ambivalence about health care, characterized by high and often unrealistic expectations, a strong sense of dependency and a critical attitude. Congruent with this view, James (1988b) drew attention to the public's assumption that there must be a reliable emergency medical response mechanism and that they have a right to such a service. Robinson (1984) explained that such public expectations are at least partly responsible for the overutilization of this emergency service, inappropriate summoning of ambulances and the misinterpretation of the level of services available. Morris and Cross (1980) determined that, of a sample of 1,000 emergency calls resulting in patients being admitted to hospital, 52% were unnecessary, and that medical education of the public was needed to reduce the number of such calls. Apart from the financial cost, these journeys also increase the danger to ambulance staff when hurrying to answer emergency calls. Solutions proffered by these authors include a financial levy on callers, medical authorization of calls, and greater public medical education, for example, through the media and first aid training in schools.

Palmer (1983) found that many emergency medical technicians enjoyed the "invisible" component of the job, which gave a sense of power and control via demonstration of technical and interpersonal skills to the public. Boyle (1997) noted the importance of the high level of trust afforded by members of the public to ambulance staff and also the high public expectations that ambulance officers are "model" citizens. She observed that "the power to calm and reassure combined with the power to alleviate pain and suffering is a powerful magnet that provides officers with a strong and meaningful work identity" (p. 226), to which may be added a strong community identity.

Physiological Measures

Ethnographic studies and self-completion measures of ambulance staff stress are complemented by those that have used physiological measures deemed to be associated with felt stress. For example, Beaton et al. (1995) used a Symptoms of Stress Inventory as one of several survey instruments to obtain information on 253 paramedics, most of whom were also firefighters. Their inventory measured somatic, behavioral, and psychological stress symptomatology. Combining their sample of paramedics with a much larger sample of firefighters, these authors found that their total sample showed highly significant differences from a comparison sample and had raised levels on apprehension/dread, gastrointestinal symptoms, sleep difficulties, throat and mouth symptoms, and intrusive/frightening thoughts. Other factors emerging from Beaton et al.'s analysis of their firefighter/paramedic sample were head/neck/facial tension, anger, indigestion/asthma, generalized anxiety, cardiopulmonary, cutaneous, headiness, itchy/rashes, exaggerated startle, agitated depression, upper respiratory, extremity tension, hunger, and nervous habits.

Using a model devised by Karasek, Russell, and Theorell (1982), Goldstein, Jamner, and Shapiro (1992) identified the job of paramedics as meeting the criteria of Karasek et al. of being inherently stressful because of a combination of being psychologically demanding and having low decision latitude. These authors cited such intrinsic job features as dealing with life-or-death decisions under potentially hazardous situations as well as repeated exposure to human tragedy and pressure to perform in ambiguous situations as being psychologically demanding. Low decision latitude results from the type of organizational factors found in studies cited in Table 1.6.3, as well as typically low levels of participation in the organization of their work and few possibilities to control work pace (Goldstein et al., 1992).

Goldstein et al. (1992) compared ambulatory blood pressure and heart rate in 30 paramedics during a 24-hour workday and a 24-hour non-workday. Despite overall stability of blood pressure readings across both days, specific situations were associated with higher blood pressure during work activity. These included being at the scene of an accident, being at the hospital, and riding in the ambulance—situations that were also associated with more negative moods (Jamner, Shapiro, Goldstein, & Hug, 1991). Goldstein et al. identified elevated blood pressure levels as being related to stress involved in confronting accident victims, handling others' grief, and responding to the threat of injury to oneself (Mitchell, 1984a; Stout, 1984).

In their study, Goldstein et al. (1992) interpreted the observed elevated blood pressure levels during certain work activities as reflecting paramedics' ability to separate stressful episodes in their working day from other work activities, identifying their distinct pattern of relatively rapid unwinding from their work activities and from acute stressful events in particular. However, the organizational context in which coping occurs can be a critical moderating factor. Boyle (1997) considered the impact of the masculine culture in the organization that she studied—which is likely to be typical of organizations in which ambulance staff work. In such a culture, officers are allowed to display "feminine" emotionality only in limited contexts. To be "men," officers are expected to be able to switch between front- and backstage emotionality at will. However, as noted by Boyle (1997), emotional switching is problematic when ambulance staff are unable to engage in emotional process work in backstage regions. She pointed out that suppression of emotional process work is important in constructing and maintaining organizational emotionality.

Mortality and Morbidity

Surveying occupational stress in the London Ambulance Service, Toombs, Quinlan, and Terry (1979) found a high rate of sickness absence, a proportion of which was deemed to be stress related. Balarajan (1989) found ambulancemen to be one of the few health worker groups in which ischemic heart disease had increased in the decade to 1981. Robinson (1986) found that symptoms of ill health were predicted by high age, divorce since ambulance employment, and poor coping skills. Mitchell and Bray (1990) noted that emergency services personnel in general experience higher divorce rates than the population norm.

Several studies have found a high incidence of burnout and job turnover among paramedics (Grigsby & McKnew, 1988; Herbison, Rando, Plante, & Mitchell, 1984; Metz, 1981; Mitchell, 1984b; Nixon, 1987). Citing Herbison et al. (1984) and Mitchell's (1984b) findings that paramedics tend to leave their occupation at a relatively young age, Goldstein et al. (1992), in their study of blood pressure among ambulance paramedics, considered that this serves to "protect" such workers from hypertension. What these authors effectively identified in this occupational group is the "healthy worker" effect, frequently documented in epidemiological studies. However, Robinson (1986) found that length of employment was negatively related to reported stress.

Consistent with this effect is S. Glendon's (1991) finding that compared with younger respondents, older ambulance workers in her study reported lower levels of stress, although this was not a strong effect. However, set against this finding was the finding that perceived fitness and perceived health correlated negatively with length of service. The picture that emerges is one of more resilient staff staying longer with the job, but nevertheless experiencing declining health as a result. In respect of self-selection of ambulance staff, James and Wright (1991) pointed out that the occupation attracts and retains people who are more capable than their peers of coping with job-related stress—another component of the healthy worker effect.

Social and Organizational Responses

Some time ago, a U.K. report recommended that, in line with other emergency service personnel, the retirement age for ambulance staff be reduced from 65 to 55 years, combined with developing improved occupational health care and phased new pension arrangements (Perry, 1988). The report was submitted to the Secretary of State and pointed to the high incidence of diseases of the musculoskeletal system as well as the greater comparative ill-health retirement of ambulance staff. However, it was neither published, nor were its recommendations implemented, presumably because of the increased superannuation and other costs that would be associated with such changes.

The issue of organizational responses to stress among ambulance staff may be considered at various levels. Although numerous palliative measures have been suggested by various authors, in-depth exploration of organizational factors that are antecedents to stress might indicate that many of these are relatively superficial attempts to deal with deep-seated stressors, which leave underlying issues unaddressed. Concomitantly, studies that explore complex underlying structural issues may have difficulty in making specific recommendations for improving the lot of ambulance staff, because of the embedded nature of issues in organizational culture.

Palliative Measures

A number of suggestions have been made about how stress may be relieved in ambulance staff. Glendon's (1991) respondents approved of the following items from a prompted list:

- Provide more gymnasium and exercise equipment.
- Provide counselling services.
- Provide better catering facilities, as well as training across a variety of fields, including:
  Preventing back injuries.
  Developing physical fitness.
  Offering stress management dealing with trauma and emergency.
  Dealing with conflict, assertiveness.
  Teaching counseling skills.
  Dealing with AIDS/hepatitis B patients.
  Coping with bereavement.

Several studies have made recommendations about reducing harmful levels of stress among ambulance personnel (Glendon, 1991; Perry, 1988; James, 1988a, 1988b; Robinson, 1986). Those that have been independently identified by at least three of these studies are:

- Reduce retirement age from 65 years.
- Offer alternative employment for older staff.
- Provide periodic health screening for employees.
- Offer confidential counselling service.

- Train in lifting techniques.
- Provide for staff's physical fitness.
- Offer stress management program.

Beaton et al. (1995) advocated a screening program to identify high-risk individuals who emerged from their survey and proposed for these groups a broadly based psychoeducational intervention program, which focused first on identifying sources and symptoms of their stress. They argued for an intervention program at both organizational and individual levels, which would include work redesign and coping skills training. Recommendations made by Beaton et al. for specific treatment modules included relaxation training —to counter apprehension, dread, and generalized anxiety; cardiovascular health—to reduce incidence of cardiovascular morbidity and mortality; conflict resolution training—to teach skills to mediate inevitable interpersonal conflicts. The authors argued that, because of numerous and diverse stress symptoms in this group, a comprehensive approach to stress management in paramedics and other at-risk groups must be more broadly based than critical incident stress debriefings. They urged further study of mediating variables that appear to be protective for a majority of paramedics and other emergency personnel, so that better preventive programs and remedial programs for high-risk individuals can be developed.

COPING STRATEGIES

General "Group Culture" Responses

Use of humor is well known among occupational groups that encounter death and mutilation as part of their work. Palmer (1983) found that black humor and language alteration were principal coping techniques when faced with horrific situations, although this technique might not be available for the most extreme situations. Thompson and Suzuki (1991) identified black humor as a coping technique among ambulance workers, which could appear macabre and insensitive to outsiders. James and Wright (1991) noted the use of humor as a therapeutic tactic for coping with difficult situations, for example, involving trauma. However, it is more difficult to imagine the effective use of humor in response to stressors that are organizational in origin. La Rocco and Jones (1978) found colleagues to be the prime social support link and buffer against negative effects of job stress for ambulance workers.

Individual Responses

Robinson (1986) found that high stress and prevalence of ill-health were associated with negative emotions and poor coping skills. Glendon (1991) found a significant relation between staff grade and coping, such that higher grades were less likely than lower grades to use emotion-focused coping strategies. Higher graded respondents in Glendon's study also reported lower overall stress levels. Glendon and Glendon (1992) found that respondents reporting high stress tended to use emotion-focused coping strategies and to report low coping effectiveness. Problem-focused coping was positively associated with effectiveness among their respondents. Results also indicated that the greater a respondent's job satisfaction, the higher was the importance of that person's job in his or her life and the more in control the person felt about coping with stress. These respondents' overall experience of stress was low, and they tended not to use emotion-focused coping strategies. Interpreting their findings in terms of a transactional approach, the authors observed that the more an individual feels in control of the work environment, the less stressful it is perceived to be. James (1988b), among others, noted that coping is dependent on degree of perceived control over a situation. For ambulance staff, experience might operate in one of two directions. Experienced stress might be increased over time because it is a learned response to situations. Conversely, perceived stress might decrease as a result of experience, because uncertainty is reduced over time. Individual ambulance workers who experience the former situation could be predicted to leave the profession early, whereas those for whom the latter situation pertains could be predicted to be more likely to remain and to cope increasingly well with typical work stressors.

Glendon and Glendon's (1992) results further indicated that the more stress experienced, the more likely that individuals adopted emotion-focused coping strategies and the less effective they deemed them to be. Emotion-focused coping does not therefore seem to reduce stress for those adopting this type of strategy. Respondents who preferred to use emotion-focused coping strategies also used problem-focused strategies and found emotion-focused strategies ineffective in dealing with stress. Respondents adopting problem-focused strategies found them to be fairly effective in coping with stress. From this evidence, there is a clear indication that problem-focused coping is more effective than emotion-focused approaches. The authors suggested that doing something about felt stress—a more concrete coping strategy than using feelings to address the stress—is congruent with ambulance staff work ethos, which demands practical measures to cope with a wide variety of situations encountered daily.

Building on the notion of problem-focused coping, Bunce and West (1994) examined innovative coping responses to occupational stress among paramedics and health service workers. The most frequently reported innovative response was "changes to working procedures." These authors showed that individuals are able to go beyond merely adapting themselves to stressful environments and that greater attention should be paid to the scope for individuals to change workplace stressors.

Bunce and West (1994) also identified a potential methodological bias in questionnaire studies of coping in that questions asking respondents how they *generally* or usually cope with stressful situations are more likely to elicit emotion-focused responses. Asking about how individuals coped in *specific* situations is more likely to result in problem-focused responses. Drawing on work by Newton (1989), these authors argued that asking respondents to recall a specific incident (coping behavior) evokes episodic memory, whereas questions relating to general coping (coping style) draws on semantic memory.

Social Support

Young and Cooper (1995a, 1995b) found that their sample of ambulance staff used a range of techniques for coping with work pressures. Compared with a normative group, they were significantly more likely to use problem-focused coping and spouse support and less likely to use other sources of social support. However, many studies of stress have reinforced the importance of social support as a coping facilitator. For ambulance staff, Boyle (1997) identified emotional process work as an important strategy for assisting ambulance staff to return to a normal emotional state. Ideally, emotional aspects of a case are discussed with a superior, peer, or spouse/partner. A dilemma for the ambulance officer identified by Boyle (1997), is that the front-stage emotional culture deems that ambulance staff should always cope. However, from an organizational perspective, maintaining emotional health is an individual responsibility. Thus, ambulance staff often have to wait until they get home to undertake this emotional process work.

Almost all Boyle's (1997) respondents admitted that they would find it difficult to do their job without the support of their spouse. They also admitted that the emotional stress of ambulance work often put significant pressure on their marital relationships. The nature of ambulance work means that staff cannot tell their spouses everything, although spouses in related professions, such as nursing, often coped better (Boyle, 1997). Thus, "spouses are the one who 'care' for the 'carers,' and as such, could be regarded as quasi employees" (p. 195). However, as Boyle noted, the organizational expectation that emotional process work should be done out of work time, which implicitly relies on spouses to provide the emotional process support, effectively absolves the organization from this responsibility.

Ambulance officers with domestic relationship problems, which can be brought on or exacerbated by the nature of their work, are liable to double jeopardy. The risk for an organization that effectively privatizes support mechanisms is that it assumes that most officers have this support available. However, if spouses withdraw their emotional support role, then this can have serious consequences for ambulance workers. There is a heavy organizational dependence on this public/private arrangement as a means of providing informal emotional support (Boyle, 1997).

EXTREME SITUATIONS
AND CRITICAL INCIDENT STRESS

In a review of the impact of events such as terrorist attacks, wars, and large-scale motor vehicle accidents, Galai (1991) emphasized understanding burnout in emergency team workers. He described an approach for preventing burnout and for providing greater emergency team efficiency in mass emergency situations. Ambulance personnel are frequently involved in the aftermath of disasters, and available literature suggests that such helpers may be hidden victims (Duckworth, 1986; Kliman, 1976) and are at risk of developing resultant emotional and psychological problems (Paton, 1989). Studies have increased understanding of critical incident effects on primary responders, such as emergency personnel, and the negative emotional, cognitive, and behavioral reactions of such personnel have been viewed as characteristics of normal people having normal reactions to abnormal situations (Mitchell, 1986: Spitzer, 1988).

The major impact of such stress is reported to occur some time after the event (James & Wright, 1991). At the scene of the incident, expertise and training take over. After an incident, ambulance workers start thinking over what happened, asking such questions as, Did I make the right diagnosis? Did I do everything I could? Winkler (1980) found that approximately 50% of personnel reported that they could not stop thinking about calls for a long time after the incident, with child trauma calls being particularly upsetting.

As "crisis workers" with repetitive exposure to duty-related trauma or "critical incidents," ambulance workers are at risk of developing secondary traumatic stress. This is defined as stress associated with helping or wanting to help a victim of trauma (Beaton & Murphy, 1993). Duty-related critical incidents tend to be unusual and unpredictable (Durham, McCammon, & Allison, 1985; Paton, 1989) and often involve applying training in unusual circumstances, dealing with unexpected or extreme events, dealing with ambiguity in regard to tasks and roles, and working with injured, mutilated or dead victims. Experience of dealing with death and mutilation has been linked with sleep disturbances, recurrent dreams, nightmares, and intrusive thoughts (Paton, Cox, & Andrew, 1989; Taylor & Frazer, 1982). Role ambiguity has been linked with feelings of depression, helplessness, and sleep difficulties (Raphael, Singh, Bradbury, & Lambert, 1983–1984), irritability, anger, and motivational changes (Duckworth, 1986), and feelings of inadequacy, frustration, and intrusive thoughts (Paton, 1989).

Identification refers to a cognitive process of emotional involvement where emergency workers see others as being similar to themselves (Brandt, Fullerton, Saltzberger, Ursano, & Holloway, 1995). Child victims seem to be particularly difficult and have been reported to evoke intense feelings of identification. Where dead victims are involved, this heightens the trauma of the disaster. Brandt et al. (1995) noted that "Feelings of helplessness and guilt for not doing enough are common in disaster workers" (p. 92).

Distancing inner thoughts and feelings from external events may be adaptive early in a trauma setting but at the cost of long-term difficulties. Thus, it has been suggested that the emotional impact of a series of exposures to critical incidents can accumulate across exposures (Corneil, 1992). Other studies document acute and chronic secondary trauma symptoms experienced by firefighters and paramedics during their careers, including intrusion, avoidance, hypervigilance, demoralization, anger, fear or physiological reactivity, marital discord, alcohol abuse, isolation, alienation, guilt, feelings of insanity, loss of control, and suicidal thoughts (American Psychiatric Association, 1987; Beaton & Murphy, 1993; Dunning & Silva, 1980; Durham et al., 1985; Markowitz, Gutterman, Link, & Rivera, 1987; McFarlane, 1988a, 1988b, 1988c, 1988d; Murphy & Beaton, 1991).

Psychological and emotional problems are common among helping professionals (Taylor & Frazer, 1982; Raphael et al., 1983–1984). Cognitive disturbances often emerge some time after a critical incident and its associated physical symptoms. Miles, Demi, and Mostyn (1984) reported cases where emotions have been repressed from between 4 and 13 months after a disaster, and these have been reported to last for long as 20 months after the stressful event (Taylor & Frazer, 1982). Sights, smells, or sounds reminiscent of the critical incident can trigger delayed responses (Dunning & Silva, 1980; Mitchell, 1984b).

The consistency of certain symptoms associated with traumatic experiences has led to their classification as post-traumatic stress disorder (PTSD), defined in the *DSM–III–R* (American Psychiatric Association, 1987) as:

1. Experience of a recognizable stressor—such as a natural or manmade disaster that would evoke significant symptoms in most individuals.

2. Re-experiencing the traumatic event in recurrent dreams, intrusive recollections, or feelings of reliving the traumatic event, prompted by some environmental stimulus.

3. Numbing of responsiveness to or involvement with the external world as shown by either diminished interest in significant activities, feelings of estrangement from others, or constriction of affective response.

4. At least two symptoms not in evidence before the traumatic experience, drawn from the following—hyperalertness, sleep disturbance,

guilt, memory or concentration difficulties, avoidance of events prompting recollection of the event, and intensification of symptoms following exposure to events resembling the trauma.

Compared with those involved in care that is removed from the scene of a disaster, personnel at the scene of the disaster are significantly more likely to report post-traumatic symptoms. Durham et al. (1985) reported that 91% of emergency workers showed at least one PTSD symptom, compared with 62% of hospital-based workers. These results suggest that the nature of the stress experienced by workers on the scene is different from that of those based in the hospital. Approximately 40% of rescue workers—many of whom were ambulance officers—attending two New South Wales multiple-fatality bus crashes experienced some degree of PTSD (Griffith & Watts, 1992).

PTSD can be severe and can negatively affect many aspects of life (Grevin, 1996). Ambulance workers' ability to make rapid and accurate diagnostic decisions can be hindered, and some individuals can be predisposed to various interpersonal and psychological problems (Grevin, 1996). The effectiveness of the service provided by ambulance personnel could be undermined if the problems faced by these workers are not identified and dealt with. Several authors argued that dealing effectively with helpers' needs involves adequate preparation through training, debriefing, and counseling services for the types of problems that may be presented (Berah, Jones, & Valent, 1984; Singer, 1982). Increased preparedness was argued to facilitate more effective planning of disaster intervention and psychological support (Paton, 1989).

Based on anecdotal reports from emergency workers during debriefings, five main sources of critical incident stress have been identified:

- Dealing with death (Dunning & Silva, 1980; Symons & Glowaski, 1988).
- Association with the victim (Miles et al., 1984; Raphael, Singh, & Bradbury, 1980).
- Difficulty in performing duties (Harris, 1988; Mitchell, 1983, 1988);
- Threat to one's life or safety (Raphael, 1984; Wraith & Gordon, 1987).
- Presence of onlookers (Lewis, 1988).

Grevin (1996) found that a significant number of paramedics appear to suffer from PTSD, as indicated by scores on the MMPI2 PK scale. Other studies illustrated that although not always in the pattern of PTSD, psychological symptoms associated with PTSD are often present. For example, 70% of personnel involved in the explosion at the Village Green complex in Greenville, North Carolina, experienced intrusive, repetitive thoughts about the disaster. Forty-four percent reported feelings of sadness related to the explosion and its aftermath, and 15% reported intrusive dreams about the disaster, with some feelings of depression and sense of disturbance when exposed to press coverage of the event (Durham et al., 1985).

Raphael et al. (1983–1984), investigating the effect of the Granville Rail disaster (Sydney, Australia) on personnel involved in the immediate rescue work, illustrated that the majority found the experience to be stressful. Twenty-five percent reported experiencing insomnia in the first month following the event; 10% found that functioning at home was affected; while 35% reported feeling more positive about their own lives as a result of the experience. These studies exemplify the complex variety of emotional states and behavioral indicators associated with ambulance and other staff involved in the aftermath of disasters.

Spitzer and Neely (1992) reported that strategies that have been introduced to help primary responders, such as ambulance staff, include psychological screening during the recruitment process, frequent rigorous training programs, education in crisis intervention skills, establishment of minimum physical endurance levels, and provision of employee assistance programs designed to promote stable functioning. Training in recognition and response has been given to staff who are negatively affected by job stress, such as supervisors and administrators (Mitchell & Bray, 1990). Although these steps are in the right direction, reports such as that by the UK Association of Chief Ambulance Officers (1990) suggest that greater attention is needed in the areas of recruitment and selection, identification of stress, development of coping strategies, and enhanced recognition of the role of accident and emergency ambulance teams.

Factors Mediating Effects

Paton (1989) suggested that increasing awareness of adverse stress reaction symptoms as normal may increase sufferers' willingness to seek support and outside aid, thereby reducing the likelihood of long-term problems developing. Factors mediating effects of trauma include prior disaster experience, appraisal of the situation, and social support (Brandt et al., 1995). Cohesive and supportive work groups seem to be associated with less occupational stress and fewer symptoms of depression (Revicki et al., 1993).

As the quality of help is dependent on understanding the psychological effects of disasters (Stewart, 1989), examining a range of disastrous events can provide information about impairment (Green, Grace, Lindy, Titchener, & Lindy, 1983). Specialized training on the job may help to reinforce and develop readiness to deal with difficult situations (Robinson & Mitchell, 1993). Mitchell and Bray (1990) argued that all firefighters and paramedics need a preventive, educational orientation to stress and stress management. However, those at risk of high stress symptoms may need more intensive and extensive interventions (Beaton et al., 1995). For example, Marmar et al.

(1996) found that high externality among emergency services personnel was predictive of avoidant coping strategies and experience of greater trauma.

Positive Effects

Disasters have also been found to have positive effects on the lives of those involved (Werner, Bates, Bell, Murdoch, & Robinson, 1992). Berah, Jones, & Valent (1984) reported that despite half the members of the volunteer health team involved in the aftermath of the Ash Wednesday fires (Australia) feeling moderate levels of sadness and depression, most claimed to have gained both personally and professionally from the experience. Werner et al. (1992), in their investigation of 25 volunteer members of the Victoria State Emergency Service, found that in general those questioned reported benefiting from the critical incident. Benefits reported were increased confidence in the ability to cope with another critical incident and increased knowledge for use in future incidents, with many indicating that their learning extended beyond operational procedures and into their personal lives. Some reported positively changing their attitudes to life, becoming more appreciative and more serious in relation to road safety. James (1988b) noted that ambulance work offers variety, challenge, responsibility, and community service. She recorded that her survey respondents report that their jobs are rewarding and that they are happy in their work. The title of Boyle's (1997) study (*Love the Work, Hate the System*) reflects both positive and negative aspects of ambulance staff work. Responsibility for others, saving and maintaining life—requiring a superior level of performance—are among the factors that contribute to ambulance staff work satisfaction.

CRITICAL INCIDENT STRESS DEBRIEFING

Critical incident stress debriefing (CISD) is a form of psychological debriefing developed to mitigate harmful effects of work-related trauma and, ultimately, to prevent PTSD in those exposed to traumatic events. Developed by Mitchell in the 1970s, CISD is one of the most widely used group intervention techniques in the world for preventing traumatic reaction, including PTSD, among high-risk professions such as emergency service personnel (Everly & Mitchell, 1992). The two main goals of CISD are to mitigate harmful effects of traumatic stress on emergency personnel and to accelerate normal recovery in people who have been exposed to a traumatic event. CISD may be used to foster the cathartic process (expression or ventilation of emotions) in a safe, supportive, and structured environment rather than a cathartic release occurring in an unstructured and chaotic environment, which could prove pathogenic (Everly, 1995). The Mitchell (1988) model is one of those in use around the world. It includes seven phases, carefully structured to flow in a nonthreatening manner. The phases begin with the cognitively oriented processing of human experience through to an emotionally oriented processing of the same experiences.

Research on crisis intervention suggests that the speed at which help is supplied after a crisis is positively correlated with recovery speed. The consensus seems to be on initiating debriefing 24 to 72 hours after the incident (Back, 1992; Mitchell, 1988). This delay between the onset of the traumatic event and initiation of treatment allows time for processing feelings that might interfere with integration of the incident (Mitchell, 1988) and allows staff to get beyond the initial stage of shock and denial (Cooper, 1995). Friedman, Framer, and Shearer (1988) found that early detection and early intervention with post-trauma reactions lead to lower costs and more favorable prognosis associated with individuals who have experienced trauma. It is universally recognized that prevention and early intervention efforts are preferable to providing ad hoc treatment of post-traumatic symptoms (Butcher, 1980; Yandrick, 1990).

Critical Evaluation

CISD has been widely used in preventing post-traumatic stress among high-risk occupational groups such as emergency services personnel. However, controversy continues over the success of the technique, and further evidence of its efficacy is needed. Anecdotal evidence for its usefulness from self-report data suggests immediate and long-term benefits of CISD, although little available empirical research evaluates outcomes. Some research has been criticized for producing little systematic evidence to support claims of success in reducing symptoms (Shalev, 1994), for not appearing to be subjected to systematic scrutiny (Orner, 1994) and for lacking follow up (Raphael, McFarlane, & Meldrum, 1994).

Studies attempting to evaluate CISD effects have also been criticized on the grounds that they typically compare a group receiving CISD with a treatment-assessment control group (Meichenbaum, 1997) without random allocation of participants. Yehuda and McFarlane (1995) noted that controlled case studies examining preventive effects of CISD, such as that of Griffith and Watts (1992), found that emergency services personnel who were debriefed following a bus crash had significantly higher levels of symptoms 6 months later. However, those not debriefed after the Newcastle earthquake showed more reductions in symptoms over a 2-year period (Kenardy et al., 1996).

In response to these criticisms, Robinson and Mitchell (1995) claimed these studies to be flawed for several reasons: lack of standardized procedures; evaluated interventions are not defined and may vary from a 5-minute chat to a 2- or 3-hour structured group discussion; studies are often assessed 1 year after the event, and 12-month anniversaries are often a tragic time in themselves; the population being debriefed is not defined, and measurement is inappropriate; debriefings are voluntary so that participants may differ from nonattenders.

Arguments have also been presented that debriefing may actually harm people. It has been suggested that talking can retraumatize some individuals and that it can be distressing to witness the distress of others, particularly if adequate processing of the event has already occurred—hence the emphasis on voluntary participation. Robinson and Mitchell (1995) stated that many of the criticisms levied at CISD are based on misconceptions of the debriefing process. Many critics have stated that if all stress–trauma signs in all people who are debriefed are not reduced then the debriefing has not been successful. However, Mitchell and Everly (1995) noted that CISD is neither a psychotherapy nor a substitute for psychotherapy. Robinson and Mitchell (1995) confirmed that the purpose of CISD is to stabilize the current situation, minimize detrimental effects, mobilize the resources of those involved, and "normalize" the experience to help restore people to functioning within their routine environments as quickly as possible.

Summarizing previous studies, Meichenbaum (1997) suggested that it appears that CISD may be useful as it provides opportunities for victims to talk to one another but that it is unlikely to provide effective treatment for ongoing, complex, persistent problems. Hiley-Young and Gerrity (1994) recognized that CISD procedures may help some disaster victims but expressed concern that there are unreasonable expectations of the usefulness of CISD among field practitioners. It is one of a number of strategies that may be required following a tragic incident. As noted by Watts (1994), "It is not possible for one session, no matter how expertly conducted and effective to resolve all issues/concerns and reactions" (p. 6).

Boyle (1997) noted that, although CISD and other components of employee assistance programs (EAPs) may help to alleviate individual stress, of themselves they do nothing to influence the context in which it is produced. She argued that, contrary to conventional wisdom, which suggests that counseling and stress management programs assist in developing a kinder, more caring culture, in their current form "such programs only serve to reinforce existing forms of emotionality and emotional culture" (p. 206).

Although this review indicates the problematic nature of CISD as a crisis intervention technique, it remains popular among high-risk professionals and is regularly asked for, and no strongly preferred alternative is generally available. CISD appears to assist in people with returning to work, reduces sick leave, and assists in recovery from trauma (Mitchell, 1983).

OCCUPATIONAL HEALTH AND SAFETY (OHS) AND LEGAL ISSUES

OHS Aspects

The many and varied situations that ambulance staff have to confront include working in difficult, dangerous, and dirty environments and dealing with people in panic, shock, and distress. In addition, as James (1988b) noted, they are subject to the risks of physical violence from unstable or violent patients, contracting various diseases from patients, being bitten by patients' pets, acute physical damage from dangerous chemicals at the scene of accidents, and long-term effects of lifting, which can result in chronic back injuries. These effects could be exacerbated in some cultures. Thus, Sparrius (1992) reported on her South African respondents being placed in physical danger, for example, as a result of racial tension, being in a crowd of rioters, as well as receiving physical or verbal abuse.

Beaton and Murphy (1993) found that paramedics expressed considerable apprehension about their personal safety. Although hard data are not readily available, violence from other parties is among the occupational hazards that particularly affect health care workers. Mezey and Shepherd (1994) identified ambulance workers as being a high-risk group. These authors identified long-term effects of such violence as including increased absenteeism, burnout, and post-traumatic stress disorder.

Legal Aspects

In a general review of the role of law in stress prevention, Leighton (1994) considered a UK case in which stress was categorized as an injury caused by employer negligence. However, as a result of conflicting ideologies over the role of law in such areas, Leighton (1994) did not conclude that legal intervention in occupational stress is likely to be widespread. Barrett (1995), in a review of English case law in respect of employer liability for emergency services employee stress induced by exposure to traumatic events, recorded the absence of case law on employers' liability for stress experienced by rescue workers, such as ambulance staff. She noted that "There is at present no recorded example of successful litigation against their employers, by those engaged in emergency work, for psychological injury caused by an isolated, but particularly traumatic experience" (p. 400). However, the author proceeded to comment that courts might have regard for the employer's total system, which could be particularly relevant if an employee's disorder could be related to cumulated experience over a period rather than attendance at one particular horrific incident. However, the onus would be on the plaintiff to provide adequate evidence of employer negligence. Such evidence might include lack of adequate training, failure to debrief after certain types of incidents, or inadequate rest periods.

Earnshaw and Cooper (1994) considered the extent to which occupational stress has the potential for giving rise to personal injury litigation in the UK. The number of cases to that time was quite low, and from a small-scale survey, most of these were settled out of court. Although the authors predicted an increase in such cases in the U.K., they pointed out that to succeed, four conditions need to be fulfilled:

1. The employee must establish that she or he has suffered a stress-induced illness—not necessarily a straightforward matter.
2. A causal link must be shown between the allegedly stressful workplace conditions and the illness—difficult if plaintiffs are questioned about other potential stressors in their lives.
3. It must be shown that the particular damage suffered by the employee was foreseeable by the employer.
4. It must be demonstrated that the employer is at fault.

Although research continues to suggest that stress-related occupational illness is likely to be increasing, Earnshaw and Cooper (1994) pointed out that the cost to employees of losing stress-related claims can be considerable. Conversely, the potential costs to employers of successful claims are also very large. On balance, it seems more likely that occupational stress incidence will continue to be reflected in absenteeism, labor turnover, and other workplace behavioral indicators, rather than in litigation.

In the case of ambulance staff, who are contracted to respond to emergency calls, both the contractor and the employer may have responsibilities to the crew under relevant sections of U.K. health and safety legislation requiring them to do what is "reasonably practicable" to protect their health and safety at work. However, a case could be decided on the issue of whether work-related stress in the case of ambulance and other emergency workers is a foreseeable risk. Barrett (1995) pointed out that "It would be difficult to argue that it is not a foreseeable risk that rescue workers might suffer work-related stress" (p. 402). Currently, there are requirements for risk assessments to be undertaken, which should include assessment of factors that could cause psychological harm to rescue workers. Where identified, steps should be taken to minimize such harm. However, because the nature of the work can be deemed to be inherently stressful and at times traumatic, the issue of what steps might be taken to protect such workers from harm arising from traumatic incidents is problematic. Barrett (1995) opined that an inspectorate that is hard pressed to deal with work-related physical injuries may be unlikely to prosecute an employer for failure to assess stressors in the work environment.

Boyle (1997) noted the additional factor of organizational change and enterprise bargaining "reforms," including linking performance, pay, and productivity, increased workload, and staff reductions. She considered the effects of ambulance staff losing conditions such as overtime payments, parental leave, and full access to workers' compensation. Loss of such access for work-related stress means that the onus of proof now falls on ambulance staff to prove that work was the sole cause of their stress. Where such provisions operate, this is very likely to result in a decrease in officially recorded rates of stress among ambulance staff.

CONCLUSIONS

Figure 1.6.1 summarizes the main factors found to be associated with stress in ambulance staff or paramedics. Broad categories of stressors include:

1. Those associated with the organizational context in which the work is undertaken.
2. Intrinsic aspects of the job that are routinely encountered.
3. Critical incidents, which are rarely encountered, but which may have dramatic consequences.

Outcomes identified in Fig. 1.6.1 may be positive, negative, or ambiguous in nature. Much of the emphasis in the literature has been on negative outcomes and on coping with these. However, although critical incidents can and do have a variety of adverse effects on those experiencing them, there is also evidence that positive outcomes are possible and that many ambulance personnel derive great satisfaction and feelings of self-worth and achievement from their work. Research effort could usefully be expended in determining how best to boost naturally occurring positive effects—even from incidents that could be categorized as "critical"—such as feelings of being more in control and improved preparation for subsequent incidents.

Marginal aspects of the model include the so-called "trauma junkie" syndrome, perhaps reflecting a more general human fascination with unusual and dramatic events, such as accidents. The propensity of traffic on the roadway opposite to an accident to slow down so that drivers can rubberneck the aftermath may reflect this motive.

A variety of factors mediate potential stressor inputs and the range of possible outcomes. How stress among any category of employee is managed by an organization is a function of its broader risk-management policy and approach, which in turn is significantly influenced by the organizational culture. A combination of organizational and individually initiated coping responses operates in any given situation. Although research has identified a range of these as being likely to lead to greater or lesser effective coping, a

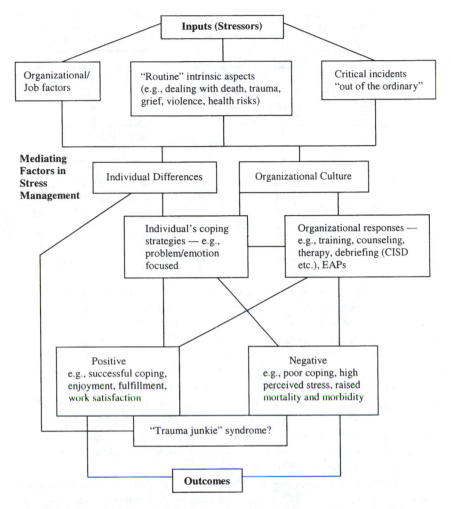

FIG. 1.6.1.   Ambulance staff/paramedic occupational stress summary model.

more inclusive model incorporating both individual and organizational factors as a basis for research into optimum coping response is still awaited.

There is also considerable scope for research investigating the interplay between individual and organizational factors relating to stress as experienced by ambulance workers. This area is little understood or appreciated at present, and adequate models need to be developed, perhaps based on job performance outcomes, that indicate how best to combine organizational support with individual initiative. Models from fields such as human resource management might usefully be used as a framework for gathering

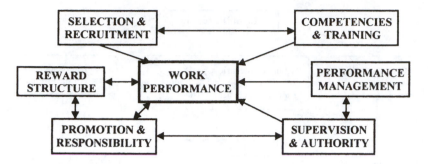

FIG. 1.6.2. Possible human resource management framework for interventions in respect of improving job performance and organizational support for ambulance workers.

and interpreting data over a range of variables. Such a model might incorporate features shown in Fig. 1.6.2.

Also generally lacking in the research literature on stress in ambulance
staff are longitudinal studies, for example, to study organizational changes,
or in the form of action research. These types of studies are more difficult
and more costly than cross-sectional surveys, but are essential if we are to
move beyond the mere counting of factors that pose as potential stressors to
ambulance personnel. At the simplest level, these could usefully be used to
experimentally assess the effectiveness of various interventions, for example, designed to reduce the worst effects of stress and to improve performance. In the longer term, a continual monitoring function could become
standard practice so that aspects of organizational climate and culture that
are known from studies described in this review to be critical to ambulance
worker performance could be improved.

In general, a greater degree of triangulation in studies of ambulance staff
is needed. Surveys of various kinds are liable to introduce bias, for example,
in overemphasizing extrinsic aspects of the job. Although these may be balanced by ethnographic studies, there have been few of these to date. Particularly lacking are intervention studies with controlled before and after designs. There is now sufficient literature on which to base field experimental
intervention studies in which changes to the work environment of ambulance staff may be more systematically analyzed. A possible constraint here is
likely to be opposition from the dominant culture of organizations employ-
ing ambulance staff.

Finally, there is a need to explore coping mechanisms more imaginatively
and more precisely. Scales that promote a bias toward either problem- or
emotion-focused coping could usefully be augmented by triangulated
methodologies, perhaps including qualitative approaches, to draw out the

full range of coping strategies adopted and the circumstances under which each is likely to be used. Research could also usefully pursue the extent to which ambulance staff are active and innovative in their responses to stressors, for example, in responding to organizational pressures, rather than adhering to an implicit model of passivity in the face of multiple stressors.

REFERENCES

American Psychiatric Association. (1987). *Diagnostic and statistical manual of mental disorders* (3rd ed., rev.). Washington, DC: American Psychiatric Association.

Association of Chief Ambulance Officers. (1990). *Ambulance 2000: Proposals for the future of the Ambulance Service.* London: Association of Chief Ambulance Officers.

Auerbach, P. S., Morris, J. A., Phillips, J. B., Redlinger, S. R., & Vaughn, W. K. (1987). An analysis of ambulance accidents in Tennessee. *Journal of the American Medical Association, 258*, 1487–1490.

Back, K. J. (1992). Critical incident stress management for care providers in the pediatric emergency department. *Critical Care Nurse, 12*(1), 78–83.

Balarajan, R. (1989). Inequalities in health within the health sector. *British Medical Journal, 299*, 822–825.

Barrett, B. (1995). Occupational stress, rescue workers and English law. *Work & Stress, 9*, 394–404.

Beaton, R., & Murphy, S. (1993). Sources of occupational stress among fire fighters/EMTs and fire fighters/paramedics and correlations with job related outcomes. *Prehospital and Disaster Medicine, 8*, 140–150.

Beaton, R., Murphy, S., Pike, K., & Jarrett, M. (1995). Stress-symptom factors in firefighters and paramedics. In S. L. Sauter, & L. R. Murphy (Eds.), *Organizational risk factors for job stress* (pp. 227–245). Washington, DC: American Psychiatric Association.

Berah, E. F., Jones, J., & Valent, P. (1984). The experience of a mental health team involved in the early phase of a disaster. *Australian and New Zealand Journal of Psychiatry, 18*, 354–358.

Boyle, M. V. (1997). *Love the work, hate the system: A qualitative study of emotionality, organisational culture and masculinity within an interactive service workplace.* Unpublished doctoral dissertation, University of Queensland, Brisbane.

Brandt, G. T., Fullerton, C. S., Saltzberger, L. Ursano, R. J., & Holloway, H. (1995). Disasters: Psychological responses in health case providers and rescue workers. *Nordic Journal of Psychiatry, 49*, 89–94.

Bunce, D., & West, M. (1994). Changing work environments: Innovative coping responses to occupational stress. *Work & Stress, 8*, 319–331.

Butcher, J. (1980). The role of crisis intervention in an airport disaster plan. *Space and Environmental Medicine, 51*, 1260–1262.

Cooper, C. (1995). Psychiatric stress debriefing: Alleviating the impact of patient suicide and assault. *Journal of Psychosocial Nursing and Mental Health Services, 33*(5), 21–25.

Corneil, W. (1992). Prevalence and etiology of post-traumatic stress disorders in firefighters. *Public Safety Personnel.* Symposium conducted at the First World Conference on Trauma and Tragedy, The Netherlands.

Cox, T., Leather, P., & Cox, S. (1988). Stress and the public services: Editorial. *Work & Stress, 3*, 277–279.

Dawson, D. (1990). *Measuring stress in the work of ambulance officers* (unpublished report). Ambulance Officer's Training Centre, Ambulance Service, Victoria, Australia.

Duckworth, D. (1986). Psychological problems arising from disaster work. *Stress Medicine, 2,* 315–323.

Dunning, C., & Silva, M. (1980). Disaster-induced trauma in rescue workers. *Victimology: An International Journal, 5,* 287–297.

Durham, T. W., McCammon, S. L., & Allison, E. J. (1985). The psychological impact of disaster on rescue personnel. *Annals of Emergency Medicine, 14,* 664–668.

Dutton, L. M., Smolensky, M. H., Leach, C. S., Lorimor, R., & Hsi, B. (1978). Stress levels of ambulance paramedics and fire fighters. *Journal of Occupational Medicine, 20,* 111–115.

Earnshaw, J., & Cooper, C. L. (1994). Employee stress litigation: The UK experience. *Work & Stress, 8,* 287–295.

Everly, G. S. (1988). *The personality profile of an emergency worker.* Paper presented at the Second International Conference on Dealing with Stress and Trauma in Emergency Services, Melbourne.

Everly, G. S. (1995). The role of the Critical Incident Stress Debriefing (CISD) process in disaster counseling. *Journal of Mental Health Counseling, 17,* 278–290.

Everly, G. S., & Mitchell, J. T. (1992, November). *The prevention of work related post-traumatic stress: The critical incident stress debriefing process.* Paper presented at the second American Psychological Association and National Institute for Occupational Safety and Health Conference on Occupational Stress, Washington, DC.

Friedman, R., Framer, M., & Shearer, D. (1988). Early response to post-traumatic stress. *EAP Digest,* 45–49.

Galai, T. (1991). Team work during emergency: A hierarchical model. *SihotDialogue Israel Journal of Psychotherapy, 5* (Feb.), 24–26.

Glendon, A. I., & Glendon, S. (1992). Stress in ambulance staff. In E. J. Lovesey (Ed.), *Contemporary ergonomics 1992: Ergonomics for industry* (pp. 174–180). London: Taylor & Francis.

Glendon, S. (1991). *Chronic stress in ambulance crews.* Unpublished master's thesis. Department of Occupational Psychology, Birkbeck College, University of London.

Goldstein, I. B., Jamner, L. D., & Shapiro, D. (1992). Ambulatory blood pressure and heart rate in healthy male paramedics during a workday and a nonworkday. *Health Psychology, 11,* 48–54.

Green, B. L., Grace, M. C., Lindy, J. D., Titchener, J. L., & Lindy, J. G. (1983). Levels of functional impairment following a civilian disaster: The Beverly Hills Supper Club fire. *Journal of Consulting and Clinical Psychology, 51,* 573–580.

Grevin, F. (1996). Posttraumatic stress disorder, ego defense mechanisms, and empathy among urban paramedics. *Psychological Reports, 79,* 483–495.

Griffith, J., & Watts, R. (1992). *The Kempsey and Grafton bus crashes: The aftermath.* Lismore, NSW, Instructional Design Solutions, University of New England.

Grigsby, D. W., & McKnew, M. A. (1988). Work-stress burnout among paramedics. *Psychological Reports, 63,* 55–64.

Hamil, N. (1991). *Sources and levels of stress in ambulance personnel and ways of coping.* Unpublished report, School of Psychology, Birmingham University, UK.

Harris, V. (1988). Working with families of emergency service staff. In *Dealing with stress and trauma in emergency services: An international conference* (pp. 26–28). Melbourne, Social Biology Resources Centre.

Herbison, R. J., Rando, T. A., Plante, T. G., & Mitchell, G. (1984). National EMS burnout survey. *Journal of Emergency Medical Services, 9,* 48–50.

Hiley-Young, B., & Gerrity, E. T. (1994). Critical Incident Stress Debriefing (CISD): Value & limitations in disaster response. *NCP Clinical Quarterly, 4.*

James, A. E. C. (1988a). *Stress in the ambulance service.* Unpublished doctoral dissertation, University of Bradford Management Centre.

James, A. E. C. (1988b). Perceptions of stress in British ambulance personnel. *Work & Stress.*
    2, 319–326.
James, A. E. C., & Wright, P. L. (1991). Occupational stress in the Ambulance service. *Journal
    of Managerial Psychology, 6*(3), 13–22.
James, A. E. C., & Wright, P. L. (1993). Perceived locus of control: Occupational stress in the
    ambulance service. *Journal of Managerial Psychology, 8*(5), 3–8.
Jamner, L. D., Shapiro, D., Goldstein, I. B., & Hug, R. (1991). Ambulatory blood pressure and
    heart rate in paramedics: Effects of cynical hostility and defensiveness. *Psychosomatic Medi-
    cine, 53*, 393–406.
Karasek, R. A., Russell, S. T., & Theorell, T. (1982). Physiology of stress and regeneration in
    job related cardiovascular illness. *Journal of Human Stress, 8*, 29–42.
Kenardy, J. A., Webster, R. A., Lewin, T. J., Carr, V. J., Hazell, P. L., & Carter, G. L. (1996).
    Stress debriefing and patterns of recovery following a natural disaster. *Journal of Traumatic
    Stress, 9*, 37–49.
Kliman, A. S. (1976). The Corning Flood Project: Psychological first aid following a natural
    disaster. In H. J. Parad, H. L. P. Resnik, & L. B. Parad (Eds.), *Emergency and disaster manage-
    ment: A mental health sourcebook.* Maryland: Charles.
La Rocco, J. M., & Jones, A. D. (1978). Co-worker and leader support as moderators of the
    stress-strain relationship. *Journal of Applied Psychology, 63*, 629–631.
Lau, B. W. K. (1988). An appraisal of lay and medical concepts of illness. *Journal of the Royal
    Society of Health, 108*, 185–187.
Leighton, P. (1994). What has the law to do with stress prevention? (Editorial). *Work & Stress,
    8*, 283–285.
Lewis, C. (1988). Red rescuer. *Royalauto, 55*, 4–5.
Mannon, J. M. (1991). *Emergency encounters: EMTs and their work.* Boston: Jones & Bartlett.
Markowitz, J., Gutterman, E., Link, B., & Rivera, M. (1987). Psychological response of fire
    fighters to a chemical fire. *Journal of Human Stress, 131*, 84–93.
Marmar, C. R., Weiss, D. S., Metzler, T. J., & Delucchi, K. (1996). Characteristics of emergency
    services personnel related to peritraumatic dissociation during critical incident exposure.
    American Journal of Psychiatry, 153(Suppl.), 94–102.
McFarlane, A. (1988a). The etiology of post-traumatic stress disorders following a natural dis-
    aster. *British Journal of Psychiatry, 152*, 116–121.
McFarlane, A. (1988b). Relationship between psychiatric impairment and a natural disaster:
    The role of distress. *Psychological Medicine, 18*, 120–139.
McFarlane, A. (1988c). The phenomenology of post-traumatic stress disorders following a
    natural disaster. *Journal of Nervous and Mental Disease, 176*, 22–29.
McFarlane, A. (1988d). The longitudinal course of post-traumatic morbidity: The range of
    outcomes and their predictors. *Journal of Nervous and Mental Disease, 176*, 30–39.
Meichenbaum, D. (1997). *Treating post-traumatic stress disorder: A handbook and practice manual
    for therapy.* Canada: Institute Press.
Metz, D. L. (1981). *Running hot.* Cambridge, MA: Abt.
Mezey, G., & Shepherd, J. (1994). *Violence in health care: A practical guide to coping with violence
    and caring for victims* (pp. 1–11). Oxford: Oxford University Press.
Miletich, J. J. (1990). *Police, firefighter, and paramedic stress: An annotated bibliography.* New York:
    Greenwood.
Miles, M. S., Demi, A. S., & Mostyn, A. P. (1984). Rescue workers' reactions following the Hy-
    att Hotel disaster. *Death Education, 8*, 315–331.
Mitchell, J. T. (1983). When disaster strikes: The critical incident stress debriefing process.
    Journal of Emergency Medical Services, 8, 36–39.
Mitchell, J. T. (1984a). The 600-run limit. *Journal of Emergency Medical Services, 9*, 52–54.

Mitchell, J. T. (1984b). Disasters leave other victims among those who try to help. *Maryland Rescue Journal, 6*, 2–3, 14–15, 39.

Mitchell, J. T. (1986, August). Critical incident stress debriefing. In R. Robinson (Ed.), *Stress and trauma in emergency services.* Proceedings from a conference dealing with stress and trauma in emergency services, Melbourne.

Mitchell, J. T. (1988). Development and functions of a critical incident stress debriefing team. *Journal of Emergency Services, 13*, 42–46.

Mitchell, J. T., & Bray, G. (1990). *Emergency services stress: Guidelines for preserving the health and careers of the emergency services personnel.* Englewood Cliffs, NJ: Prentice-Hall.

Mitchell, J. T., & Everly, G. S. (1995). *Critical Incident Stress Debriefing: An operations manual for the prevention of trauma among emergency service and disaster workers.* Baltimore, MD: Chevron.

Morris, D. L., & Cross, A. B. (1980). Is the emergency ambulance service abused? *British Medical Journal, 12 July*, 121–123.

Murphy, S. A., & Beaton, R. (1991). *Counteracting effects of trauma in everyday life: Leisure patterns among fire fighters.* Paper presented at the seventh annual meeting of the Society for Traumatic Stress Studies, Washington, DC.

Newton, T. J. (1989). Occupational stress and coping with stress: A critique. *Human Relations, 42*, 441–461.

Nixon, R. G. (1987). Burnout revisited. *Emergency Medical Services, 16*, 44–53.

Orner, R. (1994). *Intervention strategies for emergency response groups: A new conceptual framework.* Paper presented at the NATO conference on stress, coping and disaster, Bornas, France.

Palmer, C. E. (1983). A note about paramedics' strategies to deal with death and dying. *Journal of Occupational Psychology, 56*, 83–86.

Paton, D. (1989). Disasters and helpers: Psychological dynamics and implications for counseling. *Counseling Psychology Quarterly, 2*, 303–321.

Paton, D., Cox, D. E. H., & Andrew, C. (1989). *A preliminary investigation into stress in rescue workers.* RGIT Applied Social Science Research Report No. 1. Aberdeen: Robert Gordon Institute of Technology.

Perry, P. G. (1988). *The problems of long-serving ambulancemen* (Report to the Secretary of State for Health). Ambulance Council Working Party.

Queensland Parliament. (1990). *Parliamentary Select Committee of Inquiry into Ambulance Services—First Report.* Brisbane, Government Printing Office.

Raphael, B. (1984). Rescue workers: Stress and their management. *Emergency Responses, 1*, 27, 29–30.

Raphael, B., McFarlane, A. C., & Meldrum, L. (1994). *Acute interventions after traumatic events.* Unpublished manuscript, Royal Brisbane Hospital, University of Queensland.

Raphael, B., Singh, B., & Bradbury, L. (1980). Disaster: The helper's perspective. *Medical Journal of Australia, 2*, 445–447.

Raphael, B., Singh, B., Bradbury, L., & Lambert, F. (1983–1984). Who helps the helpers? The effects of a disaster on the rescue workers. *Omega, 14*, 9–20.

Revicki, D. A., Whitley, T. W., & Gallery, M. E. (1993). Organizational characteristics, perceived work stress, and depression in emergency medicine residents. *Behavioral Medicine, 19*, 74–81.

Robinson, R. C. (1984). *Health and stress in ambulance services: Report of evaluation study, Part I* (Unpublished research report). Social Biology Resources Centre, Victoria, Australia.

Robinson, R. C. (1986). *Health and stress in ambulance services: Report of evaluation study, Part 2* (Unpublished research report). Social Biology Resources Centre, Victoria, Australia.

Robinson, R. C. (1988). *Identification of sources of stress in ambulance services.* Proceedings from a conference on dealing with stress and trauma in emergency services, Social Biology Resources Centre, Victoria, Australia.

Robinson, R. C., & Mitchell, J. T. (1993). Evaluation of psychological debriefings. *Journal of Traumatic Stress, 6,* 367–382.

Robinson, R. C., & Mitchell, J. T. (1995). Getting some balance back into the debriefing debate. *The Bulletin of the APS, 17,* 5–10.

Shalev, A. Y. (1994). Debriefing following traumatic exposure. In R. J. Ursano, B. G. McCaughey, & C. S. Fullerton (Eds.), *Trauma and disaster.* Cambridge: Cambridge University Press.

Singer, T. J. (1982). An introduction to disaster: Some considerations of a psychological nature. *Aviation, Space and Environmental Medicine, 53,* 245–250.

Sparrius, S. (1992). Occupational stressors among ambulance and rescue service workers. *South African Journal of Psychology, 22,* 87–91.

Spitzer, C. (1988). The invisible toll on rescue workers. *Washington Post Health, May 10,* 13–16.

Spitzer, W. J., & Neely, K. (1992). Critical incident stress: The role of hospital-based social work in developing a statewide intervention system for first-responders delivering emergency services. *Social Work in Health Care, 18,* 39–58.

Stewart, M. (1989). Mirrors of pain. *Community Care, 748,* 121.

Stout, J. (1984). How much is too much? *Journal of Emergency Medical Services, 9,* 26–34.

Symons, I., & Glowaski, D. (1988). Stress issues relating to the County Fire Authority. In *Dealing with stress in Melbourne, Trauma in the Emergency Services: An international conference* (pp. 26–28). Social Biology Resources Centre.

Taylor, A. J. W., & Frazer, A. G. (1982). The stress of post-disaster body handling and victim identification. *Journal of Human Stress, 8,* 33–34.

Thompson, J. A., & Suzuki, I. (1991). Stress in ambulance workers. *Disaster Management, 4,* 193–197.

Toombs, F. S., Quinlan, T. P., & Terry, D. W. (1979). *Report of a survey into occupational stress factors.* London: Advisory Conciliation and Arbitration Service.

Watts, R. (1994). The efficacy of critical incident stress debriefing for personnel. *The Bulletin of the Australian Psychological Society, 6,* 7.

Werner, H. R., Bates, G. W., Bell, R. C., Murdoch, P., & Robinson, R. (1992). Critical incident stress in Victoria State Emergency Service volunteers: Characteristics of critical incidents, common stress responses, and coping methods. *Australian Psychologist, 27,* 159–165.

Winkler, R. (1980, November). *Occupational stress in Western Australian Ambulance Officers.* Summary results of a report of the 14th Convention of Ambulance Authorities, Perth, Australia.

Wraith, R., & Gordon, R. (1987). Short term responses to disaster. *Macedon Digest, 2,* 3–5.

Yandrick, R. (1990). Critical incidents. *EAPA Exchange,* January, 18–23.

Yehuda, R., & McFarlane, A. C. (1995). Conflict between current knowledge about post-traumatic stress disorder and its original conceptual basis. *American Journal of Psychiatry, 152,* 1705–1713.

Young, A. (1980). The final analysis of the recent poll among Victoria's ambulance officers. *Ambulance World.* January/February.

Young, K. M., & Cooper, C. L. (1995a). Occupational stress in the ambulance service: A diagnostic study. *Employee Counseling Today, 7*(5), 25–32.

Young, K. M., & Cooper, C. L. (1995b). Occupational stress in the ambulance service: A diagnostic study. *Journal of Managerial Psychology, 10*(3), 29–36.

1.7 Women Police: The Impact of Work Stress on Family Members

Briony Thompson
Andrea Kirk-Brown
David Brown
Griffith University

Recent studies of paramilitary organizations, such as the police service, have identified a number of possible sources of stress in the organization (Christie, 1994; Hotchkiss, 1992; Kelly & Brown, 1999). Police work has been ranked among the top five most stressful occupations (Dantzer, 1987). Much more knowledge is needed about coping with stressful work, an area of research that has lagged behind investigation of occupational stress generally (Beehr, Johnson, & Nieva, 1995). As Brown and Campbell (1990) noted, there is scarce evidence providing information on those aspects of policing that are stressful and their impact on officers. In particular, little is known about the experience of women in policing.

Much of the research on this occupational subgroup has centered around male officers. However, women police may experience qualitatively different sources of stress from male officers (Brown & Fielding, 1993). For example, in addition to the stress of operational duties, a number of researchers have noted that female police officers are exposed to possible sources of stress in the form of sexual harassment expressed as sexually oriented jokes, inappropriate touching, requests for sex, and degrading comments (Close, 1994; Daum & Johns, 1994; Heidensohn, 1994). In addition, few studies have investigated differential sources of stress for men and women. As Brown and Fielding (1993) argued, part of the difficulty in making such comparisons is in the low number of female officers (they reported a proportion of 11%). Brown and Fielding found women officers to report more stress than men when engaged in violent encounters, dealing with victims of violence, or informing a relative of a death. Nearly one half the women reported stress associated with sexual harassment as opposed to 3% of men, and although women were no more likely than men to report work–family conflict, they reported more stress associated with such conflict. Organizational and management sources of stress are, however, found to be more commonly reported than operational sources of stress at a ratio of 4 to 1 (Brown & Campbell, 1990); Brown and Campbell noted there was some indication that gender was a differentiating factor in reports of stress but failed to analyze their data by gender. Given the predominantly male samples researched, it is not known whether women officers are also more likely to report organizational and management rather than operational sources of stress and whether their experience of work stress in general parallels the findings of male samples.

In addition, little attention has been given to the possible negative transmission to family members of exposure to police organizational stress. Research indicates that occupational stressors do affect the family, but studies of spillover effects from work to home have predominantly assessed the impact of the husbands' work environment on their wives (Thompson, 1997). Alexander and Walker (1994) found that in their sample of police officers 40% admitted taking out stress on their families; unfortunately, they did not report the number of female officers in their sample, which makes it difficult to infer the extent to which these results reflect a primarily male perspective. Higher occupational demands experienced by men are correlated with dissatisfaction and distress in wives (Burke, Weir, & Duwors, 1980). Police officers experiencing high stress are likely to be more angry and uninvolved in family matters and to have unsatisfactory marriages (Jackson & Malasch, 1982). In a sample of police officers (92% male), negative effects of work demands on family (in particular, concerns about health and safety) were related to work attitudes and emotional well-being (Burke, 1994). In addition, work stress may affect the physical health and even the life expectancy of partners (Haynes, Eaker, & Feinleib, 1983). There is evidence that the direction of occupational stress transmission is primarily from men to women (Jones & Fletcher, 1993). A major limitation of models of occupational stress transmission is that little research is available on female employees in high-stress occupations such as the police service.

There are a number of reasons to assume that the experience of male officers, especially in terms of sources of stress and the impact of work stress on the family, cannot be directly applied to women officers. Thompson (1997) argued that men and women experience the work–family interface differently, with women being more susceptible to role overload and work–family conflict. Gutek, Searle, and Klepa (1991) found that controlling for the number of hours worked, women reported more work interference with family than did men, and women demonstrated a higher correlation between the number of hours spent in a role (work or family) and conflict originating in that role, than did men. In addition, Crouter (1984) found mothers reported more negative spillover between work and family than did fathers.

A key mechanism of stress transmission to family members is likely to be the management of emotion. We understand little of how emotion management in work roles influences emotional management in family roles (Wharton & Erikson, 1993) and of other mechanisms (such as social support) that may mediate the effect of occupational stress transmission from female officers to family members. Evan, Coman, Stannley, and Burrows (1993) concluded that there is a general tendency for officers to refuse to share their emotional reactions with nonpolice personnel such as their partners or families. However, women may use support mechanisms differently. For example, Etztion (1984) found that for men support in the workplace moderated the relation of work stress and burnout, whereas for women, life support (outside the workplace) was the significant moderator. The study by Alexander and Walker (1994) found that women officers were more likely than men to talk things over with family and friends; males were likely to keep things to themselves.

We are currently undertaking a study of the experience of women police officers' work-related stress and how it affects the family, with a particular emphasis on factors in the work and family environment that reduce negative spillover between work and family. We aimed to identify those factors spontaneously nominated by women officers as sources of stress and the salience or relative importance of those sources to these women. Rather than make preliminary decisions about the importance of factors based on previous research and attempt a large-scale questionnaire, we chose to use open-ended questions in a focus group methodology, which allowed women to explore in depth, in a supportive context, their experience of stress in policing and its interface with family. In line with Gutek, Searle, and Klepa's argument (1991) that work–family conflict has two dimensions, work interference with family (WIF) and family interference with work (FIW), we wished to examine not only the impact of stress in the work environment as it affects the family, but also those things in the home environment that affect the capacity to manage stress at work. Specifically, we aimed to identify the most salient sources of work stress and which of these were most likely to affect family, the effects on family, and the effect of family demands/concerns in the workplace. We also aimed to identify the factors in the work environment that either exacerbated or reduced the impact on family stresses and the factors in the home context that either exacerbated or reduced negative spillover to work. In recognition of the various family circumstances in which policewomen live, we defined family as partner, children, or parents for whom the officer was responsible or with whom the officer had a close personal relationship.

Twenty-nine women officers in operational and nonoperational roles were recruited from a metropolitan police service and nearby rural areas. Women currently represent 15% of the Queensland Police Service. Contact with potential participants was made through members of a women's advisory group in the police service who approached colleagues and invited them to participate in a group discussion exploring work stress and its relation to family. Some participants were given leave to attend during work hours; other participants attended outside working hours, usually at the end of a shift. There were two rural focus groups and five groups in the metropolitan area. In addition, one woman in a rural area was interviewed individually. Interviews lasted up to 2½ hours, but most interviews ran for about 1½ hours. All levels of the service were represented in the groups. Interviews were tape recorded, allowing for preparation of transcripts for content analysis of themes.

The questions we used to initiate discussion were:

1. What things in your work environment are most likely to cause you stress?
2. Which of these things do you think are most likely to result in stress that affects your family?
3. What are the results of such stress on your family?
4. What things about your work help you manage the stress better so that it affects your family less?
5. What things about your work environment make stress harder to manage so that it affects your family more?
6. What things about your home environment help you manage the stress better so it affects work less?
7. What things about your home environment make stress harder to manage so it affects work more?

CONTENT ANALYSIS OF FOCUS GROUP DATA

The first stage of the content analysis of the focus group data was to develop a coding taxonomy consisting of metathemes and specific themes. Two of the authors independently coded each of the eight transcripts to compile a list of metathemes. Extracted themes were then compared across the eight transcripts, with minor revisions made to produce the final 12 metathemes. Because of quantitative differences in the amount of material provided in response to each of the questions, there were varying numbers of metathemes per question. The metathemes are listed in Table 1.7.1.

Specific themes relating to each of the metathemes were then identified by using the same procedure. For example, under Question 1 we identified

TABLE 1.7.1
Content Analysis Metathemes

Question	Metatheme
1.	Operational
	Relationship/police culture
	Organizational/management
2.	Operational
	Relationship/police culture
	Organizational/management
3.	Crossover
	Spillover
4.	Factors making work impact less likely at home
5.	Factors making family impact more likely at home
6.	Factors making family impact less likely at work
5.	Factors making work impact more likely at work

three metathemes, referring to Operational themes (i.e., to do with the work itself), Relationship/police culture (references to the interpersonal and organizational context in which the job occurs), and Organizational/management (practices, policies, workload, and resourcing). Under each metatheme, we identified a number of themes, identified by key words and phrases. For example, under the Operational metatheme for Question 1, the occurrence of the key words *domestics, drunk, shot gun, criminals, safety* were regarded as identifying the theme of physical threat/danger.

The transcripts were coded by using sentences as units. The first occurrence in a sentence of any key word or phrase identifying a theme was regarded as reference to that theme. In this manner, each of the transcripts could now be coded for frequency of reference to themes. A team discussion with the authors and three raters clarified the rating process and resolved potential inconsistencies. The eight transcripts were then coded independently by the three raters. The percentage (i.e., number of references to that theme, out of total references in the metatheme) endorsement of each of the metathemes and specific themes is presented in Table 1.7.2. Where there is more than one metatheme in a question, percentages are given for the number of references to the metathemes.

CONCLUSIONS

We conceptualized sources of stress at work as of three kinds: operational (the tasks), relationship/culture (the organizational and interpersonal context), and management (practices, policies, workload, and resourcing). In a number of ways, our findings with respect to sources of stress are consistent

TABLE 1.7.2
Frequency of Endorsement for Each
of the Metathemes and Specific Metathemes

Question	Metatheme (Frequency of Endorsement)	Specific Theme	Percentage Endorsement
1	Operational (16%)	Public	11
		Physical threat/danger	8
		Exposure to trauma	17
		Physical working conditions	13
		Work schedule	19
		Legal	32
	Relationship/police culture (28%)	Lack of peer support	24
		Gender discrimination	25
		Confidentiality	18
		Interpersonal	33
	Organizational/management (56%)	Feedback/recognition	11
		Management	34
		Resources	19
		Administration (policies/demands)	9
		Workload	27
2	Operational (27%)	Danger/threat	21
		Work schedule	72
		Legal	1
	Relationship/police culture (33%)	Gender discrimination	83
		Interpersonal	17
	Organizational/management (40%)	Resources	9
		Administrative	52
		Workload	38
3	Crossover (36%)	Lack of time and support	8
		Anxiety re possible threat/danger	19
		Harassment of family	3
		Extra family responsibilities	13
		Relationship issues	57
	Spillover (64%)	Mood	38
		Negative emotions	14
		No time for family needs	14
		No time for self	7
		Don't discuss work issues	6
		Alcohol/drug abuse	3
		Discussing work at home	7
		Illness	5

Continued

TABLE 1.7.2   (Continued)

Question	Metatheme (Frequency of Endorsement)	Specific Theme	Percentage Endorsement
4	Factors that make work impact on family less likely	Supportive work–family practices	8
		Human services officer	2
		Humor	11
		Social events	10
		Physical activity	2
		Social support	51
		Work environment	5
		Workload	3
		Job satisfaction	6
5	Factors that make work impact on family more likely	Operational	21
		Resources	20
		Workload	6
		Work schedules	3
		Interpersonal	10
		Gender issues	8
		Managerial	8
		Lack of support	23
6	Factors that make family impact on work less likely	Eating/drinking	4
		Breaks	16
		Practical support from the organization	5
		Pets	3
		Humor	17
		Don't talk about work	13
7	Factors that make family impact on work more likely	Household tasks	33
		Children's needs	15
		Travel time to work	8
		Not enough time for self	28
		Lack of support	9
		Family needs	7

with other studies using predominantly male samples. Like male samples, there was evidence in this group that organizational and management sources of stress (56% of comments) were more salient than were occupational sources of stress (16% of comments); these data parallel those of Brown and Campbell (1990) collected via questionnaire, who also found reports of organizational and management stress to occur four times more frequently than reports of operational stress. In our sample, management practices and workload were primary causes of stress in this area, paralleling their findings of long work hours, overload, and management styles to be problematic. Although organizational and management issues were the most frequently noted metatheme, relationship/culture issues were also salient sources of stress; this metatheme accounted for 28% of comments overall, with interpersonal stress, lack of peer support, and gender discrimination being almost equally of concern in this category. In other studies (e.g., Brown & Fielding, 1993), operational sources of stress such as informing relatives of a death, dealing with violence or trauma, or situations of threat have been identified as key sources of stress; however, in this group legal processes and requirements were the most commonly noted theme. This theme referred to making court appearances, the strict requirements to be met, and lack of training and support in this area.

In response to Question 2, women indicated that the organizational culture affected family particularly via gender discrimination, accounting for 83% of comments in the metatheme of relationship/police culture. Women raised issues of tokenism, pressures to prove themselves via performance, being visible, hostility, undermining, and sexual comments. Given the crucial role of social support at work in moderating the impact of work stress on family, it is understandable that this issue, which women referred to in terms of lack of support, should be salient. Operational sources of stress were less likely to affect family than organizational/management and relationship/ culture sources of stress. Work schedule was the most problematic operational stress, with 72% of comments. Shifts, rosters, and surveillance work were noted to be difficult to integrate with family needs. Twenty-one percent of comments referred to danger and threat creating spillover; specifically, family members found it stressful to have officers exposed to dangerous situations. Family members found it difficult to sleep while the officer was on night duty, felt worried and anxious, and sometimes pressured officers to leave the job. Some officers also commented on harassment by the public of family members, particularly children as a result of having mothers who were police.

This type of effect has been referred to as crossover (Westman & Etzion, 1995), where strain from one family member affects another. Following Westman and Etzion's conceptualization, we distinguished crossover from spillover, where stress in one life domain causes an individual to experience stress in the other domain. Spillover effects occurred primarily in terms of negative mood (for example, anger, depression, irritability, and anxiety) and expressed emotions (crying and nightmares). Comments about mood occurred more frequently (38%) than comments about emotions (14%). Illnesses were less frequently mentioned.

Crossover was less frequently reported than spillover; the most problematic crossover effect on family members was relationship difficulties, particularly in marriages. Women reported difficulty finding time with partners and also indicated that they tended to avoid discussing work stresses at home because of feelings that family members would not understand or would be distressed by such disclosures. This strategy may, however, exacerbate relationship difficulties. Failure to share worries and concerns about work occurred in spite of the importance of social support outside the work environment for these women; family was seen to affect work less when social support was available, being clearly the most salient factor felt to reduce the likelihood of exacerbating stress at work. Social support at work was clearly the most salient factor making work stress less likely to affect family, accounting for over one half of the comments in this metatheme. The next most frequent themes were use of humor and social events, accounting for 11% and 10% of comments.

Etzion (1984) found that women seek social support from family and friends whereas men tend to seek it in the work environment. For this group of women, however, support from the family may be limited because of the nature of the work, even though emotional support is critical in moderating between work stress and burnout (Etzion, 1984). By protecting family members in this way, these women may exacerbate possible spillover effects such as mood, illness, and fatigue. Social support at work clearly plays a key role in managing stress so that it affects family less. The ability to use this type of support may however be reduced by the police culture, which emphasizes active and problem-focused methods of coping rather than social support. Alexander and Walker (1994) found that female constables talk things over more with family and friends than do males, and males are likely to keep things to themselves, suggesting that in a male-dominated culture, seeking social support may not be legitimized. Women's reports of gender discrimination also suggest that women feel unsupported in the culture as a whole, although references to supportive bosses and colleagues were common. It is therefore important to develop a better understanding of how policewomen use emotional support and its impact on their well-being. From comments in interviews friends (especially those with some experiencing of policing) and female colleagues, rather than family, seem to be primary sources of social support for these women.

In summary, our findings indicate women experience the same sources of stress as male police officers, but not surprisingly, report gender discrimination as a major source of stress, and one likely to affect family. Like male colleagues, their work stresses particularly affect partners by affecting relationships. However, social support is a primary factor reducing stress for these women and may be more crucial for women in managing stress than for male officers. We are presently conducting further studies of this population of policewomen to explore the meaning of social support in both the work and home environments and the mechanisms by which such support may reduce the negative impact of workplace stressors.

ACKNOWLEDGMENTS

The authors appreciate the support and cooperation of the Queensland Police Service in carrying out this research and particularly wish to thank those women who gave their time to organize focus groups and to participate in them. This support does not suggest that the Queensland Police Service endorses the research or its findings, and responsibility for any errors of omission or commission rest solely with the authors.

REFERENCES

Alexander, D. A., & Walker, L. G. (1994). A study of methods used by Scottish police officers to cope with work-induced stress. *Stress Medicine, 10,* 131–138.

Beehr, T. A., Johnson, L. B., & Nieva, R. (1995). Occupational stress, coping of police and spouses. *Journal of Organizational Behavior, 16,* 3–25.

Brown, J. M., & Campbell, E. A. (1990). Sources of occupational stress in the police. *Work and Stress, 4,* 305–318.

Brown, J., & Fielding, J. (1993). Qualitative differences in men and women police officers' experience of occupational stress. *Work and Stress, 7,* 327–340.

Burke, R. J. (1994). Stressful events, work–family, conflict, coping, psychological burnout, and well-being among police officers. *Psychological Reports, 75,* 787–800.

Burke, R. J., Weir, T., & Duwors, R. E. (1980). Work demands on administrators and spouse well being. *Human Relations, 33,* 253–278.

Christie, G. (1994, December). *Police perceptions of norm supporting and norm violating behaviour in communicative relationships.* Paper presented to the Australia and New Zealand Academy of Management, Wellington.

Close, D. H. (1994, April). Sexual harassment and the police environment. *The Police Chief,* 8–10.

Crouter, A. (1984). Spillover from family to work: The neglected side of the work–family interface. *Human Relations, 37e*(6), 425–442.

Dantzer, M. L. (1987). Police related stress: A critique for future research. *Journal of Police Criminal Psychology, 3*(3), 43–48.

Daum, J. M., & Johns, C. M. (1994, September). Police work from a woman's perspective. *The Police Chief,* 46–49.

Etzion, D. (1984). Moderating effect of social support on the stress–burnout relationship. *Journal of Applied Psychology, 69,* 615–622.

Evans, B. J., Coman, J., Stanley, R. O., & Burrows, G. D. (1993). Police officer's coping strategies: An Australian survey. *Stress Medicine, 9,* 237–246.

Gutek, B. A., Searle, S., & Klepa, L. (1991). Rational versus gender role explanations for work–family conflict. *Journal of Applied Psychology, 76,* 560–568.

Haynes, S. G., Eaker, E. D. & Feinleib, M. (1983). Spouse behavior and coronary heart disease in men: Prospective results from the Framingham heart study. *American Journal of Epidemiology, 118,* 1–21.

Heidenson, F. (1992). *Women in control? The role of women in law enforcement.* Oxford: Clarendon Press.

Hotchkiss, S. (1992). *Policing promotional difference: Policewomen in the Queensland Police Service.* Unpublished manuscript, Griffith University.

Jackson, S. E., & Malasch, C. (1982). After effects of job related stress: Families as victims. *Journal of Occupational Behaviour, 3,* 63–77.

Jones, F., & Fletcher, B. C. (1993). Taking work home: A study of daily fluctuations in work stressors, effects on moods and impacts on marital partners. *Journal of Occupational and Organizational Psychology, 69,* 89–106.

Kelly, S., & Brown, E. (1999). *Female perspectives in policing: A content anaylsis.* Manuscript in preparation.

Thompson, B. M. (1997). Couples and the work–family interface. In W. K. Halford & H. J. Markman (Eds.), *Clinical handbook of marriage and couples interventions.* Chichester, England: Wiley.

Westman, M., & Etzion, E. (1995). Crossover of stress, strain and resources from one spouse to another. *Journal of Organizational Behavior, 16,* 169–181.

Wharton, A. S., & Erikson, R. J. (1993). Managing emotions on the job and at home: Understanding the consequences of multiple emotional roles. *Academy of Management Review, 18,* 457–486.

1.8

Stress and Driving Performance: Implications for Design and Training

Gerald Matthews
University of Cincinnati

Paula A. Desmond
Texas Tech University

Current technological advances offer the driver access to a range of complex intelligent transportation systems. Some of these systems include navigation, route guidance, and collision-avoidance systems. Such systems provide the driver with additional information, for example, advice on headway, which the driver can use or ignore. Future technological developments promise fully automated transportation systems (FATS: Brand, 1998) in which the vehicle's braking, steering, and acceleration functions are automated.

Human factors research has sought to develop guidelines for the design of transportation systems that are based on well-established human factors principles. For example, Sanders and McCormick (1987) have highlighted the importance of compatibility between stimulus and response in relation to the system's layout and control functions. However, contemporary design approaches fail to reflect the variability and context dependence of stress reactions. An ideal transportation system should neither induce stress in the driver nor be sensitive to performance degradation if the driver should be stressed. The transactional model of driver stress (see chap. 1.1, this volume) offers a comprehensive framework for driver safety and has important implications for the design and implementation of transportation systems. In this chapter, we distinguish various potentially detrimental effects of driver stress and principles for developing practical countermeasures. The structure of this chapter is as follows. First, we provide an overview of the safety problems originating from driver stress, in which we emphasize problems associated with the introduction of new technology. Next, we examine the transactional model as a framework for driver safety, which accommodates both ecological and cognitive ergonomic perspectives. The model suggests we can seek to alter both the external environment and the person's internal cognitions. Accordingly, the next two sections discuss design implications for various transportation systems and training interventions targeted at cognitions. We conclude with some guidelines for intervention derived from the transactional model.

DRIVER STRESS AS A SAFETY PROBLEM

The statistical association between stress and accident risk has been well established. Studies of individuals who have experienced traumatic life events have revealed a greater risk of accident involvement among these individuals (Selzer & Vinokur, 1975). For example, Selzer, Payne, Westervelt, and Quinn (1967) investigated the relation between life stresses (e.g., divorce) and driving accidents in samples of drivers ranging from 50 to over 500 individuals. The study showed that the occurrence of social stresses during a 1-year period before a fatal car accident was significantly more frequent in an accident group of drivers than in a control group. Brenner and Selzer (1969) estimated that drivers who had experienced stressful events such as personal conflicts, illness, bereavement, or financial difficulties were five times as likely to cause fatal accidents as drivers not subjected to such stressful events.

Several mechanisms may be responsible for the association between stress and accident risk. Mayer and Treat (1977) demonstrated that several personal and social maladjustment measures predicted accident involvement. The authors suggested that two mechanisms might be involved in the stress–accident risk link. First, drivers who are stressed may be distracted by thoughts about their personal problems. This "cognitive interference" mechanism has been demonstrated in many studies of anxiety (e.g., Sarason, Sarason, Keefe, Hayes, & Shearin, 1986). Second, depressed drivers may lack necessary motivation in avoiding hazardous driving situations: Unwillingness to apply task-directed effort is commonly found in experimental studies of depression (Wells & Matthews, 1994).

Technology and the Changing Nature of the Driving Task

The problems of driver stress must be considered in the context of the impact of new technology on the nature and demands of the traditional driving task. The driving task currently requires the driver to manually direct the vehicle along the road, through the efficient use of the vehicle's controls, and to maintain safe interactions with other vehicles on the road. Today's driver receives safety-critical information about the driving environment through interaction with other road users. However, with the emergence of future technological advances such as automated driver systems (Hancock & Parasuraman, 1992), drivers will receive safety-critical information through the occurrence of rare or unusual events (e.g., a failure in automation). Thus, the reliability and efficiency of our transportation systems influence stressful reactions in drivers, but the cognitive demands of those systems are currently changing and will continue to change for the foreseeable future.

Attentional Overload

Matthews and Desmond (1995) identified two themes of particular relevance to driver stress in the vehicle of the future: *overload of attention* and *disruption of control*. Clearly the attentional demands of the driving task become rather different as vehicles become fitted with in-vehicle systems. Systems such as guidance systems, performance-monitoring systems, and headway warning systems place additional attentional demands on the driver because such systems present information that the driver may need to process. It is clear that in-vehicle systems provide the driver with several safety benefits. Systems that warn the driver of dangerously short headways may assist in diverting potential accidents. Similarly, performance-monitoring systems that provide feedback to the driver of fitness to drive assist in avoiding accidents caused by driver impairment. In-vehicle systems may also serve to reduce stress in drivers; guidance systems may assist in providing the driver with the most efficient route to a particular destination, reducing the frustration of congestion and the fatigue of an unduly prolonged drive. In addition, guidance systems may act to offset some of the negative consequences of monotonous driving conditions such as boredom, by providing the driver with route-related information to maintain driver interest, for example.

   However, despite the advantages of in-vehicle systems, there are a number of potentially hazardous consequences associated with such systems. One obvious problem is attentional overload (see Matthews & Desmond, 1995, for extended discussion). When the driver must process information from in-vehicle systems, the overall workload of the task increases, with potentially disastrous consequences, if the driver can no longer effectively attend to all sources of safety-critical information. Several studies have demonstrated that information displays in vehicles serve to distract the driver's attention from the primary task of driving, as shown by reduced frequency and duration of various regions of the visual scene (Lansdown, 1997). In-vehicle systems may also increase workload and stress (Lansdown, 1997). Overload of attention may be a problem for the unstressed driver, but stress factors such as worry and fatigue tend to impair functional attentional efficiency (Wells & Matthews, 1994), so that the stressed driver becomes especially vulnerable to overload.

Disruption of Control

The driver's task also changes as a consequence of more sophisticated automated systems. Systems that fully automate the vehicle's dynamics force the driver to assume the role of system monitor and may reduce the driver's overall cognitive workload. Undoubtedly the major attraction of automated driver systems is the promise of improved road safety and efficiency that they offer. Brand (1998) has discussed in detail some of the advantages of fully automated transportation systems (FATS). For example, we are all familiar with the traffic congestion that typically results from overly cautious drivers' reactions to other vehicles' brake lights. This form of traffic congestion could be avoided by a FATS (Brand, 1998).

However, the development of more intelligent transportation systems has implications for driver control and autonomy. Currently the driving task affords the driver considerable control over how he or she chooses to behave. For example, the driver may exhibit aggressive driving behavior by following closely behind another vehicle, or he or she may engage in risky driving maneuvers. In the future, the driver will have limited control over the driving task because computer software will be used to control many features of the vehicle's functions. However, a potential source of driver stress associated with such systems is that the driver may experience difficulty in relinquishing and taking over control of the vehicle. Matthews and Desmond (1995) distinguished two specific types of control problems: disruption of attentional control and disruption of effort regulation. Attentional efficiency depends not only on the imbalance between processing demands and some "capacity" or "resource," but also on the strategies used by the driver to interrogate the environment actively. As discussed by Matthews (chap. 1.1, this volume), this strategic aspect of attention is part of a wider set of self-regulative processes. The danger of new technology is that it disrupts the driver's prioritization of different information sources. Aviation psychologists are familiar with various problems resulting from suboptimal integration of information from different displays or from displays and the direct view (see Weiner & Nagel, 1989): Similar problems may be encountered in the vehicle of the future. Stress factors may interact with new technology. For example, arousing stressors tend to increase attentional selectivity, disrupting sampling strategy. Driver stress reactions may include maladaptive sampling strategies, as when an aggressive driver focuses excessively on passing the vehicle in front.

Drivers need strategies not just for sampling the traffic environment, but also for effort regulation. As with many real-world tasks, actual performance tends to be less than maximal performance. Presumably, the level of effort drivers appraise as necessary to maintain safety is less than the maximum possible effort. The danger of automated systems may be that because workload, or some aspects of workload, are reduced, the driver believes that it is safe to reduce the effort applied to the task, with potentially detrimental consequences. In fact, there is much evidence to indicate that automated systems do not necessarily result in a reduction in mental workload (see Parasuraman, Mouloua, Molloy, & Hilburn, 1996, for a review), but automation may still change the driver's beliefs about the need to invest effort in the driving task. Again, maladaptive effort regulation may also be influenced by driver stress factors (Matthews & Desmond, 1995). Both distress and fatigue seem to be associated with a reduction in task-directed effort when the driving task is relatively undemanding (see chap. 1.1, this volume). A related problem associated with in-vehicle systems and automation is one of risk homeostasis or risk compensation. If the vehicle is fitted with a collision-avoidance system, the driver may feel prepared to engage in various dangerous behaviors, such as reducing the effort directed toward maintaining safety or actively engaging in more risky or reckless driving behaviors.

A final issue to consider is the social consequences of technology on the driver. Today's driver can be characterized as an autonomous entity, in that he or she is an independent, self-directed operator of the vehicle. However, in the future the driver will become a component of an externally regulated transportation system, which will be regulated both by in-car systems and by whoever is responsible for the master plan of the traffic-flow system. Such a shift in autonomy may generate rather different stressful reactions in drivers.

TRANSACTIONAL ERGONOMICS: APPLICATION OF THE COGNITIVE-ADAPTIVE MODEL OF STRESS

Ecological and Cognitive Contexts for Ergonomics

Stanton (1995) has discussed the need for ergonomics to accommodate transactions between the operator and the environment. He pointed out that ergonomics tends to neglect contextual factors. In the absence of an explicit framework for dealing with context, results are unlikely to generalize across different contexts. Neglect of context is a weakness of conventional cognitive and engineering psychology (Norman, 1993). Hence, ergonomics should be "ecological" in situating the operator in an external environment that affords opportunities as well as imposing demands. However, ecological psychology tends to neglect the "internal context" generated by cognition. Stanton's (1995) contextual action theory (CAT) seeks to use the interaction between the objective external context and subjective internal context as the basis for practical intervention. At its simplest, CAT indicates that we may develop more adaptive human–technology interaction either by better design of the technology or by facilitating the user's coping efforts.

Stanton's (1995) ecological ergonomics are compatible with the transactional model of stress, and the model of driver stress discussed by Matthews (chap. 1.1, this volume) suggests principles for practical intervention that accommodate contextual factors. In brief, the model states that effects of environmental stressors (i.e., external context) on subjective and behavioral outcomes are mediated by cognitive stress processes of appraisal and coping in a dynamic system. The driver continually reviews the personal significance of external events and whether they call for further thought or action. Various aspects of driver stress may be identified, which relate to the driver's general mode of adaptation to external demands. For example, anxiety and emotional distress relate to concerns about personal competence, whereas aggression is associated with a focus on dealing with perceived hostility from other drivers. The individual's vulnerability to these various stress reactions is controlled by stable, contextualized self-beliefs about driving, which are expressed as personality traits such as Dislike of Driving and Aggression (Matthews, Desmond, Joyner, & Carcary, 1997). Safety consequences of driver stress reflect both "cognitive interference" associated with worry and the driver's specific choice of coping strategies, such as use of the hazardous confrontive strategies preferred by aggressive drivers.

Principles for Intervention

Combining the transactional model with Stanton's (1995) ecological perspective suggests some general principles for intervention. First, we can seek to influence the external environment, with respect to both the demands it places on the person and the opportunities or affordances it provides, keeping in mind that the appraisal of the environment is critical for stress reactions. By a *demand,* we mean a stimulus or event that requires some imperative action, such as a car in front braking sharply. By an *opportunity,* we mean an environmental attribute that permits but does not require the driver to take some goal-oriented action. Opportunities may be safety promoting, such as a gap in traffic allowing the driver to move into a slower moving lane, or potentially detrimental, such as the chance to overtake with short temporal leeway. Both demands and opportunities may influence stress processes of appraisal and coping. Demands are likely to trigger appraisals such as threat and challenge directly, whereas opportunities are more dependent on the driver's active search for pathways toward optional goals.

Curiously, current "passive" in-car systems and future automated control systems are quite differently related to levels of opportunity. In-car devices such as navigation systems extend the driver's range of choices and may tend

to introduce greater variability in driving behavior. We can subdivide opportunities into *warning systems,* which suggest to the driver options for increasing
safety, such as braking in response to a headway warning; *information systems,*
which assist with goals such as route finding (navigation aid) and hazard
detection (vision enhancement systems); and *control systems,* which allow the
driver to reduce effort (automatic cruise control) or to expand the capabilities of the vehicle (four-wheel drive). Driver performance–monitoring systems that signal loss of competence are a special case of warning systems. In
each case, the use made of the system depends on the driver's motivations.
Conversely, FATS technology, by removing control from the driver, tends to
reduce opportunities. To the extent that these opportunities relate to motives such as thrill seeking, the change is beneficial. However, because the
driver must adopt the new role of monitoring for system failure, FATS may
substitute demands for opportunities, to the extent that system failure calls
for imperative action.

The transactional analysis suggests two basic strategies for intervention.
First, we can seek to change the external context by designing the traffic environment for stress tolerance. *Traffic environment* here refers to factors such
as road layout, traffic signs, and aspects of the vehicle such as its handling
and its in-car information systems. In general, the car and the roadway
should be designed so as to minimize maladaptive stress reactions, such as
frustration and worry. Furthermore, because stress reactions are inevitable
to some extent, the traffic environment should be forgiving of the stressed
driver. Intersections and in-car navigation aids should not be so complex
that the worried driver becomes vulnerable to overload of attention. Designers must maintain awareness that traffic environments cater for drivers
whose mental states differ widely.

The second basic intervention strategy is to change the internal context,
the motivations and cognitions that the driver brings to the task. Here, there
is a continuum of targets for intervention, ranging from general factors
such as social values and personality, to specific maladaptive cognitions such
as the belief that running red lights is a worthwhile means for reducing journey time. Interventions operating at the level of the individual focus primarily on training and selection. Driver training neglects handling one's own
emotions as a skill to be acquired, and more emphasis on mood regulation
and stress-management techniques would be desirable. Organizations em-
ploying drivers may have the luxury of selecting employees on the basis of
competence. An extended view of competence might include the ability to
stay calm under the pressures of driving.

In this chapter, we focus on the individual driver, but we note also that
driver stress reactions take place against a background of social norms, and
there is scope for stress-related interventions at the societal level. Social
norms are relevant primarily to the cognitions of the individual, although

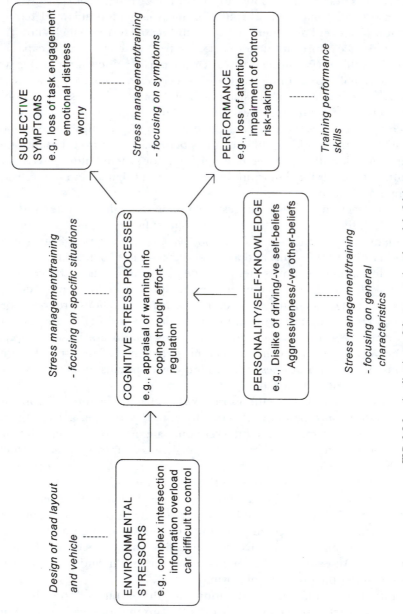

FIG. 1.8.1. Applications of the transactional model of driver stress.

they presumably influence design via designers' beliefs about acceptable levels of stress and behavioral responses to stress. In at least some Western societies, such as England, some form of rule breaking on the road may be seen as socially acceptable (e.g., speeding), whereas law breaking in other contexts is generally unacceptable (Corbett & Simon, 1991). Social norms help to shape appraisals and coping, especially perhaps in the context of aggression and frustration (see chap. 1.1, this volume): Cultural values help to define when other road users are appraised as showing hostility and the extent to which confrontive coping strategies such as gesturing or intimidation are acceptable. Social norms may be influenced through advertising and safety campaigns. In England and the United States, long-running campaigns appear to have been successful in reducing the acceptability of drinking and driving; perhaps a similar effort is required to tackle aggressive driving. Other socially influenced motives may require greater attention. For example, the achievement motive may contribute to drivers' intentions to keep driving although fatigued or to meet deadlines for journeys by speeding. The need to maintain self-esteem by appearing competent to others is potentially beneficial in encouraging drivers to attain high standards of skill, but, in anxiety-prone drivers, this motive may be associated with dangerous levels of worry.

In the next two sections, we look in more detail at, first, the implications of driver stress research for design, and second, its implications for interventions targeted at the individual driver's cognitions. Figure 1.8.1 summarizes how different types of intervention may be targeted toward the components of the Matthews (chap. 1.1, this volume) transactional model, reconfigured slightly to address the concerns of this chapter. Design-oriented interventions are targeted toward the external inputs to the cognitive system, whereas stress management and training-based interventions are variously directed toward personality (i.e., underlying self-knowledge), cognitions of specific stressful encounters, and symptoms of stress such as negative emotion.

DESIGNING FOR DRIVER STRESS

Road Engineering

The road layout itself is an important aspect of the environment. For example, badly cambered curves or intersections with poor visibility may be sources of threat. Of course, restricted visibility is dangerous because it increases the objective difficulty of hazard detection, but there may be *additional* safety costs, if drivers react by worrying about the problem. As discussed by Matthews (chap. 1.1, this volume), there is a finely balanced trade-off operative here. Threat stimuli are potentially beneficial in preventing overconfidence and eliciting safety-oriented coping, such as looking carefully at intersections, but potentially hazardous in generating worry and cognitive interference, which impairs attention. The balance appears to be tipped by personality factors, and especially by dislike of driving: The more sensitive driver may be safer in a more comforting traffic environment. Similarly, road layouts may present demands related to lane merging and traffic conflict, which present particular difficulties for aggressive drivers, and may be avoided by explicit lane prioritization, for example. Forcing drivers to compete for access to lanes is poor design.

In addition, road design controls safety-related opportunities. Traffic-calming measures seek to promote safe driving by denying opportunities for risk taking, for instance. Engineering solutions of this kind are demonstrably effective in reducing speed and may, additionally, promote greater confidence in anxiety-prone drivers. However, there may be more subtle issues that relate to frustration during driving. In some contexts, such as driving on main roads between cities, drivers expect to have opportunities to overtake. Drivers high in the stress-vulnerability dimensions we refer to as Aggression and Thrill Seeking (Matthews et al., 1997) are perhaps especially prone to expect such opportunities, and lack of opportunity is likely to frustrate them. It is an open question whether it is better to cater for these expectations by designing in at least some road sections where overtaking may be possible, and so reducing potentially dangerous frustration, or to re-educate drivers not to expect to overtake as of right. For example, the A9 Perth–Inverness trunk road in Scotland alternates normal and divided-lane sections of road, allowing periodic safe overtaking. However, the expectancy of fast driving has been blamed for accidents caused by excessive speed and overtaking on nondivided road sections. Road design should be assessed in terms of the type of coping it promotes: Encouraging confrontive coping strategies such as risky overtaking is obviously undesirable.

Informational (Traffic) Systems

Traffic information systems that are in current use include headway advisory and route-guidance systems. Through the use of radar sensors, headway advisory systems can relay information to the driver informing him or her of unsafe following distances to other vehicles. Guidance systems function to provide the driver with route-related information that the driver may pay attention to or simply ignore. In line with the transactional perspective, such systems need to be explored for their potential to give rise to negative appraisals and coping strategies in drivers. For example, in fast-moving but high-density traffic, it may be difficult to maintain a safe headway because other vehicles move into the inviting space left by the cautious driver. In such cases, frequent warnings delivered by headway advisory systems may be a source of irritation. Negative appraisals of the device may lead to its neglect in situations in which the information is more useful. Aggressive drivers exhibit a driving style that is characterized by following at short headways and frequent overtaking (e.g., Matthews, Dorn, et al., 1998), and such drivers may be especially prone to negative cognitions of safety-related warnings. At one level, of course, it is the driver's attitudes rather than the device that is at fault, and so training rather than redesign is required. However, there may be (partial) design solutions also, by giving the driver some degree of control over the operation of the device, for example, by allowing some scope for adjusting sensitivity within certain limits.

Route-guidance systems may not present an obvious problem to stress-vulnerable drivers. However, the presence of a route-guidance system may act as a source of overload for the distressed driver. Matthews (chap. 1.1, this volume) reviews evidence that Dislike of Driving is the most influential stress-related trait influencing attentional efficiency. Drivers high in Dislike are particularly vulnerable to overload of attention when the environment is stressful, and secondary task stimuli must be attended to (Matthews, 1996). Because guidance systems present additional information to the driver, it is essential that the information requires relatively little attentional capacity to assimilate. Guidance systems might be designed to reduce overload by making use of location information. For example, the display might be "grayed out" or otherwise reduced in conspicuousness when the vehicle is approaching an intersection, to reduce likelihood of distraction.

Informational (Driver) Systems

Systems have been developed to monitor drivers' performance and to provide feedback to the driver about fitness to drive (see Richardson, 1994; Ward & Fairclough, 1997). As these systems become more sophisticated, it may be possible to provide feedback to the driver on virtually any aspect of driving performance. On the surface, this possibility appears to provide potential solutions to several problems. For example, in the context of fatigue, such systems may be used to inform the driver that he or she has reached a dangerous level of fatigue and should pull over and rest. Similarly, impairment in the driver caused by alcohol or drugs may also be detected by a system of this kind. In addition to detecting driver impairment, informational systems may be used in the context of communicating to the driver when driving has become reckless or aggressive.

However, a number of concerns need to be addressed with such systems, centering on two issues (Desmond & Matthews, 1997). First, the system must validly assess impairment, and, second, whatever warning it delivers must tend to elicit a safety-enhancing response. As discussed by Matthews (chap. 1.1, this volume), both distress associated with Dislike of Driving and fatigue tend to impair driver performance when workload is relatively low, but the stressed driver effectively compensates when the task becomes more demanding. Validity of assessment varies with workload, so that the ideal system should base its assessment on low-workload driving. Such a system might use positional and navigational information to assess performance on straight roads, but not on curves or at intersections, for example. The driver's response to such systems is likely to depend on how it is appraised, and system designers should, in general, address the credibility of the system. For example, its acceptability may be enhanced to the extent that the system distinguishes different sources of impairment, such as distress versus fatigue. Fairclough (chap. 2.6, this volume) describes how alcohol and sleep deprivation generate different patterns of performance change, which might feed into appropriate warnings.

The "internal context" that the driver applies to the warning is also important. In the case of the anxiety-prone driver, high in Dislike of Driving, a system that delivers repeated warnings of poor driving performance may well serve to exacerbate emotional distress and worry. In this case, design should aim to avoid excessive negative self-appraisal, by emphasizing that any impairment is temporary rather than general, or by ensuring that the system delivers feedback to the driver only when driving performance has deteriorated to a level critical to safety. Overconfident drivers may exhibit a different problem, in that they may appraise the warning signals as a sign that they can continue to drive provided they exert more effort, when they might be better advised to cease driving. FATS may afford the possibility of terminating driving whatever the driver's wishes, leading to emotional distress but perfect safety.

In summary, the analysis here points to some fundamental difficulties in provision of warning systems, because responses to the warning may be maladaptive, especially in the case of stressed drivers. A possible but expensive solution to these difficulties may be to tailor the system to the individual user's needs. The designer might work with the user to identify potential weaknesses in driving competence and to fit an appropriate system approved by the user. Without some user involvement of this kind, it may otherwise be difficult to obtain system credibility.

Automation of Control and IHVS

Intelligent vehicle highway systems (IVHS: Hancock & Parasuraman, 1992) function to automate many features of the driver's task. In such systems, the navigation and collision avoidance systems of the vehicle are under system control. We have already discussed some of the advantages and disadvantages of IVHS. From a transactional perspective, special consideration should be given to how such technology affects drivers' appraisals and coping strategies. For distressed or fatigued drivers, IVHS that lessen their fear of danger or task difficulty may have potentially hazardous consequences. Previous research has shown that stressed drivers tend to lack confidence in their driving ability (Dorn & Matthews, 1995) and may be inclined to "switch off" from the driving task and to place complete trust in an automated driving system. Likewise, the fatigued driver may undermobilize his or her effort in the presence of an automated driving system. Indeed, previous research on driver fatigue (see Desmond & Matthews, 1997) has shown that fatigued drivers experience difficulty in regulating their effort in low-demand driving conditions, although they are able to maintain driving performance when demands are high. Desmond and Matthews suggested that the fatigued driver becomes complacent and resistant to expending further effort in the driving task when task demands are low. The implication of this work for the design of future intelligent transportation systems is that we must seek to maintain the driver's involvement in the task to offset any potential complacency problems in drivers.

The potential negative consequences of automated driver systems have been shown in empirical research. Desmond, Hancock, and Monette (1998) examined the effects of automated and manual driving on drivers' subjective states. In the study, automation of velocity and trajectory failed three times during the 40-minute drive. Both drives resulted in increased tension, depression, cognitive interference, physical and perceptual fatigue symptoms, and boredom. In addition, decreases in concentration, motivation, energy, and perceived control were found following both drives. Thus, automation may induce stressful reactions in drivers as effectively as a prolonged period of monotonous driving. Automation may also have hazardous consequences for the driver's performance. A serious problem identified with automated systems is operator complacency: De Waard, van der Hulst, Hoedemaeker, and Brookhuis (in press) reported a simulation study of an IHVS system in which around one half of drivers failed to take control of the vehicle in an emergency situation. When operators are forced to manually operate a system following a failure in automation, their performance is impaired compared with operators who perform the same task without automation (Endsley & Kiris, 1995). In Desmond et al.'s (1998) study, drivers were slower to regain efficient vehicle control following a failure in automation than drivers who were manually engaged in the driving task throughout the duration of the drive.

The findings support Hancock, Parasuraman, and Byrne's (1996) view that human-centered design approaches are superior to full automation of a task. Matthews et al. (1999) recommended that automated systems should be assessed for their effects on the three fundamental cognitive-adaptive stress syndromes discussed by Matthews (chap. 1.1, this volume): task disengagement, distress, and worry. Loss of task disengagement appears to be the most common problem with automated systems, but Desmond et al.'s (in press) study identified increased worry as a further potentially maladaptive reaction. Worry and distress may ensure when system functioning threatens personal competence, such as when the system is appraised as unreliable or when it overrides the user's control inappropriately. Discrimination of stress reactions assists the designer in focusing on maintenance of "in-the-loop" control (task disengagement), preventing episodic overload of the operator (distress), and reducing uncertainty over system operation (worry).

INTERVENTIONS FOR MALADAPTIVE COGNITIONS

The task cannot always be fitted to the person. In this section, we discuss how the person may be adapted to the task through training appraisals and coping strategies that promote safety. Such interventions are likely to be more effective than relaxation techniques that tackle symptoms rather than underlying cognitive causes of symptoms. Organizations may prefer to pre-select personnel with adaptive cognitions of driving, and we also outline strategies for selecting drivers who maintain safety in stressful environments.

Stress Management Through Training

Most researchers have accepted that (a) driving as a skill may be conceptualized at various levels (e.g., Summala, 1997), and (b) driver training tends to emphasize the "lower" levels of skill acquisition, such as vehicle maneuvering, while neglecting "higher" levels such as maintaining awareness of risk (Brown & Groeger, 1988) and regulating motivational states in the light of task demands (Delhomme & Meyer, 1997). Training needs to be directed not just toward handling of imperative demands, but toward more safety-oriented decisions when the environment affords opportunities, such as possible overtaking. In addition, we expect future research to focus more strongly on training in adults. Introduction of new technology may call for retraining of adult drivers and periodic refresher courses as the technology evolves. Some countries have also introduced voluntary or mandatory training programs for individuals convicted of traffic offenses.

Unfortunately, training directed toward management of stress reactions has been rather neglected as an explicit goal of instruction. The transactional model implies that training should be directed toward, first, the beliefs about the self and the traffic environment that influence appraisal and, second, the successful use of coping strategies in dealing with stress. Research in trainee drivers (Hatakka, 1998) shows that self-evaluations of risks, personal driving habits and skills, and so forth predict driver behavior in both cross-sectional and longitudinal data. Furthermore, self-evaluations change during driver training, but, as Hatakka (1998) stated, training generally fails to address self-perceptions and motives explicitly.

The transactional model suggests two possible foci for training: self-confidence and "social skills" for driving. Self-confidence is generally seen as desirable: A sense of self-efficacy brings various benefits such as well-being and better performance under stress (Bandura, 1986). However, in driving, self-confidence is often seen as overconfidence: Drivers may be prone to overestimate their personal control over hazardous situations (McKenna & Lewis, 1991) or the safety benefits of good technical driving skills (Hatakka, 1998). We might suspect that drivers also often overestimate their ability to drive safely in disturbed states of mind. However, lack of self-confidence may have disadvantages, too. As discussed by Matthews (chap. 1.1, this volume), the "depressive realism" of drivers high in Dislike of Driving may lead to worry and distraction from the driving task. Across individuals, Dislike of Driving may control a trade-off between worried cautiousness and serene recklessness. There may be a similar trade-off controlled by the transient state of the individual. Matthews, Quinn, and Mitchell (1998) have shown that the state of self-confidence predicts higher perceptual sensitivity on a secondary hazard detection task during simulated driving. It follows that training should encourage the benefits of both ends of the self-confidence continuum, while discouraging their costs. The goal should be an attitude of "detached realism," accepting and compensating for one's limitations as a driver without indulging in excessive worry (Matthews, Dorn, et al., 1998).

The training procedures required remain uncertain, but a first step would be to monitor self-confidence systematically during training and to develop interventions for excessively high or low levels of confidence. Mc-Kenna and Lewis (1991) described how overconfidence may be addressed by having drivers imagine themselves to blame for an accident, although they cautioned that long-term change may be difficult to effect. Under-confidence may be addressed through identifying and training specific skills that generate worry or, more generally, through applying the cognitive-behavioral techniques used in therapy for anxiety disorders (Wells, 1997). Extreme underconfidence presumably grades into driving phobia (Mathew, Weinman, Semchuk, & Levin, 1982), which requires clinical treatment.

We have attributed aggressive driving to maladaptive appraisals of other drivers as hostile and to willingness to use confrontive coping strategies. This style of cognition is clearly undesirable, and it should be identified and treated as early as possible in driver instruction. Treatment is partly a question of training behavioral "social skills" for avoiding conflict, such as defensive driving, and partly a question of training anger management and self-control. For example, Novaco's (1975) anger control treatment has subjects rehearse appropriate cognitions during four stages of a provocation sequence: preparing, confronting, and reflecting on the provocation, and coping with arousal and agitation. Aggressive drivers might be trained in dealing with provocations such as another driver's forcing them to brake sharply. However, it should be acknowledged that some instances of driving aggression and so-called road rage are signs of serious personality disorder and, as such, are difficult to treat.

In summary, training for stress requires an explicit instructional focus on self-beliefs, motives, and the driver's repertoire of coping skills. It also requires a focus on the individual driver. There may well be benefits to generalized instructional programs designed to raise awareness of dangers such as overconfidence and aggression. However, the very different stress vulnerabilities of individual drivers indicates a need to personalize instruction to some degree. Currently, only offenders tend to receive the benefits of personalized instruction, but it might profitably be extended to the general population.

Selection of Safe Drivers

Motor vehicle accidents are a significant problem for organizations. Crichton (1991) provided some 1988 UK statistics that indicate the scale of the problem in company fleets: one death or injury per 41 company cars per year, an annual claims frequency of 43.5 claims per 100 vehicles, at a mean claim cost of £251 (c. $400) per vehicle. As Crichton (1991) described, organizations should take various countermeasures, including promoting a safety culture, encouraging drivers to seek further training, and providing financial incentives for avoiding accidents. In addition, organizations can make greater use of measures related to driver behavior in personnel selection (Matthews et al., 1998). The more enlightened organization might also appraise driver behavior periodically with the intention of offering appropriate employee support and counseling.

There are various methods for selection, each of which has its advocates, but here we focus on the use of personality-like questionnaires that aim to assess individual differences in driver behavior. Personality testing in industry remains somewhat controversial, but its shortcomings often result from a failure to match the scales used to the context for job performance (Matthews, 1997). General personality measures such as extroversion and neuroticism are sometimes found to predict driver behavior criteria, but, unsurprisingly, validation coefficients are generally larger for driving-specific scales (e.g., Dorn & Matthews, 1995). Well-designed questionnaires have the considerable advantage of being reliable and valid, whereas measures of accident frequency are often of limited reliability and, hence, of limited validity.

Questionnaires measure a variety of different constructs that might be used to select safer drivers. Organizations might seek to assess dangerous behaviors directly, through using scales for errors and violations (e.g., Reason,

Manstead, Stradling, Baxter, & Campbell, 1990), or they might use measures for risk and self-perception (Hatakka, 1998). There may be advantages to using measures of cognitions rather than self-ratings of behavior, to the extent that people may have more veridical perceptions of cognitions. However, given that conceptually distinct questionnaire measures nevertheless tend to be substantially intercorrelated (e.g., Matthews et al., 1997), it is unclear a priori that any single approach will necessarily prove superior to any other: More comparative research is required.

There may be some advantages to using driver stress measures, given that the transactional model provides an explicit conceptual framework supporting various validation studies (chap. 1.1, this volume). The Driver Stress Inventory (DSI: Matthews et al., 1997) assesses five driver stress dimensions that might be used in selection. The behavioral and accident data suggest that high Aggression and Thrill seeking are liabilities in a driver, whereas high Hazard Monitoring may be advantageous. Long-distance drivers might be selected for low Fatigue Proneness, although research so far (Desmond, 1997; Matthews et al., 1997; Matthews & Desmond, 1995) shows that this dimension relates to subjective fatigue responses, but not to overall accident likelihood. Dislike of Driving also fails to relate to overall accident rate, but the validation evidence suggests, speculatively, that high Dislike drivers may be at risk when driving is particularly pressured or demanding (Matthews et al., 1998). More generally, driving might seem a less than ideal career choice for an individual of this type.

CONCLUSIONS: GUIDELINES FOR STRESS COUNTERMEASURES

Driver stress contributes to the human factors problems of designing safe transportation systems, problems that may be exacerbated by the introduction of new technology. Stress may influence the driver's processing of safety-critical information and the driver's strategies for handling the demands and opportunities of driving. As suggested by Stanton's (1994) CAT, intervening for stress is partly a question of design. The road layout and the vehicle should be designed not to elicit excessive stress. In addition, the traffic environment should be designed so that the stressed driver can still manage its demands effectively. However, design is not a complete solution to stress problems. Interventions must also be directed toward the user through training safety-enhancing appraisal and coping strategies and, in organizational settings, through using questionnaire assessment to ensure a good fit between personal characteristics and work environment. Our discussion of these principles suggests the following guidelines for tackling problems.

Distinguish Cognitive-Adaptive and Biocognitive Problems

As discussed by Matthews (chap. 1.1, this volume), "stress" phenomena require description at different levels. Driver stress research suggests that the behavioral consequences of safety follow primarily from the effects of maladaptive self-regulation on cognition and performance (the cognitive-adaptive level). However, some stress factors may influence performance via neural processes (the biocognitive level). This chapter has focused on the cognitive-adaptive level. We note in passing that the best solution to biocognitive impairments such as falling asleep, drink- and drug-related deficits, and serious age-related deficits is to keep the driver off the road, temporarily or on a permanent basis. However, compensation is sometimes possible for minor deficits, as when older drivers avoid driving at night.

Distinguish Qualitatively Different Stress Reactions

Effective intervention depends on discrimination of different stress reactions, such as emotional distress, reactive (anger-related) aggression, and fatigue. Identification of stress-related problems requires multidimensional assessment of stress reactions and choice of countermeasures related to the aspect involved. Interventions for distress should focus on maintenance of self-esteem and task-directed coping; interventions for aggression should focus on reduction of hostility toward others and confrontive coping; and interventions for fatigue should promote adaptive effort regulation and task motivation.

Design for Stress Explicitly

To some extent, good design principles, such as avoiding overload, contribute to designing systems to be stress resistant. However, the subtleties of stress reactions are such that explicit design for stress is often required. For example, minimizing overload may lead to maladaptive attentional allocation in the distressed driver and to effort reduction in the fatigued driver. Novel in-car systems should be evaluated in stressful as well as routine circumstances, paying special attention to the reactions of stress-vulnerable drivers.

Design for Variability in Workload

Research shows that stress effects on performance are highly sensitive to task demands: System weaknesses may not be apparent without sampling different workload levels. Designers should assess systems not just in stressed drivers, but also in conditions of unusually low- and high-task load. Occupational interventions should also take into account the typical workload of the driving task.

Work at the Level of the Individual Where Possible

Clearly, there is a role for group-level interventions such as advertising campaigns and driver instruction methods. However, interventions are often more effective when they are targeted at the vulnerabilities of the individual driver. Furthermore, interventions may be enhanced by engaging the active cooperation of the individual, for example, in installing warning or performance-monitoring systems or in tailoring training to the specific problems faced by the individual.

Direct Interventions Toward Explicit Criteria

Finally, all interventions should aim to meet specific outcome criteria. The over-riding criterion is normally increased safety, but other criteria such as the subjective well-being of the driver may also be important. Companies may benefit from drivers' enjoying their work more, as well as from greater safety. Occasionally, criteria may conflict: A traffic-calming intervention may be justified by its safety benefits even if it tends to be frustrating to drivers. Outcome criteria must be balanced against cost criteria: For example, the additional safety benefits deriving from a personalized intervention rather than a group intervention may or may not justify the extra expense involved.

REFERENCES

Bandura, A. (1986). *Social foundations of thought and action: A social cognitive theory.* Englewood Cliffs, NJ: Prentice-Hall.

Brand, J. L. (January, 1998). Driver out of the loop? *Ergonomics in Design,* 26–31.

Brenner, B., & Selzer, M. L. (1969). Risk causing a fatal accident associated with alcoholism: Psychopathology and stress: Further analysis of previous data. *Behavioral Science, 14,* 490–495.

Brown, I. D., & Groeger, J. A. (1988). Risk perception and decision taking during the transition between novice and experienced driver status. *Ergonomics, 31,* 585–597.

Corbett, C., & Simon, F. (1991). Police and public perceptions of the seriousness of traffic offences. *British Journal of Criminology, 31,* 153–164.

Crichton, D. (1991). *Risk management in motor vehicle fleets.* Perth: General Accident Fire and Life Assurance Corporation.

Delhomme, P., & Meyer, T. (1997). Control motivation and driving experience among young drivers. In E. Carbonell Vaya & J. A. Rothengatter (Eds.), *Traffic and transport psychology: Theory and application* (pp. 305–316). Amsterdam: Pergamon Press.

Desmond, P. A. (1997). *Fatigue and stress in driving performance.* Unpublished doctoral dissertation, University of Dundee.

Desmond, P. A., Hancock, P. A. & Monette, J. L. (1998). Fatigue and automation-induced impairments in simulated driving performance. *Transportation Research Record* No. 1628, 8–14.

Desmond, P. A., & Matthews, G. (1997). Implications of task-induced fatigue effects for in-vehicle countermeasures to driver fatigue. *Accident Analysis and Prevention, 29,* 513–523.

De Waard, D., van der Hulst, M., Hoedemaeker, M., & Brookhuis, K. A. (in press). Driver behaviour in an emergency situation in the Automated Highway System. *Transportation Human Factors.*

Dorn, L. & Matthews, G. (1995). Prediction of mood and risk appraisals from trait measures: Two studies of simulated driving. *European Journal of Psychology, 9,* 25–42.

Endsley, M. R. & Kiris, E. O. (1995). The out-of-the-loop performance problem and level of control in automation. *Human Factors, 37,* 381–394.

Hancock, P. A., & Parasuraman, R. (1992). Human factors and safety in the design of intelligent vehicle-highway systems (IVHS). *Journal of Safety Research, 23,* 181–198.

Hancock, P. A., Parasuraman, R., & Byrne, E. A. (1996). Driver-centered issues in automation for motor vehicles. In M. Mouloua & R. Parasuraman (Eds), *Automation and human performance: Theory and applications* (pp. 337–364). Lawrence Erlbaum Associates.

Hatakka, M. (1998). *Novice driver's risk- and self-evaluations: Use of questionnaires in traffic psychological research: Method development, general trends in four sample materials, and connections with behaviour.* Turku, Finland: Turku University.

Lansdown, T. C. (1997). Visual allocation and the availability of driver information. In E. Carbonell Vaya & J. A. Rothengatter (Eds.), *Traffic and transport psychology: Theory and application* (pp. 215–223). Amsterdam: Pergamon Press.

Mathew, R. J., Weinman, M. L., Semchuk, K. M., & Levin, B. L. (1982). Driving phobia in the city of Houston: A pilot study. *American Journal of Psychiatry, 139,* 1049–1051.

Matthews, G. (1996). Individual differences in driver stress and performance. In *Proceedings of the Human Factors and Ergonomics Society 40th annual meeting* (pp. 579–583). Santa Monica, CA: Human Factors and Ergonomics Society.

Matthews, G. (1997). The Big Five as a framework for personality assessment. In N. Anderson & P. Herriot (Eds.), *International handbook of selection and appraisal* (2nd ed., pp. 175–200). London: Wiley.

Matthews, G., Campbell, S. E., Desmond, P. A., Huggins, J., Falconer, S., & Joyner, L. A. (1999). Assessment of task-induced state change: Stress, fatigue and workload components. In M. Scerbo (Ed.), *Automation technology and human performance: Current research and trends* (pp. 199–203). Hillsdale, NJ: Lawrence Erlbaum Associates.

Matthews, G., & Desmond, P. A. (1995). Stress as a factor in the design of in-car driving enhancement systems. *Le Travail Humain, 58,* 109–129.

Matthews, G., Desmond, P. A., Joyner, L. A., & Carcary, B. (1997). A comprehensive questionnaire measure of driver stress and affect. In E. Carbonell Vaya & J. A. Rothengatter (Eds.), *Traffic and transport psychology: Theory and application* (pp. 317–324). Amsterdam: Pergamon Press.

Matthews, G., Dorn, L., Hoyes, T. W., Davies, D. R., Glendon, A. I., & Taylor, R. G. (1998). Driver stress and performance on a driving simulator. *Human Factors, 40,* 136–149.

Matthews, G., Quinn, C. E. J., & Mitchell, K. J. (1998). Rock music, task-induced stress and simulated driving performance. In G. B. Grayson (Ed.), *Behavioural reserach in road safety* (pp. 20–32). Crowthorne: Transport Research Laboratory.

Mayer, R. E., & Treat, J. R. (1977). Psychological, social and cognitive characteristics of high-risk drivers: A pilot study. *Accident Analysis and Prevention, 19,* 1–8.

McKenna, F. P., & Lewis, C. (1991). Illusory judgements of driving skill and safety. In G. B. Grayson & J. F. Lester (Eds.), *Behavioural research in road safety* (pp. 124–130). Crowthorne, England: Transport and Road Research Laboratory.

Norman, D. A. (1993). Cognition in the head and in the world: An introduction to the special issue on situated action. *Cognitive Science, 17,* 1–6.

Novaco, R. W. (1975). *Anger control: The development and evaluation of an experimental treatment.* Lexington, MA: Heath.

Parasuraman, R., Mouloua, M., Molloy, R., & Hilburn, B. (1996). Monitoring of automated systems. In R. Parasuraman & M. Mouloua (Eds.), *Automation and human performance: Theory and applications* (pp. 91–115). Mahwah, NJ: Lawrence Erlbaum Associates.

Reason, J., Manstead, A., Stradling, S., Baxter, J., & Campbell, K. (1990). Errors and violations on the roads: A real distinction? *Ergonomics, 33,* 1315–1332.

Richardson, J. (1994). The development of a driver monitoring system. In *Proceedings of the Conference on Driver Impairment, Fatigue and Driving Simulation.* Applecross, Western Australia: Promaco Conventions.

Sanders, M. S., & McCormick, E. J. (1987). *Human factors in engineering and design* (6th ed.). New York: McGraw-Hill.

Sarason, I. G., Sarason, B. R., Keefe, D. E., Hayes, B. E., & Shearin, E. N. (1986). Cognitive interference: Situational determinants and traitlike characteristics. *Journal of Personality and Social Psychology, 51,* 215–226.

Selzer, M. L., Payne, C. E., Westervelt, F. H., & Quinn, J. (1967). Automobile accidents as an expression of psychopathology in an alcoholic population. *Quarterly Journal of Studies on Alcohol, 28,* 505–516.

Selzer, M. L., & Vinokur, A. (1975). Role of life events in accident causation. *Mental Health and Society, 2,* 36–54.

Stanton, N. (1995). Ecological ergonomics: Understanding human action in context. In S. Robertson (Ed.), *Contemporary ergonomics 1995* (pp. 62–67). London: Taylor & Francis.

Summala, S. (1997). Hierarchical models of behavioural adaptation and traffic accidents. In E. Carbonell Vaya & J. A. Rothengatter (Eds.), *Traffic and transport psychology: Theory and application* (pp. 41–52). Amsterdam: Pergamon Press.

Ward, N. J., & Fairclough, S. (1997). Acceptance of driver status monitoring systems: Individual differences in subjective fatigue. In E. Carbonell Vaya & J. A. Rothengatter (Eds.), *Traffic and transport psychology: Theory and application* (pp. 225–235). Amsterdam: Pergamon Press.

Weiner, E. L., & Nagel, D. C. (Eds.). (1989). *Human factors in aviation.* New York: Academic Press.

Wells, A. (1997). *Cognitive therapy of anxiety disorders: A practice manual and conceptual guide.* Chichester, England: Wiley.

Wells, A., & Matthews, G. (1994). *Attention and emotion: A clinical perspective.* Hove: Lawrence Erlbaum Associates.

COMMENTARY

A Strategic Approach to Organizational Stress Management

Cary L. Cooper
Susan Cartwright
University of Manchester Institute of Science and Technology

Any organization that seeks to establish and maintain the best state of physical, mental, and social well-being of its employees needs to have policies and procedures that comprehensively address health and safety. These policies include procedures to manage stress, based on the needs of the organizations and its members, and must be regularly reviewed and evaluated.

There are a number of options to consider in looking at the prevention of organizational stress, which are termed primary (e.g., stressor reduction), secondary (e.g., stress management), and tertiary (e.g., employee assistance programs/counseling) levels of prevention and which address different stages in the stress process (Murphy, 1988).

PRIMARY PREVENTION

Primary prevention is concerned with taking action to modify or eliminate sources of stress inherent in the work environment and so reduce their negative impact on the individual. The "interactionist" approach to stress (e.g., Edwards & Cooper, 1990) depicts stress as the consequences of the "lack of fit" between the needs and demands of the individual and his/her environment. The focus of primary interventions is in adapting the environment to "fit" the individual.

Elkin and Rosch (1990) summarized a useful range of possible strategies to reduce workplace stressors: redesign the task; redesign the work environment; establish flexible work schedules; encourage participative management; include the employee in career development; analyze work roles and establish goals; provide social support and feedback; build cohesive teams; establish fair employment policies; share the rewards.

A number of general recommendations for reducing job stress have been put forward by National Institute of Occupational Safety and Health (NIOSH) in the National Strategy for the Prevention of Work Related Psychological Disorders (Sauter, Murphy, & Hurrell, 1990). A few of these recommendations are listed here:

> *Workload and work pace:* Demands (both physical and mental) should be commensurate with the capabilities and resources of workers, avoiding underload as well as overload. Provisions should be made to allow recovery from demanding tasks or for increased control by workers over characteristics such as work pace of demanding tasks.

> *Work schedule:* Work schedules should be compatible with demands and responsibilities outside the job. Recent trends toward flexitime, a compressed work week, and job sharing are examples of positive steps in this direction. When schedules involve rotating shifts, the rate of rotation should be stable and predictable.

> *Job future:* Ambiguity should be avoided in opportunities for promotion and career or skill development, and in matters pertaining to job security. Employees should be clearly informed of imminent organizational developments that may affect their employment.

> *Social environment:* Jobs should provide opportunities for personal interaction, both for purposes of emotional support and for actual help as needed in accomplishing assigned tasks.

> *Job content:* Job tasks should be designed to have meaning and provide stimulation and an opportunity to use skills. Job rotation or increasing the scope (enlargement/enrichment) of work activities are ways to improve narrow, fragmented work activities that fail to meet these criteria.

SECONDARY PREVENTION

Secondary prevention is essentially concerned with the prompt detection and management of experienced stress by increasing awareness and improving the stress-management skills of the individual through training and educative activities. Individual factors can alter or modify the way employees exposed to workplace stressors perceive and react to this environment. All individuals have their own personal stress threshold, which is why some people thrive in a certain setting and others suffer. This threshold varies between individuals and across different situations and life stages. Some key factors or "moderator" variables that influence individuals' vulnerability to stress include their personality, coping strategies, age, gender, attitudes, training, past experiences, degree of social support available from family, friends, and work colleagues.

Stress education and stress-management training serve a useful function in helping individuals to recognize the symptoms of stress and to overcome much of the negativity and stigma still associated with the stress label. Awareness activities and skills-training programs designed to improve relaxation techniques, cognitive-coping skills, and work/lifestyle modification skills (e.g., time-management courses or assertiveness training) have an important part to play in extending the individual's physical and psychological resources. They are useful in helping individuals deal with stressors inherent in the work environment that cannot be changed and have to be "lived with," such as, for example, job insecurity. Such training can also prove helpful to individuals in dealing with stress in other aspects of their life, such as nonwork related. However, the role of secondary prevention is essentially one of damage limitation, often addressing the consequences rather than the sources of stress, which may be inherent in the organization's structure or culture. These interventions are concerned with improving the "adaptability" of the individual to the environment. Consequently, this type of intervention is often described as "the Band-Aid" or inoculation approach. Because of the implicit assumption that the organization will not change but continues to be stressful, therefore, individuals must develop and strengthen their resistance to that stress. The continued demand for stress-management programs and the increasing stress levels reported in the literature (e.g., Cooper, 1995) are perhaps indicative of organizations' acceptance that stress is an inherent and enduring feature of the working environment, which has to be "coped and lived with."

TERTIARY PREVENTION

Tertiary prevention is concerned with the treatment, rehabilitation, and recovery process of those individuals who have suffered or are suffering from serious ill health as a result of stress. Interventions at the tertiary level typically involve the provision of counseling services for employee problems in the work or personal domain. Such services are provided by either inhouse counselors or outside agencies in the form of an employee assistance program (EAP). EAPs provide counseling, information, and/or referral to appropriate counseling treatment and support services (e.g., Berridge, Cooper, & Highley, 1997). Originally introduced in the United States to tackle alcohol-related problems, the concept of workplace counseling has since assumed a significantly wider focus. Such services are confidential and usually provide a 24-hour telephone contact line. Employees are able to voluntarily access these services or in some cases are referred by their occupational health function. The implementation of comprehensive systems and procedures to facilitate and monitor the rehabilitation and return to work of employees who have suffered a stress-related illness is another aspect of tertiary prevention.

There is some evidence to suggest that counseling is effective in improving the psychological well-being of employees and that it has considerable cost benefits. Based on reports published in the United States, figures typically show savings to investment rates of anywhere from 3:1 to 15:1 (Cooper & Cartwright, 1994). Such reports have not been without criticism, particularly as schemes are increasingly evaluated by the managed care companies responsible for their implementation, companies that are frequently under contract to deliver a preset dollar saving (Smith & Mahoney, 1989). However, evidence from established counseling programs, which have been rigorously evaluated, such as those introduced by Kennecott in the United States and the U.K. Post Office, resulted in a reduction in absenteeism in 1 year of approximately 60%. In the case of the British experience (Cooper & Sadri, 1991), measures taken pre- and postcounseling showed significant improvements in the mental health and self-esteem of the participating employees. However, there was no improvement in levels of employee job satisfaction and organizational commitment.

Like stress-management programs, counseling services can be particularly effective in helping employees deal with workplace stressors that cannot be changed and with nonwork-related stress (i.e., bereavement, marital breakdown, etc.), but that nevertheless tend to spill over into work life.

A COMPARISON OF INTERVENTIONS

Although there is considerable activity at the secondary and tertiary levels, primary- or organizational-level (stressor reduction) strategies are comparatively rare (Murphy, 1984). Organizations tend to prefer to introduce secondary- and tertiary-level interventions for several reasons: There are relatively more published data available on the cost-benefit analysis of such programs, particularly EAPs; those traditionally responsible for initiating interventions, the counselors, physicians, and clinicians responsible for health care, feel more comfortable with changing individuals than changing organizations (Ivancevich & Matteson, 1990); it is considered easier and less disruptive to business to change the individual than to embark on any extensive and potentially expensive organizational development program—the outcome of which may be uncertain (Cooper & Cartwright, 1994); they present a high profile means by which organizations can "be seen to be doing something about stress" and taking reasonable precautions to safeguard employees' health. This is likely to be important, not only in terms of the message it communicates to employees, but also to the external environment. This latter point is particularly important given the increasing litigation fears that now exist throughout the United States and Europe. It is not difficult to envisage that the existence of an EAP, regardless of whether an individual chooses to use it, may become an effective defense against possible legal action (Sutherland & Cooper, 2000).

Overall, evidence as to the success of interventions that focus at the individual level in isolation suggests that such interventions can make a difference in temporarily reducing experienced stress (Murphy, 1988). Generally, evidence as to the success of stress-management training is confusing and imprecise (Elkin & Rosch, 1990), which possibly reflects the idiosyncratic nature of the form and content of this kind of training. Some recent studies, which have evaluated the outcome of stress-management training, have found a modest improvement in self-reported symptoms and psychological indexes of strain (Reynolds, Taylor, & Shapiro, 1993; Sallis, Trevorrow, Johnson, Howell, & Kaplan, 1987; Sutherland & Cooper, 2000) but little or no change in job satisfaction, work stress, or blood pressure. Counseling appears to be successful in treating and rehabilitating employees suffering from stress, but as they are likely to re-enter the same work environment as job dissatisfied and no more committed to the organization than they were before, potential productivity gains may not be maximized. Firth-Cozens and Hardy (1992) suggested that as symptom levels reduce as a result of clinical treatment for stress, job perceptions are likely to become more positive. However, such changes are likely to be short-term if employees return to an unchanged work environment and its indigenous stressors. If, as has been discussed, such initiatives have little impact on improving job satisfaction, then it is more likely that the individual will adopt a way of coping with stress that may have positive individual outcomes, but negative implications for the organization, that is, taking alternative employment.

The evidence about the impact of health-promotion activities has reached similar conclusions. Research findings that have examined the impact of lifestyle and health habits have provided support that any benefits may not necessarily be sustained. Lifestyle and health habits appear to be effective in reducing anxiety, depression, and psychosomatic distress but do not necessarily moderate the stressor–strain linkage. According to Ivancevich and Matteson (1988), after a few years, 70% of individuals who attend such programs revert to the previous life-style habits.

Furthermore, as most stress-management programs or lifestyle-change initiatives are voluntary, this raises the issue as to the characteristics and health status of those participants who elect to participate. According to Sutherland and Cooper (1990), participants tend to be the "worried well" rather than the extremely distressed. Consequently, those employees who need most help and are coping badly are not reached by these initiatives. Also, sometimes access to such programs is restricted to managers and relatively senior personnel in the organization. Given that smoking, alcohol abuse, obesity, and coronary heart disease are more prevalent among the lower socioeconomic groups and that members of this group are likely to occupy positions in the organizational structure that they perceive afford them little or no opportunity to change or modify the stressors inherent in their working environment, the potential health of arguably the "most at risk" individuals is not addressed. Finally, the introduction of such programs in isolation may serve to enhance employees' perceptions of the organization as a caring employer, interested in their health and well-being, and may contribute to creating a "feel good" factor, which is unlikely to be sustained if the work environment continues to remain stressful.

Secondary- and tertiary-level interventions have a useful role to play in stress prevention, but as "stand-alone" initiatives, they are not the complete answer unless attempts are also made to address the sources of stress itself. Cardiovascular fitness programs may be successful in reducing the harmful effects of stress on the high-pressured executive, but such programs do not eliminate the stressor itself, which may be overpromotion or a poor relationship with his/her boss (Cooper & Cartwright, 1994). Identifying and recognizing the problem and taking steps to tackle it, perhaps by negotiation (i.e., a "front-end" approach), might arguably arrest the whole process. If, as has been discussed, experienced stress is related to the individual's appraisal of an event or situation, an organization can reduce stress by altering the objective situation, that is, by job redesign (Cummings & Cooper, 1979).

A further limitation of secondary- and tertiary-level interventions is that they do not directly address the important issue of control. This is particularly critical in terms of the health of blue-collar workers. Research has shown (Karasek, 1979) that jobs that place high demands on the individual but at the same time afford the individual little control or discretion (referred to as "decision latitude") are inherently stressful. Stress-management training may heighten the awareness of workers to environmental stressors that may be affecting their health, but because as individuals they may lack the resource or positional power to change them, the training may arguably even exacerbate the problem.

Again, there is not a great deal of research evidence that has evaluated the impact of primary-level interventions on employee health and well-being. However, what exists has been consistently positive, particularly in showing the long-term beneficial effects (Jackson, 1983; Kompier & Cooper, 1999; Quick, 1979).

Treatment may, therefore, often be easier than cure, but it may be only an effective short-term strategy. In focusing on the outcome or "rear end" of the stress process (i.e., poor mental and physical health) and taking remedial action to redress that situation, the approach is essentially reactive and recuperative rather than proactive and preventative.

In summary, secondary and tertiary levels of intervention are likely to be insufficient in maintaining employee health without the complimentary approach of primary/stressor reduction initiatives. Secondary- and tertiary-level interventions may extend the physical and psychological resources of the individual, particularly in relation to stressors that cannot be changed, but those resources are ultimately finite. Tertiary-level interventions, such as the provision of counseling services, are likely to be particularly effective in dealing with nonwork-related stress. Evidence researched on workplace-counseling programs indicates that approximately one quarter of all problems presented concerned relationships outside work (e.g., Berridge et al., 1997). Organizations considering counseling schemes should recognize that counseling is a highly skilled business and requires extensive training. It is important to ensure that counselors have recognized counseling skills training and have access to a suitable environment that allows them to contact this activity in an ethical and confidential manner.

A FRAMEWORK FOR ACTION: PLANNING AN INTERVENTION STRATEGY

Assess the Current Organizational Situation

A good starting point is to form a small policy or steering committee, under the leadership of a member of senior management and including personnel from the occupational health and human resource functions and union representatives. At some stage, it may be advisable to include an external consultant/agency with expertise in this field.

To formulate a policy and plan of action for addressing the issue of stress at work, the group should consider the current organizational situation in respect of the following agenda of items along the lines suggested by the U.K. government publication *Stress and Mental Health in the Workplace* (Cartwright & Cooper, 1995).

Ascertain the Existing Levels of Stress and Health in the Organization

What evidence/information do we have? What systems/procedures could we introduce to monitor these in the future? What kind of organizational culture do we have? In what ways might this contribute to stress? What is the existing role of the personnel/occupational health function in relation to stress? What health-related policies or training programs do we have at present? How adequate are our organizational policies on selection, sickness absence, and health and safety management in relation to this issue? How coordinated are they? What are the training needs of our personnel in respect of management and general life skills? Are they being adequately met? If there is no dedicated occupational health service, where do people currently go for help, and how much information can they easily get hold of? What support systems currently exist? What do we actively do to create a healthy and supportive work environment? What else could we do? What is the current level of commitment in the organization to stress-related initiatives? How could we raise the profile and importance of the issue?

Identify Organizational Stressors and Current Levels of Health by Conducting a Stress Audit

Tailoring action to suit the assessed needs of the organization is likely to be more effective than any "one size fits all" approach. Kompier and Cooper (1999) suggested that managing stress at work is more about stress prevention through primary diagnosis and intervention than about arming the individual to cope with a damaged organization. To target its resources in reducing stress in the workplace, an organization first needs answers to the following questions: What is the existing level of stress in the organization? What impact is it having on organizational functioning? Are job satisfaction and physical and psychological health better in some areas than others? How does the organization compare with other occupational groups and populations? Have we a problem, or do we anticipate a problem? If so, can we define the problem and what is causing it? What are the stressors? Are they departmental/site specific or organizational wide? Do they affect particular groups of workers, such as women as opposed to men, younger workers as opposed to old?

To obtain answers to these questions, organizations need to conduct prior diagnosis or stress audit. The Cooper and Marshall (1978) model usefully conceptualizes the sources of stress as falling in six broad categories: factors intrinsic to the job; role in the organization; relationships with others; career development and achievement; organizational structure; climate and culture; and home–work interface.

Factors Intrinsic to the Job

Each occupation has its own potential environmental sources of stress—aspects of the job itself, where it is performed, and what the task actually involves. Sources of stress intrinsic to the job include poor physical working conditions, noise, heat, bad lighting, and humidity, working long and/or unsociable hours, or shift work. Shift work has been shown to affect blood temperature, metabolic rate, blood sugar levels, mental efficiency, and motivation. It also affects sleep patterns and can place a strain on family and social life.

Work Overload/Underload

Work overload can be of two different types. *Quantitative* overload refers to simply having too much work. *Qualitative* overload refers to work that is too difficult for an individual. Problems of work overload often lead to long working hours with a resultant strain on physical and psychological health and personal relationships. It is a common myth that working longer hours means working more productively. Research has shown that working more than 48 hours per week doubles the risk of coronary heart disease. Similarly, having too little work to do leads to understimulation and boredom.

Repetitive and Understimulating Tasks

Employees who are most affected by stress at work tend to occupy jobs where there is little opportunity to exercise personal influence and control. Approximately 23% of European employees are in jobs that involve essentially repetitive tasks. Individuals working under conditions of high work pace/low decision latitude are the most at risk from stress.

Poorly Designed Equipment and Machines; Physical Danger, Chemicals, and Toxic Substances; Person–Job Mismatch

Individuals who find themselves in jobs that are ill suited to their skills, abilities, and training or that do not meet their needs and expectations are likely to experience stress. Eliminating or reducing stressors relating to factors intrinsic to the job may involve ergonomic solutions and may have implications for task redesign and work organization. Problems of work overload/underload may indicate a need to recruit, skills deficiencies, underutilization, or inappropriate selection decisions or delegation problems.

Role in the Organization

Three critical factors, role ambiguity, role conflict, and the degree of responsibility for others, are major sources of potential stress. Role ambiguity arises when an individual is uncertain about his/her work objectives and is unclear about the scope and responsibilities of the job. Role conflict exists when an individual is torn by conflicting job demands or finds him/herself having to do things that he/she does not believe in or feels comfortable about. In an organization, there are basically two types of responsibility: responsibility for people and responsibility for "things," such as budgets, plant, machinery. The stressful nature of having responsibility for others has grown in the economic climate of the 1990s, with so many industries facing cost-cutting exercises. As a result, many managers are caught between the two often conflicting goals—keeping personnel costs to a minimum and at the same time, looking after the welfare of subordinates in terms of job security and stability.

Eliminating and reducing role-related stress require clear role definitions and role negotiation.

Relationships at Work

Other people—and our varied encounters with them—can be major sources of stress and support. There are three critical relationships at work: relationships with superiors, with colleagues, with subordinates. Research studies have consistently shown that mistrust of co-workers is associated with high role ambiguity, poor communication, low job satisfaction, and poor psychological health. In organizations where there is strong social support from co-workers, the effects of job strain have been shown to be greatly reduced. Improving personal relationships in the workplace is a complex process with implications for a range of interpersonal skills training.

Career Development

Job insecurity and career development have increasingly become a source of stress during the 1980s and 1990s as more and more organizations have experienced major restructuring. Other sources of career stress include over-promotion, job relocation, and early retirement. The introduction of regular appraisals, the provision of retraining opportunities, career sabbaticals, and counseling are ways in which career stress may be reduced. As redundancy or job loss looks set to remain a feature of organizational life in the near future, the provision of outplacement facilities becomes increasingly important.

Organizational Structure and Culture

A fifth potential source of stress is simply being in the organization and the threat to the individual's freedom, autonomy, and identity that this situation poses. Different organizations develop different kinds of organizational culture, which influence the way in which work is organized and the behaviors expected of its members. Little or no participation in the decision-making process, lack of effective communication, and restrictions on behavior are examples of the types of stressors that fall into this category.

Home/Work Interface

Finally, managing the interface between work and home and its often conflicting demands, particularly in terms of time commitments, can also be a potential source of stress. With nearly 70% of married females and almost 40% of women with preschool children in the workplace in some European countries, the dual career couple and its attendant stresses are becoming key features in modern life. Managing the home/work interface can also be difficult for employees who work from home or individuals who may be experiencing financial difficulties or life crises such as bereavement.

Although the organization can arguably do little to directly alleviate the stress caused by domestic circumstances, it can offer support and perhaps provide counseling services. It can also help reduce the pressure on working parents by introducing more flexible working arrangements and adopting family-friendly employment policies.

Stress audits typically take the form of a self-report questionnaire administered on an organization-wide, site, or departmental basis. In addition to identifying the sources of stress at work and those individuals most vulnerable to stress, the questionnaire usually measures levels of employee job satisfaction, coping behavior, physical and psychological health comparative to similar occupational groups and industries. As well as directing organizational resources into areas where they are most needed and indicating the most appropriate type of intervention, audits provide a means of regularly monitoring stress levels and employee health over time and provide a baseline whereby subsequent interventions can be evaluated. The use of audits could also be extended to ascertain employee attitudes and perceived needs for secondary (stress management) and tertiary (EAPs) interventions to provide valuable information about the likely "takeup" rates of such programs before any expenditure is incurred. The Occupational Stress Indicator (OSI) devised by Cooper, Sloan, and Williams (1988) is one such instrument and is based on the Cooper–Marshall model. Many other questionnaires have been developed to assess job stress and health relationships. For example, the Occupational Stress Inventory (Osipow & Spokane, 1983) measures a wide range of job stressors, employee resources for coping with stress, and mental and physical strains. Other commonly used instruments include the Generic Job Stress Questionnaire developed by the National Institute for Occupational Safety and Health (NIOSH) and the Work Environment Scale (WES; Moos, 1981).

In smaller companies, information may be collected less formally through interviews and employee discussion groups or checklists. The agenda for such discussions/checklists might include the following main issues: job content and work scheduling; physical working conditions; employment terms and expectations of different employee groups; relationships at work; communication systems and reporting arrangements; "extra" organizational sources of stress (i.e., outside the workplace).

Another alternative is to ask employees to keep a stress diary for a few weeks in which they record any stressful events they encounter during the course of the day. Pooling this information on a group/departmental basis can be useful in identifying universal and persistent sources of stress.

Implementing and Evaluating the Intervention

Following stress assessment and problem identification, interventions need to be designed, installed, and evaluated. The intervention itself needs to be comprehensive and contain an element of stressor reduction, that is, of organizational change, in addition to any individual oriented elements.

Stressor reduction interventions require a knowledge of the dynamics of change processes in organizations, so that potential undesirable outcomes can be minimized. Stressor-targeted interventions must initially deal with the problem that organizations, like individuals, tend to resist change, and this inertia is often reinforced by the belief among many managers that the work environment does not contribute to employee distress (Cartwright, Cooper, & Murphy, 1995).

A key factor in primary prevention is the development of the kind of supportive organizational climate in which stress is recognized as a feature of modern industrial life and not interpreted as a sign of weakness or incompetence. Therefore, employees should not feel awkward about admitting to any difficulties they encounter.

Organizations need to take explicit steps to remove the stigma often attached to those with stress-related problems and to maximize the support available to staff. Some of the formal ways to accomplish this include informing employees of existing sources of support and advice in the organization, like occupational health specifically; incorporating self-development issues extending and improving the "people" skills of managers and supervisors so that they convey a supportive attitude and can more comfortably handle employee problems

Most important, there has to be demonstrable commitment to the issue of stress from both senior management and unions. This may require a move to more open communication and the dismantling of cultural norms in the organization that inherently promote stress among employees, such as cultural norms that encourage employees to work excessively long hours, take work home, or feel guilty about leaving "on time." Organizations with a supportive organizational climate are also proactive in anticipating additional or new stressors that may be introduced as a result of proposed changes, such as restructuring or new technology, and take steps to address this, perhaps by training initiatives or greater employee consultation.

Regardless of the specific intervention strategy selected, the involvement and participation of workers in the process is critical to its success. Lasting, effective change in organizations requires involvement of individuals at all levels in the organization. Once a strategy has been selected, it is important that a means of evaluating its impact are decided on from the outset.

ACKNOWLEDGMENT

This chapter is based on a European Union report for the European Foundation for the Improvement of Living and Working Conditions entitled *Stress Prevention in the Workplace: Assessing the Costs and Benefits to Organizations,* by C. L. Cooper, P. Liukennonen, & S. Cartwright, Dublin: EFILWC, 1997.

REFERENCES

Berridge, J. F. R., Cooper, C. L,. & Highley, C. (1997). *EAPs and workplace counselling.* Chichester: Wiley.

Cartwright, S., & Cooper, C. L. (1995). *Stress and mental health in the workplace: A guide for employers and organizational policymakers.* London: HMSO Department of Health.

Cartwright, S., Cooper, C. L., & Murphy, L. R. (1995). Diagnosing a healthy organization: A proactive approach to stress in the workplace. In G. P. Keita & S. Sauter (Eds.), *Job stress intervention: Current practice and future directions.* Washington, DC: APA/NIOSH.

Cooper, C. L. (1995) *Handbook of stress, medicine and health.* Boca Raton, FL: CRC Press.

Cooper, C. L., & Cartwright, S. (1994). Healthy mind, healthy organization—A proactive approach to occupational stress. *Human Relations, 47*(4), 455–471.

Cooper, C. L., & Marshall, J. (1978). *Understanding executive stress.* London: Macmillan.

Cooper, C. L., & Sadri, G. (1991). The impact of stress counselling at work. In P. L. Perrewe (Ed.), *Handbook of job stress* [Special Issue]. *Journal of Social Behaviour and Personality, 6*(7), 411–423.

Cooper, C. L., Sloan, S. J., & Williams, S. (1988). *Occupational Stress Indicator: Management guide.* Windsor, England: NFER Nelson.

Cummings, T., & Cooper, C. L. (1979). A cybernetic framework for the study of occupational stress. *Human Relations, 32,* 395–419.

Edwards, J., & Cooper, C. L. (1990). The person–environment fit approach to stress: Recurring problems and some suggested solutions. *Journal of Organizational Behavior, 11,* 293–307.

Elkin, A. J., & Rosch, P. J. (1990). Promoting mental health at the workplace: The prevention side of stress management. *Occupational Medicine: State of the Art Review, 5*(4), 739–754.

Firth-Cozens, J., & Hardy, C. E. (1992). Occupational stress, clinical treatment, change in job perception. *Journal of Occupational and Organizational Psychology, 65,* 81–88.

Ivancevich, J. M., & Matteson, M. T. (1988). Promoting the individual's health and well being. In C. L. Cooper & R. Payne (Eds.), *Causes, coping and consequences of stress at work* (pp. 267–300). New York: Wiley.

Ivancevich, J. M., & Matteson, M. T. (1990). *Stress at work.* Illinois: Scott Foresman.

Jackson, S. E. (1983). Participation in decision making as a strategy for reducing job related strain. *Journal of Applied Psychology, 68,* 3–19.

Karasek, R. A. (1979). Job demands, decision latitude and mental strain: Implications for job design. *Administrative Science Quarterly, 24,* 285–307.

Kompier, M., & Cooper, C. L. (1999). *Preventing stress, improving productivity.* London: Routledge.

Moos, R. H. (1981). *Work Environment Scale manual.* Palo Alto, CA: Consulting Psychologists Press.

Murphy, L. R. (1984). Occupational stress management: A review and appraisal. *Journal of Occupational Psychology, 57,* 1–15.

Murphy, L. R. (1988). Workplace interventions for stress reduction and prevention. In C. L. Cooper & R. Payne (Eds.), *Causes, coping and consequences of stress at work* (pp. 301–343). New York: Wiley.

Osipow, S. H., & Spokane, A. R. (1983). *A manual for measures of occupational stress, strain and coping.* Odessa, FL: Par.

Quick, J. C. (1979). Dyadic goal setting and role stress in field study. *Academy of Management Journal, 22,* 241–252.

Reynolds, S., Taylor, E., & Shapiro, D. A. (1993). Session impact in stress management training. *Journal of Occupational and Organizational Psychology, 66,* 99–113.

Sallis, J. F., Trevorrow, T. R., Johnson, C. C., Howell, M. F., & Kaplan, R. M. (1984). Worksite stress management: A comparison of programmes. *Psychology and Health, 1,* 237–255.

Sauter, S., Murphy, L. R., & Hurrell, J. J., Jr. (1990). A national strategy for the prevention of work related psychological disorders. *American Psychologist, 45,* 1146–1158.

Smith, D., & Mahoney, J. (1989, August). McDonnell Douglas Corporation's EAP produces hard data. *The Almacan,* 18–26.

Sutherland, V. J., & Cooper, C. L. (1990). *Understanding stress.* London: Chapman & Hall.

Sutherland, V. J., & Cooper, C. L. (2000). *Strategic stress management.* London: Macmillan.

1.10

The Future of Human Performance and Stress Research: A New Challenge

M. Ephimia Morphew

SPACEHAB/Johnson Engineering
NASA Johnson Space Center, Houston, Texas

Traditionally, the development of technology has taken the forefront in our efforts to sustain life underwater, in the air, in outer space, and in complex technological environments including nuclear power plants, flight decks, spacecraft, Antarctic and remote Earth stations, medical/emergency operating rooms, and air traffic control towers. This is visibly illustrated, for example, in both the space flight and aviation domains. Our efforts in human space flight were first dedicated solely toward developing the technology needed for rockets, propulsion, and engineering systems capable of delivering humans beyond Earth's atmosphere. Similarly, efforts in aviation were dedicated solely toward understanding the principles of aerodynamics and building aircraft structures capable of sustaining flight.

Achievement of these technological and engineering feats soon delivered us to an awareness of the physiological and biomedical stressors associated with operating in these environments. Myriad physiological conditions arising from space flight and high-performance flight include hypoxia, Gravity-induced Loss of Consciousness (G-LOC), Space Adaptation Sickness (SAS), bone demineralization, and cardiovascular deconditioning, among others. Accordingly, the development of biomedical and physiological countermeasures was undertaken in an effort to begin overcoming these stressors. These countermeasures have allowed us to sustain human presence in flight for increasing periods, as well as to participate in increasingly complex missions.

A new challenge has arisen, however, of which we now stand on the forefront. Our experience in long-duration space flight has revealed that it is often the human element, pertaining to poor human–technology interface design, team and interpersonal dynamics, and psychological factors that limit successful performance in space flight, rather than the purely techno-logical or extreme environmental factors that define the environment. Russ-ian experience in long-duration space flight has revealed that among the most critical problems facing humans in long-duration space flight, after the biomedical, are the psychological and psychosocial (Herring, 1997; Jdanov, personal communication, August 1996; Manzey & Lorenz, 1997; Manzey, Albrecht, & Fassbender, 1995; Morphew & MacLaren, 1997; O. Atkov, per-sonal communication, August 1996).

Just as human performance has come to be one of the most critical fac-tors in the space flight domain, so has it been revealed to be among the most critical factors in aviation operations. It has now been established that the majority of all commercial, military, aerial fire suppression, and general avi-ation aircraft accidents occur because of human and crew-related perform-ance factors (Boeing, 1994; Raymond & Moser, 1995; Ricketson, Brown, & Graham, 1980; Wiegmann & Shappell, 1997; Wiener, Kanki, & Helmreich, 1993; Yacavone, 1993).

The prominence of human performance and behavioral factors is not limited to the aviation and space flight domains, however, because similar findings are emerging in other challenging domains, including emergency medicine and anesthesia, underwater diving, firefighting, and industrial op-erations. An analysis of 2,000 critical incidents in the medical operating room, for example, revealed that 70% to 80% of medical mishaps were caused by human factors issues related to team/interpersonal interactions among the operating room team (Sexton et al., 1996; Williamson, Webb, Sellen, Runciman, & van der Walt, 1993). Other studies have revealed simi-lar findings involving the role of human error in intensive care units (Leape, 1997). Investigations of causal factors in anesthetic mishaps also revealed that an estimated 75% involve human error (Chopra, Bovill, Spierdijk, & Koornneef, 1992; Kumar, Barcellos, Mehta, & Carter, 1988). In yet another domain, analysis of fatal and serious occupational accidents in Finland re-vealed the predominant causal factors to be due to some form of human error (Salminen & Tallberg, 1996). In the underwater diving domain, find-ings are beginning to mirror those from aviation. Despite improvement in diver training, equipment design, and supporting technology, the number of diving fatalities has remained essentially unchanged (Divers Alert Net-work, 1996; Raglin, 1998). It is now becoming evident that the precipitating factors in up to 40% of both professional and recreational diver deaths are considered psychological in nature (Morgan, 1995; Raglin, 1998).

It is becoming clear through review of accident reports, empirical re-search, and operational findings, that human-related factors remain the most critical about the sustainment and optimization of human safety and functionability in these environments. These factors involve the human–technology interface, environmental, behavioral, cognitive, psychological, psychosocial, physical/physiological, and organizational issues associated with operating in high-stress conditions.

LOWERING THE BOUNDARIES BETWEEN DOMAINS

Although these and other challenging domains differ in their physical environmental attributes, the psychological factors and stressors they afford to those operating in them often remain similar. All demand performance abilities at times bordering on the limits of human capabilities and impose significant technological, psychological, psychosocial, and physiological stressors on those operating in them. All afford demanding human–technology, human-human, and human–environmental interfaces on which lives critically depend—and failure in any link of the chain can result in error, and ultimately, the loss of lives.

The psychological experience and manifestation of human panic, for example, are similar whether experienced in an aircraft emergency, diving accident, or firefighting emergency. Similarly, the effects of psychological panic on performance, regardless of the domain, produce similar decrements. Although certain factors can mediate the negative effects of stress on performance (leadership, expertise, motivation, personality, and team factors, among others; Bishop, Santy, & Faulk, 1998; Foushee & Manes, 1981; Freixanet, 1991; Heslegrave & Colvin, 1994; Kanki, Lozito, Foushee, 1989; Milgram, Orenstein, & Zafrir, 1989; Sandal, 1999; Vetter, 1994), these factors help to reduce those decrements, in whatever domain they are experienced.

Scuba Divers, Firefighters, and Athletes: A Common Thread

A prime example of the utility of unifying research and operational efforts across high-stress domains has been demonstrated by research conducted in an effort to reveal causal factors of Scuba diver fatalities that could not be meaningfully classified in terms of the cause of death (Morgan, 1995). Over 60% of all diver fatalities have been classified as caused by "drowning." This classification, however, serves as a poor indicator of what actually caused the drownings. The remaining 40% of diving fatalities have been classified as "unexplained" or owing to "undetermined" causes. The purpose of Morgan's research was to identify actual factors that led to the diver drownings and to provide insight into the remaining unclassified 40% of diver deaths.

Morgan's analysis of diver fatality reports illuminated the prevalence of a strange and persisting occurence. Deceased divers had been repeatedly found with their perfectly operating regulators (air source) removed from their mouths, with an adequate supply of air still available in their tanks. No other physiological cause of death (e.g., heart attack, stroke, injury, etc.)

aside from drowning as a result of removing the regulator from their mouths and inhaling water was found. This finding was puzzling, however, as there is no seemingly rational reason for a diver to remove a perfectly operating air source, given that doing so results in an inability to breathe and, ultimately, death. In an effort to understand this phenomenon, Morgan analyzed operational findings and conducted research in another domain in which personnel operate self-contained breathing apparatus (SCBA) under stressful conditions, namely firefighting. Operational and anecdotal experiences revealed similar findings, demonstrating that in the midst of battling a fire, firefighters, too, have been found with their SCBA removed, which resulted in their death (asphyxiation from smoke). The question remained as to why highly trained firefighters and scuba divers would remove their air source when their very lives depended on it.

To begin examining this question, a study undertaken by Morgan and Raven (1985) had 45 male firefighters undergo intensive treadmill exercise at an intensity equal to 80% of maximal oxygen consumption while wearing SCBA gear. The trait anxiety (an indication of the tendency for an individual to experience acute anxiety responses to stressors) of each firefighter was measured. Results revealed that some subjects ended the exercise session early because of a feeling of "respiratory distress" or because they "couldn't breathe" (even though this was not actually the case as measured by life-support equipment). One firefighter, who was convinced that he was suffocating, panicked and terminated the experiment by his own removal of the SCBA equipment. Interestingly, a significant percentage of the firefighters, who displayed panic and attempted to remove their air source (as a result of a feeling of suffocation, and accordingly, panic), were accurately predicted to do so by the trait anxiety measures administered in the beginning of the study.

Similar findings were replicated both in experiments conducted on scuba divers during training (Raglin, O'Connor, Carlson, & Morgan, 1996) and high-performance athletes (Kellerman, Winter, & Kariv, 1969; O'Connor, Raglin, & Morgan, 1996). Together, these findings indicated that athletes, scuba divers, and firefighters exercising to near-maximal capacity have been found to remove their SCBA under conditions of stress and panic brought on by the perception of respiratory distress/physical exertion, even though for the latter two this is the very equipment that sustains their lives in the operational environment.

Implications for Research

Deepening our understanding as well as our ability to predict human performance and behavior in stressful conditions depends, in part, on our ability to unveil the constancies of human behavior that emerges across domains and to approach human performance from a multidimensional perspective (as resulting from a culmination of factors). Perhaps one of the greatest obstacles to achieving this arises because findings from different domains pertaining to similar human psychological attributes are published in widely disbursed journals. What is needed is a forum that facilitates a comprehensive understanding of human performance and behavior across a variety of stressed conditions, in addition to addressing the factors that influence performance yet lie beyond the discipline of psychology (engineering, physiology, human factors). Such a forum should present operational and research findings that are of use, interest, and relevance to individuals tasked with optimizing performance, irrespective of their particular discipline.

Similarly, efforts are needed to facilitate an understanding between scientists and practitioners involving the many areas in the subdisciplines of psychology. Performance and behavior in stressed environments result from a culmination of psychological, cognitive, psychosocial, and human-technology interface factors. It is too often the case, however, that efforts are undertaken from the standpoint of one subdiscipline or area (e.g., perception, or cognition, or technology interface, or psychosocial), without heeding the other factors affecting performance. Without such a forum to bring these elements together, professionals are left on their own to (a) identify individually which factors outside their native domain of study are relevant, (b) search for and locate the relevant research (which lies in widely disbursed journals and publications), and (c) comprehend the findings that are often written specifically for experts in a particular area of specialty. This effort-intensive process can well be beyond the capacity of an individual researcher or scientist.

MOVING BEYOND THE LABORATORY INTO THE OPERATIONAL ENVIRONMENT

The need for stress research is becoming increasingly necessary for sustaining and improving performance in a variety of demanding environments, yet advancement in this area has a variety of unique challenges to overcome. Although laboratory and experimental research often strive to remove the complexities of the environment by isolating and examining certain variables and holding others constant, complexity is often a key defining trait of stressed operational environments. This presents a clear paradox in which the very attributes of the environment, which interact to affect performance, are removed in the laboratory setting. This is particularly critical for stressed and complex environments, in which multiple factors interact to influence performance. Additional methodologies and approaches must be developed to complement those of the laboratory, to allow for the examination of the complexity of factors mediating performance in these environments.

Because of the high level of performance demanded in these environments, the development of performance profiles to aid in understanding and predicting human performance capabilities and limitations is critical. Yet because of the methodological limitations and other difficulties associated with conducting research in operational domains, we often rely solely on laboratory experiments for building these profiles. Caution must be used, however, when using such profiles to approximate performance in actual high-stress operational environments. A variety of factors unique to operational environments exist (e.g., multiple stressors, reliance on team interactions, interface with technology, risk, complexity, situational ambiguity and uncertainty), which can lead to dissociations between actual operational findings and laboratory or experimental findings (Baddeley, 1972; Mears & Cleary, 1980; Weltman, Christianson, & Egstrom, 1970; McCarthy, 1998; Wilson, Skelly, & Purvis, 1989).

The ability of experimental laboratory research to capture the complexities of human performance in stressed and challenging operational environments is limited. This is particularly true for high-stress, high-risk environments. This fact has been recognized by a growing community of researchers in the area of Naturalistic Decision Making (NDM; Zsambok & Klein, 1997). Researchers in the NDM community have undertaken the challenge of beginning efforts toward developing methodologies for understanding human performance in the context of actual operational environments, rather than in stripped-down laboratories that remove much of the context, complexity, multidimensionality, and therefore ecological validity.

When stress research is undertaken in the laboratory setting, stress must often be induced by means of laboratory-generated "stressors," which do not psychologically or physically endanger the participant. These simulated stressors include time pressure, noise, vibration, low-grade aversive stimuli, monetary incentive, and in the case of military research, temperature extremes, and acceleration forces. Researchers have often used these elements in an effort to assess the effects of psychological and physical stress on performance in high-workload, time-pressured, high-risk, life-threatening conditions that can be experienced by pilots, astronauts, nuclear power plant operators, firefighters, military personnel, air traffic controllers, underwater divers, emergency room physicians, surgeons, emergency response personnel, and law enforcement officers among others. The importance of stress research is particularly apparent for these professionals because they are often required to perform at their cognitive, physical, emotional, psychosocial, and psychological limits, and the consequences of breakdowns in performance are severe.

The ability of laboratory experiments using these elements to study the effects of stressors on performance and behavior is limited. It has not been sufficiently demonstrated that the anxiety, fear, stress, uncertainty, risk, mental pressure, and arousal associated with performing in operational environments can be even generally approximated by laboratory-induced stressors. This is particularly true because operational environments are often characterized by multiple stressor interactions. Additionally, traditional laboratory stress research often uses inexperienced participants who have been shown to have different personality profiles, knowledge, experience, skill levels, and stress-coping abilities as compared with those who are actually selected and trained to operate in these stressful environments. Expertise and experience, for example, have been revealed to be a significant mediator of the effects of stress on performance under stressful conditions (Berkun, 1964; Falk & Bar-Eli, 1995; Fens & Jones, 1972; Hammerton & Tickner, 1967; Hancock, 1986; Lazarous & Ericksen, 1952; Mullins, Fatkin, Modrow, & Rice, 1995; O'Connor, Hallam, & Rachman, 1985; Ursin, Baade, & Levine, 1978; Weltman et al., 1970). Controlled, laboratory environments often fail to include many crucial factors that mediate stressors and stress responses (e.g., team performance/group dynamics, leadership, expertise). Correspondingly, findings concerning the effects of these laboratory stressors on performance and behavior are limited. As noted by Wilson et al. (1989), "No one has ever died in a simulator" (p. 1).

A study demonstrating this dissociation between laboratory and operational findings involved comparing pilots' responses during actual versus simulated flight emergencies (Wilson et al., 1989). Results revealed a 50% increase in pilot Heart Rate (HR) during the inflight emergency, while no marked increase occurred during the simulated flight emergency. Similarly, low Heart Rate Variability (HVR), an indicator of high stress or workload (Hancock, Meshkati, & Robertson, 1985; Heslegrave & Colvin, 1994; Wilson et al., 1989), was much more pronounced during the inflight emergencies than during the simulated (the inflight emergencies were serendipitous events in which pilot HR and HRV happened to be monitored). The notably higher HR and lower HVR found in pilots only during the actual inflight emergencies (both indicative of higher stress) were hypothesized to be due to their direct perception of a higher stress environment. Conversely, the authors hypothesized that a stress response was not observed in pilots to nearly the same degree during the simulated emergencies because pilots did not perceive the simulated emergencies as having the same level of risk and danger as the actual inflight emergencies.

In a study by Mears and Cleary (1980) that illustrates the dissociation between laboratory and operational findings, the authors were interested in examining the effect of anxiety on underwater performance. The method involved testing divers in both shallow (6m), and deep (30m) open water, both in day and night conditions. Because divers at both depths were exposed to equivalent environmental and physical factors, the authors purported that there should be no differences in performance between the 6m and 30m conditions.

Manual dexterity, cognitive performance, Heart Rate and self-rated anxiety were examined, along with other factors. Regarding manual dexterity, results revealed a 14%–30% decrement in the 6m condition, while a 45%–47% decrement was found in the 30m condition. Markedly lowered cognitive performance occurred in the 30m condition, which was accompanied by increased self-reported anxiety and higher Heart Rate. Because manual dexterity and cognitive performance should not have been affected differentially at these two depths, the authors concluded that the perception of risk and danger associated with deeper water resulted in the performance decrements observed.

Mears and Cleary's research was inspired by Baddeley's (1966) seminal study which revealed a marked difference between diver performance in the open sea, and performance in an equally pressurized chamber. Baddeley attributed the large performance decrements occurring in the open sea condition to be due to the stress associated with operating in the actual operational environment, as opposed to a chamber. Later investigations by Baddeley and Fleming (1967), and Baddeley, De Figueredo, Hawkswell, and Williams (1968), using numerous performance tests, further substantiated this finding and led to the conclusion that anxiety was the causal factor in open sea performance decrement.

A 1970 study by Weltman, Christianson, and Egstrom, which studied expert and novice underwater work performance in both a dive tank and in the open ocean, revealed a marked decrement in manual assembly time and problem-solving accuracy for the novices in the ocean condition, but not in the tank.

Pilot performance capabilities have also been shown to dissociate between laboratory and actual flight about tolerance to the acceleration forces associated with high-performance flight. Centrifuge technology is capable of simulating the precise gravity-profiles of any high-performance aircraft. Yet pilot gravity-force tolerance in actual flight is often notably lower than in the ground-based centrifuge cockpit, as indicated by the greater occurrence of gravity-induced loss of consciousness in the air (McCarthy, 1996). Similarly, a series of underwater diver studies (Baddeley, 1966; Baddeley & Flemming, 1967; Bowen, Anderson & Promisel, 1966) provided evidence for the dissociation between performance abilities in laboratory settings and actual operational environments. These studies examined underwater diver performance on identical tasks, in both laboratory (dry and wet pressure chambers, safe enclosed underwater sites near shore) and actual open sea conditions. Findings consistently revealed significant decrements in diver performance in the actual open sea conditions, which were not revealed in the land-based laboratory or in the close-to-shore protected conditions. Together, these studies illustrated that for high-stress environments in particular, laboratory research findings should be viewed as limited approximations of the performance and behavior that may be expected in actual operational environments.

HIGH-STRESS ENVIRONMENTS: WHERE TECHNOLOGY, PHYSIOLOGY, AND PSYCHOLOGY MEET

We must begin to view human performance as resulting from a culmination of interacting factors that do not heed disciplinary boundaries. Performance in the cockpit, as in firefighting operations or on battlefields, results from a culmination of the technological, physiological, and psychological factors interacting in the environment. Advancing our ability to understand, predict, and optimize human performance in such environments requires us to embrace experimentation founded on a truly interdisciplinary perspective. An example illustrative of this comes from the high-performance flight domain. One of the greatest challenges limiting human performance in the high-performance cockpit is the acceleration force associated with aerial combat maneuvering and high-performance flight. Up to 30% of all military pilots have experienced high acceleration forces leading to gravity (G)-induced Loss of Consciousness (G-LOC)—a condition that renders pilots totally incapacitated for a minimum of 15 seconds (Gawron, 1997).

Accordingly, considerable efforts have been dedicated to understanding this physiological phenomenon from a seemingly appropriate physiological perspective. Studies have been undertaken by using human centrifuges, in which the physiological mechanisms that govern the occurrence of this phenomenon were systematically examined. As our knowledge of this physiological phenomenon grew, however, a dissociation was found between pilots' physiological tolerance to G-force in the centrifuge and their tolerance in actual flight (McCarthy, 1996, 1998). After years of studying both research and operationally oriented findings related to pilot tolerance of G-force, McCarthy found that while pilots could routinely withstand over nine sustained Gz in centrifuge training on the ground, pilots in the air were still experiencing G-LOC at significantly lower G-levels. This raised two fundamental questions: What was accounting for the difference between pilot tolerance in the air and on the ground? Was this indicative of the limited ability of laboratory-based studies to provide insight into human performance in high-stress operational environments?

McCarthy attributed this discrepancy to the psychological factors that are present to a greater degree in actual flight, namely, high workload, attentional demands, and psychological stress. During inflight aerial combat maneuvering, pilots pulling high Gz are burdened with great attentional demands in addition to extremely high workload. Stress factors present in actual flight, which McCarthy thought contributes significantly to pilots' reduced tolerance to G, include risk, stress, and anxiety. These factors lower G tolerance in the air by lowering pilots' ability to devote their full attentional resources to engaging in anti-G countermeasure maneuvers. What was once thought to be a purely physiological phenomenon, is now known to be mediated significantly by psychological processes.

This inability of laboratory-based physiological research to capture the complexities of the cockpit environment in their entirety, illustrates that physiological performance in the cockpit is not dictated by physiological factors alone. We must therefore begin to view human performance and behavior from a truly multidimensional perspective. Many researchers, however, still approach problems from within the boundaries of their own discipline or area of specialization, without knowledge, awareness, or inclusion of the other factors that mediate performance. As this example illustrates, stress and other psychological factors can, and often do, mediate physiological tolerances in stressful environments. This relation is reciprocal, however, as psychologists studying pilot performance in the high-performance cockpit need a working/base level knowledge of the physiological and physical factors that influence pilot performance and mediate their stress reactions.

TOWARD INTERDISCIPLINARY EFFORTS

We are entering a new age of discovery where scientists are addressing increasingly complex problems that demand multidisciplinary methods and expertise (Bertenthal, 1998). It is vital for workers to understand and enhance human performance and behavior—with all its complexities—and to venture beyond their own area of specialty into other complimentary domains and disciplines to gain a comprehensive understanding of the properties that mediate it.

Because multiple factors inherent in stressed environments mediate performance, it is not sufficient to gain an understanding of performance simply by placing findings from individual studies (each examining an isolated effect of a single factor on stress) together to form a comprehensive understanding. Factors and dynamics present in actual environments mediate performance but are not present in the controlled laboratory environment. Efforts have been traditionally directed toward examining how isolated aspects of the environment (noise, time pressure, acceleration force, fatigue, etc.) affect certain aspects of performance including perception, cognition, human error, decision making, physiological/physical abilities, or interpersonal/team interactions. Stressed and complex operational environments, however, often present multiple stressors that interplay and collectively mediate performance. Accordingly, our ability to comprehend and predict human performance and behavior depend not only on our ability to understand how singular environmental attributes affect performance, but also how performance is mediated by a collective of real-life environmental attributes characterizing the operational environment.

OVERCOMING OUR CHALLENGES

Optimizing human performance and behavior in complex and high-stress operational environments requires viewing the operators and their environment as a system whose performance is mediated by a multitude of factors. Efforts should be taken to unite scientists conducting similar research in different high-stress domains. Although differences exist in the external environmental attributes of these domains, they often present similar psychological stressors to those operating in them. Accordingly, particular elements of human behavior in them are very often constant, stable, and predictable across domains. To facilitate recognition and understanding of this, efforts must be taken to create a forum that brings together research on stress and human performance from different high-stress domains; presents human performance and behavior issues from a multidimensional perspective; and does so in a way that researchers and practitioners, irrespective of which discipline they specialize in, can understand and use the information presented.

As we enter the new millennium, behavioral and performance research are taking their place at the forefront of science. Findings are revealing the human factors associated with operating in challenging and stressful environments to be among the most critical for maintaining the safety, health, and performance of those operating in them. Accordingly, our increased presence in these environments is dependent on our ability to deepen our understanding of human behavior in stressful conditions. In the coming millennium, we will begin to see this for the true interdisciplinary endeavor that it is.

REFERENCES

Baddeley, A. D. (1966). Influence of depth on the manual dexterity of free divers: A comparison between open sea and pressure chamber testing. *Journal of Applied Psychology, 50,* 81–85.

Baddeley, A. D. (1972). Selective attention and performance in dangerous environments. *British Journal of Psychology, 63,* 537–546.

Baddeley, A. D., & Fleming, N. C. (1967). The efficiency of divers breathing oxy-helium. *Ergonomics, 10,* 311–319.

Baddeley, A. D., De Figueredo, J. W., Hawkswell, C., & Williams, A. N. (1968). Nitrogen nar-
cosis and performance underwater. *Ergonomics, 11,* 157–164.

Berkun, M. M. (1964). Performance decrement under psychological stress. *Human Factors, 6,*
21–30.

Bertenthal, B. (1998). A new age of discovery. *APA Monitor, 29*(5), p. 17.

Bishop, S. L., Santy, P. A., & Faulk, D. (1998). Team dynamics analysis of the huautla cave diving expedition: A case study. *Journal of Human Performance in Extreme Environments, 3*(1),
37–41.

Boeing. (1994). *Statistical summary of commercial jet aircraft accidents: Worldwide operations, 1959–
1993.* (Boeing Airplane Safety Engineering Report B-210B). Seattle, WA: Boeing Commercial Airplane Group.

Bowen, H. N., Anderson, B., & Promisel, D. (1966). Studies of divers' performance during
the Sealab II project. *Human Factors, 8,* 183–199.

Chopra, V., Bovill, J. G., Spierdijk, J., & Koornneef, F. (1992). Reported significant observations during anaesthesia: A prospective analysis over an 18-month period. *British Journal of
Anaesthesia, 68,* 13–17.

Divers Alert Network. (1996). *Report on diving accidents and fatalities.* Durham, NC: Duke University Medical Center.

Falk, B., & Bar-Eli, M. (1995). The psycho-physiological response to parachuting among
novice and experienced parachutists. *Aviation, Space and Environmental Medicine, 66,* 114–
117.

Fens, W. D., & Jones, G. B. (1972). Individual differences in physiologic arousal and performance in sports parachutists. *Psychosomatic Medicine, 34,* 1–8.

Freixanet, M. G. (1991). Personality profile of subjects engaged in high physical risk sports.
Personality and Individual Differences, 12, 1087–1093.

Foushee, H. C. & Manes, K. L. (1981). Information transfer within the cockpit: Problems in
intracockpit communication. In C. E. Billings & E. S. Cheaney (Eds.), *Information transfer
problems in the aviation system* (NASA Tech. Paper 1875). Moffett Field, CA: NASA Ames Research Center.

Gawron, V. (1997). High-G environments and the pilot. *Ergonomics in Design, 5*(2), 18–23.

Hammerton, M., & Tickner, A. H. (1967). *Tracking under stress* (Medical Research Council
Rep. No. APRC 67/CS), 10(A).

Hancock, P. A. (1986). The effect of skill on performance under an environmental stressor.
Aviation, Space and Environmental Medicine, 57, 59–64.

Hancock, P. A., Meshkati, N., & Robertson, M. M. (1985). Physiological reflections of mental
workload. *Aviation, Space and Environmental Medicine, 56,* 1110–1114.

Herring, L. (1997). Astronaut draws attention to psychology, communication. *The Journal of
Human Performance in Extreme Environments, 2*(1), 42–47. (Reprinted from APS Observer,
1995, September.)

Heslegrave, R. J., & Colvin, C. (1994, September). *Selection of personnel for stressful occupations:
The potential utility of psychophysiological measures as selection tools* (Rep. No. 1035). U.S. Army
Research Institute for the Behavioral and Social Sciences, pp. 1–58.

Kanki, B. G., Lozito, S., & Foushee, H. C. (1989). Communication indexes of crew coordination. *Aviation, Space, and Environmental Medicine, 60,* 56–60.

Kellerman, J. J., Winter, I., & Kariv, I. (1969). Effect of physical training on neurocirculatory
asthenia. *Israeli Journal of Medicine, 5,* 947–949.

Kumar, V., Barcellos, W. A., Mehta, M. P., & Carter, J. G. (1988). An analysis of critical incidents in a teaching department for quality assurance: A survey of mishaps during anesthesia. *Anaesthesia, 43,* 879–883.

Lazarus, R. S., & Ericksen, C.W. (1952). Effects of failure stress on skilled performance. *Journal of Experimental Psychology, 43,* 100–105.

Leape, L.L. (1997). A systems analysis approach to medical error. *Journal of Human Performance in Extreme Environments, 4*(1), 14–20. [Reprinted from *Journal of Evaluation in Clinical Practice, 3*(3)].

Manzey, D., Albrecht, S., & Fassbender, C. (1995). Psychological countermeasures for extended manned space flights. *Journal of Human Performance in Extreme Environments, 1*(2), 66–84. [Reprinted from *Acta Astronautica, 35,* (4/5)].

Manzey, D., & Lorenz, B. (1997). Human performance during prolonged space flight. *Journal of Human Performance in Extreme Environments, 2*(1), 68.

McCarthy, G. W. (1996). G-induced loss of consciousness (GLOC): An aviation psychology challenge. *Journal of Human Performance in Extreme Environments, 1*(2), 42–43.

McCarthy, G. W. (1998, May 17–21). *Operational relevance of aeromedical laboratory research. The Aerospace Medical Association 69th Annual Scientific Meeting Program Guide* (Abstract No. 24, 57). Seattle, WA: The Aerospace Medical Association.

Mears, J. D., & Cleary, P. J. (1980). Anxiety as a factor in underwater performance. *Ergonomics, 23*(6), 549–557.

Milgram, N. A., Orenstein, R., & Zafrir, E. (1989). Stressors, personal resources, and social supports in military performance during wartime. *Military Psychology, 1*(4), 185–199.

Morgan, W. P. (1995). Anxiety and panic in scuba divers. *Sports Medicine, 20*(6).

Morgan, W. P., & Raven, P. B. (1985). Prediction of distress for individuals wearing industrial respirators. *American Industrial Hygiene Association Journal, 46,* 363–368.

Morphew, M. E., & MacLaren, S. (1997). Blaha suggests need for future research on the effects of isolation and confinement. *Journal of Human Performance in Extreme Environments, 2*(1), 52–53.

Mullins, L. L., Fatkin, L. T., Modrow, H. E., & Rice, D. J. (1995). The relationship between cognitive performance and stress perceptions in military operations. *Proceedings of the Human Factors & Ergonomics Society Annual Meeting, 39,* 868–872.

O'Connor, K., Hallam, R., & Rachman, S. (1985). Fearlessness and courage: a replication experiment. *British Journal of Psychology, 76,* 187–197.

O'Connor, P. J., Raglin, J. S., & Morgan, W. P. (1996). Psychometric correlates of perception during arm ergometry in males and females. *International Journal of Sports Medicine, 17,* 462–466.

Raglin, J. S. (1998). Psychobiological antecedents of panic in scuba diving. *Journal of Human Performance in Extreme Environments, 3*(1), 26–29.

Raglin, J. S., O'Connor, P. J., Carlson, N., & Morgan, W. P. (1996). Responses to underwater exercise in scuba divers differing in trait anxiety. *Undersea and Hyperbaric Medicine, 23,* 77–82.

Raymond, M. W., & Moser, R. (1995). Aviators at risk. *Aviation, Space and Environmental Medicine, 66*(1), 35–39.

Ricketson, D. S., Brown, W. R., & Graham, K.N. (1980). 3W approach to the investigation, analysis, and prevention of human-error aircraft accidents. *Aviation, Space and Environmental Medicine, 51,* 1036–1042.

Sandal, G. M. (1999). The effects of personality and interpersonal relations on crew performance during space simulation studies. *Journal of Human Performance in Extreme Environments, 4*(1), 43–50.

Salminen, S., & Tallberg, T. (1996). Human errors in fatal and serious occupational accidents in Finland. *Ergonomics, 39,* 980–988.

Sexton, B., Marsch, S., Helmreich, R., Betzendoerfer, D., Kocher, T., & Scheidegger, D. (1996). Jumpseating in the Operating Room. *Journal of Human Performance in Extreme Environments, 1*(2), 36.

Ursin, H., Baade, E., & Levine, S. (1978). *Psychobiology of stress: A study of coping men.* New York: Academic Press.

Vetter, C. (1994, November). Into the abyss: An ill-fated expedition into one of the world's deepest underwater caves. *Journal of Human Performance in Extreme Environments, 2*(1), 13–21. (Reprinted from *Outside Magazine.*)

Weltman, G., Christianson, R. A., & Egstrom, G. H. (1970). Effects of environment and experience on underwater work performance. *Human Factors, 12*(6), 587–598.

Wiegmann, D. A., & Shappell, S. A. (1997). Human factors analysis of postaccident data: Applying theoretical taxonomies of human error. *International Journal of Aviation Psychology, 7*, 67–81.

Wiener, E. L., Kanki, B. G., & Helmreich, R. L. (1993). *Cockpit resource management.* New York: Academic Press.

Williamson, J. A., Webb, R. K., Sellen, A., Runciman, W. B., & van der Walt, J. H. (1993). Human failure: An analysis of 2,000 incident reports. *Anaesthesia Intensive Care, 21*, 678–683.

Wilson, G., Skelly, J., & Purvis, B. (1989, October). Reactions to emergency situations in actual and simulated flight. In *Proceedings of the Aerospace Medical Panel Symposium*, 1–13. The Hague, Netherlands.

Yacavone, D. W. (1993). Mishap trends and cause factors in naval aviation: A review of Naval Safety Center data, 1986–1990. *Aviation, Space and Environmental Medicine, 64*, 392–395.

Zsambok, C. E., & Klein, G. (1997). *Naturalistic decision making.* Mahwah, NJ: Lawrence Erlbaum Associates.

PART II

Workload

THEORY

2.1

Stress, Workload, and Boredom in Vigilance: A Problem and an Answer

Mark W. Scerbo
Old Dominion University

Imagine that you have been asked to monitor a display for the occasional appearance of critical targets. The distinction between the targets and other elements on the display is subtle, yet still clearly perceptible. When you see a target, you are to press a button indicating that the target has been detected. What could be more simple?

At the outset, the task may indeed be simple. However, as time progresses, the task quickly becomes more difficult, boring, and stressful. Moreover, the ability to perform the task also deteriorates. Why is this so?

VIGILANCE AND WORKLOAD

As it turns out, this problem has been studied experimentally for nearly 50 years. Research on vigilance, or the ability of individuals to maintain attention and respond to stimuli over extended periods, began with Mackworth (1948), who sought to understand why radar operators were missing critical signals on their displays. In his initial investigation, Mackworth found that an observer's ability to detect critical signals declined over a watch-keeping session. Numerous studies since then have also shown that the quality of vigilance performance is fragile and deteriorates over time (see, for example, Davies & Parasuraman, 1982; Warm, 1984). This decrement in performance can manifest itself as drop in accuracy or an increase in response time and is referred to as the decrement function (Dember & Warm, 1979) or the vigilance decrement (Davies & Parasuraman, 1982).

The results of many recent studies on sustained attention have shown that not only does performance deteriorate, but participants also find such tasks to be very demanding (see, for example, Warm, Dember, & Hancock, 1996). For instance, Dittmar, Warm, Dember, and Ricks (1993) asked participants to monitor a display for changes in either the size or duration of stimuli. In addition, participants were asked to provide subjective estimates of the workload associated with the vigil by means of the NASA Task Load Index (TLX: Hart & Staveland, 1988). The TLX is an instrument that provides a reliable index of overall workload (test–retest correlation = .83). It includes a set of rating scales that reflect the relative contributions of six individual sources of workload: effort, frustration, performance, mental demand, physical demand, and temporal demand. The workload ratings obtained by Dittmar et al. fell at the high end of the scale and were primarily attributable to the frustration and mental demand subscales. More important, other investigators have shown a direct relation between workload and factors known to affect the quality of sustained attention. For instance, Gluckman, Warm, Dember, Thiemann, and Hancock (1988) found that more salient signals resulted in better performance and lower levels of workload. Similarly, Becker, Warm, Dember, and Hancock (1991) observed superior performance and lower workload when participants were provided with knowledge of results about their correct detections and false alarms.

STRESS AND WORKLOAD

It is important to understand that the changes made to psychophysical task parameters in these studies moderated workload, but did not result in substantially lower workload ratings. This suggests that some other aspect of the vigilance setting is contributing to the perception of workload. One possibility may be stress. Several researchers indicated that participants often report increased levels of stress on completion of a vigilance task (Galinsky, Rosa, Warm, & Dember, 1993), and others have shown that vigilance can produce physiological changes consistent with a stress response (Frankenhaeuser, Nordheden, Myrsten, & Post, 1971). Moreover, Hancock and Warm (1989) have argued that the need to maintain high levels of vigilance performance may be sufficient to produce a stress response in and of itself. If performing a vigilance task is indeed stressful, then what is the source of the stress?

A DYNAMIC MODEL OF STRESS AND VIGILANCE

Hancock and Warm (1989) have described a model of stress and sustained attention in which the rate of information and the structure of information (i.e., the meaning sought by the participant) combine to determine attentional resource capacity in vigilance. Performance is expected to be normal when the combination of information rate and structure is optimized. Minor variations in task load should have little effect on capacity demand and subsequent performance. Greater deviations from the optimal levels of these two determinants, however, require the individual to expend additional resources. Either increases or decreases beyond this "normative zone" initially result in lower levels of comfort. Further deviations, however, are likely to exceed the individual's range of psychological adaptability and to increase subjective perceptions of task load. At this point, the individual should perceive the task as stressful. Deviations beyond this point, however, compromise the individual's physiological range of adaptability and result in performance deficits.

The Hancock and Warm (1989) model suggests that psychological and physical stability requires an optimal degree of both information structure and rate of information. In most vigilance situations, however, the operator does not have control over the structure or rate of information. Thus, without the ability to control these factors, it is unlikely that an optimal level can be reached or sustained. Thackray (1981) also argued that monitoring a monotonous display produces stress because it requires the observer to maintain high levels of attention without the ability to control the events that occur.

Recently, Scerbo, Greenwald, and Sawin (1993) addressed this issue in an experiment where participants were given control over one aspect of the display. Specifically, they were permitted to adjust the rate of stimulus occurrences or event rate. Research has shown that event rate is an important parameter in vigilance and that, in general, increases in event rate result in poorer performance (see Warm & Jerison, 1984). Moreover, Parasuraman (1985) has shown that higher event rates make greater demands on attentional resources, and Galinsky, Dember, and Warm (1989) reported that workload scores on the TLX were substantially higher in the context of a high event rate.

Scerbo et al. (1993) allowed their participants to switch among five different event rates at any point during the vigil. The performance of this group was compared to a yoke control group who received the identical pattern of event rate changes, but who could not affect the event rate themselves; and a group who received a schedule of event rate changes determined at random. All participants were also asked to provide subjective estimates of workload on the NASA TLX. Scerbo et al. expected that giving participants control over the events would enable them to set an optimal rate for themselves and would therefore lead to superior performance and lower workload.

The results showed that those participants who could control the event rate outperformed those who could not. In fact, this group showed no decrement in performance even though the mean event rate for the session was fairly high at almost 40 events per minute. These performance differences, however, did not affect the workload scores. Consistent with the findings of Galinsky et al. (1989), all groups reported similar and high levels of workload.

Initially, a result such as this might seem at odds with the model offered by Hancock and Warm (1989). According to their model, giving participants control over the rate of information should help them to optimize performance and thereby lower their stress. Scerbo et al. (1993) found those who could control the event rate performed better than those who could not, but it did not affect the workload scores. Recall, however, that the Hancock and Warm model predicts optimal performance and low stress when the *combination* of information rate and information structure is optimized. Thus, it is possible that allowing participants to adjust the event rate was insufficient to relieve stress because the manipulation did nothing to affect the structure of the information. As Thackray (1981) might argue, manipulating event rate did not affect the task requirement to maintain attention throughout the vigil.

In another study, Scerbo and Sawin (1994) gave their participants the ultimate control over the need to remain alert during the course of a vigil. The participants were told that they could quit the task whenever they wanted. Although the primary goal of this experiment was to examine the nature of boredom, these investigators used a traditional vigilance task to induce boredom.

BOREDOM IN VIGILANCE

Although many definitions of boredom have been offered over the years, boredom is usually associated with feelings of increased constraint, repetitiveness, unpleasantness, and decreased arousal (Geiwitz, 1966). Hill and Perkins (1985) proposed a model of boredom with cognitive, affective, and psychophysiological components. According to this model, individuals seek stimulus variety. Thus, if faced with repetitive, unchanging stimuli, they become bored and frustrated. Moreover, this process is moderated by cognitive and affective factors. Consequently, boredom is only perceived if one construes the situation as monotonous.

Several researchers (Mackie, 1987; Scerbo, 1998; Scerbo, Greenwald, & Sawin, 1992) and countless experimental participants have commented on the boring nature of vigilance tasks. Other investigators have discussed the relation between boredom and stress (Mackie, Wylie, & Smith, 1985; O'Hanlon, 1981; Thackray, 1981). In general, high levels of boredom coupled with the need to remain vigilant are often accompanied by stress.

STIMULUS VARIETY AND SATIATION IN BOREDOM

In an effort to determine the source of task-related boredom, Scerbo and Sawin (1994) considered the role of satiation. It has been suggested that individuals become bored when they must continue to work on a task beyond the point at which they would normally reject it (Barmack, 1939). The point at which they no longer wish to continue is called the satiation point. Thus, satiation should be a precursor to boredom. To test this idea, Scerbo and Sawin asked participants to rate their perception of boredom for a given task on several Likert scales that addressed both situational characteristics and feelings associated with boredom.

To provide some means of comparison, Scerbo and Sawin (1994) had participants perform and rate two different tasks. Stimulus variety is thought to be an important element in several theories of boredom (Hill & Perkins, 1985; O'Hanlon, 1981). In general, stimuli that are very similar or activities that are highly repetitive lead to feelings of boredom. Accordingly, two tasks were selected that differed primarily in degree of stimulus variety. The first was typical of most vigilance tasks. Participants monitored the repetitive presentation of a pair of bars on their video display terminal (VDT) for occasional increases in the height of the bars. The second task required participants to monitor a kaleidoscope-like screen-saver program that drew random configurations of colored lines. Their objective was to look for patterns that were predominantly white. The participants were told to do the best that they could on each task, but if they reached a point where they no longer wished to continue they could stop.

Scerbo and Sawin (1994) argued that if a lack of stimulus variety leads to boredom then the participants should be willing to terminate the vigilance task sooner than the kaleidoscope task. The investigators did not, however, expect the boredom ratings for the two activities to differ. Instead, they argued that individuals who can terminate the activity at will should provide similar estimates of boredom for each task because, according to Barmack (1939), the participants should stop performing each task when they reach their satiation point, that is, before they become bored.

Scerbo and Sawin (1994) found that the participants were willing to spend comparable time on either task if it was their first activity. On the other hand, they spent about 60% less time on the vigilance task if it followed the kaleidoscope task. Contrary to expectations, the participants did not rate the two tasks similarly. Instead, they felt significantly more stressed, had more difficulty concentrating, and wished the task would end sooner after performing the vigilance as compared with the kaleidoscope task. Moreover, their ratings on the boredom subscale were quite high for the vigilance task, but only moderately so for the kaleidoscope task. These ratings were obtained in the absence of any order effects. Thus, this pattern of results appears to contradict Barmack (1939), who argued that individuals should stop performing a task when they became satiated, before the onset of boredom.

Scerbo and Sawin (1994) believed, however, that a methodological weakness may have contributed to the discrepant ratings for the two tasks. Specifically, if some participants felt compelled to continue to work beyond their satiation point to please the experimenter, then it is quite likely that they would report higher levels of boredom. A postexperimental survey of their participants revealed that 17% did feel constrained by the procedures but continued to work at the task even though they had become bored. Indeed, this group of participants reported higher levels of boredom on the boredom subscale than those who indicated they did not feel constrained by the procedure.

To examine further the role of constraint in boredom, Prinzel, Sawin, and Scerbo (1995) repeated the Scerbo and Sawin (1994) experiment, but in this case the participants were not allowed to terminate the task. Instead, the participants were asked to complete each task. The durations for the tasks, however, were dissimilar. Specifically, the task durations were derived from the mean session times for the four levels of task and order obtained in the Sawin and Scerbo experiment. The results from this second study showed that the vigilance as compared with the kaleidoscope task was rated significantly higher in stress, boredom, irritation, sleepiness, difficulty concentrating, desire for the task to end sooner, and significantly lower in alertness and relaxation. Moreover, Prinzel et al. found that the overall boredom ratings were higher for the vigilance task, but significantly more so when the vigil followed the kaleidoscope task. Thus, when participants could not terminate the task at will and could compare their experiences on the vigil to the more variable kaleidoscope task completed earlier, they provided very high ratings of boredom. In addition, Prinzel et al. noted that overall the boredom ratings were much higher in the constrained conditions of this study as compared with the unconstrained conditions in the Sawin and Scerbo study. Collectively, the findings from these two studies support the idea that demand characteristics, in the form of experimenter-imposed time constraints, contribute to feelings of boredom and stress in vigilance. Moreover, these findings are also consistent with Hancock and Warm's (1989) model of stress described earlier. Specifically, individuals who can terminate the task at will should experience little stress because they can quit before they ever venture outside the comfort zone. On the other hand, those who are required or feel compelled to continue working at a task after they have breeched their zone of psychological stability should report higher levels of stress.

COGNITIVE APPRAISAL OF STRESS

As noted previously, Scerbo et al. (1993) found that when participants were given control over event rate, it facilitated performance but did not reduce workload. With respect to Hancock and Warm's (1989) model, this manipulation affected the rate of information but not the structure.

In recent years, the structure or nature of information and one's interpretation of that information have become the focus of models of stress. It has been argued that stress arises from transactions between individuals and their environment (Cox, 1987; Welford, 1973). Stokes and Kite (1994) suggested that individuals must evaluate their beliefs, goals, and fears in the context of challenges, opportunities, and threats from the environment. More specifically, one must assess the demands of a given situation as well as one's own ability to meet the demands. Stokes and Kite contended that stress results from mismatches between the perception of task demands and the perception of resources needed to meet them.

Recently, Sawin and Scerbo (1995) conducted an experiment that bears on this view of stress. Traditionally, participants in vigilance research are instructed to "pay attention" and to look for "critical" signals. The investigators argued that these instructions may affect how individuals conceptualize the task and may therefore contribute to higher perceptions of stress and workload in vigilance.

Sawin and Scerbo (1995) examined this issue by asking participants to monitor a special display consisting of a computer screen encased in the shell of a microfiche reader. Thus, a standard VDT was covered with a blue filter from the microfiche reader. To enhance the impression of a unique display, a special input device was used so that the entire configuration did not resemble a standard computer system.

The participants were told that their job was to evaluate a new color display, but that unfortunately this particular unit was not operating properly. They were told that the display would flicker every so often and were asked to press the button on the special input device whenever they noticed one of these flickers. The flickers were the targets for detection and consisted of a 2 millisecond shift in luminance that occurred once per minute at random intervals in each minute.

One half of the participants were given traditional vigilance instructions emphasizing the importance of detecting critical signals. The remaining subjects, however, were told that the display emitted a special blue light that most people found quite relaxing. These participants were told to relax and enjoy the display but to press the button if they noticed a flicker. The participants monitored the display for 30 minutes. In addition, they also provided subjective ratings of boredom and workload before and after the vigil.

Sawin and Scerbo (1995) found that detection performance was comparable and declined over the vigil for all groups. Thus, the instruction manipulation had no impact on performance. The instruction emphasis did, however, affect the subjective ratings of workload. Specifically, those participants who received the traditional detection-emphasis instructions reported higher levels of workload after the vigil. By contrast, there were no differences between the pre- and postvigil workload ratings for those who received the relaxation-emphasis instructions. The instruction manipulation did not affect the overall boredom ratings, but it did impact the stress subscale. Specifically, those participants who received the detection-emphasis instructions reported similar amounts of stress before and after the vigil. On the other hand, those who received the relaxation-emphasis instructions reported lower levels of stress after the vigil. Similar results were found in another study with a slightly different task (Sawin & Scerbo, 1994).

Collectively, these results support the cognitive appraisal view of stress and show that stress and mental workload in vigilance can be influenced by how individuals interpret the situation. Thus, the source of the stress in vigilance may lie with the expectations that experimenters set for their participants. When they are asked to look for "critical signals" and to "respond as soon as you can," these instructions may place an additional burden on them and may be ultimately reflected in subsequent ratings of stress and workload. It is possible that over the course of the vigil, individuals come to believe that they no longer have the resources necessary to focus their attention on the task. This situation ultimately generates stress. By contrast, when the individuals are asked to relax, the same task no longer produces stress because, presumably, they do not perceive the task to be a drain on their resources.

These findings are also consistent with Hancock and Warm's (1989) model. In these experiments, the information rate was held constant, but the information structure was manipulated through the task instructions. The traditional emphasis on detection appeared to result in levels of psychological adaptability in which perceptions of stress and mental workload were elevated, but not to the point where performance was degraded. On the other hand, the participants who received the relaxation-emphasis instructions did not perceive the task to be high in workload and stress. It is possible that the perception of workload and stress was lowered because the instructions to relax deemphasized the need to remain alert.

CONCLUSION

The nature of sustained attention has continued to interest investigators since Mackworth (1948) conducted his seminal experiment 50 years ago.

Although much of the original interest lay with understanding the psycho-physics of vigilance, recent attention has been focused on understanding why vigilance produces such high levels of stress and workload. The research outlined in the present chapter offers one explanation. The stress in vigilance is the result of *both* task demands and boredom. The evidence for each of these contributors is reviewed in turn.

Several investigators have shown that vigilance produces high levels of workload. Furthermore, psychophysical parameters known to affect vigilance performance affect workload as well. For instance, increases in signal salience or the availability of knowledge of results about performance have been shown to decrease workload (Becker et al., 1991; Gluckman et al., 1988). Although these manipulations had a direct effect on task demands, it is important to remember that the effect merely lowered workload, but did not completely eliminate it. On the other hand, Sawin and Scerbo (1995) were able to prevent workload ratings from rising above their previgil levels by instructing participants to relax. This manipulation also affected task demands, but in this case the change in instructions had a more profound impact on workload because the need to remain alert was de-emphasized. As noted earlier, the requirement to maintain attention is one of the primary contributors to stress described by Thackray (1981). Psychophysical manipulations such as those introduced by Becker et al. and Gluckman et al. have only marginal effects on workload because they do not allay the need to remain alert. Likewise, giving participants control over the event rate as described by Scerbo et al. (1993) also had no effect on workload because it too did not relieve the demand to maintain attention.

The second contributor to stress in vigilance is boredom. As noted previously, several researchers have argued that boredom arises when individuals are required to work at a task with highly repetitive, homogeneous stimuli beyond the point where they would normally reject it. Scerbo and Sawin (1994) showed that individuals were willing to work at a task with more stimulus variety for longer intervals than a more repetitive task. Furthermore, when these individuals were allowed to terminate the task at will, they reported much lower levels of boredom and stress than did others who felt compelled to continue or who had to work for durations set by the experimenter (Prinzel et al., 1995). Thus, allowing individuals the freedom to terminate a task at will appears to be the only way to keep boredom in check. Increasing stimulus variety can prolong the time an individual is willing to work at the task, but it does not prevent boredom; it merely delays its onset. Consequently, as long as one is required to work at a boring task, the task is stressful. The stress arises from the need to combat the boredom of having to continue working beyond one's satiation point.

Recall that when Sawin and Scerbo (1995) instructed their participants to relax, the change in emphasis lowered workload ratings, but did not affect the boredom scores. Recently, Hitchcock, Dember, Warm, Moroney, and See (1997) obtained a similar pattern of results in a study examining the effects of cuing and knowledge of results on vigilance. Specifically, they found that when individuals were cued about the appearance of critical signals, it facilitated performance and lowered workload, but had no impact on boredom.

The pattern of results from these two studies should not be surprising. Changes that impact task demands in vigilance (i.e., those that affect psychophysical parameters or information structure) have no impact on the stress produced by boredom. Individuals can work at a task that they consider easy or difficult and still become bored with it. However, if they must continue working at the task beyond the point at which they would normally reject it, it becomes stressful. The stress arises not because the task demands have changed but because the individuals have become bored and cannot quit, change the task, or do anything else to relieve the monotony.

Vigilance is stressful because of the need to remain alert and combat boredom over extended periods. Vigilance tasks typically require individuals to work at a highly repetitive activity, in an understimulating and homogeneous environment, and to remain attentive for intervals determined by someone else. Simple changes to psychophysical task parameters can make the activity more or less difficult and thereby affect one's perception of workload, but they do nothing to relieve the monotony. Introducing more stimulus variety can maintain one's interest over longer intervals, but merely delays the onset of boredom. Consequently, the only viable strategy to reduce stress in vigilance, at present, appears to be giving people the freedom to stop when they become bored.

REFERENCES

Barmack, J. E. (1939). A definition of boredom: A reply to Mr. Berman. *American Journal of Psychology, 52,* 467–471.

Becker, A. B., Warm, J. S., Dember, W. N., & Hancock, P. A. (1991). Effects of feedback on perceived workload in vigilance performance. *Proceedings of the Human Factors Society 35th annual meeting* (pp. 1491–1494). Santa Monica, CA: Human Factors Society.

Cox, T. (1987). Stress, coping, and problem solving. *Work and Stress, 1*, 5–14.

Davies, D. R., & Parasuraman, R. (1982). *The psychology of vigilance.* London: Academic Press.

Dember, W. N., & Warm, J. S. (1979). *The psychology of perception* (2nd ed.). New York: Holt, Rinehart & Winston.

Dittmar, M. L., Warm, J. S., Dember, W. N., & Ricks, D. F. (1993). Sex differences in vigilance performance and perceived workload. *The Journal of General Psychology, 120,* 309–322.

Frankenhaeuser, M., Nordheden, B., Myrsten, A. L., & Post, B. (1971). Psychophysiological reactions to understimulation and overstimulation. *Acta Psychologica, 35,* 298–308.

Galinsky, T. L., Dember, W. N., & Warm, J. S. (1989, March). *Effects of event rate on subjective workload in vigilance performance.* Paper presented at the annual meeting of the Southern Society for Philosophy and Psychology, New Orleans, LA.

Galinsky, T. L., Rosa, R. R., Warm, J. S., & Dember, W. N. (1993). Psychophysical determinants of stress in sustained attention. *Human Factors, 35,* 603–614.

Geiwitz, P. J. (1966). Structure of boredom. *Journal of Personal and Social Psychology, 3,* 592–600.

Gluckman, J. P., Warm, J. S., Dember, W. N., Thiemann, J. A., & Hancock, P. A. (1988, November). *Subjective workload in simultaneous and successive vigilance tasks.* Paper presented at Psychonomic Society, Chicago, IL.

Hancock, P. A., & Warm, J. S. (1989). A dynamic model of stress in sustained attention. *Human Factors, 31,* 519–537.

Hart, S. G., & Staveland, L. E. (1988). Development of NASA-TLX (Task Load Index): Results of empirical and theoretical research. In P. A. Hancock & N. Meshkati (Eds.), *Human mental workload* (pp. 139–183). Amsterdam: North-Holland.

Hill, A. B., & Perkins, R. E. (1985). Towards a model of boredom. *British Journal of Psychology, 76,* 235–240.

Hitchcock, E. M., Dember, W. N., Warm, J. S., Moroney, B. W., & See, J. E. (1997). Effects of cueing and knowledge of results on workload and boredom in sustained attention. *Proceedings of the Human Factors and Ergonomics Society 41st annual meeting* (pp. 1298–1302). Santa Monica, CA: Human Factors and Ergonomics Society.

Mackie, R. R. (1987). Vigilance research—Are we ready for countermeasures? *Human Factors, 29,* 707–723.

Mackie, R. R., Wylie, C. D., & Smith, M. J. (1985). Comparative effect of 19 stressors on task performance: Critical literature review (what we appear to know, don't know, and should know). *Proceedings of the Human Factors Society 29th Annual Meeting* (pp. 462–469). Santa Monica, CA: Human Factors Society.

Mackworth, N. H. (1948). The breakdown of vigilance during prolonged visual search. *Quarterly Journal of Experimental Psychology, 1,* 6–21.

O'Hanlon, J. F. (1981). Boredom: Practical consequences and a theory. *Acta Psychologica, 49,* 53–82.

Parasuraman, R. (1985). Sustained attention: A multifactorial approach. In M. I. Posner & O. S. Marin (Eds.), *Attention and human performance* (Vol. II, pp. 493–511). Hillsdale, NJ: Lawrence Erlbaum Associates.

Prinzel, L. J., Sawin, D. A., & Scerbo, M. W. (1995, March). *Boredom in vigilance: Stuck in a rut.* Paper presented at the third annual Mid-Atlantic Human Factors Conference, Blacksburg, VA.

Sawin, D. A., & Scerbo, M. W. (1994). Vigilance: How to do it and who should do it. In *Proceedings of Human Factors and Ergonomics Society 38th annual meeting* (pp. 1312–1316). Santa Monica, CA: Human Factors and Ergonomics Society.

Sawin, D. A., & Scerbo, M. W. (1995). The effects of instruction type and boredom proneness in vigilance: Implications for boredom and workload. *Human Factors, 37,* 752–765.

Scerbo, M. W. (1998). What's so boring about vigilance? In R. R. Hoffman, M. F. Sherrick, & J. S. Warm (Eds.), *Viewing psychology as a whole: The integrative science of William N. Dember* (pp. 145–166). Washington, DC: APA Books.

Scerbo, M. W., Greenwald, C. Q., & Sawin, D. A. (1992). Vigilance: It's boring, it's difficult and I can't do anything about it. In *Proceedings of the Human Factors Society 36th annual meeting* (pp. 1508–1512). Santa Monica, CA: Human Factors Society.

Scerbo, M. W., Greenwald, C. Q., & Sawin, D. A. (1993). The effects of subject-controlled pacing and task type on sustained attention and subjective workload. *Journal of General Psychology, 120,* 293–307.

Scerbo, M. W., & Sawin, D. A. (1994, February). *Vigilance: How much is enough?* Paper presented at the second annual Mid-Atlantic Human Factors Conference. Reston, VA.

Stokes, A., & Kite, K. (1994). *Flight stress: Stress, fatigue, and performance in aviation.* Hampshire, England: Avebury Aviation.

Thackray, R. I. (1981). The stress of boredom and monotony: A consideration of the evidence. *Psychosomatic Medicine, 43,* 165–176.

Warm, J. S. (1984). An introduction to vigilance. In J. S. Warm (Ed.), *Sustained attention in human performance* (pp. 1–14). Chichester, England: Wiley.

Warm, J. S., Dember, W. N., & Hancock, P. A. (1996). Vigilance and workload in automated systems. In R. Parasuraman & M. Mouloua (Eds.), *Automation and human performance: Theory and applications* (pp. 183–200). Mahwah, NJ: Lawrence Erlbaum, Associates.

Warm, J. S., & Jerison, H. J. (1984). The psychophysics of vigilance. In J. S. Warm (Ed.), *Sustained attention in human performance* (pp. 15–59). Chichester, England: Wiley.

Welford, A. T. (1973). Stress and performance. *Ergonomics, 15,* 567–80.

2.2 An Autonomic Space Approach to the Psychophysiological Assessment of Mental Workload

Richard W. Backs
Central Michigan University

Berntson, Cacioppo, and Quigley (1991, 1993) formalized a theory of autonomic determinism that delineates the multiple ways in which the sympathetic and parasympathetic branches of the autonomic nervous system may respond to physical and mental events. Their theory subsumes classical autonomic theory and suggests that autonomic responses can be more useful for understanding psychological-physiological mappings when the responses are properly assessed. This chapter briefly reviews the aspects of the Berntson et al. model that are most relevant for mental workload assessment, discuss the implications for inference, and reviews some studies that support the Berntson et al. approach.

MODES OF AUTONOMIC CONTROL AND AUTONOMIC SPACE

The basic tenets of autonomic determinism relevant to mental workload assessment are the multiple neural *modes of autonomic control* (Berntson et al., 1991, p. 459) and the resulting *autonomic space* that describes a visceral effector response to neurogenic change. An autonomic control mode is a consistent pattern of sympathetic and/or parasympathetic activity to an effector organ such as the heart. The most studied and best understood response is heart rate, which is also one of the psychophysiological responses most frequently used for mental workload assessment (Kramer, 1991; Wilson & Eggemeier, 1991). Therefore, this chapter focuses solely on heart rate, with the understanding that the implications of autonomic determinism are generally applicable to all visceral responses.

A model of the functional state (f_{ij}) of an organ response (e.g., heart rate) is given in the following equation from Berntson et al. (1991, p. 471), where β is the basal functional state (e.g., heart rate with no autonomic efferent activity), c_{si} is the coupling coefficient for sympathetic activity, s_i is the sympathetic activation, c_{pj} is the coupling coefficient for parasympathetic activity, p_j is the parasympathetic activation, c_{sipj} is the coupling coefficient for interaction of sympathetic and parasympathetic activity, $s_i p_j$ is the interaction of sympathetic and parasympathetic activation (e.g., local feedback effects on one autonomic branch because of activation of the opposing branch), and e is error (which in the present model includes the effects of circulating hormones):

$$f_{ij} = \beta + c_{si} \cdot s_i + c_{pj} \cdot p_j + c_{sipj} \cdot s_i p_j + e. \tag{1}$$

Actually, the model fits heart period (the time between successive heart beats) rather than heart rate (the average number of heart beats per minute), and most recent research (including all the studies reviewed in the present chapter) uses heart period because it has better statistical properties than heart rate. For most dually innervated organs, including the heart, c_{si} and c_{pj} have opposite signs indicating that sympathetic and parasympathetic activations have opposing effects on the organ. In the case of heart period, c_{si} is negative, indicating that sympathetic activation results in shorter heart periods (i.e., increased heart rate), and c_{pj} is positive, indicating that parasympathetic activation results in longer heart periods (i.e., decreased heart rate). In practice, the model is further simplified by setting c_{sipj} to zero and assuming e to be random. A model of heart period for humans was estimated by Berntson, Cacioppo, and Quigley (1993, p. 304) and is presented here in simplified form:

$$f_{ij} = \beta - 230 \cdot s_i + 1{,}713 \cdot p_j. \tag{2}$$

In Equation 2, the effector response (f_{ij}) is heart period in milliseconds. Basal functional state (β) is an individual difference variable that can be estimated for each person and removed by subtraction. This correction is usually accomplished by obtaining an estimate of heart period during physiological and psychological rest and subtracting that estimate from the heart period obtained in the conditions of interest. The coupling coefficients for sympathetic (c_{si}) and parasympathetic (c_{pj}) activations are estimates of the dynamic range of each autonomic branch obtained from a review of the literature. Sympathetic (s_i) and parasympathetic (p_j) activations are normalized to range from zero (no input from the branch) to one (maximum input from the branch). Thus, heart period decreases 230 milliseconds (faster heart rate) with maximal sympathetic activity, and heart period increases 1,713 milliseconds (slower heart rate) with maximal parasympathetic activity.

In the classical representation of autonomic control (e.g., Cannon, 1939), sympathetic and parasympathetic activity have reciprocal effects, and the two branches are coupled (meaning that activation of one branch occurs with concomitant inhibition of the other branch). There are only two *reciprocally coupled* modes of autonomic control according to classical theory: sympathetic activation (with coupled parasympathetic inhibition) that results in shorter heart periods (i.e., increased heart rate) and parasympathetic activation (with coupled sympathetic inhibition) that results in longer heart periods (i.e., decreased heart rate).

Exceptions to the reciprocally coupled autonomic control modes for the heart have long been observed. Cannon (1939) recognized the possibility that sympathetic and parasympathetic activity may not always function in a reciprocally coupled fashion; however, these instances were considered to be anomalies. Berntson et al. (1991) were among the first to suggest that other control modes may be a routine part of autonomic functioning and to instantiate a model that accommodates multiple modes of autonomic control (see also Stemmler, Grossman, Schmid, & Foerster, 1991). There is human and/or animal evidence to support most of the eight possible modes of autonomic of control listed in Table 2.2.1. Much of the evidence for the *nonreciprocal* and *uncoupled* modes of control comes from studies that manipulated psychological factors. These other control modes may be more commonly elicited by psychologically relevant events (e.g., a conditioned stimulus signaling a noxious unconditioned stimulus) than by physiologically relevant events (e.g., a fall in blood pressure caused by orthostatic stress).

Autonomic space is a bivariate surface that represents the effector organ response as a function of sympathetic and parasympathetic activity. Figure 2.2.1 is taken from Berntson et al. (1993) and represents the autonomic space for heart period. The surface in Fig. 2.2.1 is the result of evaluating Equation 2 across the dynamic range of s_i and p_j. The units for heart period are the relative change from baseline (β) in milliseconds at point ij. Theoretically, heart period at any point ij can be reduced to the constituent sympathetic and parasympathetic activity by which it was elicited. In practice, however, independent verification of the activity of each autonomic branch is required. Various multivariate methods using multiple psychophysiological measures of cardiovascular functioning have been proposed for this purpose (e.g., Backs, 1995, 1998; Stemmler, 1993). However, the simplest method is to use psychophysiological measures that directly correspond to the underlying autonomic activity (e.g., Cacioppo et al., 1994).

Two points are needed in autonomic space to determine an autonomic mode of control. The change between these two points defines a vector in

TABLE 2.2.1
Possible Modes of Autonomic Control for Heart Rate

Control Mode	Sympathetic Response	Parasympathetic Response	Task	Study
Reciprocally coupled modes				
Sympathetic activation/ parasympathetic inhibition	⇑	⇓	Mental arithmetic, reaction time, speech stress	Berntson et al. (1994)[1]
Parasympathetic activation/ sympathetic inhibition	⇓	⇑	Aversive conditioning	Albiniak & Powell (1980)[1,5]
Nonreciprocally coupled modes				
Coactivation	⇑	⇑	Aversive conditioning	Obrist, Wood, & Perez-Reyes (1965)[2]
Coinhibition[4]	⇓	⇓		
Uncoupled modes				
Sympathetic activation	⇑	—	Reaction time	Pollak & Obrist (1988)[1,6]
Sympathetic inhibition[4]	⇓	—		
Parasympathetic activation	—	⇑	Visual illusions	Berntson, Cacioppo, & Fieldstone (1996)[3]
Parasympathetic inhibition	—	⇓	Low-intensity exercise	Robinson, Epstein, Beiser, & Braunwald (1966)[1]

[1]Verified by sympathetic and parasympathetic pharmacological blockades.

[2]Verified by parasympathetic pharmacological blockade.

[3]Not verified by pharmacological blockade.

[4]No study found to verify this control mode.

[5]Rabbits were used as subjects, and unlike humans, their conditioned cardiac response is heart rate slowing.

[6]The reason for the different modes of control during the reaction time tasks of Berntson et al. and Pollak & Obrist is not clear, but there were several important differences between studies including the sympathetic blocking agent, the reward structure of the task, and the location of measurement (hospital or laboratory, respectively).

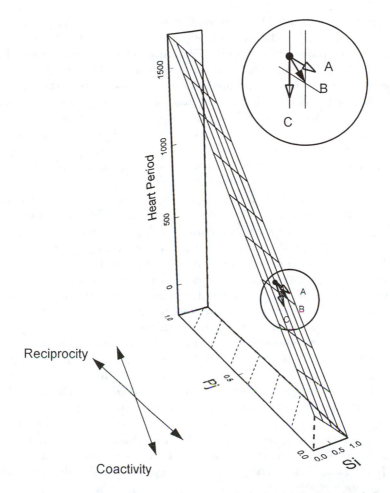

FIG. 2.2.1. Human autonomic space for heart period obtained from Equation 2. P_j and S_i are, respectively, the normalized activations of the parasympathetic and sympathetic branch inputs to the heart at point ij. P_j and S_i are scaled to be proportional to the dynamic range for each branch. Heart period is shown as the relative change from baseline (β) in ms. Uncoupled control modes are represented by change along a single axis, whereas reciprocal and coactivity control modes are represented by change along a directional vector as indicated by the arrows. Dotted lines represent isoresponse contours projected from the surface to the autonomic plane. Heart period does not change along an isoresponse contour. Inset: Example autonomic control mode vectors for uncoupled sympathetic activation (A), reciprocally coupled sympathetic activation and parasympathetic inhibition (B) and uncoupled parasympathetic inhibition (C). The filled circle is an arbitrary point in the space that was used to anchor all vectors. The vectors were computed by using Equation 2 so that each vector represents the same 100 ms decrease in heart period from the anchor point. Adapted from "Cardiac Psychophysiology and Autonomic Space in Humans: Empirical Perspectives and Conceptual Implications" by G. G. Berntson, J. T. Cacioppo, and K. S. Quigley, 1993, *Psychological Bulletin, 114,* 296–322. Copyright 1993 by the American Psychological Association, Inc. Adapted with permission.

the space that indicates the autonomic mode of control (see the example vectors in Fig. 2.2.1). The two points may be obtained from two different conditions, or they may be obtained from the same condition at two different times. For example, most of the research conducted thus far (e.g., Berntson et al., 1994) has examined vectors corresponding to the difference between heart period during resting baseline (point ij) and heart period during performance of some information processing task (point ij''). From a mental workload perspective, however, it is more important to determine the autonomic control mode corresponding to the heart period change between levels of information processing task *demand*, such as from low (point ij) to high difficulty (point ij'') versions of a task.

Two important limitations in the use of heart period (or heart rate or any other dually innervated visceral response) for making inferences about psychological phenomena such as mental workload are evident in Fig. 2.2.1. The first limitation is that equivalent heart period changes can occur as the result of many different autonomic modes of control. For example, heart period can decrease (i.e., heart rate increase) as a result of reciprocally coupled sympathetic activation, uncoupled sympathetic activation, uncoupled parasympathetic inhibition, coactivation (where sympathetic activation exceeds parasympathetic activation), or coinhibition (where parasympathetic inhibition exceeds sympathetic inhibition). This limitation is illustrated in Fig. 2.2.1 by the example vectors in the inset. The length and direction of each vector indicates a 100 millisecond decrease in heart period from the same starting point (ij) for three different control modes. The second limitation is illustrated by the so-called isoresponse contours given by the dotted lines in the autonomic plane. Note that along any one of the contours heart period does not change even though the underlying sympathetic and parasympathetic activity may change greatly. Thus, heart period (or any other dually innervated visceral response) alone is uninformative about the underlying autonomic activity.

USING AUTONOMIC SPACE TO ELUCIDATE PSYCHOLOGICAL-PHYSIOLOGICAL MAPPINGS

Cacioppo and Tassinary (1990a, 1990b) summarized the difficulties inherent in using physiological responses to make inferences about psychological processes. They classified the various relations between psychological processes and physiological responses according to two dimensions, specificity and generality, that result in four general classes of relations termed outcomes, concomitants, markers, and invariants (see Cacioppo & Tassinary, 1990a, fig. 4). Outcomes are the lowest level relation on both the specificity and generality dimensions. Outcome relations are many-to-one mappings between psychological processes and a physiological response obtained in context-bound environments. Outcomes are typically the first relation established between a psychological process and a physiological response, and in fact few psychological-physiological relations have been advanced beyond the outcome class.

For the purposes of mental workload assessment, outcome relations establish the property of *sensitivity*. O'Donnell and Eggemeier defined sensitivity as the "capability of a technique to detect changes in the amount of workload imposed on task performance" (O'Donnell & Eggemeier, 1986, p. 42-2). Outcome relations are useful if all that needs to be ascertained is that the psychological demands have changed. The outcome relations between many psychological processes and heart period are illustrative. Shorter heart period (i.e., faster heart rate) has been associated with increasing the demand on numerous psychological processes that increase mental workload. For example, heart period decreases as attention demands increase from focused to divided attention (Backs & Ryan, 1992), as working memory load increases (Backs & Seljos, 1994), and as tracking system disturbance increases (Backs, Ryan, & Wilson, 1994). However, it is important to note that the mapping between psychological processes and the physiological response does not hold in the opposite direction for outcomes. That is, merely observing decreased heart period is not sufficient to identify the psychological process that was responsible for eliciting the response.

To be useful in most application domains where mental workload needs to be assessed, the psychological–physiological relation must also have the property of *diagnosticity*. O'Donnell and Eggemeier defined diagnosticity as the "capability of a technique to discriminate the amount of workload imposed on different operator capacities or resources" (O'Donnell & Eggemeier, 1986, p. 42-43). Heart period change cannot be diagnostic for the reasons illustrated in the autonomic space in Fig. 2.2.1. To have diagnosticity with physiological responses there must be at a minimum the type of psychological-physiological relation that Cacioppo and Tassinary (1990a, 1990b) have termed a marker. A marker relation is a one-to-one psychological-physiological mapping obtained in a context-bound environment. A physiological response that is a marker is observed in the presence of one and only one psychological process. Importantly, the opposite is also true for markers, and the observation of the physiological response is sufficient to identify the psychological process that was responsible for eliciting the response.[1]

---

[1]The context-bound environmental constraint on the generality dimension for markers is not a limitation to their usefulness for mental workload assessment. Because mental workload is typically assessed in a particular application domain or in an individual, the psychological-physiological relation need hold only in that domain or that individual. It does not matter whether the relation holds in other domains or individuals, as long as this lack of generality is recognized. Of course, other markers have to be identified that do hold for other domains or individuals.

Cacioppo and Tassinary (1990a, 1990b) proposed a strategy for converting an outcome relation with a many-to-one mapping between psychological processes and a physiological response into a marker relation with a one-to-one mapping between a psychological process and a physiological response. Their strategy is to transform the physiological response to obtain a one-to-one mapping between a psychological process and a physiological response *profile* in which the individual processes are distinguished according to multiple physiological features. I contend that the first logical step for transformation of visceral responses such as heart period is to establish the mode of autonomic control for each psychological process. If consistent patterns emerge between changing demands on psychological processes and autonomic control modes, then the mode of control is more diagnostic than is heart period.

The transformation strategy is one way to begin disambiguating the heart period outcomes like those given earlier for attention, memory, and response processing demands. Figure 2.2.2 illustrates the first-stage transformation from many-to-one mappings to one-to-one mappings for these psychological (Ψ) and physiological relations (Φ). The putative autonomic modes of control for the demand on the psychological processes in Fig. 2.2.2 were obtained from principal components analyses of multiple psychophysiological cardiovascular measures. That method was described in detail in Backs (1995). Although promising, these results must remain tentative until more confirmatory, especially psychopharmacological, evidence of the autonomic control modes is available for the demand manipulations summarized in Fig. 2.2.2.

CONCLUSIONS

Of course, the transformation of a single physiological measure such as heart period to an autonomic mode of control is unlikely to result in the one-to-one mapping with psychological processes that is needed for mental workload assessment. It will undoubtedly be found that the relations between psychological processes and autonomic control modes are of the many-to-one class as the control modes for more processes are established. However, the potential power of the approach can be increased by extending the control mode transformation strategy in two ways that can be used either alone or together. The first way is to examine the control modes for multiple visceral responses instead of for a single response. The second way is to examine the control mode as it unfolds over time. Conducting these second- (Φ'') and/or third-stage (Φ''') transformations, especially for responses from related physiological systems such as cardiovascular and pulmonary, should result in patterns of autonomic control that have a greater chance of achieving marker status.

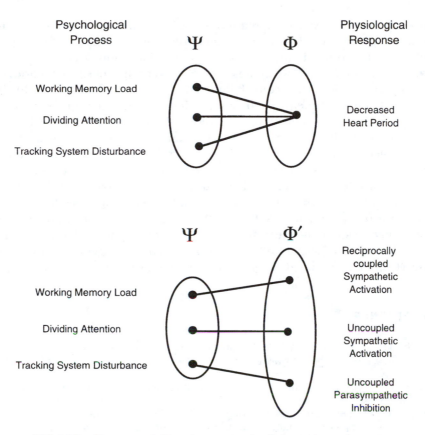

FIG. 2.2.2.   Upper panel: Many-to-one mapping between demand on psychological processes (Ψ) and the physiological response of decreased heart period (Φ). Lower panel: One-to-one mapping between demand on psychological processes (Ψ) and the transformation of decreased heart period to the purported autonomic mode of control responsible for the change (Φ').

Furthermore, Cacioppo and Tassinary (1990a, 1990b) suggested that the psychological processes can also be transformed. Higher-level constructs (Ψ') may be found that integrate several lower level processes (Ψ) and have the desired one-to-one mapping to autonomic modes of control (Φ' or Φ''). Dually transformed psychological-physiological relations of this type (i.e., Ψ' with Φ' or Φ'') may also have greater generality than relations where only the physiological response is transformed. Increasing the generality of a marker would result in the most powerful and desirable type of relation that Cacioppo and Tassinary have termed an invariant, which is one-to-one psychological-physiological mapping that is context free.

In summary, a better understanding of both Ψ and Φ transformations is required before psychophysiological responses prove useful as mental workload metrics in advanced applications such as dynamic task allocation in real-world domains. The incorporation of physiological responses into a psychological theory of human performance is a challenging task that has seldom been attempted. Two notable exceptions are the energetic theories of attention and performance by Kahneman (1973) and Sanders (1983). In the tradition of Kahneman and Sanders, it seems clear that the theoretical formulation for the higher level psychological constructs (Ψ') that delineate autonomic modes of control must be an energetic one. The theory must acknowledge that the primary role of visceral responses is in the maintenance of homeostasis and must specify how the relatively small information-processing effects are interpretable apart from the physiological background on which they are superimposed (see Gaillard & Kramer, 2000). Current research (e.g., Lenneman & Backs, 2000) is only at the initial phase of attempting to define the relevant dimensions of the higher order psychological constructs by systematically identifying the autonomic control modes for heart period across a wide range of information-processing task demands.

REFERENCES

Albiniak, B. A., & Powell, D. A. (1980). Peripheral autonomic mechanisms and Pavlovian conditioning in the rabbit (*Oryctolagus cuniculus*). *Journal of Comparative and Physiological Psychology, 94*, 1101–1113.

Backs, R. W. (1995). Going beyond heart rate: Modes of autonomic control in the cardiovascular assessment of mental workload. *The International Journal of Aviation Psychology, 5*, 25–48.

Backs, R. W. (1998). A comparison of factor analytic methods of obtaining cardiovascular autonomic components for the assessment of mental workload. *Ergonomics, 41*, 733–745.

Backs, R. W., & Ryan, A. M. (1992). Multimodal measures of mental workload during dual-task performance: Energetic demands of cognitive processes. In *Proceedings of the Human Factors Society 36th annual meeting* (pp. 1413–1417). Santa Monica CA: Human Factors Society.

Backs, R. W., Ryan, A. M., & Wilson, G. F. (1994). Psychophysiological measures of workload during continuous manual performance. *Human Factors, 36*, 514–531.

Backs, R. W., & Seljos, K. A. (1994). Metabolic and cardiorespiratory measures of mental effort: The effects of level of difficulty in a working memory task. *International Journal of Psychophysiology, 16*, 57–68.

Berntson, G. G., Cacioppo, J. T., Binkley, P. F., Uchino, B. N., Quigley, K. S., & Fieldstone, A. (1994). Autonomic cardiac control, III: Psychological stress and cardiac response as revealed by autonomic blockade. *Psychophysiology, 31*, 599–608.

Berntson, G. G., Cacioppo, J. T., & Fieldstone, A. (1996). Illusions, arithmetic, and bidirectional modulation of vagal control of the heart. *Biological Psychology, 44*, 1–17.

Berntson, G. G., Cacioppo, J. T., & Quigley, K. S. (1991). Autonomic determinism: The modes of autonomic control, the doctrine of autonomic space, and the laws of autonomic constraint. *Psychological Review, 98*, 459–487.

Berntson, G. G., Cacioppo, J. T., & Quigley, K. S. (1993). Cardiac psychophysiology and autonomic space in humans: Empirical perspectives and conceptual implications. *Psychological Bulletin, 114*, 296–322.

Cacioppo, J. T., Berntson, G. G., Binkley, P. F., Quigley, K. S., Uchino, B. N., & Fieldstone, A. (1994). Autonomic cardiac control, II: Noninvasive indices and basal response as revealed by autonomic blockade. *Psychophysiology, 31,* 586–598.

Cacioppo, J. T., & Tassinary, L. G. (1990a). Inferring psychological significance from physiological signals. *American Psychologist, 45,* 16–28.

Cacioppo, J. T., & Tassinary, L. G. (1990b). Psychophysiology and psychophysiological inference. In J. T. Cacioppo & L. G. Tassinary (Eds.), *Principles of psychophysiology: Physical, social, and inferential elements* (pp. 3–33). New York: Cambridge University Press.

Cannon, W. B. (1939). *The wisdom of the body.* New York: Norton.

Gaillard, A. W. K., & Kramer, A. F. (2000). Theoretical and methodological issues in psychophysiological research. In R. W. Backs & W. Boucsein (Eds.), *Engineering psychophysiology: Issues and applications* (pp. 31–58). Mahwah, NJ: Lawrence Erlbaum Associates.

Kahneman, D. (1973). *Attention and effort.* Englewood Cliff, NJ: Prentice-Hall.

Kramer, A. F. (1991). Physiological metrics of mental workload: A review of recent progress. In D. L. Damos (Ed.), *Multiple-task performance* (pp. 279–328). London: Taylor & Francis.

Lenneman, J. K., & Backs, R. W. (2000). The validity of factor analytically derived cardiac autonomic components for mental workload assessment. In R. W. Backs & W. Boucsein (Eds.), *Engineering psychophysiology: Issues and applications.* Mahwah, NJ: Lawrence Erlbaum Associates.

Obrist, P. A., Wood, D. M., & Perez-Reyes, M. (1965). Heart rate during conditioning in humans: Effects of UCS intensity, vagal blockade, and adrenergic block of vasomotor activity. *Journal of Experimental Psychology, 70,* 32–42.

O'Donnell, R. D., & Eggemeier, F. T. (1986). Workload assessment methodology. In K. R. Boff, L. Kaufman, & J. P. Thomas (Eds.), *Handbook of perception and human performance, Vol. II: Cognitive processes and performance* (pp. 42, 1–49). New York: Wiley Interscience.

Pollak, M. H., & Obrist, P. A. (1988). Effects of autonomic blockade on heart rate responses to reaction time and sustained handgrip tasks. *Psychophysiology, 25,* 689–695.

Robinson, B. F., Epstein, S. E., Beiser, G. D., & Braunwald, E. (1966). Control of heart rate by the autonomic nervous system. *Circulation Research, 29,* 400–411.

Sanders, A. F. (1983). Towards a model of stress and human performance. *Acta Psychologica, 53,* 61–97.

Stemmler, G. (1993). Receptor antagonists as tools for structural measurement in psychophysiology. *Neuropsychobiology, 28,* 47–53.

Stemmler, G., Grossman, P., Schmid, H., & Foerster, F. (1991). A model of cardiovascular activation components for studies using autonomic receptor antagonists. *Psychophysiology, 28,* 367–381.

Wilson, G. F., & Eggemeier, F. T. (1991). Psychophysiological assessment of workload in multi-task environments. In D. L. Damos (Ed.), *Multiple-task performance* (pp. 329–360). London: Taylor & Francis.

How Unexpected Events Produce an Escalation of Cognitive and Coordinative Demands

David D. Woods
Emily S. Patterson
Ohio State University

EXPLAINING THE CLUMSY USE OF TECHNOLOGY

Each round of technological development promises to aid the people engaged in various fields of practice. After these promises result in the development of prototypes and fielded systems, those researchers who examine the reverberations of technology change have observed a mixed bag of effects, most quite different from the expectations of the technology advocates. Often the message practitioners send with their performance, their errors, and their adaptations is one of technology-induced complexity. In these cases, technological possibilities are used clumsily so that systems intended to serve the user turn out to add new burdens that congregate at the busiest times or during the most critical phases of the task (e.g., Woods, Johannesen, Cook, & Sarter, 1994, chap. 5, Woods & Watts, 1997).

Although this pattern has been well documented in a variety of areas such as cockpit automation (Sarter, Woods, & Billings, 1997) and many principles for more effective human–machine and human–human cooperation have been developed (e.g., Norman, 1988), we have a gaping explanatory problem. There is a striking contrast between the persistent optimism of developers who before the fact expect each technological development to produce significant performance improvements and the new operational complexities that are observed after the fact. It seems quite difficult for all kinds of people in design teams to predict or anticipate operational complexities. Yet operational complexities are easy to see when the right scenarios are examined, for example, through observing incidents during practice.

Ultimately, we need to explain why this technology-induced complexity occurs so often when designers fully expect these systems to produce major benefits for the practitioners. There are many factors that could be invoked to explain this observation. Some may fall into hoary cliches about the need for human factors in the design process. Others may examine the pressures on development and developers. Here we explore one factor that contributes in part—a fundamental dynamic relation between problem demands, cognitive and coordinated activities, and the artifacts intended to support practitioners.

THE ESCALATION PRINCIPLE

On the basis of observations of anomaly response in many supervisory control domains in both simulated and actual incidents, a pattern seemed to recur. When an anomaly occurred and people began to recognize various unexpected events, there was a process of escalation of cognitive and coordinated activities. During these periods of escalating demands, we observed the penalties associated with poor design of systems that had been intended to support practitioners.

> *Escalation Principle:* The concept of escalation concerns a process—how situations move from canonical or textbook to nonroutine to exceptional. In that process, escalation captures a relationship—as problems cascade, they produce an escalation of cognitive and coordinative demands that brings out the penalties of poor support for work.

There is a fundamental relationship where the greater the trouble in the underlying process or the higher the tempo of operations, the greater the information-processing activities required to cope with the trouble or pace of activities. For example, demands for knowledge, monitoring, attentional control, information, and communication among team members (including human–machine communication) all tend to go up with the unusualness, tempo, and criticality of situations. If workload or other burdens associated with using a computer interface or with interacting with an autonomous or intelligent machine agent tend to be concentrated at these times, the workload occurs when the practitioner can least afford new tasks, new memory demands, or diversions of his or her attention away from the job at hand to the interface or computerized device per se.

The concept of escalation captures a dynamic relationship between the cascade of effects that follows from an event and the demands for cognitive and collaborative work that escalate in response (Woods, 1994). An event triggers the evolution of multiple interrelated dynamics.

There Is a Cascade of Effects in the Monitored Process

A fault produces a time series of disturbances along lines of functional and physical coupling in the process (e.g., Abbott, 1990). These disturbances produce a cascade of multiple changes in the data available about the state of the underlying process, for example, the avalanche of alarms following a fault in process control applications (Reiersen, Marshall, & Baker, 1988).

Demands for Cognitive Activity Increase as the Problem Cascades

More knowledge potentially needs to be brought to bear. There is more to monitor. There is a changing set of data to integrate into a coherent assessment. Candidate hypotheses need to be generated and evaluated. Assessments may need to be revised as new data come in. Actions to protect the integrity and safety of systems need to be identified, carried out, and monitored for success. Existing plans need to be modified or new plans formulated to cope with the consequences of anomalies. Contingencies need to be considered in this process. All these multiple threads challenge control of attention and require practitioners to juggle more tasks.

Demands for Coordination Increase as the Problem Cascades

As the cognitive activities escalate, the demand for coordination across people and across people and machines rises. Knowledge may reside in different people or different parts of the operational system. Specialized knowledge and expertise from other parties may need to be brought into the problem-solving process. Multiple parties may have to coordinate to implement activities aimed at gaining information to aid diagnosis or to protect the monitored process. The trouble in the underlying process requires informing and updating others—those whose scope of responsibility may be affected by the anomaly, those who may be able to support recovery, or those who may be affected by the consequences the anomaly could or does produce.

The Cascade and Escalation Is a Dynamic Process

A variety of complicating factors can occur, which move situations beyond canonical, textbook forms. The concept of escalation captures this movement from canonical to nonroutine to exceptional. The tempo of operations increases following the recognition of a triggering event and is synchronized by temporal landmarks that represent irreversible (or difficult to reverse) decision points.

The dynamics of escalation vary across situations. First, the cascade of effects may have different time courses. For example, an event may manifest itself immediately or may develop more slowly. Second, the nature of the responses by practitioners affects how the incident progresses—less appropriate or timely actions (or too quick a reaction in some cases) may sharpen difficulties, push the tempo in the future, or create new challenges. Different domains may have different escalation gradients depending on the kinds of complicating factors that occur, the rhythms of the process, and consequences that may follow from poor performance.

Interactions with Computer-Based Support Systems

Interactions with computer-based support systems occur in the context of these escalating demands on memory and attention, monitoring and assessment, communication and response.

In canonical (routinized or textbook) situations, technological systems seem to integrate smoothly into work practices, so smoothly that seemingly little cognitive work is required. However, cognitive work grows, and patterns of distribution of this work over people and machines grow more complex as situations cascade. Thus, the penalties for poor coordination between people and machines and for poor support for coordination across people emerge as the situation escalates demands for distributed cognitive work.

The difficulties arise because interacting with the technological devices is a source of workload as well as a potential source of support. When interacting with devices or others through devices creates new workload burdens when practitioners are busy already, creates new attentional demands when practitioners are plagued by multiple voices competing for their attention, creates new sources of data when practitioners are overwhelmed by too many channels spewing out too much competing data, practitioners are placed in an untenable situation.

As active, responsible agents in the field of practice, practitioners adapt to cope with these bottlenecks in many ways—they eliminate or minimize communication and coordination with other agents; they tailor devices to reduce cognitive burdens; they adapt their strategies for carrying out tasks; they abandon some systems or modes when situations become more critical or higher tempo. Woods et al. (1994) devoted a chapter to examples of these workload bottlenecks and the ways that people tailor devices and work strategies to cope with this technology-induced complexity. Sarter et al. (1997) summarized this dynamic for cockpit automation. Cook and Woods (1996) captured this dynamic for a case of operating room information technology. Patterson and Woods (1997) described the strategies used in one organization, space shuttle mission control, for successfully coping with escalating demands following an anomaly. In this case, practitioners who are assigned on-call responsibility invest in building a prior understanding of the mission context before problems occur to be able to come into an escalating situation more effectively should an anomaly actually occur.

AN EXAMPLE OF ESCALATING DEMANDS

To illustrate the escalation principle, consider an anomaly that occurred during the ascent phase of a space shuttle mission (Watts, Woods, & Patterson, 1996; Watts-Perotti & Woods, 1997; for a more publicized case, examine the escalating demands on mission control during the Apollo 13 accident; e.g., Murray & Cox, 1989). As shown in Fig. 2.3.1, an unexpected event produced an escalation of cognitive and coordinative demands and activities. In the figure, the escalating demands are grouped into three temporal units that roughly capture portions of the evolving event.

Several minutes into the ascent phase of the mission, one of the controllers responsible for monitoring the health and safety of the mechanical systems noticed an anomaly—an unexpected drop in hydraulic fluid in an auxiliary power unit (APU). The personnel monitoring immediately recognized that the symptoms indicated a hydraulic leak. Did this anomaly require an immediate abort of the ascent? In other words, how bad was the leak? Was it a threat to the safety of the mission? What were the relevant criteria (and who knew them, and where did they reside)? The mechanical systems controllers did a quick calculation that indicated the leak rate was below the predetermined abort limit—the mission could proceed to orbit. The analysis of the event relative to an abort decision occurred very quickly, in part because the nature of the disturbance was clear and because of the potential consequences with an anomaly at this stage for the safety of the astronauts.

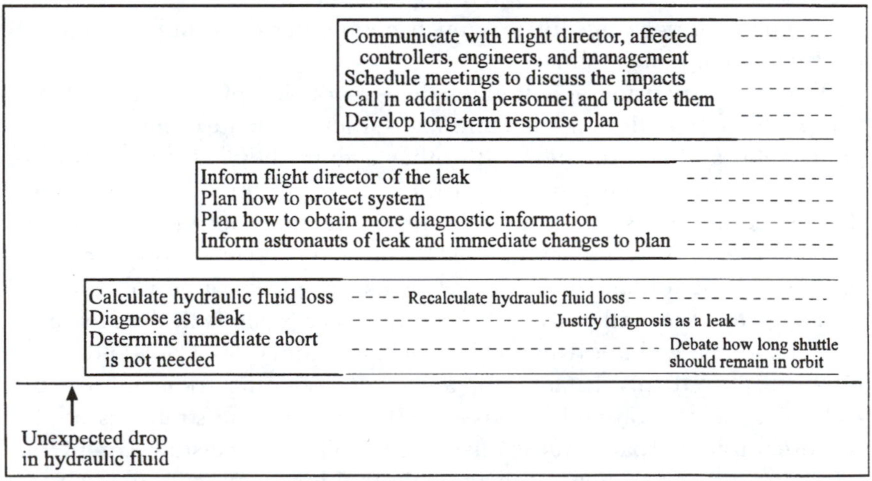

FIG. 2.3.1.   Escalation of cognitive and coordinated work following an anomaly in space shuttle mission control.

As the ascent continued, the figure points to a second collection of de-mands and activities that were intertwined and went on in parallel. The con-trollers for the affected system informed the Flight Director and the other members of the mission control team of the existence of the hydraulic leak and its severity. Because of the nature of the tools for supporting coordina-tion across controllers (voice loops), this occurred in a very cognitively eco-nomical way for all concerned (see Watts et al., 1996). The team also had to plan how to respond to the anomaly before the transition from the ascent to the orbit phase was completed. As in all safety-critical systems, planning was aimed both at how to obtain more information to diagnose the problem as well as how to protect the affected systems. This planning required resolving conflicting goals of maximizing the safety of the systems as well as determin-ing as confidently as possible the diagnosis of the anomaly. The team de-cided to alter the order in which the auxiliary power units (APUs) were shut down to obtain more diagnostic information. This change in the mission plan was then communicated to the astronauts.

A third group in the figure refers to demands and activities that occurred after the initial assessment, responses, and communications. Information about the assessments of the situation and the changed plans were available to other controllers who were or might be affected by these changes. This happened because other personnel could listen in on the voice loops and overhear the previous updates provided to the flight director and the astro-nauts about the hydraulic leak. After the changes in immediate plans were communicated to the astronauts, the controllers responsible for other sub-systems affected by the leak and the engineers who designed the auxiliary power units contacted the mechanical systems controllers to gain further information. In this process, new issues arose, some were settled, but these issues sometimes needed to be revisited or reemerged.

For example, a series of meetings between the mechanical systems con-trollers and the engineering group was called. These meetings served an important role in the process to assess contingencies and to decide how to modify mission plans such as a planned docking with the MIR space station and for re-entry. In addition, they provided opportunities to detect and cor-rect errors in the assessment of the situation, to calibrate the assessments and expectations of differing groups, to anticipate more possible side effects of changing plans (see Watts-Perotti & Woods, 1997, for a complete analysis of these functions of cooperative work in this case).

Additional personnel were called in and integrated with others to help with the new workload demands and to provide specialized knowledge and expertise. In this process, the team expanded to include an impressive num-ber of agents acting in a variety of roles and teams all coordinating their efforts.

ESCALATION HELPS EXPLAIN EPISODES
IN THE CLUMSY USE OF TECHNOLOGY

The escalation principle helps to explain some recurrent phenomena in distributed cognitive work. We briefly refer to two recurrent patterns in the impact of technology on practitioners. One is clumsy automation where automation introduced to lower workload and free up resources actually creates new bottlenecks in higher tempo and more critical situations (Sarter, Woods, & Billings, 1997). Another is how attempts to provide intelligent diagnostic systems with explanation capabilities have failed to make these artificial intelligence (AI) systems into team players (Malin et al., 1991).

CLUMSY AUTOMATION

The escalation of problem demands helps explain a syndrome, which Wiener (1989) termed "clumsy automation." Clumsy automation is a form of poor coordination between human and machine in the control of dynamic processes where the benefits of the new technology accrue during workload troughs, and the costs or burdens imposed by the technology occur during periods of peak workload, high criticality, or high-tempo operations. Despite the fact that these systems are often justified on the grounds that they help offload work from harried practitioners, we find that they in fact create new additional tasks, force the user to adopt new cognitive strategies, require more knowledge and more communication at the very times when the practitioners are most in need of true assistance. This creates opportunities for new kinds of human error and new paths to system breakdown that did not exist in simpler systems (Woods & Sarter, in press).

We usually focus on the perceived benefits of new automated systems, assuming that introducing new automation leads to lower workload and frees up limited practitioner resources for other activities (Sarter et al., 1997). Our fascination with the possibilities afforded by automation often obscures the fact that new automated devices also create new burdens and complexities for the individuals and teams of practitioners responsible for operating, troubleshooting, and managing high-consequence systems. The demands may involve new or changed tasks such as device setup and initialization, configuration control, or operating sequences. Cognitive demands change as well, creating new interface management tasks, new attentional demands, the need to track automated device state and performance, new communication or coordination tasks, and new knowledge requirements. These demands represent new levels and types of operator workload.

The dynamics of these new demands are an important factor because in complex systems human activity ebbs and flows, with periods of lower activity and more self-paced tasks interspersed with busy, high-tempo, externally paced operations where task performance is more critical (Rochlin, La Porte, & Roberts, 1987). Technology is often designed to shift workload or tasks from the human to the machine. However, the critical design feature for well-integrated cooperative cognitive work between the automation and the human is not the overall or time-averaged task workload. Rather, it is how the new demands created by the new technology interact with low-workload and high-workload periods, how they impact the transition from canonical to more exceptional situations, and especially how they impact the practitioner's ability to manage workload as situations escalate. It is these relationships that make the critical difference between clumsy and skillful use of the technological possibilities.

FAILURE OF MACHINE EXPLANATION TO MAKE AI SYSTEMS TEAM PLAYERS

The concept of escalation helps us understand why efforts to add machine explanation to intelligent systems failed to support cooperative interactions with human practitioners. Typically, expert systems developed their own solution to the problem at hand. Potential users found it difficult to accept such recommendations without some information about how the AI system arrived at its conclusions. This led many to develop ways to represent knowledge in such systems so they could provide a description of how a system arrived at the diagnosis or solution (e.g., Chandrasekaran, Tanner, & Josephson, 1989).

However they were generated and however they were represented, these explanations were provided at the end of some problem-solving activity after the intelligent system had arrived at a potential solution. As a result, they were one-shot, retrospective explanations for activity that had already occurred. The difficulties with explanations of this form generally went unnoticed. Effort was focused on building the explanation-generating mechanisms and knowledge representations. Development was directed toward contexts (or a simplified piece of a context was abstracted) where the underlying system was static and unchanging and where temporal relations were not significant. Even then a few noticed (e.g., Cawsey, 1992), in contrast to the assumptions of developers, when people engage in collaborative problem solving, they tend to provide information about the basis for their assessments as the problem-solving process unfolds to build a common ground for future coordination (e.g., Clark & Brennan, 1991; Johannesen, Cook, & Woods, 1994).

Warnings about problems with one-shot, retrospective explanations were disregarded until AI diagnostic systems were applied to dynamic situations.

Once such prototypes or systems had to deal with beyond-textbook situations, escalation occurred. The explanation then occurred at a time when the practitioner was likely to be engaged in multiple activities as a consequence of the cascade of effects of the initial event and escalating cognitive demands to understand and react as the situation evolved. These activities included generating and evaluating hypotheses, dealing with a new event or the consequences of the fault(s), planning corrective actions or monitoring for the effects of interventions, attempting to differentiate the influences due to faults and those due to corrective actions, among others.

These kinds of expert systems did not act as cooperative agents. For example, the expert systems did not gauge the importance or length of their messages against the background context of competing cues for attention and the state of the practitioner's ongoing activity. Thus, the system's output could occur as a disruption to other ongoing lines of reasoning and monitoring (Woods, 1995).

In addition, the presence of the intelligent system created new demands on the human practitioner. The typical one-shot retrospective explanation was disconnected from other data and displays the practitioner was examining. This meant the practitioner had to integrate the intelligent systems assessment with other available data as an extra task. This new task required the practitioner to shift attention away from what was currently going on in the process, possibly resulting in missed events.

Overall, the one-shot, retrospective style of explanation easily broke down under the demands of escalation. Practitioners, rather than being supported by the new systems, found extra workload during high-tempo periods and a new source of data competing for their attention when they were already confronted with an avalanche of changing data. As a result, practitioners adapted: They simply ignored the intelligent system (e.g., Malin et al., 1991, for the general pattern; Remington & Shafto, 1990, for one case).

There are several ironies about this pattern of technology change and its surprising reverberations. First, it had happened before. The same experience had occurred in the early 1980s when nuclear power tried to automate fault diagnosis with non-AI techniques. The systems were unable to function autonomously and only exacerbated the data overload that operators confronted when a fault produced a cascade of disturbances (Woods, 1994). That attempt to automate diagnosis was abandoned, although the organizations involved and the larger research community failed to see the potential to learn about dynamic patterns in human–machine cooperation.

A second irony is that to make progress in supporting human performance, efforts have moved away from autonomous machine explanations and toward understanding cooperative work and the ways that cognitive activity is distributed (Hutchins, 1995). The developers had assumed that their intelligent system could function essentially autonomously (at least on the important components of the task) and would be correct for almost all situations. In other words, they designed a system that would take over most of the cognitive work. The idea that human–intelligent system interaction required significant and meaningful cooperative work adapted to the changing demands and tempo of situations was outside their limited understanding of the cognitive demands of actual fields of practice.

IMPLICATIONS

At the beginning of the chapter, we posed a question—why is technology so often used clumsily, creating new complexities for already beleaguered practitioners?

The concept of escalation provides a partial explanation. In canonical cases, the technology seems to integrate smoothly into the work practices. The practitioners are able to process information from machine agents. The additional workload of coordinating with a machine agent is easily managed. More static views of the work environment may be acceptable simplifications for textbook situations.

The penalties for poor design of supporting artifacts emerge only when unexpected situations dynamically escalate cognitive and coordinative demands. In part, developers miss higher demand situations when design processes remain distant and disconnected from the actual field of practice. The current interest in field-oriented design techniques such as work analysis, cognitive task analysis, and ethnography reflects this state of affairs.

In part, developers misread and rationalize away the evidence of trouble created by their designs in some scenarios. This can occur because situations that escalate are relatively less frequent than canonical cases. Also, because practitioners adapt to escape from potential workload bottlenecks as criticality and tempo increase, the user hides the evidence that the system does not fit operational demands (Cook & Woods, 1996; Woods et al., 1994).

However, most important is that almost all design processes, including most human factors specialties, have missed the process of moving from canonical to exceptional that the concept of escalation captures. Supporting the escalation in cognitive and coordinated activity as problems cascade is a critical design task (Patterson, Woods, Sarter, & Watts-Perotti, 1998). To cope with escalation as a fundamental characteristic of cognitive work, one needs to design:

- How more knowledge and expertise can be integrated into an escalating situation.
- How more resources can be brought to bear to handle the multiple monitoring and attentional demands of escalating situations (Watts-Perotti & Woods, 1997).

- How to bring practitioners up to speed quickly when they are called in to support others (Patterson & Woods, 1997).

Many have noticed that scenario design is a critical activity for human-centered design processes (Carroll, 1997). Because escalation is fundamental to cognitive work, it specifies one target for scenario design. Field work techniques, such as building and analyzing corpuses of critical incidents, are needed to understand how situations move from textbook to nonroutine to exceptional in particular fields of practice, and particularly how this occurs after significant organizational or technological changes. Work is needed to identify general and specific complicating factors that shift situations beyond textbook plans (Roth & Mumaw, 1993).

Notice that the concept of escalation is not simply about problems, or about demands on cognition or on collaboration, or even simply about technological artifacts. Rather, it captures a dynamic interplay between all these factors. As a result, escalation illustrates a fundamental point distinguishing Cognitive Systems Engineering from other disciplines—joint and distributed cognitive systems are the fundamental unit of analysis for progress on understanding and designing systems of people and technology at work (Hutchins, 1995; Woods, 1998; Woods & Roth, 1988). Escalation, in particular, and distributed cognitive systems, in general, are concerned with relationships between problem demands, cognitive and coordinated activity, and artifacts.

ACKNOWLEDGMENTS

This work was supported by NASA Johnson Space Center (Grant NAGW-4560, Human Interaction Design for Anomaly Response Support, and Grant NAG 9-786, Human Interaction Design for Cooperating Automation) with special thanks to Dr. Jane Malin and her colleagues at NASA Johnson. Additional support was provided by a National Science Foundation Graduate Fellowship. Any opinions, findings, conclusions, or recommendations expressed in this publication are those of the authors and do not necessarily reflect the views of the National Science Foundation.

REFERENCES

Abbott, K. H. (1990). *Robust fault diagnosis of physical systems in operation.* Doctoral dissertation, State University of New Jersey, Rutgers.

Carroll, J. M. (1997). Scenario-based design. In M. G. Helander, T. K. Landauer, & P. Prabhu (Eds.), *Handbook of human–computer interaction* (2nd ed., pp. 383–406). Amsterdam: Elsevier Science.

Cawsey, A. (1992). *Explanation and interaction.* Cambridge, MA: MIT Press.

Chandrasekaran, B., Tanner, M. C., & Josephson, J. (1989). Explaining control strategies in problem solving. *IEEE Expert, 4*(1), 9–24.

Clark, H. H., & Brennan, S. E. (1991). Grounding in communication. In L. Resnick, J. M. Levine, & S. D. Teasley (Eds.), *Perspectives on socially shared cognition* (pp. 127–149). Washington, DC: American Psychological Association.

Cook, R. I., & Woods, D. D. (1996). Adapting to new technology in the operating room. *Human Factors, 38*(4), 593–613.

Hutchins, E. (1995). *Cognition in the wild.* Cambridge, MA: MIT Press.

Johannesen, L. J., Cook, R. I., & Woods, D. D. (1994). *Grounding explanations in evolving diagnostic situations.* (CSEL Rep. 1994-TR-03). Columbus, OH: The Ohio State University, Cognitive Systems Engineering Laboratory.

Malin, J., Schreckenghost, D., Woods, D., Potter, S., Johannesen, L., Holloway, M., & Forbus, K. (1991). *Making intelligent systems team players: Case studies and design issues.* (NASA Tech. Memo 104738.) Houston, TX: NASA Johnson Space Center.

Murray, C., & Cox, C. B. (1989). *Apollo, the race to the moon.* New York: Simon & Schuster

Norman, D. A. (1988). *The psychology of everyday things.* New York: Basic Books.

Patterson, E. S., & Woods, D. D. (1997). Shift changes, updates, and the on-call model in space shuttle mission control. In *Proceedings of the Human Factors and Ergonomics Society 41st annual meeting* (pp. 243–247). Albuquerque, NM: Human Factors Society.

Patterson, E. S., Woods, D. D., Sarter, N. B., & Watts-Perotti, J. (1998, May). *Patterns in cooperative cognition.* COOP '98, third international conference on the Design of Cooperative Systems, Cannes, France.

Reiersen, C. S., Marshall, E., & Baker, S. M. (1988). An experimental evaluation of an advanced alarm system for nuclear power plants. In J. Patrick and K. Duncan (Eds.), *Training, human decision making and control.* New York: North-Holland.

Remington, R. W., & Shafto, M. G. (1990, April). *Building human interfaces to fault diagnostic expert systems, I: Designing the human interface to support cooperative fault diagnosis.* Paper presented at CHI '90 Workshop on Computer–Human Interaction in Aerospace Systems, Seattle, WA.

Rochlin, G. I., La Porte, T. R., & Roberts, K. H. (1987, Autumn). The self-designing high-reliability organization: Aircraft carrier flight operations at sea. *Naval War College Review, 76–90.*

Roth, E. M., & Mumaw, R. J. (1993, April). *Operator performance in cognitively complex simulated emergencies.* Paper presented at the American Nuclear Society Topical Meeting on Nuclear Plant Instrumentation, Control, and Man–Machine Interface Technologies, Oak Ridge, TN.

Sarter, N. B., Woods, D. D., & Billings, C. (1997). Automation surprises. In G. Salvendy (Ed.), *Handbook of human factors/ergonomics* (2nd ed., pp. 1926–1943). New York: Wiley.

Watts, J. C., Woods, D. D., Corban, J. M., Patterson, E. S., Kerr, R., & Hicks, L. (1996). Voice loops as cooperative aids in space shuttle mission control. In *Proceedings of computer-supported cooperative work* (pp. 48–56). Boston: ACM.

Watts, J., Woods, D. D., & Patterson, E. S. (1996). Functionally distributed coordination during anomaly response in space shuttle mission control. In *Proceedings of Human Interaction with Complex Systems '96* (pp. 68–75). Dayton, OH.

Watts-Perotti, J., & Woods, D. D. (1997). *A cognitive analysis of functionally distributed anomaly response in space shuttle mission control.* (CSEL Rep. No. 1997-TR-02). Columbus, OH: The Ohio State University, Cognitive Systems Engineering Laboratory.

Wiener, E. L. (1989). Human factors of advanced technology ("glass cockpit") transport aircraft (NASA Contractor Report No. 177528). Moffett Field, CA: NASA Ames Research Center.

Woods, D. D. (1994). Cognitive demands and activities in dynamic fault management: Abduction and disturbance management. In N. Stanton (Ed.), *Human factors of alarm design* (pp. 63–92). London: Taylor & Francis.

Woods, D. D. (1995). The alarm problem and directed attention in dynamic fault management. *Ergonomics, 38*(11), 2371–2393.

Woods, D. D. (1998). Designs are hypotheses about how artifacts shape cognition and collaboration. *Ergonomics,* 41, 168–173.

Woods, D. D., Johannesen, L., Cook, R. I., & Sarter, N. B. (1994). *Behind human error: Cognitive systems, computers, and hindsight.* Dayton OH: Crew Systems Ergonomic Information and Analysis Center, WPAFB.

Woods, D. D., & Roth, E. M. (1988). Cognitive engineering: Human problem solving with tools. *Human Factors, 30,* 415–430.

Woods, D. D., & Sarter, N. (in press). Learning from automation surprises and going sour accidents. In N. Sarter & R. Amalberti (Eds.), *Cognitive engineering in the aviation domain.* Hillsdale, NJ: Lawrence Erlbaum Associates.

Woods, D. D., & Watts, J. C. (1997). How not to have to navigate through too many displays. In M. G. Helander, T. K. Landauer, & P. Prabhu (Eds.), *Handbook of human–computer interaction* (2nd ed., pp. 617–650). Amsterdam: Elsevier Science.

RESEARCH

Adaptive Control of Mental Workload

Raja Parasuraman
Catholic University of America

Peter A. Hancock
University of Minnesota

The practical importance of the concept of human mental workload was established several decades ago in the investigation of such human–machine systems as ground transportation (Brown & Poulton, 1961), air traffic control (Kalsbeek, 1965), and process control (Singleton, Whitfield, & Easterby, 1967). The theoretical development of the field can be traced to a NATO conference and the subsequent text, Mental Workload (Moray, 1979). Since that seminal volume, many studies have been conducted on the theoretical underpinnings, assessment techniques, and real-world implications of mental workload in a variety of work domains.

There has been a sequence of reviews of this literature (e.g., Andre & Hancock, 1995; Damos, 1991; Hancock & Meshkati, 1988; Hart & Wickens, 1990; Kantowitz & Campbell, 1996; Lysaght, Hill, Dick, & Wierwille, 1989; Moray, 1988; O'Donnell & Eggermeier, 1986; Parasuraman, 1990; Tsang & Wilson, 1997; Warm, Dember, & Hancock, 1996; Wickens, 1984, 1992; Wickens, Mavor, & McGee, 1997). These reviews have covered the main topics extensively, and consequently we have no wish to add yet another review to this list. Rather, in this chapter, we examine mental workload from a different perspective: that of *adaptive control,* or dynamic change to satisfy some goal. Our view of workload therefore is not as a static entity or one determined solely by the tasks imposed on the human operator, but as dynamic and multiply determined, with the changes being initiated by the human operator, the task at hand, the work environment, or some combinations of these respective factors. This perspective is predicated on the concept of humans as purposeful and goal-oriented agents (Craik, 1947; Newell &

Simon, 1972). Mental workload, according to this perspective, becomes a symptom or representation of the degree by which humans and machines (i.e., the human designers of machines) jointly and synergistically achieve their respective goals.

THE HUMAN OPERATOR AS AN ADAPTIVE AGENT

Many approaches to mental workload have assumed that it is an intervening construct that reflects the relation between the environmental demands imposed on the human operator and the capabilities of the operator to meet those demands. Workload may be *driven* by the task load imposed on human operators from external environmental sources but not deterministically so, because workload is also mediated by the individual response of humans to the load and their skill levels, task management strategies, and other personal characteristics. Although such definitions immediately suggest the notion of adaptation by the human operator, most studies have failed to explicitly examine this aspect of workload. The majority of studies have been concerned with empirical evaluations of the effects of various task load factors on various measures of mental workload—whether based on performance outcome, physiological response, or subjective report.

What is the relation between task load, mental workload, and performance? Generally, we expect that the operator's mental workload is related to task load, such that the higher the task load, the higher the operator's mental workload is likely to be. This relation is referred to as *association*. Consider the job of air traffic control (ATC). The more aircraft that have to be handled by a single controller in a given sector, the greater is the workload until, at some point, unacceptable overload results in performance failure (Arad, 1964). However, this does not necessarily mean that all controllers experience extremes of workload. Most controllers use some form of adaptive strategy to manage their performance and subjective perceptions of task involvement. Sperandio (1971) first showed that controllers handled an unexpected increase in traffic load adaptively by decreasing the amount of time they spent processing each aircraft, especially in verbal communication with the pilot. Controllers may also cease less important, peripheral tasks, thus leaving more time for active control, or alternatively they can regulate load by increasing spacing, stacking aircraft, or preventing aircraft from entering their sector.

Because ATC is a team activity, another possibility is that controllers ask colleagues to take over a particular component of the task. For example, the division of load between the radar (R)-side and data (D)-side positions is a time-honored adaptive strategy in ATC. If sector loading increases even more, these respective positions may be further subdivided. In exceptional

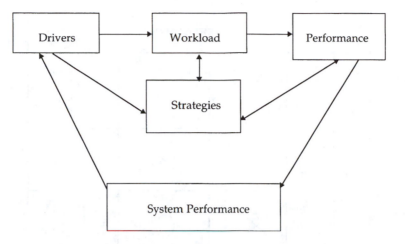

FIG. 2.4.1.   Inter-relationships between workload drivers, workload, and performance. From "Workload and Vigilance" (p. 116), by R. Parasuraman, in *Flight to the Future: Human Factors in Air Traffic Control,* 1997, Washington, DC: National Academy Press. Copyright 1997 by National Academy Press. Reprinted with permission.

circumstances, five or even six controllers may work one sector. However, subdivision of a sector has a limit, for any gains in individual controller workload are eventually outweighed by the added workload of communication and coordination between controllers (Hopkin, 1996). In general, controllers use a variety of strategies to manage workload and regulate their performance: If they do not use any of these adaptive strategies, further increases in traffic load may result in errors.

These considerations suggest that one needs to distinguish between workload *drivers* (i.e., task load and other factors in the work environment), experienced operator workload, adaptive strategies, and performance consequences (see Fig. 2.4.1). Workload can be thought of as being *driven* by various factors in the environment but also being sensitive to and therefore mediated by internal factors such as operator skill and motivation. The influence of these environmental workload drivers have been modeled (e.g., North & Riley, 1989; Rouse, Edwards, & Hammer, 1993), but must be supplemented by assessment of the human operator to measure the actual workload experienced. The human operator uses various strategies to cope with the external drivers. Human operator performance represents the joint consequences of the effects of task drivers on workload and the mediating influence of adaptive strategies.

Because humans can be very effective in their coping strategies, we do not necessarily observe association (between task load and mental workload) in

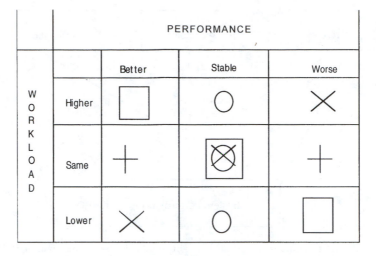

FIG. 2.4.2.   Matrix of performance and workload associations and dissociations.

all cases. Association implies that workload increases or decreases directly with external demands. However, there are a number of occasions on which *insensitivity* has been reported. In these cases, the external task load is either increased or decreased, but the experienced operator workload does not show evidence of change. Finally, there are unusual circumstances in which *dissociation* occurs. At such times, external task demand may be increasing while experienced workload is reduced, or vice versa. These unusual cases have been reported and addressed theoretically by Yeh and Wickens (1988) and Hancock (1996). Figure 2.4.2 illustrates the pattern of associations and dissociations that are possible between performance and workload. Such evidence points to the adaptive nature of human operator response and confirms that the human component cannot be considered a simple linear element of any human–machine system.

WORKLOAD MANAGEMENT

Previously (Hancock & Warm, 1989), a three-part differentiation of stress, workload, and fatigue has been presented. In this view, the objective demands of the external task represent the input load, the various strategies used to cope with this loading are termed adaptation, and the operator response level is the output. Operator performance affects system performance, which in combination with the operator's adaptive strategies can result in changes in input load (e.g., slowing down the rate of accepted aircraft by a sector controller). Thus input load, adaptive strategies, workload, and performance interact together as variables in a closed-loop system, with changes in any one variable propagating through the system. Fig. 2.4.2 illustrates these closed-loop relations.

The following relations are portrayed in Fig. 2.4.2. The first, shown by the vertical crosses (+), represents *workload insensitivity:* As performance changes, the individual adapts so that no change in experienced workload is evident. The second pattern, shown by the circles (○), indicates *performance insensitivity:* The individual experiences a change in workload while still managing to maintain a constant level of performance. The most commonly seen pattern, labeled *association,* is represented by the squares (□): As performance improves, the cost is seen in higher workload. The final pattern, indicated by the diagonal crosses (✕), shows *dissociation:* Here decreased performance results in higher workload and vice versa.

These relations provide a framework for understanding workload management and its impact on performance. As input load increases in an expected fashion, we have seen how the air traffic controller diffuses its effects by spreading the load. As a workload-management technique, load sharing can be considered to have both spatial and temporal components. Spatially dispersing load often means co-opting colleagues to reduce the rate of increase in input load, but more recently we have seen a much greater move toward machine-based task sharing. For example, the pilot may give flight control to the automation while solving a communications issue. In aviation, such sharing changes frequently, and thus automation is a dynamic adjustment. In contrast, traditional load sharing in design presents the problem of "static function allocation" to which many human factors efforts have been directed (Hancock & Scallen, 1996, 1998; Kantowitz & Sorkin, 1987). In contemporary complex systems, the spatial load sharing, that is, between the human and machine, is viewed as a dynamic and adaptive opportunity, so that the capabilities of the individual are taken as signals or symptoms of when and what tasks to switch to and from the machine component. Studies examining this type of load sharing are discussed later in this chapter. However, relatively little attention has been directed to the temporal dispersion of workload.

In the real world, many tasks can be offloaded or ignored. Unfortunately, very little systematic research has examined the conditions under which operators postpone tasks to a later time or even abandon them if they are not vital. (One exception is the very interesting work of Scerbo, 1996, who has studied the conditions under which people quit performing a boring vigilance task when given the opportunity to do so.) One reason for this lacuna are the real-world constraints versus the experimental procedures we choose to implement. Furthermore, investigators frequently use time-to-completion as an index of how well a task is being performed. Yet the same task in the real world possesses only the constraint that it be done "in time," that is before a crucial deadline. Its actual speed of completion is less relevant than its completion embedded in a sequence of other tasks. Few experimental evaluations of these context-crucial measures have been conducted, as opposed to methodologically convenient measures (for a discussion, see Hancock, 1991).

Workload management is influenced not only by the magnitude of input load but by its rate of change. One of the major environmental drivers of workload is rapid change in load (Wickens & Huey, 1993). This may take the form of unexpected increases or sudden collapse of demand. Less telling but nevertheless important are expected changes in input load. Many such work settings involve what has been termed "hours of boredom punctuated by moments of terror" (Hancock, 1997a). In some occupations such as those of professional musicians or surgeons, the schedule of demands may be known in advance, and the individual can prepare for peak performance. In other instances, however, particularly in emergency response situations, sudden changes in task load are unpredictable, and thus the transition itself becomes a stressful occasion.

AN ADAPTIVE MODEL OF WORKLOAD

The Hancock–Warm model of stress and performance has direct application to our view of workload promulgated here. Briefly, this model proposes that minor input disturbances (that is, relatively simple tasks) can be performed with little change in ongoing behavior. However, as load increases performance declines as a positive exponential at the limits of the individual's capability. Hancock and Warm (1989) showed that task performance was directly related to physiological state and that the breakdown of each at extremes bore particular resemblance. As the body itself uses physiological signals to regulate the internal environment, Hancock and Chignell (1988) suggested that the mental workload signal could be used in a comparable manner to regulate the task environment.

To try to capture these variations and the way that workload could be used as a signal to provide problem resolution, we, among others (Rouse, 1988; Scerbo, 1996), have suggested that task demand should be modified by recognition of the individual's (or group's) adaptive capacity to accept additional load. In this form of adaptive system architecture, the current level of experienced workload is used as an input to the controlling system to allocate or reallocate current task demands either away from or toward the human operator (see Hancock & Chignell, 1988). More recent versions of this concept use sophisticated elaborations of multiple inputs, together with assessment of the embedded context, to change the momentary demands. These systems have been used in aviation (Hettinger, Brickman, Roe, Nelson, & Haas, 1996) and in ground transportation (Hancock & Verwey, 1997).

OPERATOR AND SYSTEM ADAPTATION

Thus far we have considered human operators as adaptive agents whose goal is to regulate their mental workload at a manageable level. In addition to the strategies already mentioned, such as reduced processing of individual task elements (e.g., Sperandio, 1971), task prioritizing, or actual task shedding, human operators may also seek assistance from a computer or intelligent agent in workload management. Furthermore, when task shedding is not possible, temporary allocation of the task to automated control may be another and crucial adaptive strategy. The use of such forms of computer assistance or automation has been discussed under a variety of terms, including adaptive automation, adaptive aiding, adaptive interfaces, and adaptive function allocation (Byrne & Parasuraman, 1996; Hancock, Chignell, & Lowenthal, 1985; Parasuraman, 1993; Parasuraman, Bahri, Deaton, Morrison, & Barnes, 1992; Rouse, 1988; Scerbo, 1996). We use the term *adaptive automation* here to refer to cases where computer assistance is provided to the human operator at appropriate times (adaptive aiding) as well as to cases where responsibility for task performance is changed between human and computer at appropriate times (adaptive task allocation).

Adaptive automation represents an alternative to static automation in which computer assistance or task allocation between human operators and computer systems is flexible and context dependent rather than fixed. The benefits and costs of static automation have now been well documented (Parasuraman & Riley, 1997; Wiener, 1988). In a number of early theory and conceptual papers, adaptive automation was proposed to provide for regulation of operator workload and performance, while preserving the benefits of static automation (Hancock et al., 1985; Parasuraman et al., 1992; Wickens, 1992). Empirical tests of the validity of these claims have only just begun (Parasuraman, 1993; Parasuraman, Mouloua, & Molloy, 1996; Scallen, 1997; Scallen, Hancock, & Duley, 1995; see Scerbo, 1996, for a review).

Previous empirical evaluations of adaptive automation have focused primarily on the performance and workload effects of either (a) adaptive aiding (AA) of the human operator or (b) adaptive task allocation (ATA), either from the human to the machine (ATA-M), or from the machine to the human (ATA-H). Each of these forms of adaptive automation has been shown to enhance human–system performance.

Adaptive Aiding

An early study by Morris and Rouse (1986) showed that adaptive aiding in the form of target localization support enhanced operator performance in a simulated aerial search task. Benefits of AA in a more complex simulation were reported by Hilburn, Jorna, Byrne, and Parasuraman (1997), who provided air traffic controllers with a decision aid for determining optimal descent trajectories—the Descent Advisor (DA) of the Center Tracon Automation System (CTAS), an automation aid that is currently undergoing field trials at several air traffic control centers (Wickens, Mavor, Parasuraman, & McGee, 1998). Hilburn et al. (1997) found significant benefits for controller workload (as assessed by using physiological measures) when the DA was provided adaptively during high traffic loads, compared with when it was available throughout (static automation) or at low traffic loads.

Adaptive Task Allocation to the Machine

Benefits of adaptive task allocation from human to machine (ATA-M) have also been reported. In such a scheme, tasks are automated at appropriate times to allow the operator to focus attentional resources on other critical tasks. Empirical studies of this form of adaptive automation were required to ensure that workload was indeed decreased, given that previous studies have shown that automation sometimes can increase rather than reduce workload (Kirlik, 1993; Wiener, 1989). Second, any such scheme should not result in costs of returning the automated task back to manual control once the period of task overload had passed; that is, there should not be an return-to-manual performance costs. For the most part, ATA-M has been shown to benefit system performance and is not associated with significant costs (see Parasuraman, 1993; Scallen et al., 1995).

Adaptive Task Allocation to the Human

Where tasks are allocated to the machine (ATA-M), the benefits are perhaps not too surprising, given that automation can be thought of as a form of operator task shedding and hence should reduce workload under overload conditions. More surprising are studies showing that temporary allocation of a previously automated task to human control (ATA-H) can also be beneficial, particularly with respect to one form of human performance vulnerability, monitoring of automated systems. For example, a task may be automated for long periods with no human intervention. Under such conditions of static automation, operator detection of automation malfunctions can be inefficient if the human operator is engaged in other manual tasks (Molloy & Parasuraman, 1996; Parasuraman et al., 1993). The problem does not go away, and may even be exacerbated, with highly reliable automation (Parasuraman et al., 1996).

Can automation-induced monitoring inefficiency be reduced or even eliminated? One possibility is to reallocate a formerly automated task to the human operator, because an "in-the-loop" monitor performs better than one who is "out of the loop" (Parasuraman et al., 1993; Wickens & Kessel, 1979). However, this allocation strategy can clearly not be pursued generally for all automated tasks and at all times, for it would lead to excessive manual workload, thus defeating one of the principal purposes of automation. One potential solution is to allocate the automated task to the human operator for only brief periods before returning it once again to automation. The benefits of temporary allocation of a task to human control might persist for some time even after the task is returned to automation control. This hypothesis was tested in a study by Parasuraman et al. (1996). During multiple-task flight simulation, a previously automated engine-status task was adaptively allocated to the operator for a 10-minute period in the middle of a session and then returned to automatic control. Detection of engine malfunctions was better during the 10-minute block when the task was returned to human control from automation, consistent with previous reports of superior monitoring under conditions of active human control (Parasuraman, Molloy, & Singh, 1993; Wickens & Kessel, 1979). More important, however, detection performance under automation control was markedly superior in the *postallocation phase* than in the identical preallocation phase. The performance benefit (of about 66%) persisted even after the engine-status task was returned to automation, for about 20 minutes. The benefit of adaptive task allocation was attributed to this procedure's allowing human operators to update their memory of the engine-status task (see also Farrell & Lewandowsky, 2000).

Joint Adaptive Systems

Thus far we have seen that AA, ATA-M, and ATA-H can all have beneficial effects on performance-related workload. For adaptive systems to be effective, these different forms of adaptive automation need to be examined jointly in a single work domain. Furthermore, if adaptive systems are designed in a manner typical of "clumsy automation"—providing aiding or task reallocation when they are least helpful (Wiener, 1989)—then performance may be degraded rather than enhanced (see also Billings & Woods, 1994). One of the drawbacks of some flightdeck automated systems—for example, the Flight Management System (FMS)—is that they often require extensive reprogramming and impose added workload during high task-load phases of flight such as final approach and landing, while doing little to regulate workload during the low-workload phase of cruise flight.

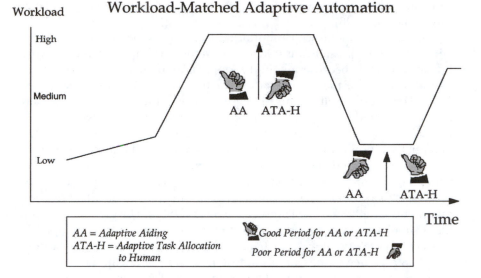

FIG. 2.4.3.   Model for workload-based adaptive aiding and adaptive task allocation. From *Automation technology and human performance: Current research
and trends* (p. 120), by R. Parasuraman, M. Mouloua, and B. Hilburn (1999),
Mahwah, NJ: Lawrence Erlbaum Associates. Copyright 1999 by Lawrence
Erlbaum Associates.

These considerations indicate that the provision of aiding or task allocation should be directly linked to the level of operator workload. Following
Hancock et al. (1985), we propose a model for effective combination of adaptive aiding and adaptive task allocation, based on matching adaptation to
operator workload. This model is illustrated is Fig. 2.4.3, which plots operator workload against work period or time. As workload fluctuates because of
variations in task load or operator strategies, high-workload periods represent good times for AA and poor times for ATA-H; the converse is true for
low-workload periods. According to this model, if AA and ATA-H are to be
combined in a adaptive system, they should be provided at the times indicated to regulate operator workload and maximize performance.

This model for adaptive automation was evaluated in a recent study in
which pilots performed multiple flight-related tasks under different conditions of adaptive automation—either AA or ATA-H (Parasuraman, Mouloua, & Hilburn, 1999). We predicted that, compared with a nonadaptive
group receiving neither AA nor ATA-H, a workload-matched adaptive group
receiving AA at high workload and ATA-H at low workload would show superior performance and reduced workload. To confirm that any observed
benefits were due to workload-matched adaptation rather than to aiding per
se, we also tested a "clumsy automation" group who received aiding at times

not matched to their workload, that is, AA at low workload and ATA-H at high workload.

Mean tracking root-mean-square error was significantly lower and subjective workload (mean NASA–TLX, or Task Load Index, score) was significantly greater during the first and last phases compared with the middle phase of the 60-minute session. Thus, the tracking difficulty manipulation was successful in inducing greater levels of mental workload during the early and late phases of the simulation, as they would be during the takeoff and landing phases of actual flight. Overall performance on all three subtasks was significantly higher in the workload-matched adaptive group than in the other two groups. In addition, subjective workload was significantly lower in this group than in the other groups. The performance benefits of workload-matched adaptation were most marked for an engine-systems monitoring task, the Engine Indicator and Crew Alerting System (EICAS). The detection rate of automation failures was significantly higher for the workload-matched adaptive group than for the other two groups. As Fig. 2.4.4 shows, the performance benefit of ATA-H persisted beyond the period of manual control (middle 20-minute phase) to the last 20-minute phase when the

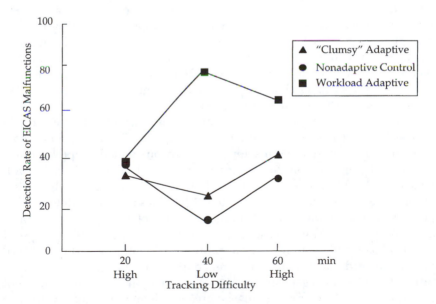

FIG. 2.4.4.   Detection rate of automation failures in the EICAS task for the workload-matched adaptive group, the "clumsy-automation" group, and the nonadaptive control group. From *Automation technology and human performance: Current research and trends* (p. 122), by R. Parasuraman, M. Mouloua, and B. Hilburn (1999), Mahwah, NJ: Lawrence Erlbaum Associates. Copyright 1999 by Lawrence Erlbaum Associates.

EICAS task was returned to automated control. This result is consistent with the previous findings of Parasuraman et al. (1996).

The workload and performance levels of the clumsy-automation adaptive group did not differ significantly from those of the nonadaptive control group. The performance levels for this group were equivalent to or in some cases lower than those of the control group. Fig. 2.4.4 shows that the monitoring performance benefits of workload-matched adaptation did not accrue to the clumsy-automation group.

These results validate a design approach to adaptive automation involving adaptation matched to operator workload (Hancock et al., 1985; Parasuraman et al., 1992). Under these conditions, adaptive aiding and adaptive task allocation both enhance performance. The results also show, however, that performance benefits are eliminated if adaptive automation is implemented in a clumsy manner.

CONCLUSIONS:
FUTURE RESEARCH ON MENTAL WORKLOAD

Our examination of human mental workload has been guided by the central concept of adaptive control. As we have seen, adaptation is clearly a fundamental characteristic of human goal-oriented behavior. Workload management can then been viewed as human adaptive response to task load. A theoretical model of workload and stress that provides a framework for understanding the overall characteristics of this human adaptability has also been discussed (Hancock & Warm, 1989).

Given that they are built by humans, machines can, and we suggest should also be designed to, exhibit some degree of adaptation. Some legitimate concerns have been raised that machines with unconstrained adaptive powers can compete with and confuse the human operator (Billings & Woods, 1994). Hence machine adaptation should be allowed only to a degree. The notion of joint adaptive systems that we discussed provides a basis for determining just how much and what kind of adaptation is beneficial to system performance. We provided empirical support for the performance effectiveness of workload-based adaptation (e.g, Parasuraman, Mouloua, & Hilburn, 1999). We also showed that "clumsy" machine adaptation can actually be counterproductive and can negate any benefits of automation. Additional studies of this type need to be carried out to examine and define the limits to effective adaptation.

As a final comment, we consider that it is important to continue to question the very concept of workload. After many years and many hundreds of empirical investigations, we still do not have a satisfactory, consensual definition of workload. Despite such concerns, there are many reasons that we should continue to understand mental workload. First, there are theoretical motivations as we seek to understand and relate this phenomenon to allied energetic and cognitive constructs such as attention, effort, and resources (Kahneman, 1973; Moray, 1967; Norman & Bobrow, 1975; Wickens, 1984). Workload is part of a more complex picture in which the full realization of human consciousness may be expected. Furthermore, detailed knowledge of mental workload has helped modify and elaborate sterile information-processing models of human behavior that neglect energetic considerations (Hockey, Gaillard, & Coles, 1986), much as cognitive neuroscience is revitalizing and allowing cognitive psychology to rediscover that minds are embodied (Gazzaniga, 1995; Parasuraman, 1998).

A second reason we seek to understand workload are the practical issues associated with human work. For most of the developed world, the nature of work has changed from being physical to primarily cognitive in nature: The Industrial Revolution of the 19th century is essentially complete, and we are entering the era of intelligent automation in the 21st century. We need to be able to assess the "load" of such information-based work, just as earlier in the century (and still in the developing countries) we needed to understand just how much physical load a worker can withstand without damage or diminution of performance capability. We wish to know, therefore, how much mental load is too much and may prove hazardous or stressful. In a similar vein, we need to know how little is too little so that individuals are sufficiently challenged to sustain useful levels of output. Some commentators have speculated that coping with underload may prove to be an even bigger challenge than overload in the coming era of automation (Hancock, 1997b; Hopkin, 1996).

A third reason is that mental workload must be assessed as a signal or indicator for use in complex technical systems to dynamically change either the nature or the demands of the work at hand. Although this may appear trivial in certain occupations, this information is vital in a number of growing realms. As virtually all forms of work involve human–computer interaction, this strategy becomes progressively more feasible and indeed crucial for complex systems. This reason alone guarantees the need for the future study of the concept of mental workload.

REFERENCES

Andre, A. D., & Hancock, P. A. (1995). Special issue on pilot workload. *International Journal of Aviation Psychology, 5,* 1–4.

Arad, B. A. (1964, May). The control load and sector design. *Journal of Air Traffic Control,* 12–31.

Billings, C. E., & Woods, D. D. (1994). Concerns about adaptive automation in aviation systems. In R. Parasuraman & M. Mouloua (Eds.), *Human performance in automated systems: Current research and trends* (pp. 264–269). Hillsdale, NJ: Lawrence Erlbaum Associates.

Brown, I. D., & Poulton, E. C. (1961). Measuring the spare "mental capacity" of car drivers by a subsidiary task. *Ergonomics, 4,* 35–40.

Byrne, E. A., & Parasuraman, R. (1996). Psychophysiology and adaptive automation. *Biological Psychology, 42,* 249–268.

Craik, K. J. W. (1947). The theory of the human operator in control systems, II: Man as an element in a control system. *British Journal of Psychology, 38,* 142–148.

Damos, D. (1991). *Multiple task performance.* London: Taylor & Francis.

Farrell, S., & Lewandowsky, S. (2000). A connectionist model of complacency and adaptive recovery under automation. *Journal of Experimental Psychology: Learning Memory and Cognition, 26,* 395–410.

Gazzaniga, M. S. (1995). *The cognitive neurosciences.* Cambridge, MA: MIT Press.

Hancock, P. A. (1991). On operator strategic behavior. In *Proceedings of the 6th International Symposium on Aviation Psychology.* Columbus: Ohio State University.

Hancock, P. A. (1996). Effect of control order, augmented feedback, input device, and practice on tracking performance and perceived workload. *Ergonomics, 39,* 1146–1162.

Hancock, P. A. (1997a). Hours of boredom, moments of terror—or months of monotony, milliseconds of mayhem. In *Proceedings of the 9th International Symposium on Aviation Psychology.* Columbus: Ohio State University.

Hancock, P. A. (1997b). On the future of work. *Ergonomics in Design, 5,* 25–29.

Hancock, P. A., & Chignell, M. H. (1988). Mental workload dynamics in adaptive interface design. *IEEE Transactions on Systems, Man, and Cybernetics, 18,* 647–658.

Hancock, P. A., Chignell, M. H., & Lowenthal, A. (1985). An adaptive human–machine system. *Proceedings of the IEEE Conference on Systems, Man, and Cybernetics, 15,* 627–629.

Hancock, P. A., & Meshkati, N. (1988). *Human mental workload.* Amsterdam: North-Holland.

Hancock, P. A., & Scallen, S. F. (1996). The future of function allocation. *Ergonomics in Design, 4*(4), 24–29.

Hancock, P. A., & Scallen, S. F. (1998). Allocating functions in human–machine systems. In R. R. Hoffman, M. F. Sherrick, & J. S. Warm (Eds.), *Viewing psychology as a whole: The integrative science of William N. Dember* (pp. 509–539). Washington, DC: American Psychological Association.

Hancock, P. A., & Verwey, W. B. (1997). Fatigue, workload, and adaptive driver systems. *Accident Analysis and Prevention, 29,* 495–506.

Hancock, P. A., & Warm, J. S. (1989). A dynamic theory of stress and sustained attention. *Human Factors, 31,* 519–537.

Hart, S. G., & Wickens, C. D. (1990). Workload assessment and predictions. In H. R. Booher (Ed.), *MANPRINT: An emerging technology* (pp. 311–334). New York: Van Rostrand & Reinhold.

Hettinger, L. J., Brickman, B. J., Roe, M. M., Nelson, W. T., & Haas, M. W. (1996). Effects of virtually augmented fighter cockpit displays on pilot performance, workload and situation awareness. *Proceedings of the 40th Annual Meeting of the Human Factors and Ergonomics Society,* Santa Monica, CA: HFES.

Hilburn, B., Jorna, P. G. A. M., Byrne, E. A., & Parasuraman, R. (1997). The effect of adaptive air traffic control (ATC) decision aiding on controller mental workload. In M. Mouloua & J. Koonce (Eds.), *Human-automation interaction: Research and practice* (pp. 84–91). Mahwah, NJ: Lawrence Erlbaum Associates.

Hopkin, D. (1996). *Human factors in air traffic control.* New York: Taylor & Francis.

Hockey, G. R. J., Gaillard, A., & Coles M. G. H. (1986). *Energetics and human information processing.* Amsterdam: Martinus Nijhoff.

Kahneman, D. (1973). *Attention and effort.* Englewood-Cliffs, NJ: Prentice Hall.

Kalsbeek, J. W. H. (1965). Mésure objective de la surcharge mentale: Nouvelles applications de la méthode des doubles taches. *Le Travail Humain, 28,* 121–132.

Kantowitz, B. H., & Campbell, B. L. (1996). Pilot workload and flight-deck automation. In

R. Parasuraman & M. Mouloua (Eds.), *Automation and human performance: Theory and applications* (pp. 117–136). Hillsdale, NJ: Lawrence Erlbaum Associates.

Kantowitz, B. H., & Sorkin, R. (1987). Allocation of function. In G. Salvendy (Ed.), *Handbook of human factors and ergonomics* (pp. 355–369). New York: Wiley.

Kirlik, A. (1993). Modeling strategic behavior in human–automation interaction: Why an "aid" can (and should) go unused. *Human Factors, 35,* 221–242.

Lysaght, R. J., Hill, S. G., Dick, A. O., & Wierwille, W. (1989). *Operator workload: Comprehensive review and evaluation of operator workload methodologies.* (Report No. 851). Fort Bliss, TX: Army Research Institute.

Molloy, R., & Parasuraman, R. (1996). Monitoring an automated system for a single failure: Vigilance and task complexity effects *Human Factors, 38,* 311–322.

Moray, N. (1967). Where is capacity limited? A survey and a model. *Acta Psychologica, 27,* 84–92.

Moray, N. (1979). *Mental workload.* New York: Plenum.

Moray, N. (1988). Mental workload since 1979. *International Reviews of Ergonomics, 2,* 123–150.

Morris, N. M., & Rouse, W. B. (1986). *Adaptive aiding for human–computer control: Experimental studies of dynamic task allocation.* (Report No. AAMRL-TR-86-005). Dayton, OH: Wright–Patterson Air Force Base.

Newell, A., & Simon, H. A. (1972). *Human problem solving.* Englewood Cliffs, NJ: Prentice Hall.

Norman, D. A., & Bobrow, D. (1975). On data-limited and resource-limited processing. *Cognitive Psychology, 7,* 44–60.

North, R. A., & Riley, V. A. (1989). A predictive model of operator workload. In G. R. McMillan (Ed.), *Applications of human performance models to system design* (pp. 81–90). New York: Plenum.

O'Donnell, R. D., & Eggermeier, F. T. (1986). Workload assessment methodology. In K. Boff, L. Kaufman, & J. Thomas (Eds.), *Handbook of perception, Volume 2: Cognitive processes and performance* (pp. 42.1–42.49). New York: Wiley.

Parasuraman, R. (1990). Event-related brain potentials and human factors research. In J. W. Rohrbaugh, R. Parasuraman, & R. Johnson (Eds.), *Event-related brain potentials: Basic and applied issues* (pp. 279–300). New York: Oxford University Press.

Parasuraman, R. (1993). Effects of adaptive function allocation on human performance. In D. J. Garland & J. A. Wise (Eds.), *Human factors and advanced aviation technologies* (pp. 147–157). Daytona Beach, FL: Embry-Riddle Aeronautical University Press.

Parasuraman, R. (1998). *The attentive brain.* Cambridge, MA: MIT Press.

Parasuraman, R., Bahri, T., Deaton, J., Morrison, J., & Barnes, M. (1992). *Theory and design of adaptive automation in aviation systems.* (Progress Rep. No. NAWCADWAR-92033-60). Warminster, PA: Naval Air Warfare Center, Aircraft Division.

Parasuraman, R., Molloy, R., & Singh, I. L. (1993). Performance consequences of automation-induced "complacency." *International Journal of Aviation Psychology, 3,* 1–23.

Parasuraman, R., Mouloua, M., & Hilburn, B. (1999). Adaptive aiding and adaptive task allocation enhance human–machine interaction. In M. Scerbo & M. Mouloua (Eds.), *Automation technology and human performance: Current research and trends* (pp. 119–123). Mahwah, NJ: Lawrence Erlbaum Associates.

Parasuraman, R., Mouloua, M., & Molloy, R. (1996). Effects of adaptive task allocation on monitoring of automated systems. Human Factors, 38, 665–679.

Parasuraman, R., & Riley, V. (1997). Humans and automation: Use, misuse, disuse, abuse. *Human Factors, 39,* 230–253.

Rouse, W. B. (1988). Adaptive aiding for human/computer control. *Human Factors, 30,* 431–438.

Rouse, W. B., Edwards, S. L., & Hammer, J. M. (1993). Modeling the dynamics of mental workload and human performance in complex systems. *IEEE Transactions on Systems, Man, and Cybernetics, 23,* 1662–1671.

Scallen, S. F. (1997). *Performance and workload effects for full versus partial automation in a high-fidelity multitask system.* Unpublished doctoral dissertation, University of Minnesota, Minneapolis.

Scallen, S. F., Hancock, P. A., & Duley, J. A. (1995). Pilot preference for short cycles of automation in adaptive function allocation. *Applied Ergonomics, 26,* 397–403.

Scerbo, M. S. (1996). Theoretical perspectives on adaptive automation. In R. Parasuraman & M. Mouloua (Eds.), *Automation and human performance: Theory and applications* (pp. 37–63). Mahwah, NJ: Lawrence Erlbaum Associates.

Singleton, W. T., Whitfield, D., & Easterby, R. S. (1967). *The human operator in control systems.* London: Taylor & Francis.

Sperandio, J. C. (1971). Variation of operator's strategies and regulating effects on workload. *Ergonomics, 14,* 571–577.

Tsang, P., & Wilson, J. (1997). Mental workload. In G. Salvendy (Ed.), *Handbook of human factors and ergonomics* (2nd ed., pp. 417–449). New York: Plenum.

Warm, J. S., Dember, W. N., & Hancock, P. A. (1996). Vigilance and workload in automated systems. In R. Parasuraman & M. Mouloua (Eds.), *Automation and human performance: Theory and applications* (pp. 183–200). Hillsdale, NJ: Lawrence Erlbaum Associates.

Wickens, C. D. (1984). Processing resources in attention. In R. Parasuraman & D. R. Davies (Eds.), *Varieties of attention* (pp. 63–102). Orlando, FL: Academic Press.

Wickens, C. D. (1992). *Engineering psychology and human performance* (2nd ed.). New York: HarperCollins.

Wickens, C. D., & Huey, B. M. (1993). *Workload transition.* Washington DC: National Academy Press.

Wickens, C. D., & Kessel, C. (1979). The effects of participatory mode and task workload on the detection of dynamic system failures. *IEEE Transactions on Systems, Man, and Cybernetics, 9,* 24–34.

Wickens, C. D., Mavor, A., & McGee, J. (1997). *Flight to the future.* Washington DC: National Academy Press.

Wickens, C. D., Mavor, A., Parasuraman, R., & McGee, J. (1998). *The future of air traffic control: Human operators and automation.* Washington, DC: National Academy Press.

Wiener, E. L. (1988). Cockpit automation. In E. L. Wiener & D. C. Nagel (Eds.), *Human factors in aviation* (pp. 433–461). San Diego, CA: Academic Press.

Wiener, E. L. (1989). *Human factors of advanced technology ("glass cockpit") transport aircraft.* (Report No. 177528). Moffett Field, CA: Ames Research Center.

Yeh, Y. Y., & Wickens, C. D. (1988). Dissociation of performance and subjective measures of workload. *Human Factors, 30,* 111–120.

Assessment of Drivers' Workload: Performance and Subjective and Physiological Indexes

Karel A. Brookhuis
Dick de Waard
University of Groningen

There are many reasons that the measurement of drivers' mental workload has great interest these days and will increasingly enjoy this status in the near future. Accidents are numerous, seemingly ineradicable, very costly, and in fact largely attributable to the victims themselves, human beings. Human errors in the sense of imperfect perception, insufficient attention, and inadequate information processing are the major causes of the bulk of the accidents on the road (Smiley & Brookhuis, 1987; Treat et al., 1977). Although both low and high mental workloads are undoubtedly basic conditions for these errors, an exact relation between mental workload and accident causation is not easily established, let alone measured in practice. Brookhuis, Van Winsum, Heijer, and Duynstee (1999) discriminated between underload and overload, the former leading to reduced alertness and lowered attention, the latter to distraction, diverted attention, and insufficient time for adequate information processing. Both factors have been studied in relation to driver impairment; however, the coupling to accident causation is not via a direct link (see also Brookhuis et al., 1999). Criteria for *when* impairment is below a certain threshold, leading to accidents, have to be established. Only then can accidents and mental workload (high or low) be related, in conjunction with the origins, such as information overload, boredom, fatigue, or factors such as alcohol and drugs. The traffic environment and traffic itself will only gain in complexity, at least for the present, with the rapid growth in numbers of automobiles and telematics applications. Aging plays a role in the interest in the measurement of drivers' mental workload these days and will increasingly do so in the near future.

COGNITIVE PROCESSING AND EFFORT

At least 25 years ago, Kahneman (1973) defined mental workload as directly related to the proportion of the capacity an operator spends on task performance. The measurement of mental workload is the specification of that proportion (De Waard & Brookhuis, 1997; O'Donnell & Eggemeier, 1986), in terms of the costs of the cognitive processing, which is also referred to as mental effort (Mulder, 1980). Mental effort is similar to what is commonly referred to as doing one's best to achieve a certain target level, even to "trying hard" in case of a strong cognitive processing demand. The concomittant changes in effort do not easily show in some of the driving-performance measures because drivers are inclined to actively cope with changes in task demands by adapting their driving behavior (Cnossen, Brookhuis, & Meijman, 1997). However, they are apparent in self-reports of the drivers and a fortiori in the changes in certain physiological measures.

Mulder (1986) discriminated between two types of mental effort, the mental effort devoted to the processing of information in controlled mode (computational effort) and the mental effort needed to apply when the operator's energetical state is affected (compensatory effort). Computational effort is exerted to keep task performance at an acceptable level, for instance, when task complexity level varies or tasks are added to the primary task. In case of (ominous) overload, extra computational effort could forstall safety hazards in this way. Compensatory effort takes care of performance decrement in case of, for instance, fatigue up to a certain level. Underload by boredom, affecting the operator's capability to deal with the task demands, might be compensated as well. If effort is exerted, whether computational or compensatory, both task difficulty and mental workload are increased. Effort is a voluntary process under control by the operator, whereas mental workload is determined by the interaction of operator and task. As an alternative to exerting effort, the operator might decide to change the (sub)goals of the task. Adapting driving behavior as a strategic solution is a well-known phenomenon. For example, overload because of an additional task, such as looking up telephone numbers while driving, is demonstrated to be reduced by lowering vehicle speed (see De Waard, Van der Hulst, & Brookhuis, 1997).

MEASUREMENT TECHNIQUES
IN THE FIELD OF TRAFFIC RESEARCH

There are three global categories distinguished in this field: measures of task performance, subjective reports, and physiological measures (see also Brookhuis, 1993; O'Donnell & Eggemeier, 1988; Wierwille & Eggemeier,

1993). The first and by far the most frequently used category of measures is based on techniques of direct registration of the operator's capability to perform the driving task at an acceptable level, that is, with respect to an acceptably low accident likelihood. These measures of task performance are directly related to vehicle handling, that is, lateral and longitudinal vehicle control, such as steering and car following. Subjective reports of driving performance are of two kinds: observer reports, which are mostly given by experts, and self-reports by the drivers. The value of the first exists by virtue of strict protocols that limit variation as produced by personal interpretation; the second exists mainly by virtue of validation through multiple applications in controlled settings. Well-known examples of the latter are the NASA–Task-Load Index (NASA-TLX: Hart & Staveland, 1988) and the Rating Scale Mental Effort (RMSE: Zijlstra, 1993). Finally, physiological measures are a natural type of workload index because work demands physiological activity by definition. Physical workload and also mental workload have a clear impact on heart rate and heart rate variability (Mulder, 1980, 1986, 1988). Mental workload might increase heart rate and decrease heart rate variability at the same time. Other measures of interest are the activity of the brain (Brookhuis, 1993; De Waard & Brookhuis, 1991) and of certain facial muscles (Jessurun, 1997).

Examples of all three categories are given for different types of mental workload in this chapter, with their limitations and a view to specific forms of effort exerted.

MEASURES OF TASK PERFORMANCE

The (theoretical) objective of collecting performance measures is to register the capability of the driver to perform the driving task in such a way that the likelihood of safely arriving at destination is optimal. Optimal performance on the task of driving a motor vehicle from A to B is available from detailed task analyses of driving behavior like the one composed as early as 1970 by McKnight and Adams (1970a, 1970b). McKnight and Adams described the task of driving a motor vehicle in such detail that strictly applying the prescriptions of their normative analysis in theory leads to nothing less than perfect performance. In practice, the driver performs less adequately than theoretically feasible. This is no problem, however, because the available space for variation in driving behavior is normally sufficient to allow for ample deviation from the optimum. The problem lies in assessing criteria for the determination of unacceptable deviation from the optimum with respect to traffic safety (Brookhuis, 1995). In other words, at what level of which parameter does the observer or investigator decide that driving performance is unacceptably poor in the sense of accident likelihood. The (practical) objective of collecting performance measures is to determine whether driving behavior parameters are within performance margins (Wickens, 1984; see also De Waard & Brookhuis, 1997). Driving-performance measures are based on indexing lateral and longitudinal vehicle control, with ample margin between perfect (the norm) and borderline safe.

One of the most frequently used parameters for primary task performance is the amount of variability in lateral control, the standard deviation of the lateral position (SDLP). SDLP is closely related to the likelihood of leaving the traffic lane and getting involved in an accident (Allen & Stein, 1987). The amount of swerving increases if the energetic state of the driver is suboptimal, which could be the result of the use of sedative drugs (e.g., Brookhuis, 1995; Brookhuis, De Vries, & De Waard, 1993; O'Hanlon, Haak, Blaauw, & Riemersma, 1982), of being fatigued, or of driving under the influence of low amounts of alcohol (De Waard & Brookhuis, 1991). Extremely high additional task demands, such as the earlier mentioned telephone number lookup task (De Waard, Van der Hulst, & Brookhuis, 1999a), are also reflected in increased swerving behavior. However, safety margins in lane control can be very large, as relatively wide motorway lanes allow for swerving. An increase in SDLP of 0.03 metres may thus be statistically significant, but it can well be irrelevant. As a matter of fact, in dual task performance, SDLP has sometimes been found to *decrease* compared with single task performance (cf. Brookhuis, De Vries, & De Waard, 1991), whereas in this condition an increase would be expected. In these conditions, an increase in task demands has led to the mobilization of effort and increased alertness resulting in improved performance. Longitudinal control is the other primary task in keeping the vehicle safely on the road. Apart from speed, headway to cars in front must also be regulated by the driver. Headway control is usually assessed in terms of chosen time distance or time headway. Adaptation of behavior to maneuvers of other cars requires perception and attention. Reaction time to these maneuvers has been shown to be a good indicator of driver performance (Brookhuis, De Waard, & Mulder, 1994). Both increased task demands (handling a car phone) and decreased capacity as a result of the intake of a sedative drug delayed drivers' response to lead-car speed changes (Brookhuis, De Vries, & De Waard, 1991; Brookhuis et al., 1993).

There is a useful distinction between two methods or techniques of measuring driving performance, that is, primary task and secondary task methodology. Performance on an added, secondary task is sometimes used to assess mental workload (e.g., Verwey & Veltman, 1996). The idea behind this secondary task methodology is that both the primary and the secondary tasks tap the same limited resources. Although during primary task performance the driver still has "spare" capacity (Brown & Poulton, 1961), this capacity is required when performing a secondary task. As more than full capacity might be needed for performance in this case, either the primary or the added-task performance suffers. At the same time, this very same effect points to the main problem of the secondary task technique: It is impossible to find out how drivers deal with the trade-off between the primary and secondary tasks. Do they give (continuous?) priority to one task over the other? Are there additional "central" costs for coordination of resource assignment to the two tasks? Moreover, the secondary task is frequently unnatural, unnecessary for safety control, and clearly "added." De Waard (1996) has argued that for the assessment of drivers' mental workload, embedded secondary tasks are the best choice. Embedded tasks are tasks that are performed during normal system operation, but are distinct from the function that is under assessment (Eggemeier & Wilson, 1991). In particular, in naturalistic environments and well-trained tasks such as car driving, embedded tasks offer the possibility for nonintrusive secondary performance assessment. Brookhuis et al. (1991), Fairclough, Ashby, and Parkes (1993), and Lansdown (1997) have used the number and duration of glances in the rearview mirror as a secondary task performance measure. Brookhuis et al. (1991), for instance, found less glances in the rearview mirror under conditions of high mental workload, that is, while driving on busy roads, and when handling a carphone.

Subjective Measures

There are two types of subjective measures, self-reports and (expert) judgments. The most frequently used self-reports of mental workload are the SWAT and NASA-TLX (see Eggemeier & Wilson, 1991). These are multidimensional scales, so that ratings on different dimensions have to be given. Although these different ratings can give insight into the origin of mental workload, it is not always easy for subjects to differentiate between the dimensions. For many subjects, the meanings of the TLX-dimensions of effort, mental workload, and physical workload are not very different. Moreover, whenever an overall rating of workload is required, unidimensional scales have been found to be superior (Hendy, Hamilton, & Landry, 1993). The RSME (Zijlstra, 1993) is an example of such a scale, on which mental effort is indexed on a continuous line. On the scale, the amount of invested effort is rated, instead of the earlier-mentioned more abstract concepts as workload. In a flight simulator study, the RSME has been found to be more sensitive to mental workload than is the TLX (Veltman & Gaillard, 1996), while De Waard (1996) has reported sensitivity of the RSME to both compensatory (energetic) and computational (controlled processing) effort.

A disadvantage of self-reports is that people are sometimes not aware of internal changes. Brookhuis et al. (1993) found that subjects in a driving test after antihistamine administration performed significantly worse than under placebo; however, they were not aware of reduced alertness and only slightly aware of impaired performance, but they did not indicate more effort to compensate. The main advantage of self-report ratings is that they allow those concerned to express what they experienced. The fact that self-report ratings are easy to obtain and that no esoteric knowledge is required probably also help to make them very popular.

Expert observers should be able to give a quality rating of driving performance on a series of safety-related aspects, assessing the measures per driving situation. Several preconditions are necessary (see also Verwey, Brookhuis, & Janssen, 1996). To ensure the reliability of the expert opinions, inter-rater variability should be determined. If this variability is high, data are useless. The expert observers should provide detailed quality opinions about driving according to a predetermined list of driving quality variables, using standardized techniques for assessing erroneous and unsafe driving behavior. Finally, to avoid any prejudice, judgments are to be made from video.

Physiological Measures

The final category of measures to be treated here for registering mental workload is the measurement of physiological parameters. Probably the most frequently applied measure in applied research is the electrocardiogram (ECG). The time between successive R-waves, the interbeat-interval time (IBI), as well as variability in the IBI are the prime measures. Variability in heart rate (HRV) can be computed in the time domain and is standardized by dividing the standard deviation of the IBI by the average IBI. HRV can also be analyzed in the frequency domain. When frequency analyses are performed on the IBI, the signal is decomposed into components that can be associated with biological mechanisms (Kramer, 1991). A frequency band that has been identified as sensitive to mental effort is the window between 0.07 and 0.14 hertz (the "0.10 Hz component" related to fluctuations in blood pressure), confusingly referred to as both low-frequency band (Berntson et al., 1997) and mid-frequency band (Mulder, 1992). IBI (or heart rate) has been found to be generally sensitive to both driver alertness level and computational effort, whereas the 0.10 hertz component is not sensitive to compensatory effort but exclusively to computational effort (De Waard & Brookhuis, 1997; Wiethoff, 1997). Although self-reports usually (and best) are collected after completion of a task, registration of physiology during task performance can reveal changes during performance without task interruption. Examples of changes in heart rate and 0.10 hertz heart-rate variability during task performance are given in Figs. 2.5.1 and 2.5.2. In the figures, the average heart rate of 22 subjects is computed over 30-second intervals and is displayed in "steps" of 10 seconds. Data were taken from a simulator study in which subjects "drove" through different road environments

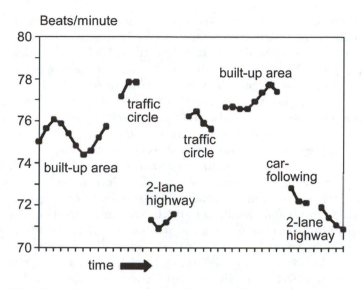

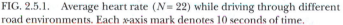

FIG. 2.5.1.   Average heart rate ($N = 22$) while driving through different road environments. Each x-axis mark denotes 10 seconds of time.

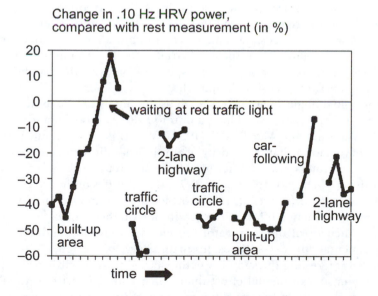

FIG. 2.5.2.   The 0.10 Hz component of heart rate variability, averaged over 22 subjects and compared with a resting level of 0. The same road environments as in Fig. 2.5.1 are depicted.

while a tutoring system monitored law compliance and provided subjects with feedback if a violation was detected (for further details on the experiment, see De Waard, Van der Hulst, & Brookhuis, 1999b). The effect of driving through different road environments can be clearly seen; driving through a traffic circle coincides with increased heart rate, whereas driving on a two-lane highway is accompanied by the slowest heart rate. In Fig. 2.5.2, heart-rate variability of the 0.10 hertz component is displayed, compared with a rest measurement (when subjects sat quietly in the car). The 0.10 hertz component is suppressed when drivers invest computational mental effort, which can be seen again when two-lane roadway driving is compared with driving through a traffic circle. Interestingly, compared with average heart rate (Fig. 2.5.1), more details are visible when driving in a builtup area. Standing still and awaiting a traffic light to turn green increases heart rate variability, indicative for reduced mental effort. This study, as well as other research (Veltman & Gaillard, 1993), shows that heart rate and HRV can serve as indexes of mental effort during task performance. Jorna's conclusion (Jorna, 1992) that HRV is sensitive only to major task differences, such as rest versus task or single task versus dual task performance, is therefore not supported.

A moving average of heart rate and heart rate variability (as in Figs. 2.5.1 and 2.5.2) is obtained by making use of the so-called profile technique (Mulder, 1992) The technique has proved to be very useful in reflecting changes in these parameters over relatively short times. Quite frequently, a time window of 30 or 40 seconds is used to calculate parameters, after which the window moves 10 seconds, resulting in a continuous, fairly smooth course in heart rate (variability). Nevertheless, in traffic, the lower limit of 30 seconds is still relatively long for some task aspects. Moreover, the bandwidth in which the measure is sensitive is limited.

Other physiological parameters, such as the phasic activity of task-irrelevant muscles, might be useful in addition to ECG. Van Boxtel and Jessurun (1993), Waterink and Van Boxtel (1994), Waterink (1997), and Jessurun (1997) have shown that the activity of three facial muscles reflects mental effort in laboratory tasks (Waterink, 1997) and tentatively also in on-the-road driving (Jessurun, 1997; Zeier, 1979). Amplitudes of the frontalis muscle, the corrugator supercilii, and the orbicularis oris inferior are increased with increased mental effort. However, in one on-the-road experiment, no increased activity of the corrugator muscle was found in conditions of increased mental effort (De Waard, Jessurun, Steyvers, Raggatt, & Brookhuis, 1995). An interesting question is whether facial muscle activity reflects another dimension of mental effort than, for instance, heart rate variability and subjective ratings. In view of the potential to measure and register EMG activity over very short times, these measures deserve more attention in future applied research.

Additionally, certain EEG parameters are certainly potentially very useful as an index for mental workload. Frequency analysis of brain activity renders a picture of distribution of activity across frequency bands, serving as an indication for driver state with respect to alertness and vigilance in different driving conditions (Brookhuis, 1995). Event-related potentials of the brain might serve as an index for (spare) capacity in conditions of different mental workload (Kramer, 1991). A Problem with EEG measurement is the low signal-to-noise ratio, forcing high amplification of uncontrollable signals, which are always present in a working environment such as a motor vehicle in traffic.

Finally, blood assays of compounds such as catecholamines give insight into long-term effects of persisting mental workload, which might have repercussions on health. However, although this might be important with respect to workload of professional drivers in the long run, it is considered outside the scope of this chapter.

DRIVER PERFORMANCE, TASK DEMANDS, AND MENTAL WORKLOAD

Of major importance is how performance, subjective ratings, and physiology relate to one another. Not all measures are sensitive in the same bandwidth, which leads to "dissociation" of measures (e.g., Eggemeier & Wilson, 1991). Measures dissociate whenever one measure indicates an increase in workload while another does not. Dissociation of measures is probably rather an advantage than a problem. The effort concept made clear that not all increases in mental workload have to become overt in deteriorated task performance. In this sense, disagreement between self-reports and task performance may provide *more* information, as it indicates successful, active driver interference in keeping the level of performance intact (De Waard, 1996; see also Muckler & Seven, 1992).

Although the investment of effort has the advantage that task performance remains at a certain target level, there are costs for this achievement. Short-lasting effort investment is probably without health consequences and is one of the advantages of human flexibility to deal with changes in demands. However, prolonged, continuous effort compensation can be a threat to good health, as it has been suggested that repetitive activation of the cardiovascular defense response (i.e., computational effort; see Mulder, 1986) may lead to hypertension (Johnson & Anderson, 1990).

It may be clear that in the assessment of driver mental workload, effort plays a central role. Effort is a voluntary process, and it is invested by the driver whenever deemed necessary, for instance, in conditions where task demands increase. If no effort is invested in these conditions, performance drops. As individuals differ in capability, *when* effort is invested also differs. This is in accord with the notion that the difficulty of a certain task, and thus mental workload, is not the same for all. Novice drivers frequently "have to try harder" than experienced drivers. The quest for the generally acceptable level of workload ("the workload redline"; see, e.g., Wierwille & Eggemeier, 1993) is therefore doomed to fail. Mental workload is determined in the interaction between operator and task. Driver capability differs between drivers, and also in drivers, for instance, as a result of the energetic state of the operator. Subjective appraisal of workload demand may also affect the level of workload, as top performance (e.g., swerve as little as possible) is not always required. The only way to assess how drivers deal with a certain task is to take multiple measures, which can reveal level of performance, applied strategy, effort investment, and compensation.

CONCLUSION

The measurement of drivers' mental workload offers opportunities and pitfalls, as illustrated in De Waard's model of driver performance, demands, and mental workload (De Waard, 1996). Although stability of primary measures of driving performance over time is what the drivers' goals are, the conditions are variable and sometimes strongly demanding and require effort in variable "amounts" that at times are beyond capacity. The accident proneness that follows such conditions is the (for the time being irrefutable) rationale for the measurement of drivers' mental workload.

REFERENCES

Allen, R. W., & Stein, A. C. (1987, June). *The driving task, driver performance models and measurement.* Paper presented at the second international symposium on medicinal drugs and driving performance, Maastricht, Netherlands.

Berntson, G. G., Bigger, J. T., Eckberg, D. L., Grossman, P., Kaufmann, P. G., Malik, M., Nagaraja, H. N., Porges, S. W., Saul, J. P., Stone, P. H., & Van der Molen, M. W. (1997). Heart rate variability: Origins, methods, and interpretive caveats. *Psychophysiology, 34,* 623–648.

Brookhuis, K. A. (1993). The use of physiological measures to validate driver monitoring. In A. M. Parkes & S. Franzén (Eds.), *Driving future vehicles* (pp. 365–377). London: Taylor & Francis.

Brookhuis, K. A. (1995). Driver impairment monitoring system. In M. Vallet & S. Khardi (Eds.), *Vigilance et transports: Aspects fondamentaux, dégradation et préventation* (pp. 287–297). Lyon, France: Presses Universitaires de Lyon.

Brookhuis, K. A., De Vries, G., & De Waard, D. (1991). The effects of mobile telephoning on driving performance. *Accident Analysis and Prevention, 23,* 309–316.

Brookhuis, K. A., De Vries, G., & De Waard, D. (1993). Acute and subchronic effects of the H1-histamine receptor antagonist ebastine in 10, 20, and 30 mg dose, and triprolidine 10 mg on car driving performance. *British Journal of Clinical Pharmacology, 36,* 67–70.

Brookhuis, K. A., De Waard, D., & Mulder, L. J. M. (1994). Measuring driving performance by car-following in traffic. *Ergonomics, 37,* 427–434.

Brookhuis, K. A., Van Winsum, W., Heijer, T., & Duynstee, M. L. (1999). Assessing behavioural effects of in-vehicle information systems. *Transportation Human Factors, 1,* 261–272.

Brown, I. D., & Poulton, E. C. (1961). Measuring the spare "mental" capacity of cardrivers by a subsidiary task. *Ergonomics, 4,* 35–40.

Cnossen, F., Brookhuis, K. A., & Meijman, T. (1997). The effects of in-car information systems on mental workload: A driving simulator study. In K. A. Brookhuis, D. de Waard, & C. Weikert (Eds.), *Simulators and traffic psychology* (pp. 151–163). Groningen: Centre for Environmental and Traffic Psychology.

De Waard, D. (1996). *The measurement of drivers' mental workload.* (Published doctoral thesis). Haren, Netherlands: Traffic Research Centre, University of Groningen.

De Waard, D., & Brookhuis, K. A. (1991). Assessing driver status: A demonstration experiment on the road. *Accident Analysis and Prevention, 23,* 297–307.

De Waard, D., & Brookhuis, K. A. (1997). On the measurement of driver mental workload. In J. A. Rothengatter & E. Carbonell Vaya (Eds.), *Traffic and transport psychology* (pp. 161–171). Amsterdam: Pergamon.

De Waard, D., Jessurun, M., Steyvers, F. J. J. M., Raggatt, P. T. F., & Brookhuis, K. A. (1995). The effect of road layout and road environment on driving performance, drivers' physiology and road appreciation. *Ergonomics, 38,* 1395–1407.

De Waard, D., Van der Hulst, M., & Brookhuis, K. A. (1999a). The detection of driver inattention and breakdown. In P. Albuquerque, J. A. Santos, C. Rodrigues, & A. Pires da Costa (Eds.), *Human factors in road traffic, II: Traffic psychology and engineering* (pp. 102–107). Braga, Portugal: University of Minho.

De Waard, D., Van der Hulst, M., & Brookhuis, K. A. (1999b). Elderly and young drivers' reaction to an in-car enforcement and tutoring system. *Applied Ergonomics, 30,* 147–157.

Eggemeier, F. T., & Wilson, G. F. (1991). Performance-based and subjective assessment of workload in multi-task environments. In D. L. Damos (Ed.), *Multiple-task performance* (pp. 207–216). London: Taylor & Francis.

Fairclough, S. H., Ashby, M. C., & Parkes, A. M. (1993). In-vehicle displays, visual workload and usability evaluation. In A. G. Gale, I. D. Brown, C. M. Haslegrave, H. W. Kruysse, & S. P. Taylor (Eds.), *Vision in vehicles* (Vol. IV, pp. 245–254). Amsterdam: North-Holland.

Hart, S. G., & Staveland, L. E. (1988). Development of NASA-TLX (Task Load Index): Results of experimental and theoretical research. In P. A. Hancock & N. Meshkati (Eds.), *Human mental workload* (pp. 139–183). Amsterdam: North-Holland.

Hendy, K. C., Hamilton, K. M., & Landry, L. N. (1993). Measuring subjective workload: When is one scale better than many? *Human Factors, 35,* 579–601.

Jessurun, M. (1997). *Driving through a road environment.* (Published doctoral thesis). Haren, Netherlands: Traffic Research Centre, University of Groningen.

Johnson, A. K., & Anderson, E. A. (1990). Stress and arousal. In J. T. Cacioppo & L. G. Tassinary (Eds.), *Principles of psychophysiology* (pp. 216–252). Cambridge: Cambridge University Press.

Jorna, P. G. A. M. (1992). Spectral analysis of heart rate and psychological state: A review of its validity as a workload index. *Biological Psychology, 34,* 237–257.

Kahneman, D. (1973). *Attention and effort.* Englewood Cliffs, NJ: Prentice-Hall.

Kramer, A. F. (1991). Physiological metrics of mental workload: A review of recent progress. In D. L. Damos (Ed.), *Multiple-task performance* (pp. 279–328). London: Taylor & Francis.

Lansdown, T. C. (1997). Visual allocation and the availability of driver information. In J. A. Rothengatter & E. Carbonell Vaya (Eds.), *Traffic and transport psychology* (pp. 215–223). Oxford: Pergamon Press.

McKnight, A. J., & Adams, B. B. (1970a). *Driver education task analysis, Volume I: Task descrip-*

tions: Final report (Contract FH 11-7336). Alexandria, VA: Human Resources Research Organization.

McKnight, A. J., & Adams, B. B. (1970b). *Driver education task analysis, Volume II: Task analysis methods: Final report* (Contract FH 11-7336). Alexandria, VA: Human Resources Research Organization.

Muckler, F. A., & Seven, S. A. (1992). Selecting performance measures: "Objective" versus "subjective" measurement. *Human Factors, 34,* 441–455.

Mulder, G. (1980). *The heart of mental effort.* Doctoral thesis, University of Groningen, Groningen, Netherlands.

Mulder, G. (1986). The concept and measurement of mental effort. In G. R. J. Hockey, A. W. K. Gaillard, & M. G. H. Coles (Eds.), *Energetics and human information processing* (pp. 175–198). Dordrecht, Netherlands: Martinus Nijhoff.

Mulder, L. J. M. (1988). *Assessment of cardiovascular reactivity by means of spectral analysis.* Doctoral thesis, University of Groningen, Groningen, Netherlands.

Mulder, L. J. M. (1992). Measurement and analysis methods of heart rate and respiration for use in applied environments. *Biological Psychology, 34,* 205–236.

O'Donnell, R. D., & Eggemeier, F. T. (1986). Workload assessment methodology. In K. R. Boff, L. Kaufman, & J. P. Thomas (Eds.), *Handbook of perception and human performance, Volume II, 42: Cognitive processes and performance* (pp. 1–49). New York: Wiley.

O'Hanlon, J. F., Haak, T. W., Blaauw, G. J. & Riemersma, J. B. J. (1982). Diazepam impairs lateral position control in highway driving. *Science, 217,* 79–81.

Smiley, A., & Brookhuis, K. A. (1987). Alcohol, drugs and traffic safety. In J. A. Rothengatter & R. A. de Bruin (Eds.), *Road users and traffic safety* (pp. 83–105). Assen, Netherlands: Van Gorcum.

Treat, J. R., Tumbas, N. S., McDonald, S. T., Shinar, D., Hume, R. D., Mayer, R. E., Stansifer, R. L., & Castellan, N. J. (1977). *Tri-level study of the causes of traffic accidents.* (Report DOT-HS-034-3-535-77 (TAC)). Bloomington: Indiana University.

Van Boxtel, A., & Jessurun, M. (1993). Amplitude and bilateral coherency of facial and jaw-elevator EMG activity as an index of effort during a two-choice serial reaction task. *Psychophysiology, 30,* 589–604.

Veltman, J. A., & Gaillard, A. W. K. (1993). Indices of mental workload in a complex task environment. *Neuropsychobiology, 28,* 72–75.

Veltman, J. A., & Gaillard, A. W. K. (1996). Pilot workload evaluated with subjective and physiological measures. In K. Brookhuis, C. Weikert, J. Moraal, & D. de Waard (Eds.), *Aging and human factors* (pp. 107–128). Haren, Netherlands: Traffic Research Centre, University of Groningen.

Verwey, W. B., Brookhuis, K. A., & Janssen, W. H. (1996). *Safety effects of in-vehicle information systems.* (Report TM-96-C002). Soesterberg, Netherlands: TNO Human Factors Research Institute.

Verwey, W. B., & Veltman, H. A. (1996). Detecting short peaks of elevated workload: A comparison of nine workload assessment techniques. *Journal of Experimental Psychology: Applied, 2,* 270–285.

Waterink, W. (1997). *Facial muscle activity as an index of energy mobilization during processing of information: An EMG study.* Doctoral thesis, Tilburg University, Delft, Netherlands: Eburon.

Waterink, W., & Van Boxtel, A. (1994). Facial and jaw-elevator EMG activity in relation to changes in performance level during a sustained information processing task. *Biological Psychology, 37,* 183–198.

Wickens, C. D. (1984). Processing resources in attention. In R. Parasuraman & D. R. Davies (Eds.). *Varieties of attention* (pp. 63–102). London: Academic Press.

Wierwille, W. W., & Eggemeier, F. T. (1993). Recommendation for mental workload measurement in a test and evaluation environment. *Human Factors, 35,* 263–281.

Wiethoff, M. (1997). *Task analysis is heart work.* (Published doctoral thesis). Delft University of Technology. Delft, Netherlands: Delft University Press.

Zeier, H. (1979). Concurrent physiological activity of driver and passenger when driving with and without automatic transmission in heavy city traffic. *Ergonomics, 22,* 799–810.

Zijlstra, F. R. H. (1993). *Efficiency in work behavior. A design approach for modern tools.* (Published doctoral thesis). Delft University of Technology. Delft, Netherlands: Delft University Press.

2.6 Automation and Workload in Aviation Systems

Mustapha Mouloua
University of Central Florida

John Deaton
CHI Systems, Inc.

James M. Hitt, II
University of Central Florida

The aviation industry has grown remarkably over the last decades. From the Wright brothers' triumph in 1903 to Charles Lindbergh's legacy of being the first to fly his airplane across the Atlantic Ocean, we now can fly the same trip in less than 3 hours by certain supersonic aircraft (Concorde). Furthermore, the Boeing jumbo jets (767 and 777) and the Airbus (320 and 340) series can now accommodate a large number of travelers across the seven continents. Looking into the future, we can expect to see aircraft designed much like the blended wing body proposed by Boeing and NASA (Weingarten, 1998). This aircraft, expected to be in operation around 2015, will carry 800 passengers and travel at speeds 25% greater than the Concorde (over 1,500 mph). All these innovations have been made possible by the rapid technological progress during the second half of the 20th century.

Today, an increased number of automated systems have changed the size of the aircrew from three to two members and have made flight possible under virtually any weather conditions. Automation has generally been implemented in efforts to reduce operational costs, reduce operators' workload and fatigue, allow more precise flight paths and fuel utilization, increase safety, and allow faster and more precise control of multiple simultaneous tasks. However, several human performance problems have also emerged, which appear to be related to the user's interaction with automation technology. These problems include: (a) a reduction in the operator's system awareness, (b) an increase in monitoring workload, and (c) degradation in manual flying abilities. A reduction in manual control skills appears

particularly critical, as degradation in this skill limits the ability of the operator to quickly resume accurate manual control of a process following automation failure. Although these issues are widely available in today's literature
(for an indepth review, see Mouloua & Koonce, 1997; Mouloua & Parasuraman, 1994; Parasuraman & Mouloua, 1996), in the present chapter, we focus
on workload issues as they relate to aviation domains by providing a snap-
shot of current findings. We examine automation technology as it is conceived and applied to several aviation systems and extensively review and
evaluate these findings in terms of recent theories, their applications, and
future trends.

HUMAN MONITORING AND VIGILANCE

Research on vigilance has shown that monitoring performance and vigilance of tasks requiring the detection of low-probability events are degraded
after prolonged periods of watch (Davies & Parasuraman, 1982). Research
with complex automated tasks requiring low-probability events has also
shown similar results (Parasuraman, Molloy, & Singh, 1993). Operators' detection of automation failures was degraded when automation was static and
not adaptive or responsive to operators' task demands and workload. In the
so-called static or "conventional" automation, the function allocation of a
task always remains fixed between the operator and the system. Automation-
induced monitoring inefficiency is one of the negative effects of highly automated systems for which human operators are not well suited (Parasuraman,
1987). In two different studies, Parasuraman and his colleagues reported
performance decrements in the detection of automation failures by using
a flight simulation task. Both pilots and nonpilots exhibited lower levels of
detection failures under automation mode. When performance was carried
out under manual control, detection performance was substantially higher.
The results show that monitoring inefficiency represents one of the performance costs of long-term static automation. This cost developed shortly
after 10 minutes of performance under automation control. It was suggested that performance cost under full automation mode might be due to
the fact that under such automation conditions, pilots and nonpilot participants fail to allocate full attention to the automated task. A possible countermeasure to automation-induced inefficiency lies in the form of adaptive automation, or automation that is implemented dynamically in response to
changing task demands placed on the operator. In adaptive systems, the division of labor between the human operator and the computer systems is not
fixed, but varies from time to time as task demands change. For example, a
pilot may choose to perform a routine engine-monitoring task manually
while cruising at 30,000 feet, but may prefer to have this task automated dur-

ing the demanding landing phase. That is, the allocation of a task or a function between the operator and the system is flexible and responsive to operators' performance and level of workload (Parasuraman, Bahri, Deaton, Morrison, & Barnes, 1992). Since the mid-1970s, adaptive automation technology was regarded as a viable design option (Rouse, 1977). It was suggested that such adaptive automated tasks are superior to the so-called static automated tasks because they improve situational awareness, regulate workload, improve vigilance in high-risk environments, and help to maintain manual control skills.

There have been few empirical studies designed to demonstrate the superiority of the claims proposed by the advocates of adaptive automation. Parasuraman, Mouloua, and Molloy (1996) examined the effects of adaptive task allocation on monitoring performance. Given that both pilots and nonpilots are relatively inefficient in monitoring automation failure for a task that is automated for a long period (Parasuraman et al., 1993), adaptive task allocation in the form of allocating a previously automated task to a pilot resulted in improved performance. The complacency effects, defined as the over-reliance on an automated system, were diminished with adaptive task allocation, and benefits, as measured by performance on a monitoring task, reached 62% above baseline measures. Function allocation is a viable design option that requires extensive human monitoring in automated tasks such as industrial inspection, piloting, and air traffic control. The results of these studies supported the claim that conventional (static) automation exacts a cost in terms of pilot performance. Furthermore, the empirical data suggested that the problem of automation-induced monitoring inefficiency can be remedied by properly designed adaptive systems. The findings also provided strong empirical evidence for the superiority of adaptive automation over static automation technology, at least for one aspect of performance—the ability of humans to monitor an automated complex system for extended periods.

PILOT INTERACTION WITH COCKPIT AUTOMATION

Modern cockpits or "glass cockpits" in transport aircraft have resulted in a series of unprecedented concerns in the aviation community. Wiener (1985) clearly illustrated the major concerns that have resulted from the implementation of advanced flight decks. Although cockpit automation has promised to compensate for some of the biological limitations of pilots (e.g., physical workload, fatigue, stress, etc.), it has also resulted in increased mental workload as we indicated earlier in this chapter. However, the major concern at this point is not to rehash some of these problems but instead to stress the importance of each individual factor on safety, efficiency, and system reliability. It is well established in the literature that pilots do not interact well with highly automated cockpits. However, there is no alternative to remedy these deficiencies. Sarter and Woods (1994) pointed out the importance of keeping the pilots in the loop to increase situation awareness. Often, when automated systems fail, pilots cannot detect automation failures because they are not aware of the system state and subsequently cannot revert to manual control in time. The current training also does not allow them to practice with false alarms or automation failures that may happen to the aircraft suddenly and in unpredictable ways. Several subjective studies have reported pilots' attitudes toward increased automation in the cockpit. Funk, Lyall, and Niemczyk (1997) examined 418 documents containing citations of flight deck automation problems and concerns from various sources. A total of 1,635 citations were reported in 150 of the documents that have been analyzed. In addition to the various articles surveyed, Funk, Lyall, and Niemczyk (1997) reported 368 citations in 246 Aviation Safety Reporting Systems reports. In each of these reports, flight deck automation was clearly a contributing factor to some major problems. All these problems point to the poor automation design, complexity, poor pilot-automation interface design, incompatibility with ATC system, and cultural problems. The findings of this study as well as those of several other researchers all point to the need for a more human-centered approach to designing advanced automation in aviation systems (Billings, 1997; Mouloua & Koonce, 1997; Mouloua & Parasuraman, 1994; Parasuraman & Mouloua, 1996).

ALERTING SYSTEMS IN THE COCKPIT

Advanced technologies have resulted in an increased number of alarms/warning systems in the modern cockpit. Aircrews often have to process an almost overwhelming amount of information pertaining to the aircraft operating in varied flight environments. Beyond direct sensing, information is commonly conveyed to the flight crew through displays, gauges, alarms, alerts, and warnings. Although these applications may have different meanings, they have often been used interchangeably to refer to alerting systems in the cockpit. This is the term we use throughout the chapter. These warnings/alarms are aimed at detecting faults or errors in hardware and software to prompt pilots about any potential breakdowns and systems malfunctions. However, several behavioral problems have also resulted from poor interaction between the pilot and the alerting systems. Understanding how pilots interact and respond to various alerting systems in the cockpit is essential for system efficiency, safety, and reliability. Among the problems that alerting systems may cause in the cockpit are pilot confusion, misinterpretation, inattention, poor discrimination, loss of situation awareness, and increased workload (Fitts & Jones, 1947; Gilson, Deaton, & Mouloua, 1996; Roscoe,

1968; Sanders & McCormick, 1993; Wiener & Nagel, 1988). To date, there have been widespread developments in aircraft and its systems. Modern avionics have by far surpassed their predecessors, today's aircraft having reached the highest possible altitudes at the highest possible speed. Such accomplishments come with the addition of a multitude of complex gauges and alerting systems that were originally poorly designed. Early investigations of such deficiencies were based on the attentional attributes of the medium (e.g., intensity of sound or light, pitch, color, type, size, and location; for an indepth review, see Sanders & McCormick, 1993; Stanton, 1994; Wickens, 1992; Wiener & Nagel, 1988). However, only recently has research started to focus on the growing number of displays and alarms in the human interface system because of the problems that have resulted in mishaps, mission aborts, and premature maintenance.

A study by Tyler, Gilson, and Mouloua (1996) reviewed and categorized 10 years of U.S. Naval Safety Center Accidents/Incident/Hazard Reports, covering all Navy and Marine Corps aircraft that were involved between January 1984 and February 1994. The results showed that 59% (851 out of 1,450) of the reports indicated the installed warning system activated when there was no hazard or primary system malfunction; 18% (258 cases) were cases where erratic instrument indications were themselves valid indications of failing aircraft subsystems; and 16% (237 cases) were missed alarms documented as times when a hazard was present and the appropriate warning system failed to activate. These categories were further analyzed by both aircraft type and aircraft system. False "fire warnings" contributed the most to the false alarms category (333); erratic indications associated with the F-18 inertial navigation system (INS) was the largest single contributor in the "valid alarms" category (86); stuck or frozen A-6 analog display indicators (ADI) accounted for the majority of the "failed to activate" reports (182; for an indepth review, see Tyler et al., 1996).

These results indicate that the proliferation of alerting systems has led to an increased likelihood of false indication, confusion, and multiple alarms that consequently increase the operator's inability to accurately diagnose the underlying causes of the alarms/alerts and to select the appropriate response. What are the implications of these navy aircraft findings? Current naval aviation senior leadership have imposed an institutional philosophy that assumes that all cockpit alarms are valid. Accordingly, pilots are trained to execute the prescribed emergency procedure every time the alarm activates. The results of such a philosophy may result in the following: Naval aircraft and crews are put at risk by responding to false alarms; missions are canceled unnecessarily, often en route, because of false alarms; and resources are wasted in troubleshooting and "cleaning up" after false alarms. These data clearly point to the fact that aircrew reactions to false alarms need to be addressed because of a lack of guidelines in current training approaches to these problems. Traditional interface design and training procedures do not control for all these mentioned problems. As a result, missions are unnecessarily aborted, various uncertainties detract from critical mission tasks, or in the worst case, avoidable accidents resulting in loss of life and equipment occur. These anomalies and their consequences could be eliminated with redesigned alerting systems and training procedures as documented by Gonos, Mouloua, Gilson, McDonald, Deaton, and Shilling (1996). In their study, Gonos and his associates examined pilots' attitudes toward these alerting systems via structured interviews and questionnaires. Results indicated that pilots are not well suited to interact with current alerting systems because of either a lack of training or inappropriate alert/alarm designs.

PILOT WORKLOAD

Increased automation in the cockpit has placed more demands on pilots' limited capacity to process the multitudes of displays of modern glass cockpits. The Airbus series (320 and 340) and the modern Boeing 777 are now equipped with an overwhelming number of computerized systems. It is well established that the major aviation industries such as Boeing and Airbus use workload measurement for the certification of their aircraft (see later section in this chapter for details of Airbus workload certification). Online, real-time physiological and subjective workload assessment techniques are used. Heart rate variability (HRV) and evoked response potentials (ERP) techniques are now widely used as indexes of mental workload.

The importance of studying pilot workload was evidenced by a special issue of the *International Journal of Aviation Psychology* published in 1995 (edited by Anthony D. Andre & Peter A. Hancock). This edition included six papers that examined pilot workload from a different perspective. This is proof of the varied nature of pilot workload and the many research efforts aimed at trying to decipher the mystery of the measurement of workload. The papers included attempts to measure workload via physiological (Backs, 1995), subjective (Becker, Warm, Dember, & Hancock, 1995), and performance measures (Andre, Heers, & Cashion, 1995). Attempts such as this help shape future research in workload analysis as well as to stimulate discussion on the topic. It is interesting to note how many of the same questions the research community addressed in the early workshops on workload (Moray, 1979; Williges & Wierwille, 1979) remain unanswered.

LABORATORY AND FIELD STUDIES EXAMINING AUTOMATION AND WORKLOAD

As automation has seen ever-increasing use in aviation systems, there has been a corresponding increase in the number of laboratory and field studies to investigate the effects of automation on mental workload. Early researchers cautioned designers of automated systems not to ask whether their systems can be automated, but rather to ask *should* they be automated (Wiener & Curry, 1980). Technology was no longer the driving force in automation as human-centered design became important. Designers now had to include human limitations and capabilities into the design of aircraft systems (see Billings, 1997).

Aircraft companies made grand claims of how automated systems would reduce pilot workload inside the cockpit. Wiener (1988) even speculated that McDonnell Douglas might have never been able to sell the MD-80 if it required a three-person crew because of labor costs. Field studies of the MD-80 revealed pilots did indeed reveal an overall reduction in workload but not as large as previously stated (Wiener, 1985). Much of the automation research often used questionnaires to gain pilots' attitudes toward the use of automation. Wieners' work with B-757 pilots also indicated that pilots thought workload is not changing in absolute level but that a redistribution of workload is occurring. The automated systems are demanding a greater amount of monitoring of systems (cognitive activity) and a decrease in manual control (psychomotor). McClumpha, James, Green, and Belyavin (1991) examined questionnaire data from over 1,300 European pilots about cockpit automation. Results indicated that pilots felt workload was decreased in automated cockpits when compared with nonautomated cockpits. Results also revealed that those flying glass cockpits perceived lower workload levels than those flying with flight management systems (FMS). Research by Rudisill (1994, 1995) has shown evidence that pilots thought automation often reduces workload at periods of low workload and increases workload at periods of increased workload (takeoffs and landings). Rudisill also reported pilots often turn off the automated systems during periods of high workload. These results paralleled the findings of Wiener (1988).

In 1991, Arnegard and Comstock (1991) introduced the Multi-Attribute Task (MAT) Battery. This task battery was created to examine operator performance and workload in a multitask environment. Tasks in the MAT include monitoring gauges, a tracking task, auditory communications, a predictive scheduling window, and a resource management task. These tasks can all be performed in manual or automated mode. Over the last 7 years, many versions of the MAT have been adopted by various research labs (i.e., Catholic University, University of Minnesota, Mankato State University, and Old Dominion University) for their particular research interests. Several government agencies (NASA and the Naval Air Warfare Center—Aircraft Division) have also contributed resources and funding for many of the studies in this area. A review of the literature reveals that many laboratory studies have used part or all of the tasks in the MAT battery. We now describe several of these studies in an effort to understand the various ways the MAT battery (and its related versions) can be used to test workload and automation issues.

Research at Catholic University has used a modified version of the MAT battery in an attempt to define parameters for use in adaptively automated systems. A team of scientists headed by Parasuraman has investigated issues such as automation induced "complacency" (Parasuraman et al., 1993), the monitoring of automated systems (Parasuraman, Mouloua, & Molloy, 1994, 1996; Parasuraman, Mouloua, Molloy, & Hilburn, 1996), and psychophysical approaches to adaptive task allocation (Byrne & Parasuraman, 1995). More recent research by these scientists has examined adaptive automation in air traffic control (ATC) systems (Hilburn, Jorna, Byrne, & Parasuraman, 1997; Hilburn, Jorna, & Parasuraman, 1995). Results have shown that mental workload can be significantly decreased with use of adaptive systems.

Research at two universities in Minnesota (University of Minnesota and Mankato State) has also examined a number of issues related to automation and workload. A main thrust of the research from these researchers examines the effect of operator-initiated automation and differences in anticipated versus unanticipated task load. Using a revised version of the MAT battery called MINUTES (see Harris, Hancock, Arthur, & Manning, 1992), these researchers concluded that with sufficient training operators experience less workload when automation is user initiated. They also warned designers of the hazards associated with allowing system operators to control the level of system automation outright (Harris, Hancock, Arthur, & Caird, 1995). Automation control is itself a source of demand, adding yet another demand on attention and information-processing resources. They also concluded that automation does not decrease workload at low task levels, but invoking automation at high task-load levels may not only increase workload but also decrease the operators' trust and reliance in use of the automated system. A study by Harris, Goernert, Hancock, and Arthur (1994) examined how workload and performance were altered when automation was induced at the beginning of high task-load periods. They found when operators were warned before increases in task load, performance did not differ between operator- and system-induced automation. In conditions in which operators were not informed of upcoming increases in task load, errors in resource management were greater in operator-initiated automation than when the system initiated the automation itself. They suggested adaptive automation might be best used in system environments with very dynamic workload changes.

Research efforts at Old Dominion University and NASA-Langley, led by Mark Scerbo and Alan Pope, respectively, are currently investigating using psychophysical measures (EEG) as the criterion for initiating adaptively automated systems. Using various ratios of electroencephalogram (EEG) wavelengths (alpha, beta, and theta) developed from research originally used in biofeedback literature for sufferers of attention-deficit hyperactivity disorder (see Lubar, 1991; Lubar, Swartwood, Swartwood, & O'Donnell, 1995), these researchers used these same ratios in an attempt to measure operators' overall mental state of awareness. A study by Pope, Bogart, and Bartolome (1995) tested various EEG ratios to determine what ratio produced results that best matched the expected results of feedback control systems. They acknowledged the ratio of beta / (alpha + theta) as being the best candidate ratio for examining automation issues in simulated flight. Prinzel, Freeman, Scerbo, and Mikulka (1997) also did further testing of EEG ratios. Their results paralleled the findings of Pope et al.

Determining the best method for changing modes of automation (using EEG) in adaptively automated systems became a major research question. Prinzel, Scerbo, Freeman, and Mikulka (1995) showed that with the use of a closed-loop biofeedback system it was possible to use EEG as a criterion for achieving an acceptable level of engagement moderation. This study replicated the findings of Pope et al. (1995) and also illustrated that the system was sensitive to changes in task load (represented by a significant increase in task allocations during periods of multitasking). (Readers interested in literature related to EEG and workload are advised to read studies by Sterman [see Sterman et al., 1993], Wilson [see Backs, Ryan, Wilson, & Swain, 1995; Wilson, Monett, & Russell, 1997], and Hicks [see Hicks, 1991].)

Thus far, we have merely attempted to discuss the myriad of research questions and theoretical demands, which have arisen (with respect to mental workload) with the increased use of automation in today's aviation systems. We have also discussed the methodologies of mental workload, and the most recent passages have briefly reviewed several of the research labs conducting relevant research in this area. In the last section, we turn our focus to four aviation systems that have all been designed with some level of automation and that have considered workload in their design and implementation.

PILOT ASSOCIATE (PA) PROGRAM

In 1986, the Defense Advanced Research Projects Agency (DARPA, now called ARPA) initiated the Pilot Associate (PA) program. The purpose of the program was to demonstrate the application of artificial intelligence to aid pilots in tactical aircraft (see Rouse, Geddes, & Hammer, 1990). It is in tactical aircraft that we see pilots stressed to their limits. That is, pilots experience physical stress because of high gravity forces and mental stress from extremely rapid changes in the combat situation with the need to respond rapidly and accurately to hostile forces. Decision aiding is one method of overcoming some of these potential difficulties.

As aircraft increase in technological complexity, the need for some sort of decision aiding increases dramatically. The PA system was intended to automate many of the functions traditionally left to the pilot, thus helping the pilot perform his/her mission. In the case of the PA, however, intelligent automation was used to help the pilot overcome limitations and enhance pilot abilities. The purpose was not merely to automate more of the tasks performed by the pilot, as research has shown that this does not necessarily help the pilot. Traditional automation results in the pilot being a monitor of automation, a task that humans do not perform well.

Providing the pilot information via the pilot–vehicle interface (PVI) was the primary function of the PA system. Traditional interfaces often overwhelmed the pilot with information, much of it irrelevant, and forced the pilot to filter and integrate the data into a useful form. The associate system, on the other hand, generated and managed the display of information to provide appropriate information at the optimal time and in the optimal format. Some of the information generated by the associate system included judgments about the situation and recommended responses.

The PA system could, with pilot approval, perform actions on behalf of the pilot. Conditions of task overload resulted in the allocation of low-importance tasks to automation. These tasks would be the output of the recommended responses produced earlier by the PA. The process of intelligent automation, through control of selected aspects of the aircraft, allowed the pilot to remain in control through authorization of the conditions under which specific tasks could be automated.

In specific cases, the associate system may take the initiative within the limits imposed by the pilot. For example, it may change displays or present recommendations to the pilot on its own initiative. The PA is more like an electronic crew member than traditional automation is. As a result, this necessitates new types of knowledge in the design of the interaction between intelligent automation and human operators of complex systems (i.e., tactical aircraft).

An associate system must obey certain rules of interaction, much like a human crew member. The primary rule is that the pilot is in charge and may do whatever he/she chooses without any restriction from the PA system. The associate must follow the pilot's preferences. However, the associate system does have the same access as the pilot to aircraft systems. Furthermore, the associate system must interact with the pilot at rates that do not overload the pilot. Finally, plans proposed by the associate system may be accepted or rejected by the pilot, or even ignored, with predictable outcomes.

Several technical challenges needed to be overcome during the evolution of the PA program. The first issue concerns communication among distributed knowledge-based systems. PA had five subsystems that would need to communicate in some coordinated fashion. Skeptics doubted whether a distributed knowledge-based program could actually work as needed. Another challenge concerned real-time performance of artificial intelligence (AI) software. Because AI software usually has a rather large average time to execute, it was soon realized that the associate system would have to be coded in a conventional language and run on avionics processors in real time. A third challenge was acquiring knowledge for an aircraft with capabilities that were different from what pilots were currently flying. In fact, the domain experts used during the design of the PA system had no experience with the crew-station concepts that were proposed.

A major focus of the PA system was the intelligent interface (PVI). The intelligent interface used inputs, models, and knowledge bases to make several critical decisions: What information should be displayed to the pilot? Which tasks would be allocated to automation and automatically executed at the proper time? Which pilot actions or inaction are errors? The models developed to answer these questions needed to describe both the intentions and the workload of the pilot.

The subsystems of the PVI are the intent inferencer, the resource model, the error monitor, the adaptive aider, and the plan proposer. In general, these subsystems are responsible for interpreting the pilot's plans and goals, estimating pilot workload (used to determine the amount of information filtering required and also by the adaptive aider to assess how much help the pilot needs to successfully complete critical tasks), determining if pilot actions or inactions have serious negative consequences, determining which tasks should be allocated to automation, and organizing various plans into coherent proposals for presentation to the pilot. It is not the intent of this chapter to describe these subsystems in detail, although the next section discusses the resource model subsystem as it is relevant to the topic addressed in this chapter (see Hammer & Small, 1995).

THE RESOURCE MODEL

As mentioned earlier, the resource model estimates pilot workload to implement intelligent declutter and by the adaptive aider to make task allocation decisions. During the initial stages of PA development, the resource model was based on a psychophysical measurement model of workload by Gopher and Braune (1984) and Wickens' multiple resource theory (Wickens, 1984). Gopher and Braune used a power function model as seen in psychophysics to model the contributions of individual tasks to overall workload. Wickens' multiple resource theory hypothesized that human information processing has multiple perceptual, cognitive, and motor dimensions. The psychophysical and multiple resource theories were combined in this earlier PA model.

It soon became apparent that there were several problematic issues associated with this earlier resource model. As a result, a different theoretical approach was used. The Subjective Workload Assessment Technique (SWAT) (Reid, Potter, & Bressler, 1989) and the modified Cooper–Harper task demand scales were used to model workload. A neural network (Shewhart, 1992) replaced psychophysical power functions as the combining method (see other chapters in this book for a more detailed description of these assessment tools).

Both SWAT and the modified Cooper–Harper models were used to produce outputs. In general, both of these approaches use verbal reports by pilots as inputs to workload models. Because verbal inputs are impractical for an associate system, the workload models used situation descriptions already available within the associate system. The Cooper–Harper model used knowledge-engineered attributes associated with pilot intentions, while the SWAT method used a neural network that was connected with all the inputs that might have some effect on workload.

Problems did arise, however, with the modified resource model described previously. For one, the neural network model required considerable amounts of data for training the network to cover the multidimensional space. There was a tendency to overestimate workload as a result. After more complete data were collected and used to train the network, results were more satisfactory

Tomorrow's pilots will most likely face even bigger challenges against more capable threats while flying aircraft having even more airframe and avionics capability. Designers of such new aircraft need to provide onboard support systems, like PA, to enable pilots to do those things they can do best, fly and fight. The more routine functions like avionics systems control will be allocated to an electronic crew member. The benefits provided by such a system include increased mission effectiveness by providing the right information at the right time so that the pilot can make the right decisions. With a significant reduction in workload levels, the pilot can concentrate on formulating the tactics required to defeat the enemy and complete the mission.

HEALTH AND USAGE MONITORING SYSTEMS (HUMS)

Today's budget constraints have prevented the military from procuring new aircraft to replace an aging fleet. Moreover, the services are constantly seeking new ways of minimizing maintenance and repair costs while preserving high standards of safety. Traditionally, scheduled aircraft maintenance has been performed on the basis of service time, which often does not coincide with actual maintenance needs. In response to these concerns and as a result of progress made in the area of mechanical diagnostics, advanced sensor and analysis technologies are now being introduced to a small subset of operational helicopters, providing significant contributions to subsystem performance, workload reduction and state assessment and helping with maintenance management. The new diagnostic capability has been called HUMS (Health and Usage Monitoring Systems). A key assumption made by researchers in this field is that if the operators and maintainers of complex mechanical systems could accurately determine the health of their equipment, then important issues such as workload, safety, and affordability could be favorably affected. The availability of information about system status, trends, impending breakdowns, and actual failures provided by these advanced sensor and analysis technologies has brought us a step closer to realizing the benefits of *condition-based maintenance,* enhancing flight safety and affordability by focusing maintenance and troubleshooting activities on subsystem current states rather than on the basis of subcomponent "flight hours" in service (e.g., see Nickerson, 1994; Nickerson & Hall, 1995).

Clearly, HUMS systems have considerable potential for alerting the aircrew of inflight developments of mechanical faults. The HUMS system can potentially alert aircrews in several distinct ways to help reduce high workload levels during emergency operations. The most basic form of aiding informs the aircrew that a problem actually exists and provides the locus and type of fault. HUMS can then provide predictions (within a certain accuracy level) as to how long a subsystem will function before failure. HUMS can provide corroborative evidence to the aircrew to verify that the identified fault is real rather than a false alarm. The impact of malfunctions on other system components as well as on mission success factors can also be assessed. HUMS can provide reminders to the aircrew at later points in a mission when an earlier identified fault becomes especially significant. Finally, HUMS can offer specific action line recommendations that will help the aircrew in performing time critical procedures. This helps overcome some basic memory limitations in moments of high workload.

Recent research on mechanical diagnostics technology has revealed the possibility of using advanced mechanical diagnostics technology capabilities in real-time environments. Current work sponsored primarily through the Office of Naval Research (ONR) has focused on mechanical diagnostics for the CH-46 aircraft, with particular emphasis on HUMS instrumentation of the rotorcraft's drive train, by using various sensor and signal-processing components. Although it is clear that advanced sensors and signal processors are capable of generating data that are relevant to real-time aircrew decisions, it is not clear how to present the appropriate information in simple and straightforward terms, in a form that is directly pertinent to aircrew decisions (see Deaton et al., 1997a, 1997b; Deaton & Glenn, 1998a, 1998b; Nickerson & Chamberlain, 1995).

HUMS technology, if it is to be successful, must assist the aircrew in formulating accurate situation diagnoses and in identifying viable courses of action. It goes without saying that pilots will need to be convinced that by automating tasks that previously had been the purview of personnel on-board the aircraft, overall mission effectiveness is increased. This is by no means an easy matter. The reliability of such a system is critical; otherwise operational personnel ignore the recommendations provided by HUMS systems. The benefits of offloading critical diagnostic and prognostic tasks to an "intelligent crew assistant" can potentially minimize pilot workload during times of increased stress, as is the case when an emergency evolves during the course of a mission.

ACKNOWLEDGMENTS

We are indebted to the insightful comments and assistance of David Abbott, Richard Gilson, Dennis Vincenzi, and Haydee Mesa of the University of Central Florida. The present chapter was partially supported by a research contract from Litton/Tasc of Orlando, FL. The views expressed in this chapter are those of the authors and do not represent the organizations with which they are affiliated.

REFERENCES

Andre, A. D., Heers, S. T., & Cashion, P. A. (1995). Effects of workload preview on task scheduling during simulated instrument flight. *International Journal of Aviation Psychology, 5*(1), 5–24.

Arnegard, R. J., & Comstock, J. R. (1991). Multi-Attribute Task Battery: Applications in pilot workload and strategic behavior research. In *Proceedings of the Sixth International Symposium on Aviation Psychology* (pp. 1118–1123), Columbus: Ohio State University Press.

Backs, R. W. (1995). Going beyond heart rate: Autonomic space and cardiovascular assessment of mental workload. *International Journal of Aviation Psychology, 5*(1), 25–48.

Backs, R. W., Ryan A. M., Wilson, G. F., & Swain, R. A. (1995). Topographical EEG changes across single-to-dual task performance of mental arithmetic and tracking. In *Proceedings of the Human Factors and Ergonomics Society 39th annual meeting* (p. 953). Santa Monica, CA: HFES.

Becker, A. S., Warm, J. S., Dember, W. N., & Hancock, P. A. (1995). Effects of jet engine noise and performance feedback on perceived workload in a monitoring task. *International Journal of Aviation Psychology, 5*(1), 49–62.

Billings, C. E. (1997). *Aviation automation: The search for a human centered approach*. Mahwah, NJ: Lawrence Erlbaum Associates.

Byrne, E. A., & Parasuraman, R. (1995). *Psychophysiological approaches to adaptive task allocation*. (Tech. Rep. No. CSL-A-95-1). Cognitive Science Laboratory, Catholic University of America.

Davies, D. R., & Parasuraman, R. (1982). *The psychology of vigilance*. New York: Academic Press.

Deaton, J., & Glenn, F. (1998a). The development of an automated monitoring system interface associated with aircraft condition. In *Proceedings of the third Automation Technology and
Human Performance conference*. Hillsdale, NJ: Lawrence Erlbaum Associates.

Deaton, J., & Glenn, F. (1998b). The development of specifications for an automated monitoring system interface associated with aircraft condition. *International Journal of Aviation
Psychology, 9*(2), 175–187.

Deaton, J., Glenn, F., Federman, P., Nickerson, G. W., Byington, C., Malone, R., Stout, R.,
Oser, R., & Tyler, R. (1997a). Aircrew response procedures to in-flight mechanical emergencies. *Proceedings of the 41st annual meeting of the Human Factors and Ergonomics Society*. Albuquerque, NM: Human Factors and Ergonomics Society.

Deaton, J., Glenn, F., Federman, P., Nickerson, G. W., Byington, C., Malone, R., Stout, R.,
Oser, R., & Tyler, R. (1997b). Mechanical fault management in navy helicopters. *Proceed-
ings of the 41st annual meeting of the Human Factors and Ergonomics Society*. Albuquerque, NM:
Human Factors and Ergonomics Society.

Fitts, P. M., & Jones, R. E. (1947). *Analysis of factors contributing to 460 "pilot error" experiences
in operating aircraft controls*. (Memorandum Rep. TSEA-4-694-12, Aeromedical Laboratory). Wright Patterson Air Force Base, OH: AARMRL. Reprinted in H. W. Sinaiko
(Ed.), *Selected papers on human factors in the design and use of control systems*. New York: Dover,
1961.

Funk, K. H., Lyall, E. A., & Niemczyk, M. C. (1997). Flightdeck automation problems: Perceptions and reality. In M. Mouloua & J. M. Koonce (Eds.), *Human–automation interaction: Research and practice* (pp. 29–34). Hillsdale, NJ: Lawrence Erlbaum Associates.

Gilson, R., Deaton, J., & Mouloua, M. (1996, October). Development of training strategies for
alarm diagnosis in the cockpit. *Ergonomics in Design*, 12–18.

Gonos, G., Mouloua, M., Gilson, R., McDonald, D., Deaton, J., & Shilling, R. (1996). *Experimental design definition to evaluate system operator response to alerts and alarms*. (Tech. Rep. No.
HFRL-96-04). Orlando: University of Central Florida.

Gopher, D., & Braune, R. (1984). On the psychophysics of workload: Why bother with subjective measures. *Human Factors, 26*(5), 519–532.

Hammer, J. M., & Small, R. L. (1995). An intelligent interface in an associate system. *Human/
Technology Interaction in Complex Systems, 7*, 1–44.

Harris, W. C., Hancock, P. A., Arthur, E., & Caird, J. K. (1995). Performance, workload, and
fatigue changes associated with automation. *International Journal of Aviation Psychology,
5*(2), 169–185.

Harris, W. C., Hancock, P. A., Arthur, E., & Manning, C. (1992). *Minnesota universal task evaluation systems (MINUTES)*. (Tech. Rep. HFRL 92-02). Minneapolis, MN: University of Minnesota, Human Factors Research Laboratory.

Hicks, M. R. (1991). *EEG indicators of mental workload: Conceptual and practical issues in the development of a measurement tool*. (AGARD Report No. CP-490). Safety network to enhance performance degradation and pilot incapacitation. France: NATO.

Hilburn, B., Jorna, P. G. A. M., Byrne, E., & Parasuraman, R. (1997). The effect of air traffic
control (ATC) decision aiding on controller mental workload. In M. Mouloua & J. Koonce
(Eds.), *Human–automation interaction: Research and practice* (pp. 84–91). Mahwah, NJ: Lawrence Erlbaum Associates.

Hilburn, B., Jorna, P. G. A. M., & Parasuraman, R. (1995). The effect of advanced ATC strategic decision aiding automation on mental workload and monitoring performance: An
empirical investigation in simulated Dutch airspace. In *Proceedings of the Eighth International Symposium on Aviation Psychology* (pp. 387–391). Columbus: Ohio State University.

Lubar, J. F. (1991). Discourse on the development of EEG diagnostics and biofeedback for at-
tention-deficit/hyperactivity disorders. *Biofeedback and Self-Regulation, 16*(3), 201–225.

Lubar, J. F., Swartwood, M. O., Swartwood, J. N., & O'Donnell, P. H. (1995). Evaluation of the effectiveness of EEG neurofeedback training for ADHD in a clinical setting as measured by changes in T.O.V.A. scores, behavioral ratings, and WISC-R performance. *Biofeedback and Self-Regulation, 20*(1), 83–99.

McClumpha, A. J., James, M., Green, R. G., & Belyavin, A. J. (1991). Pilots' attitudes towards cockpit automation. In *Proceedings of the Human Factors and Ergonomics Society 35th annual meeting* (pp. 107–111). San Francisco: HFES.

Moray, N. (1979). *Mental workload, theory and measurement*. New York: Plenum.

Mouloua, M., & Koonce, J. M. (1997). *Human–Automation interaction: Research and practice*. Mahwah, NJ: Lawrence Erlbaum Associates.

Mouloua, M., & Parasuraman, R. (1994). *Human performance in automated systems: Current research and trends*. Hillsdale, NJ: Lawrence Erlbaum Associates.

Nickerson, G. (1994, October). *Conditioned-based maintenance: A system perspective*. Paper presented at the 1994 ASME/STLE International Tribology conference and exhibition, Lahaina, HI.

Nickerson, G., & Chamberlain, M. (1995, August). *On-board diagnostics: More than a red light on a panel*. Paper presented at the Future Transportation Technology conference, Irvine, CA.

Nickerson, G., & Hall, D. (1995, June). *Research imperatives for condition-based maintenance*. Paper presented at the Eighth international conference on Condition Monitoring and Diagnostic Engineering Management, Kingston, Ontario, Canada.

Parasuraman, R. (1987). Human–computer monitoring. *Human Factors, 29*, 695–706.

Parasuraman, R., Bahri, T., Deaton, J., Morrison, J., & Barnes, M. (1992). *Theory and design of adaptive automation in aviation systems* (Progress Rep. No. NAWCADWAR-92033-60). Warminster, PA: Naval Air Warfare Center, Aircraft Division.

Parasuraman, R., Molloy, R., & Singh, I. L. (1993). Performance consequences of automation-induced "complacency." *International Journal of Aviation Psychology, 3*, 1–23.

Parasuraman, R., Mouloua, M., & Molloy, R. (1994). Monitoring automation failures in human–machine systems. In M. Mouloua & R. Parasuraman (Eds.), *Human performance in automated systems: Current research and trends* (pp. 45–49). Hillsdale, NJ: Lawrence Erlbaum Associates.

Parasuraman, R., & Mouloua, M. (1996). *Automation and human performance: Theory and applications*. Mahwah, NJ: Lawrence Erlbaum Associates.

Parasuraman, R., Mouloua, M., & Molloy, R. (1996). Effects of adaptive task allocation on monitoring of automated systems. *Human Factors, 38*(4), 665–679.

Parasuraman, R., Mouloua, M., Molloy, R., & Hilburn, B. (1996). Monitoring of automation systems. In R. Parasuraman & M. Mouloua (Eds.), *Automation and human performance: Theory and applications* (pp. 91–115). Mahwah, NJ: Lawrence Erlbaum Associates.

Pope, A. T., Bogart, E. H., & Bartolome, D. S. (1995). Biocybernetic system evaluates indices of operator engagement in automated task. *Biological Psychology, 40*(1–2), 187–195.

Prinzel, L. J., III, Freeman, F. G., Scerbo, M. W., & Mikulka, P. J. (1997). A biocybernetic system for examining psychophysiological correlates for adaptive automation. In *Proceedings of the Human Factors and Ergonomics Society 41st annual meeting* (p. 1378). Santa Monica, CA: HFES.

Prinzel, L. J., III, Scerbo, M. W., Freeman, F. G., & Mikulka, P. J. (1995). A bio-cybernetic system for adaptive automation. In *Proceedings of the Human Factors and Ergonomics Society 41st annual meeting* (pp. 1365–1369). Santa Monica, CA: HFES.

Reid, G., Potter, S., & Bressler, J. (1989). *Subjective Workload Assessment Technique (SWAT): A user's guide*. (AAMRL Tech. Rep. AAMRL-TR-89-023). New York: American National Standards Institute.

Roscoe, S. N. (1968). Airborne display for right and navigation. *Human Factors, 10*, 321–332.

Rouse, W. B. (1977). Human computer interaction in multi-task situations. *IEEE Transactions on Systems, Man, and Cybernetics, 7*, 384–392.

Rouse, W. B., Geddes, N. D., & Hammer, J. M. (1990). Computer-aided fighter pilots. *IEEE Spectrum, 27*(3), 38–41.

Rudisill, M. (1994). Flight crew experience with automated technologies on commercial transport flight decks. In M. Mouloua & R. Parasuraman (Eds.), *Human performance in automated systems: Current research and trends* (pp. 203–211). Hillsdale, NJ: Lawrence Erlbaum Associates.

Rudisill, M. (1995). Line pilots' attitudes about and experience with flight deck automation: Results of an international survey and proposed airlines. In *Proceedings of the eighth international Symposium on Aviation Psychology* (pp. 288–298). Columbus: Ohio State University Press.

Sanders, M. S., & McCormick, E. J. (1993). *Human factors in engineering and design.* New York: McGraw-Hill.

Sarter, N. B., & Woods, D. D. (1994). Pilot interaction with cockpit automation, II: An experimental study of pilots' mental model and awareness of the Flight Management System (FMS). *International Journal of Aviation Psychology, 4,* 1–28.

Shewhart, M. (1992). A neural-network-based tool. *IEEE Spectrum, 29*(2), 6.

Stanton, N. (1994). *Human factors in alarm design.* Basingstoke, England: Taylor & Francis.

Sterman, M. B., Kaiser, D. A., Mann, C. A., Suyenobu, B. Y., Beyma, D. C., & Francis, J. R. (1993). Application of quantitative EEG analysis to workload assessment in an advanced aircraft simulator. In *Proceedings of the Human Factors and Ergonomics Society 37th annual meeting* (pp. 118–121). Santa Monica, CA: HFES.

Tyler, R., Gilson, R., & Mouloua, M. (1996). *False alarms in naval aircraft: A review of Naval Safety Center mishap data.* (Tech. Rep. No. HFRL-96-01). University of Central Florida, Orlando.

Wickens, C. D. (1984). Processing resources in attention. In R. Parasuraman & D. R. Davies (Eds.), *Varieties of attention* (pp. 63–102). New York: Academic Press.

Wickens, C. D. (1992). *Engineering psychology and human performance.* New York: HarperCollins.

Wiener, E. L. (1985). *Human factors of cockpit automation: A field study of flight crew transition.* (NASA Contractor Rep. CR-177333). Warrendale, PA: Society of Automotive Engineers.

Wiener. E. L. (1988). Cockpit automation. In E. L. Wiener & D. C. Nagel (Eds.), *Human factors in aviation* (pp. 433–461). San Diego, CA: Academic Press.

Wiener, E. L., & Curry, R. E. (1980). Flight-deck automation: Promises and problems. *Ergonomics, 23*(10), 995–1011.

Wiener, E. L., & Nagel, D. C. (1988). *Human factors in aviation.* San Diego, CA: Academic Press.

Weingarten, T. (1998, February 23). Beyond the Concorde, and other fantasy flights. *Newsweek, Vol. CXXXI,* pp. 8, 12.

Williges, R. C., & Wierwille, W. W. (1979). Behavioral measures of aircrew mental workload. *Human Factors, 21,* 549–574.

Wilson, G. F., Monett, C. T., & Russell, C. A. (1997). Operator functional state classification during a simulated ATC task using EEG. In *Proceedings of the Human Factors and Ergonomics Society 41st annual meeting* (p. 1382). Santa Monica, CA: HFES.

2.7 Causes, Measures, and Effects of Driver Visual Workload

Terry C. Lansdown
The Transport Research Laboratory

This chapter reviews the literature related to the causes, measurement, and effects of driver and visual workload. It discusses the performance of the human visual system and its limitations on driving. Experimental methods for the measurement and interpretation of visual behavior are considered. Relevant research that illustrates the workload implications of basic vehicle control, use of conventional instruments, and interaction with advanced driver information systems is highlighted. The need for context-relevant measurement techniques is highlighted for evaluation of workload situations where impairment in visual system may result in greater accident risk.

CAUSES OF VISUAL WORKLOAD

It has been argued that vision is the largest single resource available to the driver and is the major information-processing input in driving (Sabey & Staughton, 1975; Wierwille, 1993a). The ability to attain and process environmental and in-vehicle visual information may be compounded by the driver's individual characteristics. For example, it has been shown that when compared with experts, novice drivers adopt different strategies when obtaining and manipulating visual information (Mourant & Rockwell, 1972). Experienced drivers were reported to maintain lane position by using peripheral vision, whereas novices tended to sample the visual scene by using foveal fixations. Thus, it is suggested that expert drivers are experiencing reduced workload with respect to novices and consequently have more available attention for other roadway events.

Safe performance of the driving task requires that the driver have sufficient available resources for assessment of the visual environment and maintenance of lateral and longitudinal position in the roadway (Boyce, 1981). The skills required to perform this task have been identified as visual information acquisition (scanning); perceptual-motor coordination; anticipation and assessment of the traffic situation; risk estimation; setting safety margins; and balancing speed and caution (Duncan, Williams, & Brown, 1991). The experienced driver automates many of these processes, such that under normal conditions the multiple task demands of driving remain in their information-processing capacity (Hancock & Parasuranman, 1992). However, excessive driver visual workload leads to distraction from the primary task. The distraction may come from the road environment (e.g., a complex road network in an unfamiliar city) or from an in-vehicle device (e.g., a visually demanding congestion warning system). Imposed workload has been stated to be inversely proportional to reserve information-processing capacity (Wickens, 1984).

MEASURES OF VISUAL WORKLOAD

Methodologies that have been used by human factors practitioners to evaluate driver information systems have been highlighted (Hughes, 1989). These include the following visual workload measures: performance indicators (e.g., search and response times), verbal protocol, occlusion of the visual field, and eye movement analysis.

A distinction can be made between measures and their interpretation. For example, vision behavior may be recorded and interpreted as mean glance duration. This section concentrates on measures that can be recorded rather than on their interpretation.

Direct Measurement of Visual Workload

Direct measures of visual workload are considered here as those that record and analyze the allocation of visual behavior throughout the field of view, rather than infer visual workload from other metrics. Verbal protocols (Ericson & Simon, 1993) may be considered in this category. They have been used in the measurement of drivers' visual workload by numerous researchers (Hughes & Cole, 1986; Renge, 1980; Schraagen, 1990). However, no studies have been found to demonstrate the validity of this metric in the driving context. The study of eye movements provides another key mechanism for consideration of drivers' visual behavior. It involves two processes, eye movement data collection and eye movement analysis. Methods for the collection of visual behavior include occlusion studies (Rockwell, 1972; Senders, Kristofferson, Levision, Dietrich, & Ward, 1967), eye tracking methods (e.g., limbus tracking, pupil tracking, corneal reflex, search coil, electro-oculography and Purkinje image tracking; Carr, 1988), or concurrent/post hoc transcription of glance behavior.

Indirect Measurement of Visual Workload

Performance measures such as lane deviation may be considered as indirect indexes of visual workload and can be classified as primary or secondary tasks. Primary tasks are those that the driver must competently perform to maintain vehicle control. Zwahlen, Adams, and DeBald (1988) used position in lane as an index of primary task performance. The similar measure time to line (road edge) crossing has been proposed as a valid predictor of driver task performance (Godthelp, 1986; Godthelp, Milgram, & Blaauw, 1984). Lane deviations are a function of visual distraction from the roadway and driver visual-motor coordination. Thus, under circumstances of high visual workload from an in-vehicle device, the driver may be distracted from the forward view for an extended time with the consequence that the available reaction time to roadway events is reduced. Secondary task measures include backward counting and/or stimulus reaction time measurement. Secondary task measures provide a practical method to evaluate spare mental capacity (Schlegel, 1993). For example, driving in heavy traffic has been shown to degrade secondary task performance (Brown & Poulton, 1996, Peackock & Karowowski, 1993). However, the intrusion of a secondary task on the primary one must be carefully considered. Primary task performance in low workload conditions has been shown to decrease with addition of secondary activities (Wierwille & Gutmann, 1978); however, this effect was not found under high workload conditions.

EYE MOVEMENT INTERPRETATION

Regardless of the data collection method, numerous measures of eye movement can be calculated, including glance frequency, glance duration, and total glance duration (to an identified target location during the task duration). In the published literature, mean and standard deviation data are commonly presented for the metrics highlighted earlier. However, it has been argued (Lansdown, 1997) that the presentation of measures of central tendency and variance may not be the most appropriate measures of visual workload. Presentation of, for example, the 90th percentile may be more reflective of the implications of extended distraction from the forward view, particularly for glance duration measures. Additional measures that have been reported in the literature (Wierville, 1993a) include fixation probability (the likelihood that a particular location is fixated upon during task performance) and link value probability (the strength of association between two identified locations).

A practical problem in the interpretation of empirical research on driver visual workload is the lack of consistent definitions of measures adopted. All measures are obtained from the same fundamental visual behavior, that is, glances and transitions between glances, which in turn are composed on fixations and saccades. It is clear that the specific apparatus and experimental approach may dictate the calculation of higher order metric from these basic components, for instance, a single glance duration may reasonably be considered to encompass a transition from Region A to Region C; the dwell time on Region C (including the fixations and saccades in the area of interest); and the transition back to Region A. However, it may reasonably be argued that transition from Region C to Region A should be associated with the next glance to Region A. Three points are important to note from this: The basic building blocks remain the same regardless of interpretation; specific experimental aims may reasonably result in different calculations of metrics commonly considered as the same; and the absolute magnitude of study-specific differences may not be important (i.e., coupling of transition with dwell times targets may not radically shift total glance duration times; e.g., 0.56 seconds vs. 0.67 seconds both impose far lower visual workload on the driver than a glance of 3.77 seconds). In an attempt to clarify the basic building blocks of visual behavior, Fig. 2.7.1 is presented.

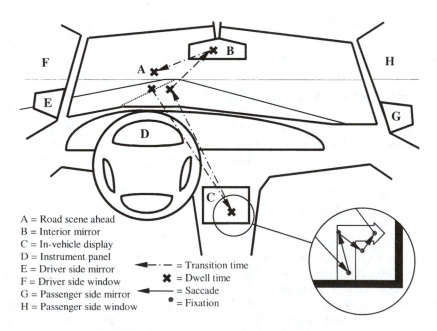

FIG. 2.7.1.   Basic components of driver visual behavior.

LEVELS OF VISUAL WORKLOAD

It has been estimated that over 90% of the information received by the driver is visual (Sabey & Staughton, 1975). Regardless of the specific percentage, vision has been ranked as the single most important source of information for the driver (Wierwille, 1993a), and "overloading of the visual pathway" (Wall, 1992) is a key area for accident prevention research. Furthermore, the forward visual scene has been described as the primary information-gathering source, and therefore all other information must be secondary (Wierwille, 1993b). Connoly (1968) reported that many of today's traffic conditions impose unreasonable workloads on the driver's perceptual resources. The introduction of advanced driver information systems may further reduce the driver's capacity to safely contend with road infrastructure and environmental situations.

In the following section the observable effects of driver visual workload are explored under a three-stage taxonomy, that is, lateral and longitudinal control, interaction with conventional systems, and interactions with advanced information systems.

VISUAL REQUIREMENTS
FOR BASIC VEHICLE CONTROL

The focus of expansion (Gibson, 1980) has been suggested as a primary target to aim the vehicle at during straight-lane driving. However, Gordon (1966) argued that aiming of the vehicle at the focus of expansion is too simplistic an approach. The point is illustrated by using curvilinear motion (cornering) where the geometry of the bend is uncertain and liable to change. The ability to detect the external environment may be mediated by the visual characteristics of the roadway.

Visual Sampling Strategies

It has been stated: "The only way that the driver can gather detailed visual information from sources at different positions is to move the foveal resource about in time, that is, to sample or time-share. This fact [time sharing] has profound implications for the design of in-vehicle displays" (Wierwille, 1993a, p. 134). If the display demands foveal attention, by definition it must distract the driver from the primary task of observing the forward scene (Wierwille, 1993a). However, it has been suggested (Bhise, 1971) that the foveal region has increased information-processing speed per unit time when compared with the peripheral regions, such that the driver may be able to assimilate information more quickly from the fovea. It is relevant to note that when recording driver visual behavior, as Rockwell (1972) argued, the point of regard may not indicate the location of the obtained information.

Spare Visual Capacity

It has been suggested (Hughes & Cole, 1986) that between 30% to 50% of drivers' visual scanning may be unrelated to driving. Sender et al. (1967) was one of the first investigators to employ occlusion as a methodology to consider the relation of drivers' visual behavior to their task performance. The results suggested that when less frequent glances to the forward view imposed, they resulted in reduced speed, as did shorter glances to the forward view. Conversely, with increased fixed speed and presumably workload, the drivers sampled the roadway more frequently. For given occlusion and glance frequency times, a roadway with more severe curves resulted in lower speed. Senders et al. suggested that when the drivers are faced with a greater attentional demand than they feel able to contend with, they slow down, look more frequently, or look for a longer period, thereby adjusting information uptake to an acceptable level. Senders et al. proposed that the methodology adopted in their report "will allow an objective measure, based on driver behaviour, of the attentional demand of a segment of road" (p. 32).

Rockwell (1972) considered the role of peripheral vision and the concept of spare visual capacity. The research suggested that fixations in driving tend to be less than 6° travel from the focus of expansion, with more than 90% located within 4°. The distribution of fixations reported seems consistent with that observed by other researchers (Cole & Hughes, 1988). According to Rockwell (1972, p. 322), "Most interstate driving requires less than 50% of a driver's perceptual capability." Drivers were said to frequently sample irrelevant information (e.g., "signs which are covered," p. 322). Indeed, drivers may not be processing information from such task-irrelevant objects at all. Rockwell suggested that fixations to objects other than those that may be considered necessary for driving, such as a lamppost, may be used to anchor visual perception for subsequent peripheral information. Spare resources provide an opportunity for the implementation of driver-information systems. The challenge remains how to use spare periods when the driver can assimilate information while not overloading the visual resource required to maintain safe control of the vehicle at other times.

Visual Attentional Processes

During driving, an individual may be gazing at the forward view (the focus of regard) by not consciously attending to any single object or person (attention attractant). Hughes and Cole (1986) suggested a distinct difference between attention and focus regard. They conducted a study in which 25 subjects reported features that attracted their attention during a suburban drive of 21.9 kilometers (13.6 mi). A laboratory study, also with 25 subjects, performed the same task by using a video film of the road routes. The subjects were required to make concurrent verbal reports. The results of the study showed that 30% to 50% of attention was directed to objects "not related to driving" (e.g., advertising). Furthermore, they stated: "Eye fixations may in fact be a poor indicator of what is noticed" (p. 376). It is suggested that this could reflect subjects' spare visual capacity. Interestingly, approximately, 15% to 20% of attention was attributed to traffic control devices (e.g., traffic information signs and lights). It is argued that the drivers could not attend to all traffic information devices with this development of visual resources. The road trial experiment contained a condition where the subjects were passengers. No substantial differential effect on attentive behavior was observed during the passenger and driver conditions. On this basis, the authors suggested that "the visual information presented by the movie film is sufficient to generate attentive processes characteristic of driving" (p. 377).

Observations were shown to be significantly greater in the shopping center. It is possible that the total number of discrete elements per visual angle per unit time was higher in this regard or that the relevance/consequence of ignoring cues in this region was greater. In support of the latter hypothesis, there was no difference in the number of reports not relating to driving. Specifically, the subjects were not distracted more by nondriving objects in the shopping center than in residential or arterial driving. They were reporting driving relevant objects more frequently.

WORKLOAD AND VARIATION IN VISUAL BEHAVIOR

Hughes (1989) suggested that variations of fixation duration (considered here as glance) are dependent on task type, task importance, the nature of the information processing, and the overall workload. Hughes' research is cited as an example where, as the visual complexity and number of stimuli in the roadway are increased (in a driving simulator), drivers were observed to change fixation frequency. This resulted in fixation (or glance) duration reduction from 440 to 409 milliseconds. It was found that in a visually rich environment (shopping center) successive fixations to the road surround were 60% compared with 51% in simulated residential driving. Hughes suggested that visual properties (i.e., visual clutter or bits of information per area per unit time) and driver information goals are both important determinants in the distribution of attention in visual scanning.

Hughes (1989) explored strategies to change the distribution of drivers' visual scanning. During normal (or undirected) driving in a vehicle simulator, subjects were observed to glance to the left of the road (Australian research, left-hand lane driving) 18% of the time. In a target-detection condition, the proportion of fixations to the left increased to 41%. As might be expected, the proportion of glances to the focus of expansion decreased correspondingly, from 25% to 14%. The situation investigated was somewhat artificial, in that such dramatic changes in visual behavior would be unlikely to occur on public roads. However, the findings show that drivers adopt specific visual-scanning strategies that influence the distribution of visual behavior throughout the forward scene.

Published literature on drivers' visual system has discussed their ability to control the vehicle and to avoid obstacles on a moment-to-moment basis. Displays that enrich the information available to drivers ease the process of vehicle control by enabling anticipation (e.g., engine frequency and speedometer readings provide, in addition to the already available optic-flow information, cues about safe velocities to negotiate corners). Therefore, it is important to consider the information sources that support drivers' strategic and maneuvering activities, especially to ensure that they support and do not overload drivers.

VISUAL ACTIVITY WITH CONVENTIONAL IN-VEHICLE INFORMATION

All cars provide the driver with information about the status of the vehicle, for example, vehicle speed, fuel level, and engine temperature. This information is typically presented to the driver by using analog indicators or hazard telltales (lights that function only during a warning condition). Research has shown that drivers employ a rational, visual, adaptive sampling strategy when performing vehicle-related tasks (Wierwille, 1987, as cited in Wierwille, 1993a). Wierwille (1993a) suggested that visual tasks are conducted by a series of exchanged glances to the instrument and to the forward view, as can be seen in Fig. 2.7.2. Wierwille stated that "normal drivers develop this time sharing (sampling) strategy, because it is the only way they can meet competing visual requirements of driving per se and carrying out in-vehicle tasks" (p. 136).

The visual impact of conventional controls was investigated in a study that compared stalk- and panel-mounted devices (Moussa-Hamouda & Howard, 1977). Five stalk and three panel controls were compared. The results suggest that glance duration was slightly longer at stalk controls, and glance frequency increased as reach distance increased. Glance duration was defined by Rockwell (1988) as "the time off the roadway to attend to a target (e.g.,

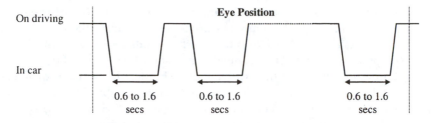

FIG. 2.7.2.   Sampling model of in-vehicle task performance. Reprinted from W. W. Wierwille, *Demands on Driver Resources Associated With Introducing Advanced Technology Into the Vehicle* (p. 103), with permission from Elsevier Science.

mirror, stereo, speedometer, etc.)" (p. 319). Data reported in his research concentrate on mean glance duration. Mean number of glances (often reported as glance frequency) is considered to be a much more sensitive measure of driver differences than average glance duration. Average glance duration is reported to be "about the same" for successive glances, whereas mean number of glances may vary from one glance to four or five glances to complete a variety of in-car entertainment system tasks. Mean data may hide the long glances that could arguably be those glance durations most reflective of unreasonable workload imposed by a driver-information system. This issue is explored in more detail in Lansdown and Fowkes (1998).

Rockwell (1988) reported that glance duration remains consistent across a variety of in-vehicle tasks. Average glance durations for three experiments were very similar (1.27 secs to 1.42 secs) and were shown to be not significantly different across the studies. Standard deviations were also reported to be "remarkable consistent, but rather large." The large range of the data suggests that use of mean glance duration may not, in fact, be the most reliable measure for the assessment of driver visual workload. Furthermore, it contradicts Wierwille's (1993a) assertion that glance duration varies little (see Fig. 2.4.2).

There is no consensus in the research on the duration of glances. Some researchers suggested that glance durations vary little during in-vehicle displays' use (Rockwell, 1988; Wierwille, 1993a). However, studies have demonstrated significant differences in glance durations during the use of complex in-vehicle systems (Fairclough, Ashby, & Parkes, 1993; Lansdown & Fowkes, 1998). Data reported by Rockwell (1988) demonstrate that qualitatively different tasks present different visual workloads. Mean glance duration for drivers' scanning to the driver's left mirror and incar radio was 1.1 seconds (SD = 0.3) and 1.4 secs (SD = 0.5), respectively. It can be seen that the distribution of glances to the radio is larger, for both the magnitude and frequency, than to the left mirror.

The maximum visual distraction duration for use of a driver-information system has been sought by many researchers. Rockwell (1988) suggested that a visual workload (glance duration) of 3 seconds is unreasonable. However, this point is mediated by the need to consider the context of the glance (e.g., traffic density at the time of glance). Whether this is mean or maximum single glance duration is not reported. The concept of a *2-second rule* (i.e., that drivers avoid glancing away from the forward view for more than 2 seconds) is presented, presumably, as a self-controlling mechanism for drivers' visual sampling. Rockwell (1988) stated that "when complex displays require glance duration beyond the 90th percentile, most drivers are clearly facing special visual workload problems" (p. 322).

VISUAL WORKLOAD AND THE INTRODUCTION OF INTELLIGENT TRANSPORT SYSTEMS

Current developments in electronic control and communication systems have made possible the introduction of complex displays of information in the vehicle. One example is route guidance, with many prototype systems produced in recent years. These reflect a wide range of approaches to information presentation in the vehicle. Some systems have used relatively simple symbolic, static information displays; others have applied multicolored scrolling maps. Regardless of the approach to design, the question raised is whether the introduction of advanced information displays unnecessarily distracts the driver from the primary task (i.e., perceiving the road and controlling the vehicle safely through it). To operationalize this question, the scope of visual behavior measures must be quantified.

Information Acquisition

Driver fixation times have been found to be longer when extracting information from alphanumeric signs (529 ms) than from symbolic displays (312 ms; Mori & Abdel-Halim, 1981). Hughes (1989) suggested that fixation times and measures of performance can be used for the development of visual information. Mori and Abdel-Halim (1981) presented data indicating that only 11% of driver fixations to road signs were long enough to extract the information contained in them. The data show a shortfall between the visual workload of the task and the human operator's ability to meet it (Hughes, 1989). However, the relevance of the traffic information of the driver was not related to the duration of glances.

Dingus, Antin, Hulse, & Wierwille (1989) investigated the visual demands of conventional driving tasks with the additional use of a computerized moving map display. The experiment employed an instrumented vehicle on public highways. Results indicated that visual workload (the sum of glance durations into the car) varied widely across the tasks; total glance duration was shown to have a minimum of 0.78 seconds (speedometer) and a maximum of 10.63 seconds (establishing the road name of the next turning); single glance durations varied from 0.62 to 1.66 seconds; and mean glance frequency ranged from 1.26 to 6.64 glances.

The study reinforced the representation of visual sampling that emerges from the literature (i.e., the glance durations were subject to some variation but remained within a relatively small range, and glance frequency appeared more representative of increases in task complexity). According to Wierwille (1993a, p. 137), "Most importantly, drivers do not, on the average, allow their single glance times to exceed about 1.6 second, even for complex information-gathering tasks." Wierwille advocated that single mean glance durations of 1.25 seconds are acceptable, and shorter are preferred. Glance frequencies of six or less are considered acceptable, particularly if the mean single glance duration is less than 1.25 seconds. Although this appears intuitively reasonable, no basis was presented for the proposal of these limits. Wierwille defended these assertions, stating that visual workload research is lacking and further work is required to form a more valid basis for such judgments. Research should aim to assist design and to keep glance durations to a minimum. The consequences of high glance frequency and extended glance duration were not discussed. It may be that poor design is reflected by high glance frequency, but a single 3-second glance is considerably more hazardous than 6 half-second glances.

Driver Performance and Visual Workload

Identification of a metric that is reflective of the impact of an information system of the driver has been the aim of many researchers. Zwahlen et al. (1988) considered further the issue of maximum time away from the forward view, presented by Rockwell (1988). The much-cited paper on the introduction of an in-vehicle touch screen display into the vehicle considered the driver's ability to maintain position in a lane as a measure of safety. The subjects were required to drive along a disused runway at 40 miles per hour while attempting to use a simulation of a device. The lane deviations were obtained from a paint dropper and analyzed post hoc.

Zwahlen et al. (1988) related increased lane deviations to different road widths. They stated: "One important measure of driver performance is the amount by which an automobile's path deviated from the centre of the lane while the driver is operating the CRT touch panel since even small deviations of a few feet may prove fatal" (p. 337). The authors reported (based on the obtained vehicle standard deviations): "If a six foot wide car were traveling in a 12 foot wide straight lane under ideal conditions (sunny, calm, dry pavement) at 40 mph then there would be a 3% chance of the vehicle laterally deviating out of the lane (either to the right or left) while the driver was operating the CRT touch panel. If the lane were reduced to 10 feet then the probability of the vehicle laterally exceeding the lane increased to 15%" (p. 341). They stated that the percentages were based on best case, and under more demanding circumstances the latter exceedance would be more likely. Although small deviations in lane may prove extremely hazardous, it is suggested that the environmental constraints of the roadway influence the drivers' propensity to deviate in lane. It is hypothesized that the negative consequences of poor control influence the effort the driver invests in maintaining position in lane. That is, narrow lanes result in better maintenance of road position than wide lanes, particularly during busy traffic.

A conceptual model of driver-information acquisition was proposed (see Fig. 2.7.3). The model was developed from previously published research (Senders et al., 1967). There are evident similarities between this model and that proposed by Wierwille (1993a). The consideration of road/traffic information in relation to visual behavior and available working memory makes the model particularly worthy of note.

The model was used to develop a tentative design guide for the design of driver-information systems (see Fig. 2.7.4). Zwahlen et al. (1988) hypothesized that "if more than three looks are required inside the vehicle to obtain a specific chunk of information during a relatively short period of time [4 seconds in Fig. 2.7.4] then the task becomes uncomfortable for the driver, since at the fourth look inside the vehicle an inadequate amount of road

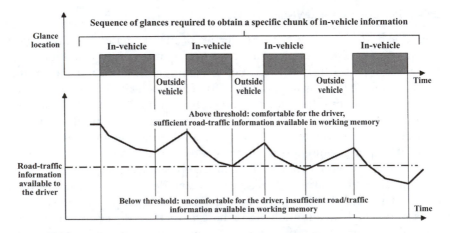
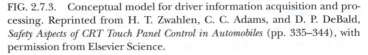

FIG. 2.7.3. Conceptual model for driver information acquisition and processing. Reprinted from H. T. Zwahlen, C. C. Adams, and D. P. DeBald, *Safety Aspects of CRT Touch Panel Control in Automobiles* (pp. 335–344), with permission from Elsevier Science.

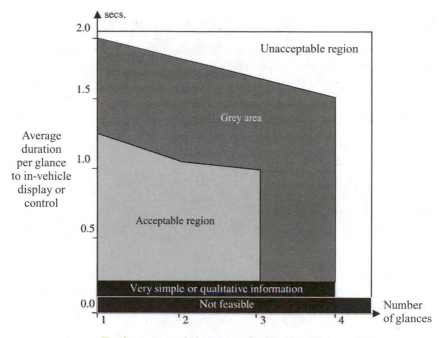

Total sequence of glances required inside vehicle to obtain a
specific chunk of information from a display or to operate a control

FIG. 2.7.4.   Proposed tentative design guide to be used when designing
sophisticated in-vehicle displays or CRT touch panel controls and/or applications. Reprinted from H. T. Zwahlen, C. C. Adams, and D. P. DeBald,
Safety aspects of CRT touch panel control in automobiles, pp. 335–344. Copyright
1988, with permission from Elsevier Science.

and traffic information is stored in the visual working memory of the driver"
(p. 337). However, the statement is presumably based on Fig. 2.7.3, and this
does not adequately consider the context in which glances into the vehicle
occur. For example, a visually barren environment would be unlikely to
overload the visual and cognitive resources of the driver. However, the
model provides insight and could perhaps be applied in a worst case (i.e.,
busy inner-city multilane junction maneuvers). Regardless, the model requires some statement of the context in which it may be used. It is unfortunate that the lack of ecological validity of the test environment is not discussed in the paper. The model has much appeal in the simplicity of its
application. The guide requires further refinement to define what is meant
by each chunk of information and whether average glance duration is more
appropriate than maximum glance duration. Additionally, the method of
translating results to the design guide is not stated.

Conventional and Navigation System Visual Workloads

Wierwille, Hulse, Antin, and Dingus (1988a) conducted two experiments to investigate the attentional demands of an in-vehicle navigation system. The first study compared the navigation system's visual workload with conventional in-vehicle tasks. All tasks were considered (both a conventional and navigation system) to determine which had high attentional demands. The total time spent glancing at tasks was ranked to determine the most visually demanding. The total glance times that were high for navigation tasks were roadway name (determine the road name on the route, 10.63 secs [SD = 5.80]); roadway distance (to the next turning, 8.84 secs [SD = 5.20]); and cross street (establish the name of the next street, regardless of relation to intended route, 8.63 [SD = 4.86]).

High single glance times were reported for the following navigation tasks: destination direction (which direction for travel, 1.20 secs [SD = 0.73]); correct direction (decide if the heading is correct, 1.45 secs [SD = 0.67]); heading (establish general heading, 1.30 secs [SD = 0.56]); zoom level (adjust the display to the correct setting with respect to the destination, 1.40 secs [SD = 0.65]); roadway name (1.63 secs [SD = 0.80]); roadway distance (1.53 secs [SD = 0.65]); and cross street 1.66 secs [SD = 0.82]).

The experimenters stated that the information presented by the moving map navigation system was "somewhat" more complex than the conventional in-vehicle tasks. Evidence was reported of a change in imposed workload as a consequence of the availability of the required information. Wierwille et al. (1988a) showed that when the subjects needed to zoom in or out (i.e., when the information required was not immediately available) the mean total glance duration doubled. The results were based on the three most visually demanding navigation system tasks. It is important to note that several of the conventional tasks investigated were shown to present high visual workload, namely total glance time; radio tuning (7.60 secs [SD = 3.41]); engaging cruise control (4.82 secs [SD = 3.80]); and powered mirror adjustment (5.71 secs [SD = 2.78]).

Under some circumstances, use of advanced driver information systems may demand unreasonable attention, potentially compromising safety. Researchers (Fairclough et al., 1993; Lansdown, 1996; Wierwille, Hulse, Fischer, & Dingus, 1988b) have argued that drivers are able to rationally adapt their behavior to compensate for the additional demands imposed. Wierwille et al. (1988a) presented data in support of this hypothesis. Higher traffic density was stated to produce more conservative driving. The finding was based on greater attention allocated to the driving and less to navigational tasks. Wierwille et al. (1988a) suggested that these features support the theory that drivers' behavior changes as a result of task demand (i.e., as the task becomes difficult, more attention is apportioned to control the vehicle).

Wierwille et al. (1988b) suggested the objective attentional demand increased as the proportion of driving related to glances increased. Glances to a navigation system were reported to decrease with a corresponding increase in objective attentional demand. Additionally, as the anticipated attentional demand increased, the proportion of time spent glancing at the navigation system was reduced and the subjects' visual scanning to the driving-related regions increased.

In a second experiment, unanticipated attentional demand was manipulated as the independent variable. Incident analysis was employed on three routes of approximately 8 miles of "industrial plants, and shopping centres, and along main streets of small towns" (Wierwille et al., 1988b, p. 2.663). Traffic density was reported as moderate to heavy. It was argued that because *incidents* are unpredictable, the attentional demand of such events must be largely unanticipated. The same subjects were used as those in Experiment 1. Visual behavior was recorded online by an experimenter seated in the rear of the vehicle. This was performed by depressing a button when the eyes were stationary and releasing it when fixation changed. Although some error checking was performed by the addition of different tones corresponding to eye movements, the accuracy of the approach is uncertain. However, such a method is likely to capture the broad distribution of visual behavior and therefore to differentiate between normal duration of glances and potentially dangerous extended duration glances.

Both studies presented evidence that the driver is able to adapt visual sampling to compensate for changes in task workload. A model of the adaptation process based on the findings of the authors is as follows: In both *anticipated* and *unanticipated* attentional demands, drivers shift a proportion of visual scanning from navigation systems to the forward central view; under conditions of increased *anticipated* attentional demand, the drivers' forward view sampling rate increases; and for increased unanticipated attentional demand, visual sampling rate to the forward view decreased and the glance duration increased.

The safety of public road trials precludes maximal loading of the driver. Consequently, the data presented support rational adaptation for submaximal task demands. However, one would expect the driver to be able to deal with these task demands. A significant problem remains in the prediction of the point at which task complexity exceeds the driver's ability to contend with it and of which information systems will demand more visual attention from the driver than they can safely provide.

Distribution of Visual Scanning

Task workload has also been shown to disrupt visual scanning. Data taken from Antin, Dingus, Hulse, and Wierwille (1988) were used by Wierwille

(1993b) to illustrate the task demand in differentiating driver visual behavior. Results demonstrated shifts in the distribution of visual scanning between normal driving, use of a paper map to navigate, and use of a computerized moving map display. The data suggest that the introduction of a navigation task results in changes in the proportions of visual scanning. Subjects' use of a computerized moving map was shown to demand radically more visual attention than the paper map, 33% and 7%, respectively. The computerized moving map was demonstrated to shift visual attention from the roadway (center and off center) while the other normal scanning behavior was hardly affected. The probability of glancing at the instrument was reduced in the same manner as use of a paper map (i.e., from 3% to 2%).

Assessment of Visual Workload

To ensure that information presented by a system is effectively and easily acquired by the operator, Hughes (1989) argued that this can be achieved by investigation of the information sources required for the task (which may be established, for example, by task analysis) and the effort required in acquisition of the visual information (which can be influenced by the allocation of functions between the driving task and the in-vehicle device). Consider whether visual workload limits driving performance or overloads the driver's capabilities. Visual demands that limit task performance may still be acceptable under some circumstances, for example, conditions of low mental workload. If visual workload can be determined to overload capabilities, such systems are clearly unacceptable for introduction into vehicles.

SUMMARY: VISUAL DEMAND AND ADVANCED DRIVER INFORMATION SYSTEMS

Time sharing between vehicle control and driver-information system use is fundamental to the imposed visual workload on the driver. The driver must be able to extract the required information in an acceptable glance duration. Many in-vehicle tasks have been shown to be reasonable in terms of the visual workload they impose (Rockwell, 1988; Wierwille et al., 1988b). However, task-related variability in visual workload has been demonstrated, and some tasks have been assessed as highly visually demanding (Dingus et al., 1989).

CONCLUSIONS

Literature relevant to the visual workload imposed by the introduction of driver information systems into vehicles has been described in this chapter.

I have presented evidence in support of foveal time sharing during information extraction and of increases in the available resources of the driver with task experience. However, attentional resources have been shown to be affected by environmental, traffic, and individual factors. Research was outlined for the measurement of driver visual workload. Rarely has this been compared with control data, and the methodologies adopted have not been described in sufficient detail to facilitate replication of the experimental work. Several models of in-vehicle sampling have been proposed, but all current models fail to fully represent driver visual behavior by using advanced information systems. Research into operator workload has been argued to fall into three contexts: workload prediction (i.e., development of metrics for the quantification of operator task demands); assessment of workload imposed by equipment; and assessment of operator experienced workload (Wickens, 1984, p. 390). A large body of visual workload literature has been published in recent years. However, there has been no clear definition of terms and measure for the assessment of operator workload. Consequently, it remains difficult to compare literature to develop a comprehensive representation of driver visual workload.

REFERENCES

Antin, J. F., Dingus, T. A., Hulse, M. C., & Wierwille, W. W. (1988). The effects of spatial ability on automotive navigation. In *Trends in Ergonomics/Human Factors, 5*, 241–248.

Bhise, V. D. (1971). *Relationship of eye movements to conceptual capabilities to visual information acquisition in automobiles*. Unpublished doctoral dissertation, Ohio State University.

Boyce, P. R. (1981). Driving. *Human factors in lighting* (pp. 170–203). London: Applied Science.

Brown, I. D., & Poulton E. C. (1961). Measuring the spare "mental capacity" of car drivers by subsidiary auditory task. *Ergonomics, 21*, 221–224.

Carr, K. T. (1988). *Survey of eye movement measuring equipment*. London: Applied Vision Association.

Cole, B. L., & Hughes, P. K. (1988). Drivers don't search: They just notice. In D. Borgan (Ed.), *Visual search* (pp. 407–417). London: Taylor & Francis.

Connoly, P. L. (1968). Visual considerations: Man, the vehicle, and the highway. *Highway Research News, 30*, 71–74.

Dingus, T. A., Antin, J. A., Hulse, M. C., & Wierwille, W. W. (1989). Attentional demand requirements of an automobile moving-map navigation system. *Transportation Research, 23*(4), 310–315.

Duncan, J. D., Williams, P., & Brown, I. (1991). Components of driving skill: Experience does not mean expertise. *Ergonomics, 34*(7), 919–937.

Ericson, A., & Simon, H. A. (1993). *Protocol analysis: Verbal reports as data*. New York: MIT Press/Bradford Books.

Fairclough, S. H., Ashby, M. C., & Parkes, A. M. (1993). *In-vehicle displays, visual workload and usability evaluation*. In A. G. Gale, I. D. Brown, C. M. Haslegrave, H. W. Kruysse, & S. P. Taylor (Eds.), *Vision in Vehicles* (Vol. IV, pp. 245–254). Amsterdam: North-Holland.

Gibson, C. P. (1980). Binocular disparity and head-up displays. *Human Factors, 22*(4), 435–444.

Godthelp, H. (1986). Vehicle control during curve driving. *Human Factors, 28*(2), 211–221.

Godthelp, H., Milgram, P., & Blaauw, G. J. (1984). The development of a time-related measurement to describe driving strategy. *Human Factors, 26*(3), 257–268.

Gordon, D. A. (1966). Perceptual basis of vehicular guidance. *Public Roads, 34,* 53–68.

Hancock, P. A., & Parasuraman, R. (1992). Human factors and safety in the design of intelligent vehicle-highway systems (IVHS). *Journal of Safety Research, 23,* 181–198.

Hughes, P. K. (1989). Operator eye movement behaviour and visual workload in aircraft and vehicles. In *The 25th annual conference of the Ergonomics Society of Australia: Ergonomics, technology and productivity* (pp. 97–105). Canberra: Australian Academy of Science.

Hughes, P. K., & Cole, B. L. (1986). What attracts attention when driving? *Ergonomics, 29*(3), 377–391.

Lansdown, T. C. (1996). How is visual behaviour affected by changes in the availability of advanced driver information? In T. Rothengatter & E. Carbonell (Eds.), *International conference on Traffic and Transport Psychology, Valencia, Spain* (pp. 215–223). London: Pergamon Press.

Lansdown, T. C. (1997). *Visual demand and the introduction of advanced driver information systems into road vehicles.* Unpublished doctoral dissertation, Loughborough University, Loughborough, England.

Lansdown, T. C., & Fowkes, M. (1998). An investigation into the utility of various metrics for the evaluation of driver information systems. In A. G. Gale (Ed.), *Sixth international conference on Vision in Vehicles, University of Derby* (pp. 215–224). Elsevier Science, North-Holland.

Mori, M., & Abdel-Halim, M. H. (1981). Road sign recognition and on-recognition. *Accident Analysis and Prevention, 13,* 101–115.

Mourant, R. R., & Rockwell, T. H. (1972). Strategies of visual search by novice and experienced drivers. *Human Factors, 14*(4), 325–335.

Moussa-Hamouda, E., & Howard, J. M. (1977). *Human factors requirements for fingertip reach controls.* (Tech. Rep. No. DOT-HS-803-267). Washington, DC: U.S. Department of Transportation.

Peacock, B., & Karowoski, W. (1993). *Automotive ergonomics.* London: Taylor & Francis.

Renge, K. (1980). The effects of driving experience on a driver's visual attention. An analysis of objects looked at using the "verbal report" method. *International Association of Traffic Safety and Sciences Research, 4,* 95–106.

Rockwell, R. H. (1972). Eye movement analysis of visual information acquisition in driving. *Australian Road Research Board Proceedings, 6*(3), 316–331.

Rockwell, T. H. (1988). Spare visual capacity if driving—Revisited. In A. G. Gale, M. H. Freeman, C. M. Haslegrave, P. Smith, & S. P. Taylor (Eds.), *The second international conference on Vision in Vehicles, Nottingham, England* (pp. 317–324). Elsevier Science, North-Holland.

Sabey, B. E., & Staughton, G. C. (1975). Interacting roles of road environment, vehicle and road user in accidents. In *Proceedings of the fifth international conference of the Association for Accident and Traffic Medicine, London,* pp. 1–17.

Schlegel, R. E. (1993). Driver mental workload. In B. Peacock & W. Karwowski (Eds.), *Automotive ergonomics* (pp. 359 – 382). London: Taylor & Francis.

Schraagen, J. M. C. (1990). *Strategy differences in map information use for route following in unfamiliar cities: Implications for in-car navigation systems.* (Final IZF 1990 B-6). Soesterberg, Netherlands: TNO Institute for Perception.

Senders, J. W., Kristofferson, A. B., Levision, W., Dietrich, C. W., & Ward, J. L. (1967). The attentional demand of automobile driving. *Highway Research Record, 195,* 13–15.

Wall, J. G. (1992). Vehicle safety—What are the needs? In XXIV FISITA Congress Technical Papers: Safety, the vehicle & the road: Part 1. C389/470: 925201. I. Mech. E., London, pp. 9–19.

Wickens, C. D. (1984). *Engineering psychology and human performance* (2nd ed.). New York: HarperCollins.

Wierwille, W. W. (1993a). Demands on driver resources associated with introducing advanced technology into the vehicle. *Transportation Research* (C), *1*(2), 133–142.

Wierwille, W. W. (1993b). Visual and manual demands of in-car controls and displays. In B. Peacock & W. Karowowski (Eds.), *Automotive ergonomics* (pp. 299–320). London: Taylor & Francis.

Wierwille, W. W., & Gutmann, J. C. (1978). Comparison of secondary and primary task measures as a function of simulated vehicle dynamics and driving conditions. *Human Factors, 20*, 233–244.

Wierwille, W. W., Hulse, M. C., Antin, J. F., & Dingus, T. A. (1988a). Visual attentional demand of an in-car navigation display system. In A. G. Gale, M. H. Freeman, C. M. Haslegrave, P. Smith, & S. P. Taylor (Eds.), *Second international conference on Vision in Vehicles, Nottingham, England* (pp. 307–316). Amsterdam: Elsevier Science, North-Holland.

Wierwille, W. W., Hulse, M. C., Fischer, T. J., & Dingus, T. A. (1988b). Strategic use of vision resources by the driver while navigating with an in-car navigation display system. In XXII FISITA congress technical papers: Automotive systems technology: The future (pp. 2.661–2.675). Warrendale, PA: Society of Automotive Engineers.

Zwahlen, H. T., Adams, C. C., & DeBald, D. P. (1988). Safety aspects of CRT touch panel controls on automobiles. In A. G. Gale, M. H. Freeman, C. M. Haslegrave, P. Smith, & S. P. Taylor (Eds.), *Second international conference on Vision in Vehicles, Nottingham, England* (pp. 335–344). Amsterdam: Elsevier Science, North-Holland.

PRACTICE

The Value of Workload in the Design and Evaluation of Consumer Products

Anthony D. Andre

Interface Analysis Associates, San Jose, California

In theory, workload is a powerful, attractive, and useful concept, even as it applies to the consumer product design field. Yet the concept of workload is often not invoked at all or is not used to its fullest advantage by product designers and evaluators, simply because it has never specifically been defined for this particular context. It is argued here that the concept of workload can serve as a unique platform for discussing and addressing the issues of effort, time-pressure, anxiety, and frustration in the context of designing and evaluating the user interface of a given product or procedure. This chapter first discusses both the positive and negative aspects of workload, and then describes how this concept can effectively be applied to many human-centered issues that are typically ignored in the product design and evaluation process. To this end, a checklist of workload dimensions is provided, along with guidelines for considering the economic value of these dimensions.

THE CONCEPT OF WORKLOAD

Workload is like an old family tale, passed down from generation to generation; always told in an enlightening and convincing manner with a slight touch of morality. The story goes like this:

> When evaluating a given task, environment, or system, we typically focus on overt performance, that is, the ultimate success or failure of the activity. Our ability to detect changes in performance are often gross in nature, in that only when a failure occurs do we know that performance has declined (e.g.,

373

an aircraft accident). However, it is likely that before any failure occurs, there are many (measurable) implicit changes taking place—increased effort, stress, or frustration experienced by the operator. Thus, while performance may sometimes appear to be constant, the workload experienced by the operator may be increasing. Clearly, then, if we could see or measure the workload, we might better be able to predict a failure before it happens.

Figure 2.8.1 shows a hypothetical plot of the workload concept. Note that from Time t through t5, performance (solid black line) is relatively stable, staying between 17 and 20 units. But at Time t6 performance sharply declines, or falls off—this is the point of observable failure (or red-line). In contrast, the workload curve (dashed line) provides a completely different picture. Here, we see that the workload experienced by the operator, while not showing any overt performance detriment, steadily increases over the time period t–t5. One could easily argue, and many have, that knowledge of the workload underlying performance could predict, and therefore prevent, the performance breakdown shown at Time t5-t6.

THE FAILURE OF WORKLOAD

I have always been attracted to this concept of workload; on paper it makes so much sense. But, in reality, looking at the impact (or lack thereof) of workload measures on "real-world" systems design and training decisions, my attraction turns to disappointment. There are several reasons why I think workload has failed as a useful tool for the system designer or evaluator in the aviation and process control domains where the concept is most popular, and I argue that these same issues, along with others unique to the product design domain, underlie why the concept is not often applied to the design or evaluation of consumer products:

1. There is reason to believe that one of the most critical and dramatic premises of workload—that performances suddenly collapses at a catastrophic rate—may not be a common occurrence (see Moray & Liao, 1988). Clearly, if we can measure a more gradual degradation in performance then the value of workload as a correlated measure is severely diminished.

2. The concept of workload originated out of the aviation and process control domains, where large-scale, continuous, complex and dynamic systems prevail. It is more difficult to see how this same concept applies to more simplified and discrete processes, such as someone opening a jar lid or a computer user selecting a software icon.

3. There are few, if any, documented cases where design decisions were based on workload measures instead of performance measures, when the two were in conflict. If we only pay attention to workload when it correlates

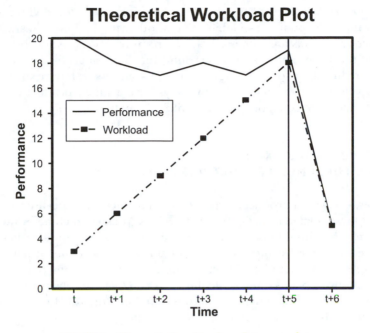

FIG. 2.8.1.   Theoretical workload–performance plot.

perfectly with overt performance, then workload becomes a redundant and somewhat vacuous concept.

4. The objective (physiological) and subjective measures of workload are often considered as two different ways to measure the same thing; and due to their simplicity in administration and analysis, most researchers favor to use subjective ratings (Hart & Staveland, 1988; Moroney, Biers & Eggemeier, 1995). I strongly disagree with this widely held approach and believe that the physiological and subjective measures are, more often than not, mutually exclusive in nature. For example, an operator's heart rate while performing a task and their subsequent rating of the mental effort involved in performing the task, represent different responses to the performance of that task, and therefore should be viewed as nonredundant sources of information.

5. Much of the workload research has been too focused on the study of "load" at the neglect of the study of "work" (Andre & Hancock, 1995). Clearly, "how" we go about interacting with a product is as important as the result of that interaction, if for no other reason that the two are necessarily related. In this sense, performance can be defined as "the end," whereas workload describes the "means to the end."

These are but a few of the problems associated with the definition and use of the concept of workload; no doubt, other methodological and measurement problems are discussed elsewhere in this book. Notwithstanding, we use this information as a background for discussing how workload has been used in the product design domain, its success or failure in that use, and finally, to suggest more effective ways for consumer product designers and evaluators to conceptualize and utilize the concept.

DOES WORKLOAD EXIST
IN THE CONSUMER PRODUCT DOMAIN?

If you speak with a product designer or usability tester, or peruse the literature of these related domains, you will be hard pressed to come by the term *workload*. This should not be surprising, because, as suggested earlier, both the theory and subsequent research of the workload concept has mainly been directed at complex, dynamic and continuous systems (e.g., aircraft, process control plants). Yet, while the term *workload* is rarely mentioned (or known to product designers and testers) the basic theme of the concept is alive and well—under a variety of different names and auspices:

> *Usability testers* have long used facial expressions as an indicator of a user's emotion or process state—frustration, concentration, confusion, surprise, etc. (see Fig. 2.8.2)—in addition to collecting subjective data on various workload-related dimensions (effort, ease-of-use, stress).
>
> *Product designers* often consider the emotional and psychological response of users when conceiving the look and feel of a product. Their goal often is to create designs that appear facile and reduce the subjective workload experience.

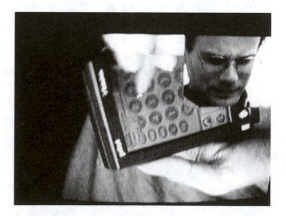

FIG. 2.8.2. Usability test photo. The user's face is shown in the upper right corner; the product he is interacting with fills the remainder of the screen.

Ergonomists are always considering the physical and physiological effects of products and environments. One of their main objectives is to limit the potential for pain or injury caused by over-exertion or stress.

Taken together, these brief descriptions allude to the fact that the concept of workload does apply, and indeed is applied, to the design and evaluation of consumer products. Note, however, that it is not a de facto consideration by any means. In fact, a relatively small amount of product designers and evaluators consider workload issues, largely because the concept has never been specifically defined for this particular context. I strongly believe that workload can serve as a useful guiding concept for product designers and evaluators, but only if the concept can be defined and applied within the lexicon and issues of the consumer product domain. Moreover, the purpose of using the workload concept should not be to study workload per se (that is yet another problem with previous approaches); rather, its use must directly serve the needs of the product designers and evaluators. To this end, the remainder of this chapter will attempt to articulate a set of guidelines for incorporating workload into the product design and evaluation process.

APPLYING WORKLOAD TO PRODUCT DESIGN AND EVALUATION

Defining Workload

In order to employ workload as a useful concept in the product design domain, we must first define it.

> *Workload is a hypothetical construct that represents the cost incurred by a human operator to achieve a particular level of performance.* (Hart & Staveland, 1988, p. 140)

There are many definitions of workload, and most invoke a multidimensional concept (Derrick, 1988). For example, the NASA TLX scale (Hart & Staveland, 1988) involves measuring workload along six dimensions: mental demand, physical demand, temporal demand, performance, effort and frustration. In contrast, the SWAT technique (Reid & Nygren, 1988) measures workload along only three dimensions: time load, mental-effort load, and psychological-stress load. Still others define workload as a psychophysiological response (e.g., heart rate) to a given task or situation (Backs, 1995).

Stepping back from the operational definitions of workload noted above, it is useful to first develop a conceptual definition of workload as it applies to the product design and evaluation process. To do this, we first discuss what can be, or is, typically considered or measured by designers and usability testers. Of course, we can easily see if a user successfully interacts with a product—whether it be opening a jar lid, finding a menu item in a software program, setting the time on a VCR, or understanding the meaning of a sign or instruction. And we can further qualify the success (or failure) of the interaction in terms of time, path, hesitation, micro-errors, and so forth. Indeed, these observational measures of user–product interaction are the foundation of usability testing. Equally considered in the evaluation process is the subjective comments and preferences of the users. Stating whether or not the task/product was "easy to use," "intuitive," "understandable," or "preferred" are common subjective usability dimensions.

Given this seemingly well-rounded approach to measuring the usability of a product, one might ask "what's missing?" The answer to this question lies in recalling the fundamental goal of usability testing: that is, to either determine if one product is better (more usable) and/or preferred over another. In other words, while typical usability evaluations can determine performance and preference differences between two products, or determine that a given interface provides a suitable level of performance or preference, they most often do not why/how the observed differences occur. To this end, it is argued here that, anytime we engage to learn if one product is more usable, we should attempt to determine the underlying causes for the differences observed. After all, if we don't learn how the various aspects of the products evaluated contribute, positively or negatively, toward their respective usability, then we can never apply this knowledge to the design of future products. Arguably, in order to learn why relative differences, or even absolute levels of performance, are observed we must gain an understanding of *how* the user interacts with, and responds to, the product in question.

In consideration of both the original definitions of workload and the requirements of usability testing discussed above, we can conclude that workload can best serve the product designer or evaluator by encompassing usability dimensions beyond performance and user preference. More specifically, workload as a concept must aid the product designer or evaluator in developing a more holistic definition of usability and subsequently, in determining design attributes that directly support the entire set of relevant usability dimensions for a given product. Accordingly, we can now define workload as it applies to the consumer product domain.

Workload refers to the various physical, physiological, psychological and emotional responses required of, and emanated from, users as they interact with a product.

Clearly, this definition of workload refers to the more "emotional" measures of usability, in contrast to the typical measures of behavioral usability (time on task, error rates)/performance and preference (see Logan, 1994; Logan, Augaitis, & Renk, 1994).

TABLE 2.8.1
Consumer Product Workload Dimensions

Category	Dimension	Measurement Method
Physical	Physical effort	Subjective
	Twisting/Reaching	Observation, Instrument, Subjective
	Dexterity	Observation, Subjective
	Force	Instrument, Subjective
	(Dis)Comfort	Instrument, Subjective
Physiological	Pain/Sensation	Observation, Subjective
	Heart rate	Instrument
	Temperature	Instrument
	Metabolic rate	Instrument
Psychological	Cognitive demand	Subjective
	Perceptual demand	Subjective
	Memory demand	Subjective
	Locus of control	Subjective
	Familiarity	Observation, Subjective
	Predictability	Observation, Subjective
Emotional	Stress/Anxiety	Observation, Instrument, Subjective
	Frustration	Observation, Subjective
	Intrigue	Observation, Subjective
	Excitement	Observation, Subjective

THE WORKLOAD CHECKLIST

Table 2.8.1 provides a checklist of workload dimensions and measures that apply to most consumer products. The dimensions are grouped into four main categories: physical, physiological, psychological and emotional. Taken together, these dimensions represent a comprehensive assessment of *both* the required and evoked responses of the user while interacting with a product. The dimensions also vary in the ways they can be measured: (a) real-time (or video) observations of body movements, facial expressions and verbal comments, (b) various measurement devices or (c) subjective surveys. Note that not all dimensions and measures apply to all products or people, and those that do apply will vary in their importance and intrusiveness. For example, a designer might be interested in the user's emotional reaction to their product, whereas a lawyer for the same company may be more interested in the physical and physiological responses as they pertain to legal or safety issues.

Performance Versus Preference

When attempting to evaluate workload it is important to consider the relationship, or lack thereof, between performance (or objective workload measures) and subjective workload/preference measures. It is sometimes the case that users preferences for, and ratings of their interaction of, one product versus another dissociate from observed or measured performance (Andre & Wickens, 1995). For example, in a recent usability study the author objectively measured the repeated failure of a user to successfully operate a hand-held electronic device (in fact, the test administrator had to eventually tell the user how to perform the function). The observations of, and verbal comments made by, the user clearly revealed a high level of frustration, effort and even anger. Yet, when later asked to rate the usability of product, the user gave a highly positive rating. The practical implications of such a finding is critical to the commercial viability of the product(s) being evaluated: Even though the user failed to successfully perform with the product, she is inclined to provide a positive endorsement of this product to others, and to potentially purchase this product herself.

While performance–preference dissociations can sometimes be predicted or explained (see Andre & Wickens, 1995), they are often difficult to interpret. For example, in this case, the user might have rated usability in terms of how easy the product was to use only after they were instructed on how to perform various tasks. Another explanation could be that while they were aware that they had difficulty figuring out how to use the product, they are attracted to some other aspect(s) of the product and therefore want to provide positive ratings.

The lessons to be learned is that, when possible, usability evaluators should attempt to measure both the objective and subjective responses of users for each of the workload dimensions of interest. When a dissociation is found, and can't be easily explained, it is useful to sit down with the user and review the objective and subjective data. In most cases they will be able to explain how or why the dissociation occurred.

THE ECONOMICS OF WORKLOAD

Testing a product is often somewhat different from evaluating the design of a product (Barfield, 1993). The former activity involves determining that the specified goal (function) of the product was achieved. So, if I was to design a phone that allowed one to enter names and numbers into an electronic address book, then I would test to see this function actually worked, and if users could interact with this function in an accurate and timely manner. Keep in mind that the same or different designers could come up with many different designs for such a product, all that achieve the same design objective (that users can create address book entries). However, in evaluating these different designs, we witness and therefore implicitly judge not only the specified design objective, but also the nonspecified attributes of the product (Barfield, 1993). Here, we see that many, seemingly peripheral product attributes play a large role in the overall success, satisfaction and safety of the user.

I have argued for, defined, and listed a set of workload dimensions that expand product evaluations beyond performance and user preference measures. However, the benefit of this approach is severely limited without the establishment of an a priori workload value (economic) system. After all, if performance measures are always given a greater value than workload measures, then the latter are of little practical significance. Recall that this is one of the main limitations of the current concept of workload, discussed earlier in this chapter.

The key to developing a value system for workload is to list the set of economic consumer responses/issues that product manufacturers are ultimately concerned with. These are, in no particular order:

Initial purchase/use. Which workload factors underlie the user's initial purchase decision? What aspects of a product's design lead users to feel that they won't perform well, or safely?

Repeat use. Which workload factors underlie the user's repeated usage of a product? How are these factors different from those that support the initial purchase/use decision?

Successful performance. Which workload factors define successful performance? How sensitive are users to various performance levels. What level of performance is necessary?

Safe performance. Which workload factors underlie the safety of the product?

References. Which workload factors contribute to a level of satisfaction that prompts the user to promote the product to others?

Next, one must determine which issues are effected by which workload dimensions. If we only concentrate on the factors that underlie successful performance (3) or references (5), we might be inclined to give little value to any or all workload dimensions. If, however, we consider the other issues as well, then we naturally assign economic value to various workload dimensions. For instance, looking at Table 2.8.1, it is easy to see how any adverse physiological response (pain, heart rate, etc.) might undermine the safety of the product and why these workload dimensions should therefore be given a high value, where such effects are likely to happen and/or have catastrophic economic consequences.

In summary, there are several factors that, together, determine the economic success of any consumer product or system. It is important for the product designer and evaluator to determine the relative importance of these economic factors, to map each performance, preference and workload dimension to these factors, and to assign economic values to each dimension based on their potential impact.

SUMMARY

Workload is a somewhat infamous concept most familiar to the aviation and process control communities. While consumer product designers and evaluators have long considered some dimensions of workload, the value of the concept has been limited by the lack of a conceptual rationale, operational definition and economic system specific to the consumer product domain. Further, whereas in the past, human factors researchers and product evaluators were concerned with measuring the resources *required* by the product, the approach outlined here emphasizes the resources, responses and reactions *evoked* by the user. To this end, it is perhaps a more truly "user-centered" approach to the concept of workload.

In this chapter I have defined workload along various categories, dimensions, and measures, which together provide a more comprehensive and emotional measure of a product's usability. Further, I have proposed a set of five economic factors to which each workload dimension can be assigned and valued. Although the workload categories and dimensions proposed here are not new, it is hoped that by re-applying this powerful and useful concept to the consumer product domain, designers and evaluators will be better equipped to predict, measure, and optimize the functional, physical, physiological, psychological, and emotional aspects of our interaction with consumer products.

ACKNOWLEDGMENTS

Thanks to Chris Wickens for first introducing me to the concept of workload, and to Sandy Hart, Neville Moray and Peter Hancock for expanding my interest in, and knowledge of, the topic. Thanks also to Kara Andre for her editorial support. Interface Analysis Associates is a human factors consulting firm specializing in product design and usability testing (http://www.interface-analysis.com).

REFERENCES

Andre, A. D., & Hancock, P. A. (1995). Special issue editorial. *International Journal of Aviation Psychology, 5*(1), 1–4.
Andre, A. D., & Wickens, C. D. (1995). When users want what's not best for them. In *Ergonomics in design* (pp. 10–14). Santa Monica, CA: Human Factors & Ergonomics Society.

Backs, R. (1995). Going beyond heart rate: Autonomic space and cardiovascular assessment of mental workload. *International Journal of Aviation Psychology, 5*(1), 25–48.

Barfield, L. (1993). *The user interface, concepts & design.* Reading, MA: Addison-Wesley.

Derrick, W. L. (1988). Dimensions of operator workload. *Human Factors, 30*(1), 95–110.

Hart, S. G., & Staveland, L. E. (1988). Development of NASA-TLX (task load index): Results of empirical and theoretical research. In P. A. Hancock & N. Meshkati (Eds.), *Human mental workload* (pp. 139–183) New York: Elsevier Science (North Holland).

Logan, R. J. (1994). Behavioral and emotional usability: Thomson Consumer Electronics. In M. Wiklund (Ed.), *Usability in practice.* Cambridge, MA: Academic Press.

Logan, R., Augaitis, S., & Renk, T. (1994). Design of simplified television remote controls: A case for behavioral and emotional usability. *Proceedings of the Human Factors and Ergonomics Society, 36,* 365–369.

Moray, N. P., & Liao, J. (1988). *A quantitative model of excess workload, subjective workload estimation, and performance degradation* (Rep. No. EPL-88-03). Urbana-Champaign: University of Illinois.

Moroney, W. F., Biers, D. W., & Eggemeier, F. T. (1995). Some measurement and methodological considerations in the application of subjective workload measurement techniques. *International Journal of Aviation Psychology, 5*(1), 87–106.

Reid, G. B., & Nygren, T. E. (1988). The subjective workload assessment technique: A scaling procedure for measuring mental workload. In P. A. Hancock & N. Meshkati (Eds.), *Human mental workload* (pp. 185–218). Amsterdam: North Holland.

Workload and Air Traffic Control

Brian Hilburn
Peter G. A. M. Jorna
National Aerospace Laboratory NLR

Although the global air traffic control (ATC) system represents one of the most complex human–machine systems ever developed, ATC upgrades are having a difficult time keeping pace with the ever-increasing demand for air transportation. By even conservative projections, air traffic is expected to rise by some 40% over the current decade. Unfortunately, increasing "sectorization" of the airspace—simply carving up the total airspace into smaller regions—does not solve this problem, as increased intersector communication and coordination demands are likely to deny any net gain in system throughput. The goal of designers, therefore, has been to create future ATC systems that permit the *individual* controller to handle more aircraft at present or better levels of safety. This goal is reflected in current efforts to develop new automated tools (e.g., decision-aiding *conflict probes*), interfaces (e.g., *datalink* communication modes), and operational concepts (e.g., *free flight*, in which aircraft assume greater responsibility for route selection and separation assurance). Historically, one of the chief design criteria for such developments has been the mental workload they impose on the controller.

Air traffic control is accomplished by a tightly coordinated network of individual controllers, who are responsible for different segments of the airspace. The ATC structure in the United States, for instance, includes six different classes of airspace, characterized by altitude, proximity to terminal areas, visibility, and so on (Nolan, 1990). The controller's job involves a variety of tasks, including the following (Danaher, 1980):

- Observing aircraft (either directly or via computer-generated displays).
- Operating display controls.
- Making data entries.

- Processing and updating flight progress information.
- Communicating with both aircraft and ground-based agents.
- Coordinating with co-workers.
- Selecting/revising plans and strategies.

How this collection of tasks influences an individual controller's work-load is not a straightforward question, and this difficulty is reflected in the variety of task load indexes that has historically been associated with ATC.

TASK LOAD INDEXES AND WORKLOAD

Task load (i.e., the demand imposed by the ATC task) is generally distin-guished from workload (i.e., the controller's subjective experience of that demand). A number of studies have attempted to identify task load indexes for ATC. Of many prospective task load indexes, the number of aircraft under control (i.e., traffic load) has shown the clearest predictive relation to workload measures (Hurst & Rose, 1978; Stein, 1985). Traffic load by itself, however, does not accurately capture the total load imposed by the airspace. Other factors that contribute to overall airspace complexity include number of traffic problems (Kalsbeek, 1976); number of flight altitude transitions (Murphy & Cardosi, 1995); mean airspeed (Hurst & Rose, 1978); aircraft mix (as it relates to differences in aircraft performance envelopes); vari-ations in directions of flight (Wyndemere, 1996); proximity of aircraft and potential conflicts to sector boundaries (Wyndemere, 1996); and weather (Scott, Dargue, & Goka, 1991; Mogford, Murphy, & Guttman, 1993). At-tempts have been made to develop objective measures of airspace complex-ity (Rodgers, Mogford, & Mogford, 1995; Wyndemere, 1996), by integrating the influence of various airspace, operational and traffic factors. Arad (1964) enumerated the following additional airspace factors that contribute to ATC task load: Sector flow organization coefficient; mean airspeed; sector area; and mean aircraft separation.

Airspace factors are clearly not the only contributors to ATC task load. Such other considerations as the ATC position (e.g., oceanic versus ter-minal; Wickens, Mavor, & McGee, 1997) and the controller interface (in-cluding both the visual display and the data entry system) are critical in determining a controller's task load. They do not always do so, however, in predictable or beneficial ways. As research from other domains has demon-strated, a system's interface itself can impose additional task demands. For instance, automated tools can have the unintended effect of raising task load (Kirlik, 1993; Selcon, 1990). The potential for such situations appears increasingly likely as more sophisticated "advisory" types of decision aids emerge in ATC. By presenting the controller the additional tasks of consid-

ering the system's advice and comparing the system's solutions to those he/
she must continue to generate (if he/she is to remain "in the loop"), such
decision-aiding automation may paradoxically force an additional task on
the controller (Hilburn, Jorna, & Parasuraman, 1995) or lead the controller
to feel "driven" by the system (Whitfield, Ball, & Ord, 1980).

The link between ATC task load and workload is a causative (albeit indirect) one that is colored by a number of internal factors. In the past, attempts to assess ATC workload have sometimes equated measures of direct
task performance (such as time to perform discrete ATC tasks) with workload. Such observable workload (Cardosi & Murphy, 1995), however, provides only a partial picture of the workload experienced by a controller. For
instance, a controller's observable performance cannot always convey the
cognitive task demands—such as planning, decision making, and monitoring—imposed by ATC. Factors such as skill, training, experience, fatigue,
and other "stressors" all mediate the relation between task demands and
workload experienced by a controller. Furthermore, strategy plays an especially important part in determining a controller's workload. In very basic
system demands, there are few constraints on how a controller should handle air traffic (Cardosi & Murphy, 1995; Reason, 1988). As a result, the system
can accommodate various control strategies without suffering a negative
outcome (i.e., a loss of separation or worse a collision). How a controller
chooses to prioritize tasks or the strategies used to respond to workload fluctuations (e.g., shedding or deferring tasks and deciding which tasks to handle first) both influence the controller's workload. Figure 2.9.1 depicts a
simplified schematic of the relation between ATC task load and controller
workload. Many other factors (e.g., time pressure, motivation, effort) are
omitted from this figure.

ATC is commonly held to be a demanding job. Not surprisingly, the assessment of ATC workload has generally focused on the possible overload

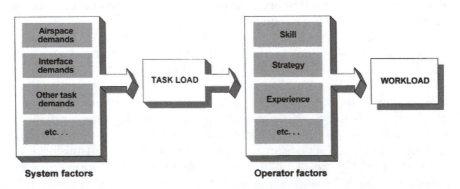

FIG. 2.9.1.   Task load and workload determinants in ATC.

condition. Indeed, operational data support the view that ATC overload poses a very real threat to air safety. Endsley and Rodgers (1996), for instance, demonstrated a relation between high levels of task load and operational errors. Although design efforts usually focus on the task overload condition, operational and theoretical evidence reveals that *under*load poses at least as large a threat to air safety (Danaher, 1980; Jorna, 1993; Stager & Hameluck, 1990). A review of ATC incidents in Canada (Stager, 1991), for example, showed that most occurred during low or moderate traffic load and normal traffic complexity. Similar data have emerged from studies of U.S. ATC operational errors (Redding, 1992). The suggestion has been made that controllers can adapt to heavy traffic peaks, but become error prone as traffic lightens (Fowler, 1980).

WORKLOAD MEASUREMENT TECHNIQUES IN ATC

Over the years, various workload measures have been used to assess ATC. These have included subjective, behavioral (both primary and secondary task), and physiological measures. Table 2.9.1 provides some examples of ATC workload measures that have been used in the past.

TABLE 2.9.1
Workload Measures in ATC: Some Empirical Studies

Subjective
 NASA-TLX (Brookings, Wilson, & Swain, 1996; Hooijer & Hilburn, 1996)
 Air Traffic Workload Input Technique (ATWIT: Leighbody, Beck, & Amato, 1992)
 Subject matter expert/Over-the-shoulder ratings (Schaffer, 1991)
 Instantaneous Self Assessment (ISA) technique (Eurocontrol, 1997; Whittaker, 1995).

Behavioral
 Number of control actions (Mogford, Murphy, & Guttman, 1993)
 Communications efficiency (Leplat, 1978)
 Communication time, message length (Stein, 1992)
 Flight data management (Cardosi & Murphy, 1995)
 Intersector coordination (Cardosi & Murphy, 1995)

Psychophysiological
 EEG, EMG, and EOG (Costa, 1993)
 Heart rate measures (Brookings et al., 1996; Hooijer & Hilburn, 1996; Laurig et al., 1971)
 Eye blink rate (Brookings et al., 1996; Stein, 1992)
 Respiration (Brookings & Wilson, 1994)
 Biochemical activity (Costa, 1993; Zeier, 1994)
 Pupil diameter (Hilburn, Jorna, & Parasuraman, 1995)
 Eye scanning randomness (*entropy:* Hilburn, Jorna, & Parasuraman, 1995)
 Visual fixation frequency (Hilburn, 1996; Stein, 1992)

A handful of subjective measures has been used with respect to ATC. Of these, the most accepted current measure appears to be the multidimensional NASA–Task-Load Index (NASA-TLX: Hart & Staveland, 1988). Other instruments that have been used with ATC include the unidimensional Rating Scale for Mental Effort (RSME: Zijlstra & van Doorn, 1985). One subjective instrument, the Air Traffic Workload Input Technique or ATWIT (Leighbody, Beck, & Amato, 1992; Stein, 1985) has been specifically tailored for use with ATC tasks. Another tool, the Instantaneous Self-Assessment or ISA (Whittaker, 1995), cues controllers for unidimensional workload ratings every 2 minutes.

Overall, ATC workload evaluations have tended to rely mainly on subjective measures. Although the use of subjective measures is attractive (they are inexpensive and easily collected), researchers should remain mindful of their limitations. This admonition seems especially relevant to the ATC domain, in which highly skilled operators (with possibly idiosyncratic strategies) are asked to give subjective reports. Such operators can be unable (or, for reasons of job security, unwilling) to give accurate reports. Furthermore, such reports may be subject to individual biases, preconceptions, or memory limitations. These and other factors limit their usefulness in cross-task, or cross-controller, comparisons.

Despite potential limitations, subjective assessment techniques also have some very strong potential benefits for ATC assessments. There are at least two clear reasons for considering their use. First, subjective techniques can be sensitive to task demands below the threshold for performance impairment and thus might be more sensitive than behavioral measures (and less costly or complex than psychophysiological measures) for assessing the critical ATC underload condition. Second, rapid technological advances are inviting a revolutionary overhaul of the basic process of air traffic control. This is reflected in the current development of new automated tools, interfaces, and operational concepts. One interesting aspect of such advances is that their successful introduction relies increasingly on controller acceptance. If an optional-use tool, for instance, is not perceived as useful, controllers are likely to develop sophisticated ways for bypassing its use. For cases in which subjective acceptance can limit the introduction of new tools or procedures, the use of subjective measures seems especially critical.

Again, many studies have attempted to relate overt behavioral measures (e.g., total radio communication time) directly to controller mental workload. This use of behavioral measures is attractive at first glance, given that they can often be related directly to operational performance. A number of ATC-relevant tasks can be used as embedded indicators of ATC workload. For instance, the communication demands of a new ATC system (operationalized as, say, total microphone key press time per hour) appear to relate directly to both how hard the controller is working and how efficient the system can be expected to perform. As in other domains, however, behavioral measures can sometimes prove insensitive in ATC. Compensatory strategies—such as vectoring aircraft into a holding stack to "buy time"—can enable controllers to allocate additional attention to a given task and thus maintain task performance. As a result, task load fluctuations are not necessarily reflected in observable performance changes. Sperandio (1971) showed that controllers can marshal a variety of idiosyncratic strategies to maintain task performance in the face of changing task demands.

Cross-task workload comparisons (e.g., communication versus intersector coordination workload) can be difficult with primary task behavioral measures. This is a large potential stumbling block to their use, given the variety of individual tasks—such as monitoring, planning, coordinating, and communicating—involved in ATC (Hopkin, 1979). Although behavioral measures often enjoy high face validity and can be easy and natural to collect, the researcher should recognize that they, like subjective measures, have potential pitfalls with respect to ATC. For instance, problems with sensitivity (e.g., at low task-load levels) or diagnosticity (in the multitask ATC situation) can limit their usefulness for evaluation of ATC workload.

The use of physiological measures for ATC is largely subject to the same ledger sheet of potential costs and benefits as for other domains. Their potential diagnosticity and sensitivity benefits (even at low task-load levels; Braby, Harris, & Muir, 1993) must be weighed against possible drawbacks such as monetary costs, required technical expertise, and increased need for experimental control of environmental factors. Furthermore, Cardosi and Murphy (1995) noted that psychophysiological measures are subject to large individual differences across controllers—a traffic load increase, for instance, might result in a large heart rate change for one controller, yet no such change for another controller. As a result, a within-subjects experimental design is generally preferable.

Brookings, Wilson, and Swain (1996) evaluated a series of physiological workload measures and the NASA TLX subjective measure by using military air traffic controllers, who performed a series of ATC scenarios on the commercially available TRACON computer-based ATC simulation. They demonstrated a significant main effect of traffic difficulty (operationalized as 6, 12, or 18 aircraft under control) on eye blink rate, respiration rate and amplitude, electroencephalogram (EEG) activity (across the four major bands, from 1.1 to 24.9 Hz) and NASA-TLX measures. Traffic complexity (a function of aircraft mix, flight type mix, and pilot compliance), on the other hand, demonstrated significant effects only on EEG activity.

Hilburn (1996) used a battery of psychophysiological, performance and subjective measures to assess the impact of automation assistance and traffic load manipulations on controller workload. Traffic load manipulation demonstrated a significant effect on most of the psychophysiological mental workload measures. The use of pupil diameter and heart rate variability (HRV), which both varied fairly consistently with the number of aircraft under control, seemed especially encouraging. Pupil diameter of highly experienced controllers increased (i.e., indicated workload increased) steadily with number of aircraft, even though the presence of more onscreen elements (aircraft) corresponded to a greater number of illuminated pixels (which, everything else being equal, would have been expected to cause a *decrease* in pupil diameter). This pattern, which has been replicated in later studies, is depicted in Fig. 2.9.2.

As a group, eye tracking measures of mental workload (either blink/pupil related, or point-of-gaze derived) have not always yielded unambiguous results. The potential for data artifacts as well as inherently large individual differences are certainly to blame in some cases. As an example of the former, Stein (1992) demonstrated that 5 to 10 minutes are required for an observer's scan pattern (both the magnitude and duration of saccades) to stabilize. Other potential sources of data artifacts include fatigue, age, and ambient illumination level. Through careful experimental design, however, these risks can be minimized in ATC simulations. For example, realistic ATC simulations do not preclude controlled lighting. Furthermore, it is realistic for a controller to acclimate to the traffic pattern (by over-the-shoulder monitoring) for 5 to 10 minutes before assuming control after a crew shift change.

Much of the past research on workload assessment has taken place in the aviation cockpit. In this context, some of the typical concerns with psychophysiological measures—such as their potential complexity, intrusiveness,

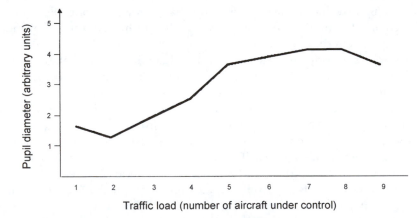

FIG. 2.9.2.   Pupil diameter, as a function of traffic. Data are from *Strategic Decision Aiding: A Comprehensive Bio-Behavioral Assessment of Human Interaction With Real-Time Decision Aids* (Report) by B. Hilburn, 1996, Brussels, Belgium: North Atlantic Treaty Organization (NATO).

and data artifacts—are understandable. Luckily for the ATC researcher, however, many of the potential problems can be minimized simply through experimental design. When used correctly, these measures can provide valuable and unique information (Jorna, 1998)—information not available through either subjective or behavioral measures. For instance, one of the most appealing attributes of physiological measures is their potential temporal resolution. Pupil diameter, as an example, offers a theoretical resolution on the order of 500 to 700 milliseconds. Similarly, evidence from cockpit simulations (Jorna, 1997) shows that "event-related" analysis of heart rate data also promises greater temporal resolution than is generally associated with either behavioral or subjective measures.

CONCLUSION

There is clearly no single ideal workload metric for all situations, indeed not necessarily for even a single task situation. All measures have advantages and limitations that should be understood, if data from them are to be properly analyzed and interpreted. In general, experience recommends a comprehensive approach to workload measurement, one that combines psychophysiological, subjective, and performance measures (Cardosi & Murphy, 1995; Hart & Wickens, 1990; Hilburn et al., 1995). In this way, various classes of workload measures can provide converging evidence and a fuller understanding of ATC workload. For instance, Hilburn, Bakker, Pekela, and Parasuraman (1997) demonstrated that, although controllers tended to query traffic more under simulated free flight than under conventional "controlled" flight (in which ATC remained responsible for separation assurance and route selection), this result actually seemed to represent more of a strategy than a workload difference: A battery of psychophysiological indicators pointed toward controller workload *reductions,* not increases, under free flight.

ATC workload assessment has generally focused on mental, as opposed to physical, workload. Regardless of workload measure, researchers should bear in mind the potential for physical artifacts in ATC workload data. Although ATC tends to impose cognitive rather than physical load, it is important to bear in mind that physical demands (e.g., keyboard entry) can also indirectly contribute to the mental workload of ATC (Wickens et al., 1997). Experimental design should control for such possible artifacts.

Besides simply focusing on how to measure ATC workload, the researcher should also stay mindful of the higher level *application* of such data—that is, how should workload data be used to drive ATC system development efforts? In this regard, at least one caveat seems in order: In ATC as in a number of other domains, the issue of fun must be taken seriously. Air traffic

controllers are highly skilled operators, and a large measure of their job satisfaction derives from handling challenging traffic situations. Exercising the
creativity and intelligence that characterize Rasmussen's (1981) *knowledge-
based* level of behavior is a genuine source of satisfaction for air traffic controllers. To deprive the operators of such complex systems of the opportunity to skillfully apply their craft runs a substantial risk of limiting user acceptance. As in other domains, therefore, the goal of ATC system designers
should be to develop systems capable of ensuring optimal, as opposed to
simply low, workload.

REFERENCES

Arad, B. A. (1964, May). The controller load and sector design. *Journal of Air Traffic Control*,
   12–31.
Braby, C. D., Harris, D., & Muir, H. C. (1993). A psychophysiological approach to the assessment of work underload. *Ergonomics, 36*, 1035–1042.
Brookings, J. B., Wilson, G. F., & Swain, C. R. (1996). Psychophysiological responses to
   changes in workload during simulated air traffic control. *Biological Psychology, 42*(3), 361–
   377.
Cardosi, K. M., & Murphy, E. D. (1995). *Human factors in the design and evaluation of air traffic
   control systems* (DOT/FAA/RD-95/3). Cambridge, MA: U.S. Department of Transportation, Volpe Research Center.
Costa, G. (1993). Evaluation of workload in air traffic controllers. *Ergonomics, 36*, 1111–1120.
Danaher, J. W. (1980). Human error in ATC system operations. *Ergonomics, 22*, 535–545.
Endsley, M., & Rodgers, M. (1996). Attention distribution and situation awareness in air
   traffic control. In *Proceedings of the 40th annual meeting of the Human Factors Society* (pp.
   82–85).
Eurocontrol. (1997). PD/1 final report annex A: Experimental design and methods. (Report
   PHARE/NATS/PD1-10.2/SSR;1.1). Brussels, Belgium: Author.
Fowler, F. D. (1980). Air traffic control problems: A pilot's view. *Human Factors, 22*, 645–653.
Hart, S. G., & Stavelend, L. E. (1988). Development of the NASA-TLX (Task Load Index): Re-
   sults of empirical and theoretical research. In P. A. Hancock & N. Meshkati (Eds.) *Human
   mental workload* (pp. 139–184). Amsterdam, NL: Elsevier.
Hart, S., & Wickens, C. D. (1990). Workload assessment and prediction. In H. R. Booher
   (Ed.), *MANPRINT: An emerging technology, advanced concepts for integrating people, machine,
   and organization.* New York: Van Nostrand Reinhold.
Hilburn, B. (1996). *The impact of ATC decision aiding automationm on controller workload and human–machine system performance.* (NLR Tech. Pub. TP96999).
Hilburn, B., Bakker, M. W. P., Pekela, W. D., & Parasuraman, R. (1997, October). The effect
   of free flight on air traffic controller mental workload, monitoring and system performance. In *Proceedings of the Confederation of European Aerospace Societies (CEAS) 10th European
   aerospace conference.* Confederation of European Aerospace Societies.
Hilburn, B., Jorna, P. G. A. M., & Parasuraman, R. (1995, April). The effect of advanced ATC
   strategic decision aiding automation on mental workload and monitoring performance:
   An empirical investigation in simulated Dutch airspace. In *Proceedings of the eighth international symposium on Aviation Psychology.* Columbus: Ohio State University.
Hooijer, J. S., & Hilburn, B. G. (1996). *Evaluation of a label oriented HMI for tactical datalink com-*

munication in ATC. (NLR Tech. Pub. TP 96676 L). Amsterdam: National Aerospace Laboratory NLR.

Hopkin, V. D. (1979). Mental workload measurement in air traffic control. In N. Moray (Ed.), *Mental workload: Its measurement and theory.* New York: Plenum.

Hurst, M. W., & Rose, R. M. (1978). Objective job difficulty, behavioral response, and sector characteristics in air route traffic control centres. *Ergonomics, 21,* 697–708.

Jorna, P. G. A. M. (1993). The human component of system validation. In J. A. Wise, V. D. Hopkin, & P. Stager (Eds.), *Verification and validation of complex systems: Human factors issues.* Berlin: Springer-Verlag.

Jorna, P. G. A. M. (1997). Pilot performance in automated cockpits: Event related heart rate responses to datalink applications. In *Proceedings of the ninth international conference on Aviation Psychology.* Columbus: Ohio State University.

Jorna, P. G. A. M. (1998). Automation and free(er) flight: Exploring the unexpected. In *Proceedings of the third international conference on Automation Technology and Human Performance.*

Kalsbeek, J. W. H. (1976). Some aspects of stress measurement in air traffic control officers at Schiphol Airport. *Symposium on stresses of air traffic control officers* (pp. 39–42). Manchester: University of Manchester Department of Postgraduate Medical Studies.

Kirlik, A. (1993). Modeling strategic behavior in human–automation interaction: Why an aid can and should go unused. *Human Factors, 35,* 221–242.

Leighbody, G., Beck, J., & Amato, T. (1992). An operational evaluation of air traffic controller workload in a a simulated en route environment. *37th annual Air Traffic Control Association conference proceedings* (pp. 122–130). Arlington, VA: Air Traffic Control Association.

Leplat, J. (1978). The factors affecting workload. *Ergonomics, 21,* 143–149.

Mogford, R. H., Murphy, E. D., & Guttman, J. A. (1993). The use of direct and indirect techniques to study airspace complexity factors in air traffic control. In *Proceedings of the first Mid-Atlantic Human Factors Conference* (pp. 196–202). Norfolk, VA: Old Dominion University.

Murphy, E. D., & Cardosi, K. M. (1995). Issues in ATC automation. In K. M Cardosi & E. D. Murphy (Eds.), *Human factors in the design and evaluation of air traffic control systems* (pp. 219–264). (DOT/FAA/RD-95/3). Cambridge, MA: U.S. Department of Transportation, Volpe Research Center.

Nolan, M. S. (1990). *Fundamentals of air traffic control.* Belmont, CA: Wadsworth.

Rasmussen, J. (1981). Models of mental strategies in process plant diagnosis. In J. Rasmussen & W. Rouse (Eds.), *Human detection and diagnosis of system failures.* New York: Plenum.

Reason, J. (1988). Cognitive under-specification: Its varieties and consequences. In B. Baars (Ed.), *The psychology of error: A window on the mind.* New York: Plenum.

Redding, R. E. (1992, October). Analysis of operational errors and workload in air traffic control. In *Proceedings of the Human Factors Society 36th annual meeting* (pp. 1321–1325), Atlanta, GA.

Reiche, D., Kirchner, J. H., & Laurig, W. (1971). Evaluation of stress factors by analysis of radiotelecommunication in ATC. *Ergonomics, 14,* 603–609.

Rodgers, M. D., Mogford, R. H., & Mogford, L. S. (1995). Air traffic controller awareness of operational error development. In *Proceedings of the second annual Situation Awareness Conference.* Daytona Beach, FL: Embry Riddle Aeronautical University.

Schaffer, M. (1991, January–March). The use of video for analysis of air traffic controller workload. *Journal of Air Traffic Control,* 35–37, 40.

Scott, B. C., Dargue, J., & Goka T. (1991). Controlled and automatic information processing. 1: Detection, search, and attention. *Bulletin of the Psychonomic Society, 18,* 207–210.

Selcon, S. J. (1990). Decision support in the cockpit. Probably a good thing? *Proceedings of the Human Factors Society 34th annual meeting* (pp. 46–50). Santa Monica, CA: Human Factors Society.

Sperandio, J. C. (1971). Variation of operators' strategies and regulating effects on workload. *Ergonomics, 14,* 571–577.

Stager, P. (1991). The Canadian automated air traffic system (CAATS): An overview. In J. A. Wise, V. D. Hopkin, & M. L. Smith (Eds.), *Automation and systems issues in air traffic control* (pp. 39–46). Berlin: Springer-Verlag.

Stager, P., & Hameluck, D. (1990). Ergonomics in air traffic control. *Ergonomics, 33*(4), 493–499.

Stein, E. S. (1985). *Air traffic controller workload: An examination of workload probe.* (Report FAA/CT-TN90/60). Atlantic City, NJ: Federal Aviation Administration Technical Center.

Stein, E. S. (1987). Where will all our air traffic controllers be in the year 2001? In *Proceedings of the 4th symposium on Aviation Psychology* (pp. 195–201). Columbus: Ohio State University.

Stein, E. S. (1988). *Air traffic controller scanning and eye movements—A literature review.* (DOT/FAA/CT-TN 84/24). Atlantic City Airport, NJ: Federal Aviation Administration Technical Center.

Stein, E. S. (1992). *Air traffic control visual scanning.* (FAA Technical Report DOT/FAA/CT-TN92/16). Washington, DC: U.S. Department of Transportation.

Whitfield, D., Ball, R. G., & Ord, G. (1980). Some human factors aspects of computer-aiding concepts for air traffic controllers. *Human Factors, 22,* 569–580.

Whittaker, R. A. (1995). *Computer assistance for en-route (CAER) trials programme: Future system 1 summary report.* (CS Rep. 9529). London: Civil Aviation Authority National Air Traffic Services.

Wickens, C. D., Mavor, A. S., & McGee, J. P. (Eds.) (1997). *Flight to the future: Human factors in air traffic control.* Washington, DC: National Academy of Sciences, National Research Council.

Wyndemere Inc. (1996). *An evaluation of air traffic control complexity.* (Final Rep. No. NAS2-14284). Boulder, CO: Author.

Zeier, H. (1994). Workload and psychophysiological stress reactions in air traffic controllers. *Ergonomics, 37*(3), 525–539.

Zijlstra, F. R. H., & van Doorn, L. (1985). *The construction of a scale to measure subjective effort.* (Tech. Rep.). Delft, Netherlands: Delft University of Technology, Department of Philosophy and Social Sciences.

2.10 Secondary-Task Measures of Driver Workload

Barry H. Kantowitz
University of Michigan Transportation Research Institute

Ozgur Simsek
Battelle Human Factors Transportation Center

Driving a car in the 20th century is complex. Excessive information inside the vehicle and odd road geometry contribute to driver workload. In this chapter, we first define workload and discuss several theoretical issues that underlie workload measurement. We then review several studies that provide insight on how to evaluate driver workload. We focus on secondary-task measures of workload and summarize the state of research in two areas: workload implications of in-vehicle information systems and workload implications of highway design.

Despite all efforts to create a safe driving environment, accidents do happen. Most researchers would agree that accident risk is related to the mental workload of the driver. Indeed, Harms (1986) showed that driver workload on 100-meter segments of roadway covaried with the number of reported traffic accidents on these segments. Therefore, it is reasonable to expect that keeping workload at an acceptable level would reduce the number and severity of traffic accidents. This chapter discusses the impact of both new in-vehicle technology and highway design on driver workload.

Workload measures fall into one of the following categories: objective performance measures (primary and secondary task), subjective measures, and physiological measures. Although all of these are extensively used to assess driver workload, we primarily focus on secondary-task performance measures for two reasons. First, we believe, on the basis of decades of aviation research on pilot workload, that this technique provides an effective measure of workload. Second, theoretical assumptions underlying the secondary-task technique are complex and greatly influence how data can be interpreted. Research on driver workload requires appropriate control conditions and levels of independent variables that are not always included. Our review of empirical efforts emphasizes methodological issues in hopes that future research will make optimal use of secondary-task techniques.

DEFINING AND MEASURING DRIVER WORKLOAD

There have been many reviews of workload, so many that one researcher has called for a review of reviews, but it is beyond the scope of this article to provide a review of reviews. Instead, we take the narrower position of merely reviewing the senior author's theoretical position on workload evaluation, which has remained fairly constant over the last 2 decades. Only a brief recapitulation is given here; readers wanting tendentious details are referred to original sources starting with Kantowitz and Knight (1976) and progressing to Kantowitz (1985, 1987) and Kantowitz and Campbell (1996).

The most important point in defining mental workload is that workload is a theoretical construct. Workload cannot be directly observed but must be inferred from changes in performance. This makes it an intervening variable. Kantowitz (1988) has defined workload as an intervening variable, similar to attention, which modulates the tuning between the demands of the environment and the capabilities of the organism.

Thus, any methodology used to measure workload must make certain theoretical assumptions, whether or not the researchers using the methodology are aware of these assumptions (Kantowitz, 1992). For example, the common methodology of using subjective rating scales to measure workload is fraught with peril for those who are unaware of the theoretical limitations of psychometrics, as has been amply explained by Nygren (1991).

Although the measurement of pilot workload in the aviation domain has been generally successful (e.g., Kantowitz & Campbell, 1996; Kantowitz & Casper, 1988), this does not guarantee that the underlying assumptions of workload measurement hold in the domain of ground transportation. The measurement of driver workload by using secondary tasks goes back at least to the early work of Brown (1962), and many others have attempted to measure driver workload (e.g., Schlegel, 1993). However, a key assumption of secondary-task methodology is that the primary (driving) task is uninfluenced by addition of the secondary task. This assumption tends to hold for airplane pilots who are highly trained to fly the plane first (e.g., Bortolussi, Kantowitz, & Hart, 1986), but this assumption must be evaluated anew when vehicle drivers become the test population (Kantowitz, 1992). The important point stressed here is that the minimum requirement for testing this assumption is a single-task control condition where the primary task is performed alone. Similarly, decrements in the secondary task can not be established without a single-task control condition where the secondary task is performed alone. Furthermore, it is also desirable to investigate more than

TABLE 2.10.1
Summary of Lessons Learned (Kantowitz, 1992)

1. Performance is not workload. The additional complexity of an intervening variable is worth its cost.
2. Workload methodology carries theoretical assumptions that the practitioner must understand.
3. Tools for proper workload measurement are available, but must be used. Convenience is not a good substitute for scientific measurement criteria.
4. Planning and decision making are vital components that have great potential to alter workload.
5. There is no universal secondary task.
6. There is no universal dependent variable for scoring the primary task, although some kind of root mean square error is often useful.
7. No single research environment, e.g., laboratory, field, or simulator, is sufficient to study workload by itself.
8. It is difficult to get the two tribes of human factors, scientists and practitioners, to agree on useful measures of workload.
9. It is possible to get so involved in studying workload that the construct becomes reified into an end rather than a means.
10. It is possible to want a practical measure of workload so desperately as to abandon most reasonable scientific criteria.

one level of secondary task (Kantowitz & Knight, 1976). Without such control conditions, one cannot evaluate the theoretical assumptions that underlie objective measurement of driver workload.

Table 2.10.1 shows 10 lessons learned in the measurement of aviation workload that are salient for researchers interested in driver workload. These lessons can help researchers navigate between the Scylla of ignoring previous aviation research and the Charybdis of inappropriate rote application of these techniques.

INTELLIGENT TRANSPORTATION SYSTEMS AND DRIVER WORKLOAD

All manner of new in-vehicle systems are rapidly becoming commercially available, including navigation and route guidance systems, collision avoidance systems, and enhanced convenience and entertainment systems. Although designers intend that these devices should improve safety, it is possible that the variety and number of these vehicular enhancements might reduce safety by increasing driver workload. Thus, it is important to evaluate driver workload in the presence of this new technology. More and more studies are being published that attempt such evaluation. However, as noted in the preceding section, the best workload evaluations are sensitive to the theoretical nature of the workload construct and contain many helpful control conditions. The studies reviewed in this section are arranged in increasing order of their approximation to a perfect workload study containing every necessary and desirable control condition. Not all researchers have the resources and the desire to attain such methodological nirvana in every study; even studies that are incomplete from the perspective of workload methodology may attain their authors' goals. For example, a finding that insertion of a secondary task produced poorer primary-task driving performance has important implications for safety, even though such changes in primary-task performance make it more difficult to evaluate workload. Studies reviewed here represent the best efforts so far in evaluating effects of in-vehicle technology; many poor studies on this topic have been reviewed but are not discussed herein.

Tables 2.10.2 and 2.10.3 summarize the results of studies reviewed, indicating the workload measure, effects of independent variables on single- and dual-task performance, and whether the workload measure was influenced in dual-task conditions by the other task. These tables are a useful propaedeutic device for comparing studies and reminding readers about requisite control conditions.

Noy (1989) conducted an experiment to investigate the effects of auxiliary task load, resource structure, and driving load on driving performance; and to test/demonstrate a methodology that could be used to evaluate the intrusiveness of an intelligent display on driving. Subjects drove in a moving-base simulator and performed cognitive tasks on a cathode-ray tube display that was located on the instrument panel to the right of the driver. The two display tasks, a spatial perception task and a verbal memory task, were designed to place differential demands on cognitive resources. Display task difficulty and driving difficulty were manipulated within subjects.

Although there was ample evidence that driving was influenced by insertion of a secondary task, the complementary data for secondary-task performance were not reported, being beyond the scope of the paper: "While these results may be relevant to display designers, it is not considered relevant from the road safety perspective" (p. 53). This perspective is understandable, but it is unfortunate that the remaining results from an excellent experiment were not presented. Indeed, even from a safety perspective, it might be helpful to obtain results on possible trade-offs between driving and secondary-task performance. However, because these results were collected and analyzed, interested readers might try obtaining them directly from the author. There are other studies, too dismal to mention, where the authors did not even collect the secondary-task data.

Verwey and Veltman (1996) investigated a primary onroad driving task combined with a visual detection secondary task. In addition, a third task, called the loading task, required either perception of a pattern of light-emitting diodes (L-counting task) or a continuous memory task. These loading

TABLE 2.10.2
Summary of Primary-Task Workload Measures

| Study | Workload Measure | Main Effect | | Influenced By At Least One Secondary Task? |
		Driving Only	Dual Task	
Noy (1989)	Standard deviation of lateral position	Not reported	Curve type	Yes
	Lane exceedance ratio	Not reported	None	No
	Time to line crossing	Curve type*	Curve type	Yes
	Headway	Not reported	Secondary task difficulty	Yes**
	Velocity	Not reported	None	No
	SD velocity	Not reported	None	Yes
Verwey & Veltman (1996)	Speed	None	Not reported	No
	Steering reversal rate	Presence of loading task	Not reported	Yes
Verwey (1991)	Speed	Driving situation, traffic density	Driving situation, traffic density	No
	Standard deviation of speed	Driving situation, traffic density	Driving situation, traffic density	No
	Steering wheel action rate	Driving situation, driving experience	Driving situation, driving experience	Yes

*The graphs suggest that there is a difference; however, relevant statistical tests are not reported.

**Driving-only condition was not statistically different from the mean of the dual-task conditions; it was not compared with each of the four dual-task conditions. However, because there was a significant effect of secondary task difficulty, and headway in the driving-only condition was lower than in any of the dual task conditions, it is likely that the headway was influenced by at least one level of the secondary task.

tasks, although not intended to represent any particular in-vehicle intelligent transport system (ITS), share some of the important cognitive demand characteristics of actual ITS devices. Vehicle speed was not influenced by addition of the loading tasks, whereas vehicle steering intervals decreased when loading tasks were added. Vehicle speed was not altered when the secondary task was added, but steering reversal rate was influenced by the secondary task, making it a less desirable measure. The L-counting task was performed less accurately when combined with the secondary task. Primary-task dependent variables were not reported separately for single- and dual-task conditions (i.e., their fig. 2 does not distinguish between the presence or absence of the secondary task). Performance of the secondary task alone (without driving) was not reported. Although this experiment contains

TABLE 2.10.3
Summary of Secondary-Task Workload Measures

Study	Workload Measure	Main Effect		Influenced by Primary Task?
		Secondary Task Only	Dual Task	
Noy (1989)	Perception task reaction time	Not reported	Yes	Not reported
	Memory task reaction time	Not reported	None	Not reported
Verwey & Veltman (1996)	Percentage detected targets (visual detection)	Not reported	Presence, type, and duration of loading tasks	Not reported
Verwey (1991)	Visual detection (reduction of % detected due to driving)	Not reported	Not reported	Yes
	Visual addition (reduction of % correct due to driving)	Not reported	Not reported	Yes
	Auditory addition (reduction of % correct due to driving)	Not reported	Not reported	Yes

many interesting results including physiological and subjective workload measures, the omission of some control conditions and the use of only a single level of the secondary task make its findings difficult to fully interpret in terms of secondary-task workload methodology. Unreported cells in Tables 2.10.2 and 2.10.3 always indicate an incomplete experiment design. Although it is clear that the loading tasks performed while driving produced secondary-task decrements, the absence of a secondary-task–only control condition does not allow assessment of the workload produced by driving itself. It seems reasonable to guess that such a control condition would have exhibited 100% target detection, but these data are not presented. Furthermore, in dual-task conditions without loading, performance on the secondary task is close to 100%, indicating a possible ceiling effect. Thus, having a more difficult level of the secondary task might have been quite useful; as argued previously in this chapter, having two levels of secondary-task difficulty is often prudent. It may well be that the secondary task in this experiment would have been insensitive to workload effects of driving only and that the decrements observed in secondary-task performance are entirely due to demands of the loading tasks rather than to any demands of the driving task. Nevertheless, this is an important experiment that shows it is possible to evaluate driver workload on the road.

Another onroad study conducted by Verwey (1991) did include multiple levels of the secondary task as well as a secondary-task–only control condition. The visual detection secondary task required subjects to detect a number presented on a dashboard-mounted display. The visual addition condition required the subjects to add 12 to the presented number and to respond vocally. The auditory addition secondary task also required adding 12 but used an auditory display. Driving speed and its standard deviation were not influenced by adding secondary tasks, although steering-action rate was influenced. Unfortunately, difference scores (reduction of secondary-task performance in dual-task conditions relative to the single-task control) were used to measure secondary-task performance. This is a poor statistical technique (Cronbach & Furby, 1970) because, among other flaws, it assumes that each score is perfectly reliable. It would have been much better to analyze single- versus dual-task performance as an independent variable. It also results in unreported cells in Table 2.10.3 (e.g., one cannot compare the different levels of the secondary task alone). However, it was clear that secondary task decrements were obtained with interesting differences between experienced and inexperienced drivers.

Although these experiments, as well as others not reviewed, show that secondary-task methodology can be useful in measuring driver workload, we do not believe that researchers are obtaining the maximum amount of information from their experiments. More care need be paid to testing, analyzing, and reporting single-task control conditions, and research would also benefit from including at least two levels of difficulty or complexity for both secondary and primary tasks.

ROAD GEOMETRY AND DRIVER WORKLOAD

There has been less research relating workload to road geometry than to in-vehicle devices, but this topic has been of concern to civil engineers and efforts have been increasing. Road geometry refers to the physical aspects of a highway such as radius of curvature, lane width, and shoulder width (Neuman, 1992). Highway engineers understand that accident rates can be related to road geometry (for example, a run-off-the road accident is more likely at a sharp curve than at a shallow curve) and are trying to relate driver workload to accidents (e.g., Krammes & Glascock, 1992).

A typical human factors approach to driver workload uses secondary tasks to evaluate workload associated with geometric road features. For example, Kantowitz (1995) used a driving simulator to study three levels of road curvature: hard curves (radius of curvature from 150–550 ft), easy curves (600–1000 ft radius), and straight segments (called tangents by highway engineers). Driving scenarios contained only one type of road geometry. Four secondary tasks were investigated: reaction time (RT) to reading a tachometer, RT to a single letter indicating compass direction, digit perception that required immediate recall of a seven-digit number presented auditorally, and digit retention that required delayed recall after 30 seconds. Every driving scenario that contained secondary tasks had a matched scenario where no secondary tasks were presented; this was a blocked single-task (driving only) control condition. Traffic density in the opposite lane was either light (55 vehicles per scenario) or heavy (125 vehicles). Although a traffic engineer would not regard this difference as a strong manipulation of traffic density, it proved sufficient for experimental purposes. Results showed that the tachometer RT task was a sensitive index of workload imposed by road geometry, whereas the digit perception task was a sensitive index of workload imposed by traffic density. The remaining two secondary tasks were not influenced by any independent variables.

Comparisons with the primary task performed alone versus with the secondary tasks showed that for all primary-task workload measures (lane position and its standard deviation, vehicle speed and its standard deviation, and steering rate standard deviation) except steering rate, driving was not influenced by the addition of a secondary task. This, of course, is a necessary condition for interpreting secondary-task data as workload indicants.

Both substantive and methodological conclusions can be drawn from this study. First, road conditions do impose driver workload as measured by a secondary task. Second, not all secondary tasks are appropriate for measuring driver workload. *There is no universal secondary task.* Third, not all primary-task performance measures are suitable for combining with a secondary task.

A similar experiment performed on the same type of driving simulator did not have such clear results. Moss and Triggs (1997) used a sophisticated attention-switching secondary task and obtained the unusual result that secondary-task performance was superior during dual-task conditions. The authors attributed this unexpected finding to increased arousal in dual-task conditions. No difference in attention-switching RT was found between straight and curved road segments. However, primary-task performance measured by the standard deviation of lateral position worsened for curved segments during dual-task conditions. This violates the key assumption discussed earlier: Inserting the secondary task should not alter primary-task performance. This violation makes it difficult to interpret their secondary-task results. The decreased switching RT may be an artifact of trade-offs between primary and secondary tasks: Worse driving performance may facilitate better attention switching, rather than being due to changes in arousal.

To elaborate this point, let us consider what the curve secondary-task performance would be if the curve driving performance remained uninfluenced by the attention-switching task. Two outcomes are possible. Most probably, we would observe some decrement in the curve secondary-task performance in the dual-task condition compared to the current outcome. With a big enough decrement, curve secondary-task performance in the dual-task condition would be the same as or worse than in the single-task condition. In that case, arousal would not be a necessary explanation on curves.

Another possibility is no change in curve secondary-task performance in the dual-task condition compared to the current outcome. In this case, the secondary-task performance in the dual-task condition would be the same on both straight sections and curves. Other research has shown some secondary tasks to be effective measures of driver workload imposed by road geometry (Kantowitz, 1995). Thus, this second possible outcome would mean that the attention-switching task is not sensitive to changes in workload induced by road geometry and, therefore, is not an effective measure.

Thus, regardless of which possible outcome might occur, the arousal interpretation offered by Moss and Triggs (1997) may not be the only explanation of their results. In dual-task research, it is always vital to examine the constancy of primary-task performance.

A typical civil engineering approach to driver workload uses subjective ratings to evaluate geometric features. For example, Messer, Mounce, and Brackett (1979) developed criticality ratings for 10 geometric features such as horizontal curves, bridges, and shoulder-width changes. Workload for a feature is defined as:

$WL_n = Rf \times S \times E \times U + C \times WL_{n-1}$.

WL_n = workload value for feature n.

Rf = workload potential rating for feature n.

S = sight-distance factor.

E = feature expectation factor.

U = driver unfamiliarity factor.

C = feature carryover factor.

Wl_{n-1} = workload value for the preceding feature.

More recent work examined the difference between the workload rating of a feature and a moving average workload, which was termed yaw (Woolridge, 1994). Greatly increased accident rates were associated with high yaw values.

Human factors experts have used a similar approach based on objective measures of road geometry. Hulse, Dingus, Fischer, and Wierwille (1989) defined driver workload as:

$Q = .4A + .3B + .2C + .1D.$

Q = driving workload.

$A = 20 \log2 (500/Sd)$, where Sd = sight distance.

$B = 100Rmax/R$, where R = radius of curvature.

$C = -40So + 100$ where So = distance of closest obstruction to road.

$D = -36.5W + 267$, where W = road width for two lanes.

Correlations between Q and ratings of drivers' abilities to take their eyes off the road in a field study supported the workload construct, given that ability to scan the road is an appropriate workload index. A further test of Q was conducted in a driving simulator by Green, Lin, and Bagian (1994). Only the correlation between Q and mean lateral position (−0.43) was statistically significant; correlations with the standard deviation of position, mean speed, mean yaw, and their standard deviations were low. A nonsignificant correlation of 0.32 was obtained between Q and subjective ratings of workload on a 10-point scale. Subjective workload ratings had significant correlations with mean lateral position (−0.37) and standard deviation of lateral position (−0.42). Although the generally low correlations obtained in this study may in part be attributed to limitations of the driving simulator, they also raise questions about procedures used to measure driver workload. The weak link between theoretical models of workload and operational definitions (see Kantowitz, 1990, on measurement, and Kantowitz & Fujita, 1990, on the distinction between parameter estimation and curve fitting) is a major obstacle in these attempts to study driver workload.

A more theoretical approach (Senders, 1970; Senders, Kristofferson, Levison, Dietrich, & Ward, 1967) was derived from information theory and used visual occlusion as an index of driver attention. As the information-processing demands of driving increased, drivers would be less able to tolerate being unable to see the road. Van Der Horst and Godthelp (1989) found that median voluntary occlusion times increased with increasing lane width and decreased with increasing vehicle speed. In a vehicle field study, Krammes et al. (1995) found that workload estimates based on occlusion worked well for degree of curvature, but they did not obtain statistically significant results for curve deflection angle or its interaction with curvature. (A highway curve can be defined geometrically by radius of curvature and either length of curve or deflection angle, which is the angle formed by joining the opposite ends of a curve to the center of the curve.) Indeed, they obtained the result that deflection angles of 90 degrees had lower workload values than angles of 45 degrees; the authors noted that this result is counterintuitive.

The interpretation of visual occlusion data depends on the model of attention and workload being used. For example, a multiple-resource model that postulated a separate source of capacity for visual information might reach a different explanation from a single-resource model with one pool of total capacity (Kantowitz, 1987). Visual occlusion may be more effective as a manipulation (e.g., an independent variable) of driver workload than as a measure (e.g., a dependent variable) of driver workload. As a manipulation, occlusion requires the same single- and dual-task control conditions as discussed earlier in this chapter if driver workload is being measured. We have been unable to find a study that manipulated both road geometry and occlusion time while including a secondary task (presumably auditory or tactile). Without such studies, findings of experiments with visual occlusion may be consistent with models of driver workload, but they lack the converging operations necessary to measure driver workload construed as an intervening variable. Thus, the deflection angle result summarized earlier may reflect methodological limitations in measuring driver workload rather than an odd limitation in driver performance. It would be interesting to repeat the Krammes et al. experiment using converging operations, such as secondary-task measures of workload.

One study that did include converging operations was performed on rural roads in the Netherlands (DeWaard, Jessurun, Steyvers, Raggatt, & Brookhuis, 1995). Three theoretical models were invoked, including a mental load model based on mental effort. This effort was measured by recording heart rate. The model predicted that narrower smooth road tracks would increase effort, and this prediction was supported by lower power in the 0 to 10 hertz component of heart rate variability and lower heart rate variability coefficients on experimental sections of road relative to control sections. This study is an excellent demonstration of how theory can be used to achieve practical goals, in this case reducing speeding. The models predicted that changes in the experimental roadways would cause drivers to slow down, in part because their workload had increased. Results were consistent with predictions, revealing that road geometry controls driver performance, at least partly because of its effects on driver mental workload.

In summary, it seems likely that effects of road geometry on driver behavior are mediated by changes in driver workload. The most promising techniques for measuring these changes start with models of workload rather than with regression equations or simple formulas based on geometric roadway features.

SUMMARY

Workload is defined as an intervening variable that modulates the tuning between the demands of the environment and the capabilities of the organism (Kantowitz, 1988). Workload, being a theoretical construct, cannot be directly observed but must be inferred from changes in performance. Therefore, any methodology used to measure workload must make certain

theoretical assumptions. For example, when using secondary-task methodology, a key assumption is that the primary (driving) task is uninfluenced by
addition of the secondary task. Testing this assumption requires a single-task
control condition where the primary task is performed alone. Similarly, a
secondary-task–only control condition is necessary in order to measure decrements in secondary-task performance. Furthermore, it is also desirable to
investigate more than one level of secondary task. Without such control conditions, one cannot evaluate the theoretical assumptions that underlie objective measurement of driver workload.

Workload evaluations in the driving environment have related workload
to both in-vehicle information systems and road geometry. Although these
studies show that secondary-task methodology can be useful in measuring
driver workload, we do not believe that researchers are obtaining the maximum benefit from their experiments. More care needs to be given to testing, analyzing, and reporting single-task conditions. Research would also
benefit from including at least two levels of difficulty or complexity for both
secondary and primary tasks. We believe that the most promising workload
techniques are based on theoretical models of workload, rather than regression equations or simple formulae based on characteristics of the driving
environment.

REFERENCES

Bortolussi, M. R., Kantowitz, B. H., & Hart, S. G. (1986). Measuring pilot workload in a mo-
    tion base trainer. *Applied Ergonomics, 17,* 278–283.
Brown, I. D. (1962). Measuring the spare mental capacity of car drivers by a subsidiary audi-
    tory task. *Ergonomics, 5,* 247–250.
Cronbach, L. J., & Furby, L. (1970). How should we measure change, or should we? *Psycholog-
    ical Bulletin, 74,* 68–70.
DeWaard, D., Jessurun, M., Steyvers, F. J. J. M., Raggatt, P. T. F., & Brookhuis, K. A. (1995). Ef-
    fect of road layout and road environment on driving performance, drivers physiology and
    road appreciation. *Ergonomics, 38,* 1395–1407.
Green, P., Lin, B., & Bagian, T. (1994). *Driver workload as a function of road geometry: A pilot ex-
    periment* (UMIRI-93-39/Great Lakes No. GLCTTR 22-91). Ann Arbor, MI: University of
    Michigan. Transportation Research Institute. Great Lakes Center for Truck and Transit
    Research.
Harms, L. (1986). Drivers' attentional responses to environmental variations: A dual-task real
    traffic study. In A. G. Gale, M. H. Freeman, C. M. Haslegrave, P. Smith, & S. P. Taylor
    (Eds.), *Vision in vehicles* (pp. 131–138). Amsterdam: North-Holland.
Hulse, M. C., Dingus, T. A., Fischer, T., & Wierwille, W. W. (1989). The influence of roadway
    parameters on driver perception of attentional demand. *Advances in Industrial Ergonomics
    and Safety 1,* 451–456.
Kantowitz, B. H. (1985). Stages and channels in human information processing: A limited
    analysis of theory and methodology. *Journal of Mathematical Psychology, 29,* 135–174.

Kantowitz, B. H. (1987). Mental workload. In P. A. Hancock (Ed.), *Human factors psychology* (pp. 81–121). Amsterdam: North-Holland.

Kantowitz, B. H. (1988). Defining and measuring pilot mental workload. In J. R. Comstock, Jr. (Ed.), *Mental-state estimation 1987* (pp. 179–188). Hampton, VA: National Aeronautics and Space Administration, Scientific and Technical Information Division.

Kantowitz, B. H. (1990). Can cognitive theory guide human factors measurement? In *Proceedings of the Human Factors Society 34th annual meeting* (pp. 1258–1262). Santa Monica, CA: Human Factors and Ergonomics Society.

Kantowitz, B. H. (1992). Heavy vehicle driver workload assessment: Lessons from aviation. In *Proceedings of the Human Factors Society 36th annual meeting* (pp. 1113–1117). Santa Monica, CA: Human Factors and Ergonomics Society.

Kantowitz, B. H. (1995). Simulator evaluation of heavy-vehicle driver workload. In *Proceedings of the Human Factors and Ergonomics Society 39th annual meeting* (Vol. 2, pp. 1107–1111). Santa Monica, CA: Human Factors and Ergonomics Society.

Kantowitz, B. H., & Campbell, J. L. (1996). Pilot workload and flightdeck automation. In R. Parasuraman & M. Mouloua (Eds.), *Human performance in automated systems* (pp. 117–136). Mahwah, NJ: Lawrence Erlbaum Associates.

Kantowitz, B. H., & Casper, P. A. (1988). Human workload in aviation. In E. Wiener & D. Nagel (Eds.), *Human factors in aviation* (pp. 157–187). New York: Academic Press.

Kantowitz, B. H., & Fujita, Y. (1990). Cognitive theory, identifiability and human reliability analysis (HRA). *Reliability Engineering, and System Safety, 29,* 317–328.

Kantowitz, B. H., & Knight, J. L. (1976). Testing tapping timesharing, II: Auditory secondary task. *Acta Psychologica, 40,* 343–362.

Krammes, R. A., Brackett, R. Q., Shafer, M. A., Ottesen, J. L., Anderson, I. B., Fink, K. L., Collins, K. M., Pendleton, O. J., & Messer, C. J. (1995). *Horizontal alignment design consistency for rural two-lane highways.* (Report No. FHWA-RD-94-034). Washington, DC: Federal Highway Administration.

Krammes, R. A., & Glascock, S. W. (1992). Geometric inconsistencies and accident experience on two-lane rural highways. *Transportation Research Record 1356,* 1–10.

Messer, C. J., Mounce, J. M., & Brackett, R. Q. (1979). *Highway geometric design consistency related to driver expectancy, Volume III—Methodology for evaluating geometric design consistency.* (Report No. FHWA/RD-81/037). Washington, DC: Federal Highway Administration.

Moss, S. A., & Triggs, T. J. (1997). Attention switching time: A comparison between young and experienced drivers. In I. Noy (Ed.), *Ergonomics and safety of intelligent driver interfaces* (pp. 381–392). Mahwah, NJ: Lawrence Erlbaum Associates.

Neuman, T. R. (1992). Roadway geometric design. In J. L. Pline (Ed.), *Traffic engineering handbook Institute of Transportation Engineers* (4th ed., pp. 154–203). Englewood Cliffs, NJ: Prentice-Hall.

Noy, Y. I. (1989). Intelligent route guidance: Will the new horse be as good as the old? In *Proceedings of the IEEE Vehicle Navigation and Information Systems Conference (VNIS 89;* pp. 49–55). Piscataway, NJ: IEEE.

Nygren, T. E. (1991). Psychometric properties of subjective workload measurement techniques: Implications for their use in the assessment of perceived mental workload. *Human Facrtors, 33,* 17–34.

Schlegel, R. E. (1993). Driver mental workload. In B. Peacock & W. Karwowski (Eds.), *Automotive ergonomics* (pp. 359–382). London: Taylor & Francis.

Senders, J. W. (1970). The estimation of operator workload in complex systems. In *Systems Psychology* (pp. 207–216). New York: McGraw-Hill.

Senders, J. W., Kristofferson, A. B., Levison, W. H., Dietrich, C. W., & Ward, J. L. (1967). *Attentional demand of automobile driving.* (Report 1482). Washington, DC: U.S. Department of Commerce, Bureau of Public Roads.

Van Der Horst, R., & Godthelp, H. (1989). Measuring road user behavior with an instrumented car and an outside-the-vehicle video observation technique. *Transportation Research Record 1213*, 72–81.

Verwey, W. B. (1991). *Towards guidelines for in-car information management: Driver workload in specific driving situations*. (Report No. IZF 1991 C-13). Soesterberg, Netherlands: TNO Institute for Perception.

Verwey, W. B., & Veltman, H. A. (1996). Detecting short periods of elevated workload: A comparison of nine workload assessment techniques. *Journal of Experimental Psychology: Applied, 2*, 270–285.

Woolridge, M. D. (1994). Design consistency and driver error. *Transportation Research Record 1445*, 148–155.

2.11 Evaluating Safety Effects of In-Vehicle Information Systems

Willem B. Verwey

TNO Human Factors Research Institute
Institut für Arbeitsphysiologie an der Universität Dortmund

In recent years, there has been a considerable increase in the research and development of modern technology in road transport. From the beginning, many people expressed their concern that this technology, known as advanced transport telematics (ATT) or intelligent transport systems (ITS), would jeopardize traffic safety (e.g., Hancock & Parasuraman, 1992; Parkes & Ross, 1991). Reasons for this concern include that drivers might be distracted at critical moments and that, in less critical situations, they devote too much attention to in-vehicle displays.

The present chapter describes a field study that addresses the development of a method for assessing safety effects of in-vehicle information systems as caused by distraction and information overload. That is, what should a safety assessment method look like, and what measures should be used? I believe that safety effects of in-vehicle information systems can be assessed only when normal drivers are observed in situations that are common to them. This implies that safety assessment should take place in a car in normal traffic. Safety assessment in driving simulators is not possible. Of course, driving in an experimental study is never completely normal; a device has to be handled which is relatively unfamiliar. Proper experimentation requires a certain degree of control by the experimenter, and various requirements must be fulfilled before safety related measures can be registered. Still, I consider the use of vehicles that are instrumented for this purpose a reasonable compromise (Verwey, Burry, & Bakker, 1995).

Given that safety assessment of in-vehicle information systems involves an instrumented vehicle, what are the indicators for safety? By now, there is a tradition of two methods for safety assessment. The first includes registration of various objective performance measures. If, relative to a control condition, any of the performance measures is found to change when the driver uses some in-vehicle system, the typical conclusion is that safety might be jeopardized. This method suffers from the problem that most driving-performance measures, such as steering frequency and speed variations, have little direct relation with safety because they often do not take the driving environment into account; a bit more swerving on an empty road need not be dangerous. The other method for assessing safety is having "expert drivers" judge the safety of driving (e.g., Carsten, 1995). These experts might be driving instructors or highly experienced professional drivers. This method is likely to yield clearer results, but it remains unclear to what extent these results are affected by these experts' opinions about the system and by errors in their safety judgments. Given the problems with both methods, I decided to combine objective measurements and subjective safety judgments in one experiment to compare their utility. If certain objective and subjective indicators covaried, this would support their value for safety assessment.

Another purpose of this chapter is contributing to the development of ergonomic guidelines for in-vehicle information systems. There seems to be some consensus that in-vehicle information should be presented in the auditory modality so that the visual channel is not overloaded. Preventing visual overload is certainly important, but several studies have demonstrated that listening to speech may be detrimental for driving performance too (e.g., Alm & Nilsson, 1994; Briem & Hedman, 1995; Brookhuis, de Vries, & De Waard, 1991). In some situations, visual information might be preferable because the information, for instance, geographical information, is better suited for visual presentation. In addition, visual information can be attended at the moment the driver has time to do so. Speech messages can be easily missed and for that reason may tempt drivers to pay less attention to driving in demanding driving situations. This raises the question of whether, for some types of messages, visual information is preferred over auditory information.

A second question with respect to ergonomic guidelines is whether drivers should be allowed to carry out relatively complex control actions while driving. Even though the moment that these actions are being performed is usually determined by the driver, this does not necessarily imply that drivers postpone control of the system to situations that are more appropriate. Especially with relatively complex control procedures, the driver may be tempted to first finish the procedure when driving suddenly requires more attention. If it could be demonstrated that controlling an in-vehicle information system reduces driving safety in more demanding driving situations, even when there is little time pressure for the driver to perform the task, then this would indicate that drivers do not allocate their attention in safe ways.

These questions were addressed in a field study in which drivers interacted with a simulated in-vehicle information system. The system was a so-called radio data system–traffic message channel (RDS-TMC) system that is expected to come on the market in a few years. It provides up-to-date traffic congestion information with respect to a selected set of roads, either as soon as new information becomes available at the traffic control center or when the driver consults the system. The system is particularly interesting for the present study because it presents fairly complex visual or speech information and needs relatively complex control actions. For example, most speech versions of RDS-TMC systems can filter out congestion information from irrelevent areas, and this filter must be programmed via a fairly demanding procedure.

Even though safety evaluations with RDS-TMC prototypes are scarce, preliminary results compiled by Katteler (1995) and based on driver self-observations in large-scale demonstration projects suggest that operating RDS-TMC applications is more attention demanding than listening to RDS-TMC messages. Listening to RDS-TMC messages was not considered more distracting than listening to traffic messages broadcasted in a conventional way but, as noted before, this does not necessarily mean that there is no negative safety effect associated with it.

In the present experiment, 12 experienced drivers drove a fixed route through the city of Amersfoort in the Netherlands. Congestion information was presented by the RDS-TMC system via a map display or a speech message at right turns and intersections and while driving at straight inner-city roads. Earlier research showed that driver workload was especially high at turns and intersections (Verwey, 2000). It was the participants' task to determine whether useful information was presented to them. In addition, participants programmed the system while driving in the mentioned situations. Driving performance, looking behavior, and safety ratings given by a driving instructor were obtained while the participants were carrying out the three RDS-TMC tasks and, in part, compared with single-task control conditions.

METHOD

Participants

Twelve drivers who had their driver's license for at least 5 years participated. Their driving experience varied between 5 and 30 years (average 19 years), and between 5,000 and 35,000 kilometers per year, with the exception of one participant who drove only 1,000 kilometers per year (average 14,000 km/yr). The participants were between 25 and 52 years old with an average of 42. There were 5 women and 7 men. They were selected from the institute's participant pool. All participants were unfamiliar with the city where the experiment took place and had driven the experimental route only once before in an earlier experiment.

Experimental Route.   The experimental route was in the Dutch city of Amersfoort. In quiet traffic, the route took about 20 to 25 minutes to drive. Four different types of driving situations were distinguished: Turning right from a main road into a minor road while taking bicycles and mopeds between the vehicle and the curb into account; approaching a general rule intersection, which, in the Netherlands, implies that motorized vehicles coming from the right have priority; approaching a yield intersection from the minor road, implying that all traffic at the main road has right of way, motorized vehicles as well as bicycles; driving straight ahead at a major inner city artery. The experimental route contained three instances of each of these four situations, making up a total of 12 experimental situations. The order in which the experimental situations were encountered was as follows: general rule intersection, yield intersection, straight ahead, right turn, yield intersection, straight ahead, right turn, right turn, general rule intersection, general rule intersection, and yield intersection.

RDS-TMC Tasks.   Participants carried out three types of RDS-TMC tasks. In two of these tasks, participants were asked to indicate whether there was congestion on a predefined route. In half the instances, the correct response was "yes"; in the remaining ones, it was "no." The third task was a filter programming task.

In the first congestion presentation task, the *map* condition, a computer-generated voice told participants their would-be current location and their would-be destination. These locations consisted of cities in a range of about 60 kilometers from Amersfoort. Following this context information, a map was displayed on the color display that was built into the dashboard of the instrumented vehicle (Fig. 2.11.1). This map showed major roads and a number of cities. The map was artificial in that the city names were not at their usual positions, but were associated with alternative positions. This mapping was changed at each trial to mimic a situation in which drivers do not know the environment and must search the map for the relevant cities. Congestion was indicated by red road segments at the otherwise green road network. For example, the instruction was "You are driving from Utrecht to Apeldoorn. Is there any congestion?" (i.e., at the connecting roads). This instruction was supported by presenting both names on the left corner of the display to reduce the workload associated with keeping the current location and the destination in mind. After a 2 to 7 second pause, the map display was presented. Participants were instructed to verbally state whether or not there was congestion on the roads connecting the current location and des-

FIG. 2.11.1. Example of a display in the map condition.

tination. Responses were registered by the experimenter and were later analyzed for correctness.

In the second task, the *speech* condition, the context information was followed by a digitized voice reading aloud five messages in a format similar to that of most RDS-TMC applications. Table 2.11.1 presents an example. As is the case in reality, this task assumes that the driver knows the location of the cities in the area. Because only larger cities were used and all participants were from the region, the participants in this study were expected to be able to perform this task.

TABLE 2.11.1
Example of a Trial in the Speech Condition

Context information
   "You are driving from Utrecht to Apeldoorn. Is there congested traffic?"

Display
   Utrecht → Apeldoorn

RDS-TMC speech messages
   "At the A1, Amersfoort Apeldoorn, slow and congested traffic."
   "The A2, Vianen Utrecht, is closed due to traffic works."
   "At the A27, Utrecht Almere, congestion near Blaricum."
   "At the A28, Utrecht in the direction of Amersfoort, congestion due to an accident."
   "At the A50, Apeldoorn Arnhem, congestion in both directions due to traffic works."

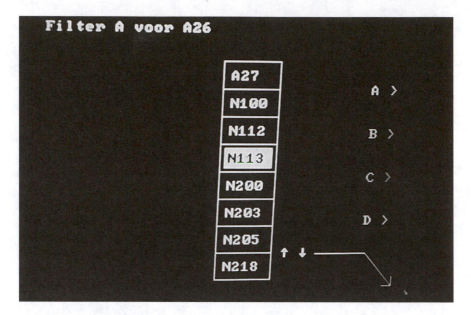

FIG. 2.11.2.   Example of a display in the filter-programming condition.

The third RDS-TMC task involved programming a filter on the system. This *filter programming* task started with an instruction given by a digitized voice, for example, "Adjust Filter C for A26." Then, participants searched with two (arrow) push buttons through a list of 34 alphabetically ordered road numbers presented on a display at the center part of the dashboard. Eight road numbers were visible at the same time; the cursor position was always in the center of the display and was highlighted. An example display is presented in Fig. 2.11.2. As soon as the proper road number was highlighted, one of four filter buttons at the right-hand side of the display (representing Filters A, B, C, and D) was pressed.

Each participant performed each of these three tasks twice in each of the four driving situations. Across all participants, the same type of task was performed equally often at each specific situation.

Apparatus and Data Registration

The experiment was carried out in one of the institute's instrumented cars (Verwey et al., 1995). This car, a Dodge Ram van with dual controls, contains an IBM 486 personal computer and has various possibilities for measuring driving behavior and generating stimuli.

Auditory stimuli were presented by a digitized, prerecorded male voice on a speaker behind the participant and were clearly audible under all driving conditions. Presentation of the stimuli was controlled by the onboard computer. Visual stimuli were presented on a backlit liquid crystal display (LCD) mounted on the center panel of the instrumented car next to the speedometer. The visual angle between the normal fixation point on the road ahead and the screen was about 27° horizontally and about 27° vertically. The eye–screen distance was about 90 centimeters for an average participant of about 1.80 meters in height. Text was clearly readable for all participants.

During the experiment, the participants' answers were registered by the experimenter. Objective measures such as speed, stimuli presented, lane position, and actions of the participants in the programming task were registered at a sample rate of 10 hertz.

Procedure

Participants were first introduced to the aim of the study and then filled out a written consent form. Next, they were familiarized with the instrumented car and the RDS-TMC tasks they were to perform. It was explicitly stated that they were to drive safely and that they would earn a bonus of 50 cents for each correct answer on the RDS-TMC tasks as long as they would not drive in an unsafe manner. After the instruction, they drove with the vehicle for about half an hour without performing an additional task. Next, while remaining still, they carried out each of the three RDS-TMC tasks several times until it was clear that they mastered these tasks sufficiently well.

Subsequently, participants drove to the starting point of the experimental route. They were guided along the experimental route by the experimenter to prevent them from being burdened by navigation. All participants drove this route four times, two times in the control condition and two times in the experimental condition. Half the participants started with two control drives at the experimental route followed by two experimental drives. This was reversed for the other half of the participants. After completion of the first experimental drive, baseline performance on the various tasks was obtained in a standing-still situation.

Following each of the 12 experimental driving situations, the experimenter, who was an experienced driving instructor, rated the safety of driving in that situation on a number of three-point rating scales. Half the participants drove in the morning (10 AM to 12:30 PM), whereas the other half drove in the afternoon (2:30 PM to 5 PM).

Alongside the experimental route, just in front of each experimental situation, inconspicuous markers were attached to light poles. These markers were used by the technician in the back of the car to trigger the onboard system that determined which task instruction was given to the participant. This ensured proper and consistent timing of the RDS-TMC tasks. Each task was performed while driving in the situations specified before.

Data Analyses and Design

Subjective safety ratings were given by the experimenter who was also a driving instructor. These items are presented in Table 2.11.2 and were derived from a standardized list and validated in earlier work (De Gier, 1980). A distinction was made between items referring mainly to vehicle control and items indicating interactions with other traffic participants. Each item was scored as being *satisfactory, acceptable, or unsafe.* An unsafe rating was given when a critical part of the task was not carried out at all (e.g., not looking at priority traffic) or in a poor, unsafe manner (e.g., braking exceptionally hard or swerving).

Table 2.11.3 presents an overview of all dependent variables. Video recordings were made of the participants' heads and were later analyzed by one person. Because intersections had been selected where priority traffic could be seen only when the vehicle was close to the intersection, looking to crossing traffic was marked by a clear head movement. In right turn and yield situations, this allowed derivation of the hypothetical deceleration of the instrumented vehicle in case the driver suddenly detected a vehicle with priority at the moment of looking right and had to stop before entering the crossroad. (This usually did not occur.) In some straight driving situations, TLC and line crossing were assessed with a special-purpose lane-tracking device (Van der Horst & Godthelp, 1989).

The data were analyzed with a Task (task versus control) × Instance (3) × Replication (2) design for each type of situation. In addition, the various dependent variables were analyzed across all situations with a Task (2) × Situ-

TABLE 2.11.2

Items Used by the Experimenter/Driving Instructor to Rate
at a 3-Point Scale Driving Safety at the Control and Maneuver Levels
of Driving for Each of the Situation Types

Type of Situation	Items Related to Vehicle Control	Items Related to Maneuver Tasks
Straight driving	1. Braking and decelerating 2. Course keeping	3. Adapting speed to other traffic 4. Headway 5. Anticipation in general
Right turn	1. Braking and decelerating 2. Course keeping	3. Looking over the shoulder 4. Hampering ongoing traffic
General rule intersection	1. Braking and decelerating 2. Course keeping	3. Judging the situation in advance 4. Looking at priority traffic 5. Giving priority
Yield intersection	1. Braking and decelerating 2. Course keeping	3. Judging the situation in advance 4. Looking at priority traffic 5. Giving priority

TABLE 2.11.3
Dependent Variables per Driving Situation (cf. Verwey, 1996)

Situation	Measure
Straight driving	Subjective ratings
	Speed
	Standard deviation of speed
	Frequency of critical time-to-line crossings (TLC)
	Line crossing-frequency
	Steering interval
	Proportion high decelerations (> 3 and > 2 m/s^2)
	Glance frequencies and durations
	In-vehicle task accuracy
Right turn	Subjective ratings
	Frequency of looking over the shoulder
	Hypothetical emergency deceleration
	Proportion high decelerations (> 3 and > 2 m/s^2)
	In-vehicle task accuracy
General rule intersection	Subjective ratings
	Looking at priority traffic
	Frequency of critical time-to-intersection (TTI)
	Proportion high decelerations (> 3 and > 2 m/s^2)
	In-vehicle task accuracy
Yield intersection	Subjective ratings
	Frequency of critical time-to-intersection (TTI)
	Hypothetical emergency deceleration
	Proportion high decelerations (> 3 and > 2 m/s^2)
	In-vehicle task accuracy
Remaining still	Response times
	In-vehicle task accuracy

Note.   Subjective ratings are presented in more detail in Table 2.11.2.

ation (4) × Instance (3) × Replication (2) design. For frequency data, a logarithmic model was used in which Rating (satisfactory/acceptable versus unsafe) was an additional variable.

RESULTS

Objective and Subjective Measures of Driving

General.   Table 2.11.4 shows the number of driving situations at which driving behavior was judged unsafe by the driving expert with respect to at least one of the various ratings. Driving was considered significantly more unsafe when performing any of the additional tasks than in the control condition [map versus control $\chi^2(1) = 7.0$, $p < .01$; speech versus control $\chi^2(1) =$

TABLE 2.11.4
Frequency (and Percentage) of Unsafe Judgments and
Satisfactory/Acceptable Judgments Across 12 Driving Situations

	Unsafe	Satisfactory/Acceptable	Total
Map	10 (10.4%)	86 (89.6%)	96 (100%)
Speech	14 (14.6%)	82 (85.4%)	96 (100%)
Filter	28 (29.2%)	68 (70.8%)	96 (100%)
Control	10 (3.5%)	278 (96.5%)	288 (100%)

15.2, $p < .001$; filter programming versus control $\chi^2(1) = 53.3$, $p < .001$]. Behavior was judged significantly more unsafe in the filter condition than in the map and the speech condition [$\chi^2(1) = 10.6$, $p < .001$; $\chi^2(1) = 6.0$, $p < .05$, respectively]. No difference was found in the number of unsafe ratings in the map and speech conditions [$\chi^2(1) = 0.8$, $p > .20$].

Across all situations, there were two types of driving behavior that were repeatedly considered unsafe. First, course keeping was generally rated more often unsafe in the task conditions than in the control condition [$\chi^2(1) = 17.6$, $p < .001$]. Planned comparisons showed that this was caused by course keeping in the filter-programming condition [$\chi^2(1) = 32.2$, $p < .001$] and not by course keeping in the map and speech conditions [$\chi^2(1) = 0.4$, $\chi^2(1) = 1.8$, $ps > .10$, respectively]. In the map condition, course keeping was judged to be unsafe twice (out of 96 ratings = 2.1%), in the speech condition this happened 6 times (6.3%), and in the filter programming condition course keeping was considered unsafe in 22 cases (22.9%). In the control condition, it occurred only once (out of 288 ratings = 0.3%).

The second type of driving behavior that was repeatedly judged unsafe when using the RDS-TMC system was braking/deceleration. The major cause for these judgments was that participants did not sufficiently anticipate forthcoming braking actions rather than that they actually braked too hard. Braking/deceleration was considered more often unsafe in the task conditions than in the control condition [$\chi^2(1) = 10.2$, $p < .01$]. Braking behavior was considered unsafe 7 times (out of 96 = 7.3%) in the map condition, 4 times (= 4.2%) in the speech condition, and 18 times (= 18.8%) in the filter condition. In the control condition, poor braking occurred 6 out of the 288 times (= 2.1%). Planned comparisons of the individual in-vehicle tasks showed that unsafe ratings were given more often in the filter than in the control condition [$\chi^2(1) = 11.2$, $p < .001$]. The map and speech tasks did not yield more unsafe braking/deceleration ratings than the control condition [$\chi^2(1) = 1.0$, $\chi^2(1) = 0.8$, $ps > .20$, respectively].

Right Turns. At right turns, drivers were supposed to look across their right shoulders to detect any bicycles and mopeds between the curb and the

TABLE 11.2.5
Subjective Ratings on Looking Right
over the Shoulder When Turning Right

Rating	In-Vehicle Tasks	Control
Satisfactory	45	55
Acceptable	18	17
Unsafe	9	0

vehicle. Expert ratings indicate that participants frequently did not show this behavior in the task conditions $[\chi^2(1) = 6.9, p < .01]$ (Table 2.11.5).

This reduction of looking in the proper direction was associated with a marginally significant increase in the number of times that mopeds and bicycles were hampered $[\chi^2(1) = 3.7, p < .06]$. Ongoing traffic was hampered 6 times (out of 72) in the task conditions (map: 1 = 1.4%, speech: 2 = 2.8%, filter programming: 3 times = 5.2%), and not at all in the control condition. Course keeping significantly deteriorated while turning right in the task condition $[\chi^2(1) = 5.4, p < .05]$. Finally, safety of the braking actions was judged to suffer from performing an in-vehicle information system task $[\chi^2(1) = 15.2, p < .001]$.

According to the video analysis, participants did not look over their right shoulders on 46 out of 69 occasions in the task conditions (= 67%), whereas in the control condition this was the case for 38 out of 67 cases [= 57%; $\chi^2(1)$ = 1.4, $p > .20$]. The difference in these figures compared with the subjective ratings can be attributed to the fact that participants often used their mirrors to check for any mopeds or bicycles. This could not always be detected from the video. The percentages of hypothetical decelerations over 2 and 3 m/s^2 in case of another car's taking right of way did not differ significantly in task and control conditions.

General Rule Intersections.   When approaching the general rule intersection, course keeping was judged as being poorer in the task conditions: Unsafe ratings were given 12 times (= 17%) in the task conditions, and 0 times in the control condition $[\chi^2(1) = 17.2, p < .001]$. Closer inspection of the data showed that out of the 12 occurrences of unsafe course keeping, 10 were in the filter-programming condition $[\chi^2(1) = 4.7, p < .10]$. Task condition had a marginally significant effect on safety judgments of braking and deceleration behavior at general rule intersections $[\chi^2(1) = 3.4, p < .07]$: Whereas 7 (9.7%) unsafe situations occurred in the task condition, there were 2 (2.8%) such situations in the control condition.

Video analyses showed that participants tended to look less in the direction of priority traffic in the task conditions (46 out of 72 = 64%) than in the control condition [56 out of $72 = 78\%$; $\chi^2(1) = 3.4$, $p < .07$]. The percentages of hypothetical decelerations over 2 and 3 m/s² did not differ significantly in task and control conditions.

Yield Intersections.   When approaching yield intersections from the secondary road, the expert ratings indicated that braking/deceleration was affected by in the in-vehicle tasks [$\chi^2(1) = 11.3$, $p < .001$] (Table 2.11.6). No other differences between conditions were found.

Straight Driving.   Expert judgments showed that participants swerved more in the task conditions than in the control condition [$\chi^2(1) = 5.5$, $p < .05$] (Table 2.11.7).

With respect to objective measures, driving performance was analyzed for the period that the driver was engaged in the in-vehicle tasks (i.e., between presentation of the message and the response). For the control condition, driving performance was analyzed for the same road segments as in the task conditions. Analysis of objective measures of steering indicated that the average interval between subsequent steering movements was lower in the task conditions, as compared with the control condition [from 0.91 s to 0.56 s; $F(2,22) = 29.5$, $p < .001$]. Subsequent Tukey testing showed that this effect was caused by the filter condition alone [map and speech: $ps > .20$, filter programming: $p < .001$]. Speed, standard deviation of speed, and frequency that time-to-line crossing was below the critical value of 1.1 seconds (see Verwey, 1996) did not differ in task and control conditions.

TABLE 2.11.6
Subjective Ratings on Braking and Decelerating
While Approaching Yield Intersections

Rating	In-Vehicle Tasks	Control
Satisfactory	21	48
Acceptable	41	22
Unsafe	10	2

TABLE 2.11.7
Subjective Ratings on Course Keeping
While Driving Straight

Rating	In-Vehicle Tasks	Control
Satisfactory	51	69
Acceptable	23	1
Unsafe	8	1

TABLE 2.11.8
Various Measures of Display-Looking Behavior While Driving Straight

Condition	Glance Frequency (Range)	Average Time per Glance (s) (Range)	Total Glance Time (s)	Percentage Glances > 1 s	Percentage Glances > 2 s
Map	5.2 (3–12)	1.5 (1.0–2.7)	7.0	51	18
Filter	6.8 (4–10)	1.5 (1.0–2.2)	10.2	65	24

The data suggest that safety was negatively affected in the task conditions, especially in the filter condition. Therefore, it is important to know for how long and how often participants looked at the RDS-TMC display. Table 2.11.8 shows various measures of looking behavior that seem to be related to the situation that behavior starts being unsafe.

Overview of Driving Safety Results.   The results show that, according to safety judgments by the driving expert, safety was affected by the in-vehicle tasks and clearly more so for filter programming than for the map and speech tasks. In the filter-programming task, these effects were largely caused by poor course keeping and (poor anticipation for) braking and decelerating. Significant effects in the objective measures were found only as an increase in steering frequency when carrying out in-vehicle tasks at the straight inner-city road and a tendency to look less to the right at the general rule intersection in the filter condition. With respect to the development of a method for safety evaluation, it is interesting to note that according to the expert, interactions with other cars (adapting speed to other traffic, distance to heading traffic, anticipation in general, giving priority, watching priority traffic) were hardly affected by the use of the RDS-TMC system. Bicycles and mopeds may have sometimes been endangered in the right-turn situations.

In-Vehicle Task Performance

To assure that the in-vehicle tasks had been properly carried out, performance on these tasks was analyzed, too. Table 2.11.9 presents the occurrence of errors on the three RDS-TMC tasks. These data were analyzed with a driving versus remaing still × map versus speech × satisfactory/acceptable versus unsafe logarithmic model. This analysis showed that significantly more errors had been made with speech messages than with maps [$\chi^2(1) = 15.4$, $p < .001$], but that the map versus speech × driving versus remaining still interaction did not reach significance [$\chi^2(1) = 0.9$, $p > .20$]. This suggests that the difference between the map and speech presentation modes was related to task difficulty and not to interference with driving.

TABLE 2.11.9
Distribution of Errors in Map, Speech, and Filter-Programming Tasks
in 4 Different Driving Situations (of 24 per cell)
and While Remaining Still (of 48)

	Map	Speech	Filter	Total
Right turns	3 (13%)	7 (29%)	6 (25%)	16 (22%)
General rule intersections	0 (0%)	9 (38%)	6 (25%)	15 (21%)
Yield intersections	0 (0%)	4 (17%)	3 (13%)	7 (10%)
Straight driving	1 (4%)	1 (4%)	2 (8%)	4 (6%)
Remaining still	3 (6%)	8 (17%)	1 (2%)	12 (8%)
Total	7 (5%)	29 (20%)	18 (13%)	54 (13%)

CONCLUSION

The present study sheds some light on two questions: How can safety be measured, and how should the human–machine interface of in-vehicle information systems be designed? As to the first issue, it turns out that subjective measures were far more sensitive to what happened when the driver was using the in-vehicle information system than were the objective measures of driving. Subjective ratings that showed the clearest effects, at least in the most demanding task condition (i.e., the filter condition), were the ones on course keeping and (anticipating) braking and decelerating. Ratings on the other scales also showed effects of the in-vehicle tasks, but not as consistently across the various situations. Of the objective measures, only steering frequency at straight inner-city roads increased with filter programming, and participants tended to look less to the right at general rule intersections. The other measures listed in Table 2.11.3 appeared not to be sensitive measures for reductions in driving safety. The fact that increased steering frequency coincided with reduced safety ratings on the straight road segments suggests that steering frequency is an indicator for safety reductions in such situations.

These subjective and objective indications for unsafe behavior are remarkable because participants had been instructed to postpone in-vehicle tasks if they thought safety would be affected. This suggests that participants could not sufficiently predict the safety effects of their interactions with the in-vehicle system. From a safety point of view, it might have been better if they had waited until the vehicle was in a less demanding driving situation.

The expectation that certain subjective and objective indicators for safety effects would covary was not fulfilled. The finding that safety was clearly affected according to subjective measures, and hardly according to the objective measures, might either mean that subjective measures were influenced by misjudgments and biases of the driving expert or that the objective measures were not sensitive enough. In the absence of a safety criterion, this issue is hard to settle. However, informal reports by the driving expert and the technician underline that dangerous situations did occur, even when the objective measures did not demonstrate this. A few times, the driving instructor was even required to take control. I think that, despite all problems associated with subjective ratings, this method does yield useful indications for reductions in safety.

The fact that hardly any objective indications for safety effects were found is probably caused by the fact that in field studies, objective measures are subject to variations caused by momentary traffic conditions and individual differences. For example, it was noted that drivers sometimes did not look over their right shoulders at right turns. However, it turned out that driving was perfectly safe because drivers had used their mirrors long in advance and started driving near the curb to keep bicycles and mopeds from riding between the vehicle and the curb. Objective measures do not clearly reflect such alternative behaviors that can not always be anticipated by the researcher. Subjective judgments are probably much less subject to these variations in measurements because the human observer acts as a noise filter: Unimportant events that affect objective measures do not affect subjective measures. Despite all problems associated with subjective safety ratings, I conclude that expert judgments on standardized rating lists should be preferred for safety evaluations of in-vehicle information systems.

If we accept that the present subjective safety results are reliable reflections of "true" safety effects, then it is also possible to propose some guidelines with the respect to in-vehicle information systems. The data make sufficiently clear that programming an RDS-TMC filter affects driving safety to a considerable extent. Handling an in-vehicle information system while driving appears to be a dangerous task, even when there is no (explicit) time pressure to finish the task. It seems that participants underestimated the safety effects of filter programming. The reason that filter programming was so detrimental to driving safety may lie in the visual as well as in the manual workload of filter programming (Verwey, 2000; Wickens, 1984). The findings with the filter task suggest that handling a system that requires a sequence of actions and of readings from a display is not acceptable from a safety point of view. The glance data indicate that this holds for tasks that take total viewing times as long as 10 seconds.

Despite these serious safety effects of filter programming, it should not be forgotten that driving behavior was also judged to be unsafe three to four times as often with a visual map display and with speech messages as when driving without an in-vehicle task. Notably, the safety effects in the speech condition did not significantly differ from those in the map condition. It has not been demonstrated before that the safety effects of mentally loading tasks may actually be comparable to those of visually loading tasks. An explanation for that fact that speech had such an unexpectedly large effect on safety is that drivers could not pay attention to the additional task when they wanted to. With visual messages, this is usually much less of a problem. Hence, in the design of in-vehicle information systems, the choice between visual and speech messages is one that still needs careful consideration. The high error rates in the speech condition while remaining still demonstrate that for the present type of information, speech information was simply less suited than visual information. We can learn from the present study that visual information may sometimes be preferable over speech messages. However, the data also show that the visual messages we used and that required as much as 7 seconds total looking time affected traffic safety. This safety effect is not acceptable, and total looking times should in any case remain below 7 seconds.

Taken together, the present results suggest that evaluating negative safety effects of in-vehicle information systems should be based on the use of a standardized safety rating list that should focus on control aspects like swerving and braking and decelerating. Steering frequency might be a useful objective safety measure while driving straight. With respect to the design of in-vehicle information systems, the data show that for some types of information, visual information is better suited from a safety point of view than are speech messages. Finally, driving safety is jeopardized when the interaction between the driver and the system involves 10 seconds or more total looking time and when inspection of a visual display alone lasts 7 seconds or more total looking time.

ACKNOWLEDGMENT

This research was funded in part by the Dutch Ministry of Transport, Public Works, and Water Management.

REFERENCES

Alm, H., & Nilsson, L. (1994). Changes in driver behaviour as a function of handsfree mobile phones: A simulator study. *Accident Analysis & Prevention, 26,* 441–451.

Briem, V., & Hedman, L. R. (1995). Behavioral effects of mobile telephone use during simulated driving. *Ergonomics, 38,* 2536–2562.

Brookhuis, K. A., de Vries, G., & De Waard, D. (1991). The effects of mobile telephoning on driving performance. *Accident Analysis and Prevention, 23,* 309–316.

Carsten, O. (1995). A framework for safety evaluation. In J. P. Pauwelussen & H. B. Pacejka (Eds.), *Smart vehicles* (pp. 419–427). Lisse, Netherlands: Swets & Zeitlinger.

De Gier, J. J. (1980). *Evaluatie van geneesmiddelen in het verkeer* [Evaluation of drugs in real driving situations]. Doctoral thesis (in Dutch), University of Utrecht, Netherlands.

Hancock, P. A., & Parasuraman, R. (1992). Human factors and safety in the design of intelligent vehicle-highway systems (IVHS). *Journal of Safety Research, 23,* 181–198.

Katteler, H. (1995). *Acceptance and impacts of RDS/TMC traffic information: Results of the ATT cross-project collaborative study.* Cord (Drive V2056) deliverable AC16. Nijmegen, Netherlands: Instituut voor Toegepaste Sociale Wetenschappen.

Parkes, A. M., & Ross, T. (1991). The need for performance based standards in future vehicle man–machine interfaces. *Advanced ATT in road transport* (Vol. 2, pp. 1312–1321). Amsterdam: Elsevier.

Van der Horst, A. R. A., & Godthelp, H. (1989). Measuring road user behavior with an instrumented car and outside-the-vehicle video observation technique. *Transportation Research Record, 1213,* 72–81.

Verwey, W. B. (1996). *Evaluating safety effects of in-vehicle information systems: A detailed research proposal.* (Report TM-96-C045). Soesterberg, Netherlands: TNO Human Factors Research Institute.

Verwey, W. B. (2000). On-line driver workload estimation. Effects of road situation and age on secondary task measures. *Ergonomics, 43,* 187–209.

Verwey, W. B., Burry, S., & Bakker, P. J. (1995). Investigating driver support systems in real traffic with the Instrumented Car for Computer Aided Driving (ICACAD). *Analise Psicologica, 3,* 279–286.

Wickens, C. D. (1984). Processing resources in attention. In R. Parasuraman & D. R. Davies (Eds.), *Varieties of attention* (pp. 63–102). London: Academic Press.

COMMENTARY

The Human Capacity for Work: A (Biased) Historical Perspective

John M. Flach
Wright State University

Gilbert Kuperman
Air Force Research Laboratory, Wright–Patterson Air Force Base

It might be useful to set the discussion of stress, workload, and fatigue in an historical context. This is particularly important when old constructs (e.g., workload, human factors) are being challenged by new constructs (e.g., situation awareness, cognitive systems engineering [CSE]). Do these new constructs represent radical paradigm shifts, or do they reflect incremental progress in our understanding of the relations between humans and technology? This is not a neutral account of the history, but an account that is strongly biased by our personal perceptions of where the field has been and by our beliefs about where it ought to be going. It is hoped that our opinions about the trajectory of ideas stimulates debate and prompts others to articulate their own personal histories. We believe that discussions at this level help to ensure that the next step (whether radical shift or incremental progression) is at least not a step backward.

WORKING SMARTER

One important root for current approaches to humans and work was the scientific-management approach of Taylor, Gilbreth, and others (e.g., see Konz, 1987). This work was inspired by the technologies that enabled mass production (as well as by the economic opportunities offered). The key contribution of these men was the importance that they gave to measurement and standardization. First, they helped to establish "time" or "rate" (units/time) as a fundamental ruler for judging the "best way" or "the "right way." Armed with this ruler, they had a basis for controlled comparisons by which the best way to shovel coal or to lay bricks could be empirically determined. Furthermore, these controlled experiments provided a basis for shifting the responsibility for choosing the "work method" from the individual worker or artisan to the manager. It became management's responsibility to provide resources needed to determine the best way and to provide training and incentives to get the workers to comply. Also, assembly lines needed to be designed. Because of the number of people involved, the time and money were not available to let assembly lines naturally evolve to a stable configuration. To a large extent, the stability had to be designed into the system (the line needed to be "balanced") if it was to function properly. Thus, the time of component operations needed to be predictable. This led to increasing standardization of methods and tools. Thus, the industrial engineer's or efficiency expert's job was to determine the "best way" through empirical tests on component operations; to organize those components into a balanced configuration to produce safe, efficient, and stable processes; and to implement training and incentive programs so that workers would comply with the standards on which the design was developed.

In addition to the introduction of empirical methods, an enduring theme that can be traced to Taylor and Gilbreth is that increased effort (working harder) was not the only path to increased productivity. For example, Taylor showed that it was possible to increase productivity by mandating rest breaks. People began to recognize that it was possible to increase productivity by "working smarter."

Rochlin (1993, 1997) presented an interesting discussion of some of the negative side effects of the scientific-management approach and some reasons that it might be a mistake to extrapolate this approach to distributed information systems. These side effects arise from the conflict between efficiency and reliability. The checks and safeguards that protect systems against erroneous decisions and actions create friction that reduces the efficiency of other correct actions. A system that gives the manager the power to micromanage activities makes it easier to implement both good and bad ideas. Thus, Rochlin introduced the notion of "essential friction." That is, friction, in terms of checks and balances, may well be essential to the stability of distributed information-management systems. Other negative side effects of the scientific-management approach include deskilling workers and disenfranchising them from a sense of ownership or responsibility for the quality of the product as a whole.

HUMAN ERROR

Whereas the prime mover for Taylor and Gilbreth was efficiency, World War II (and the technology of war) and later the development of nuclear power plants caused increased concern about issues associated with safety. Efforts to determine the "right way" shifted from an emphasis on "time" to an increased concern about "human error" as a "cause" of potentially catastrophic system failures. The Human Engineering Division in the Air Force's Armstrong Laboratory was an early center for studying human performance in the aviation domain. Fitts and Jones' (1947a, 1947b) analyses of pilot error in operating aircraft controls and of reading and interpreting aircraft instruments and Christensen's (1947) analysis of navigation were early examples of studies of cognition in the wilds of aviation. Fitts and Jones' analyses were based on retrospective verbal reports from pilots. Analyses of the errors reported by pilots inspired laboratory studies of such things as shape coding of controls (Jenkins, 1947) and alternative designs for altimeter displays (Grether, 1949), which eventually led to design innovations. Christensen's (1947) field observations of navigators earned him membership in the Pole Vaulter's Club as the first civilian to fly over the North Pole in an air force aircraft. These field observations were followed by experimental evaluations (Christensen, 1949), which led to the design of improved plotters that became the air force standard.

Similar work was being done in England by researchers such as Bartlett and Craik who studied the effects of stress and fatigue on skilled performance in the "Cambridge cockpit" (designed around a Spitfire cockpit). The Cambridge program made many important "contributions in the areas of aircrew selection and training, the effects of sleep loss and fatigue, and various aspects of visual perception and display design" (Edwards, 1988, p. 7). The combination of field observations with empirical laboratory studies set the flavor for "human engineering" analyses to come. However, the balance between field observations and empirical evaluations achieved in this early work is a standard that is rarely achieved today.

The work of Fitts and Bartlett (and their colleagues) is often recognized as a critical step toward the development of the formal disciplines of human factors (engineering psychology or ergonomics). Many people currently active in the field proudly trace their intellectual roots to either Fitts or Bartlett. Several important themes emerged from this era. First, the work in these laboratories further demonstrated the promise of an empirical, experimental program for investigating the "best way." Second, the emphasis on achieving the "best way" shifted from the managers of the work to the designers of the tools (technology). It became clear that the probability of human error was influenced by the design of tools. Finally, the most important theme was that the errors were often traced to failures to process information. There was an increased appreciation for the mental or information processing aspects of work (e.g., perception, memory, and decision making).

Research on human error in nuclear power got great impetus as a result of the accident at Three Mile Island. Early work (e.g., technique for human error rate prediction; THERP: see Miller & Swain, 1987) characterized human error in terms of error probabilities and analyzed the work environment in terms of performance-shaping factors (PSFs) that could potentially increase or decrease the probability of human error. These PSFs reflected general design principles (e.g., consistency in the layout of instruments) that had their origin in the early "human engineering" analyses of Fitts and his colleagues.

However, because of the high degree of automation in nuclear power production, attention began to shift from the perception/action-based errors (slips) that dominated Fitts and Jones' analyses to errors of intention (mistakes). In the automated world of process control, the operator was less a real-time controller and more a supervisor—whose principal function was to detect and diagnose faults. In this domain, much of the interesting "activity" was below the surface. Thus, it was difficult to make inferences about the factors that influenced performance from observing activities alone. Also, it was difficult to characterize the problem-solving processes in terms of probabilities. Verbal protocol analysis became a critical supplement to field observations, and qualitative models of the strategies of the human problem solver began to replace the more quantitative chronometric and probability-based analyses.

RISØ National Laboratory was an important center for early research on human fault diagnosis (Vicente gives a brief account of the early years at RISØ, 1997). Rasmussen and Jensen's (1973; see also Rasmussen, 1986; Rasmussen, Pejtersen, & Goodstein, 1994) field study of electronic trouble shooting was an important early study. It relied extensively on verbal protocols. These studies were conducted in the electronics shop at RISØ. Audio recordings were made of technicians as they diagnosed problems with scientific instruments that were brought to them for repair (e.g., multichannel analyzers, oscilloscopes, TV receivers, etc.). The technicians were not asked to introspect about their thought processes, but were asked to "merely tell what they were trying to accomplish, what they were doing and/or observing, and with which instruments" (Rasmussen et al., 1994, p. 86). Analysis of the protocols allowed multiple strategies for diagnosis to be identified and described. An important aspect of this work was that the strategies most commonly adopted by the technicians (e.g., topographic search) were not the ones that were predicted by abstract analysis of problem solving (e.g., hypothesis and test). Rasmussen (1986) considered whether the strategies of the technicians were rational:

> The experiment discussed here clearly indicates that the task is defined by the technicians primarily as a search to find where the fault originates in the system. They are faced with a system which they suppose has been working properly, and they are searching for the location of the discrepancy between normal and defective states. They do not see the task as a more general problem-

solving task in order to understand the actual functioning of the failed system and thus to explain why the system has the observed faulty response. . . . What is rational depends upon the performance criteria adopted by the person. Normally a reasonable criterion for a maintenance technician is to locate the fault as quickly as possible, and only in special circumstances will his criterion be that of minimizing the number of measurements. . . . From this point of view, the procedures found in our records are rational since in most cases the faults were found within very reasonable times. . . . The system designer with his theoretical background may quite naturally value as rational the "elegant" deductive procedure which is informationally very efficient and based upon few observations, but this criterion is not the appropriate one on the basis of which performances in real life maintenance work can be judged. (pp. 55–56).

Despite the productivity and influence of the work at RISØ, much of the work on human factors and human–machine systems tended to emulate the experimental methods and laboratory paradigms derived from the work of Fitts and the Applied Psychology Unit at Cambridge. Through the 1970s and 1980s, there was a heavy emphasis on experimental rigor. Field studies such as the work on electronic trouble shooting at RISØ were exceptions rather than the rule. Most of the work on human performance and cognition used abstract experimental paradigms (e.g., sine wave tracking, probe reaction time tasks, probability judgments, visual search, syllogisms, etc.). These paradigms were assumed to reflect general, fundamental aspects of human performance, and there was relatively little concern about representativeness. Naturalistic field studies were rarely reported in the technical literature. Chronometric analyses dominated the literature, and verbal protocol analyses were generally regarded with deep skepticism.

INFORMATION OVERLOAD

Attention and workload became central themes for laboratory-based empirical work in the 1970s and 1980s (e.g., Gopher & Donchin, 1986; Hancock & Meshkati, 1988; O'Donnell & Eggemeier, 1986). The problem of Three Mile Island and increasingly problems in aviation and other domains were viewed as problems of "too much data." Measuring "mental workload" became a critical challenge for researchers. Important design questions focused on: How many people were needed to meet the data processing demands of a job (e.g., is a single-operator attack helicopter feasible)? How can automation be used to off-load processing demands from the human operator (e.g., can the co-pilot be replaced by an automated pilot's associate)? How can the data be packaged (integrated in graphical displays or distributed over modalities) to minimize the workload?

Theoretical work on attention attempted to differentiate between structural and resource constraints on information processing. For those who framed attention in structural terms, the critical question was to discover the locus of the bottleneck (early or late). Wickens' (e.g. 1983, 1984) multiple resource model included features of both styles of models. Attention was modeled as a resource, but there were structural constraints on how this resource could be allocated across stages of processing. The multiple resource model provided logical post hoc accounts for multiple task performance measured in the laboratory, but it has had limited value for predicting performance in complex natural tasks where there are rich semantic links across the various component tasks.

Applied work on attention focused on the issue of workload measurement. The central problem was to come up with reliable, relatively noninvasive measures that could be used in design trade-off studies to evaluate alternative interfaces or function allocation decisions and/or that could be used as real-time measures to provide the basis for dynamic adaptation and augmentation to prevent potentially catastrophic information overloads. Consideration was given to primary task measures (e.g., tracking remnant), secondary tasks (e.g., probe reaction time), physiological measures (e.g. heart rate variability), and subjective reports (e.g., the NASA-Task Load Index [TLX]). Active research continues in all areas of workload measurement. However, in practice, subjective measures are used most frequently. Despite evidence for dissociation between subjective reports and other measures (e.g., see O'Donnell & Eggemeier, 1986), subjective reports seem to be nearly as sensitive and reliable as anything else, and they tend to be far easier to implement.

VISUALIZATION

In the mid-1980s, in part stimulated by a failure to achieve consensus on a reliable ruler for measuring workload and in part inspired by advances in graphical displays and successes such as the graphical user interface on personal computers, the characterization of the problem began to shift. The problem appeared to be less a limitation in data processing and more a problem of understanding (visualization). During this period, it became increasingly clear that the capacity to process information depended critically on the representation of that information. Extracting information from a table of numbers might be overwhelmingly difficult, whereas extracting that same information from a graph might be virtually effortless. However, it was not simply the "form" of the representation, but the correspondence between the form and the meaning of the process (work) being represented that was critical (e.g., Bennett & Flach, 1992; Bennett, Nagy, & Flach, 1997; Flach, 1999; Flach & Bennett, 1996; Rasmussen & Vicente, 1989; Sanderson, Flach, Buttigieg, & Casey, 1989; Woods, 1991).

In the aviation domain, this shift in emphasis from data processing to understanding was evident in the emergence of a new construct—situation awareness (e.g., Endsley, 1995; Flach, 1995; Flach & Rasmussen, in press). Here the inspiration came from observations of striking differences in how individuals responded to complex information-processing demands such as those that might be associated with combat. One individual is consistently overwhelmed and lost, whereas another responds instinctively and almost effortlessly. Both individuals are facing the same situation. However, one appears to have a superior internal representation—better situation awareness. In other words, one pilot is more expert!

During this period, Klein and his colleagues (e.g. Klein, Calderwood, & Clinton-Cirocco, 1986) began studying decision making in natural environments. They began by "observing and taking protocols from urban fireground commanders about emergency events that they had recently handled" (Klein, 1993, p. 139). Klein (1993) discovered:

> The fireground commanders' accounts of their decision making do not fit into a decision-tree framework. The fireground commanders argued that they were not "making choices," "considering alternatives," or "assessing probabilities." They saw themselves as acting and reacting on the basis of prior experience; they were generating, monitoring, and modifying plans to meet the needs of the situations. We found no evidence for extensive option generation. Rarely did the fireground commanders contrast even two options. We could see no way in which the concept of optimal choice might be applied. Moreover, it appeared that a search for an optimal choice could stall the fireground commanders long enough to lose control of the operation altogether. The fireground commanders were more interested in finding actions that were workable, timely, and cost effective. (p. 139)

As a result of observations of decision making in natural environments, Klein developed his "recognition-primed decision model," which emphasized the importance of perception and the "zeroing-in" aspects of the decision problem. Equally important, however, was the fact that Klein's work helped to stimulate wider interest in the problem of "expertise" and "naturalistic decision making" (e.g., Klein, Orsanu, Calderwood, & Zambok, 1993) and helped to legitimize verbal protocol analysis and field observations as essential to a complete understanding of human performance in complex work environments.

In 1995, Hutchins' book *Cognition in the Wild* was published. This book summarized a comprehensive field (ethnographic) study of ship navigation that began in the 1980s. Hutchins chose the title "in the wild" to distinguish between "the laboratory, where cognition is studied in captivity, and the everyday world, were human cognition adapts to its natural surroundings. I hope to evoke with this metaphor a sense of an ecology of thinking in which human cognition interacts with an environment rich in organizing resources" (p. xiv). Hutchins' description of how variations in map design shaped the computation problem, making some problems easier and others harder, provided a clear demonstration of the fundamental link between the "right way" and the "right representation."

The realization that the functionality of a representation (internal or external) relates to its correspondence to problem semantics (to a situation) has opened up the problem of work analysis. Whereas the emphasis of the information-processing approach was on internal human-processing constraints (channel capacity, bottlenecks, or resource limitations), there is now increasing interest in constraints in the situation. There is growing appreciation that the search for the "right way" has to be framed in terms of both an analysis of human information processing (awareness) and an analysis of the semantics intrinsic to the work domain (situation). In other words, the problem of the "right way" involves a distributed information-processing system that spans the human and the work ecology. This is the focus of cognitive systems engineering (CSE)—to understand the semantics of situations and to discover how awareness can be facilitated through the engineering of representations. These representations can be engineered through training (internal representations) or through the design of visualization tools (external representations).

The works of Rasmussen, Klein, and Hutchins also helped to show that it was possible for researchers to discover general principles and broad insights into human cognition by delving deeply into the particulars of specific work domains. These works illustrated that the image of human cognition molded from general, laboratory paradigms did not always hold up in the wild. The behavior of technicians, fire-ground commanders, and navigators often did not conform to expectations based on logical analyses. Furthermore, the "nonsense" laboratory tasks derived from the logical analyses of information processing tended to miss the essence of the natural tasks. Semantic and contextual aspects that were critical to performance in the natural world tended to be abstracted out of the laboratory tasks.

In summary, the safety tradition was inspired by field observations of natural phenomena (e.g., accidents). However, early researchers (Fitts' human engineering group and Bartlett's applied psychology unit) were extremely innovative in extracting key features of the natural phenomenon that could be studied in the laboratory by using rigorous controls and quantitative measures of performance. The success of the laboratory paradigms tended to overshadow the phenomenon-based insights that inspired them. As a result, the 1970s and 1980s were dominated by a paradigmatic information-processing approach to human performance where the laboratory paradigms became the phenomena of interest. Field studies were rare, and it was generally assumed that natural tasks could be understood as a combination of the laboratory tasks (e.g., landing an airplane might be considered a combination of tracking, visual search, and decision-making tasks).

However, pockets of phenomena-centered research were maintained. Rasmussen, Klein, and Hutchins are prominent representatives of researchers who were not seduced by the power of laboratory paradigms. These researchers maintained a phenomenon-centered science. In part, this was motivated by the practical need to address problems in natural systems like nuclear power plants. But it was far more than this. It was also motivated by a sense that human cognition in the wild was fundamentally different from cognition observed in captivity; a sense that the context-free logical puzzles and toy worlds of the cognition laboratory were not representative of the demands faced in natural work domains; and a sense that performance in the laboratory, where human limitations were easily catalogued and the fallibility of human reasoning seemed obvious, was not representative of the creative, adaptive nature of people in their fields of expertise. Rasmussen, Klein, and Hutchins were not alone. However, their work was prominent and helped draw attention to earlier insights of people such as deGroot (e.g., Vicente & Wang, 1998), Brunswik (Hammond, 1993; Kirlik, 1995), and Gibson (Flach, Hancock, Caird, and Vicente, 1995). Gradually, there seems to be growing appreciation for the importance of ethnographic observations for a basic understanding of cognition. Researchers are beginning to consider ways to scale the methods of observation and analysis to the complexity of natural phenomena, as opposed to cutting a phenomenon into pieces that fit conveniently into laboratory paradigms. This is the challenge that has stimulated the new discipline of CSE—to understand the holistic and emergent properties of work in a way that generalizes to design—to achieve expertise by design.

CONCLUSION

Table 2.12.1 summarizes this discussion in terms of four "eras" of work on human–machine systems. These eras are not necessarily distinct historical epochs, but represent themes that, although interleaved in time, are distinctive in terms of the technological changes that are addressed. Each era has framed the problem differently, and each era has focused on a different metric for comparing and evaluating potential solutions.

In the first era, the motivating technology was the assembly line. The key problems were efficiency, standardization of methods, and line balancing. The primary performance metric was production rate—how many units per minute could be produced. With the advances in technology arising during World War II and later with the development of nuclear power, focus shifted from efficiency to safety or reliability. The defining issue was human error and how to reduce the probability of system failure. The primary performance metric was error probability. However, as nuclear process research continued and as other systems (like civil aviation) became more complex,

TABLE 2.12.1
Four Eras of Work on Human-Machine Systems

ERA	Key Technologies	Defining Problem	Measurement
Scientific management	Assembly line Mass production	Efficiency Standardization Line Balancing "Too much variability"	Rate Pieces/Minute
Human Error	Advanced military systems Nuclear power	Reliability: "Too many human errors"	Error Rate BHEP (basic human error probability)
Information overload	Nuclear Power Civil Aviation Automated systems	Attention (bottle-necks, limited resources) "Too much data"	Mental Workload (subj., physiol., prim. & sec. task)
Visualization	Graphical interfaces	Situation awareness, representations "Too little understanding"	Stability? (timing and adaptability)

the focus began to shift to the explosion of data that operators were being asked to deal with. Increasingly, system failures were being attributed to limitations of human information processing. The defining issue became attention, and numerous metrics (subjective, physiological, primary task, and secondary tasks) for measuring mental workload were developed. This is still a major concern in human factors. As graphical displays became available, it began to be evident that, with the appropriate representation, large amounts of data could be processed very efficiently. Thus, the focus is now shifting to the problem of understanding or situation awareness—how to integrate complex data in a way that facilitates understanding. This is the era of data visualization.

Note that there is a question mark in the measurement column associated with the era of visualization. This is the challenge that we are now faced with—how do you measure cognitive work (i.e., the potential for creative problem solving, discovery, innovation)? How do you measure the effectiveness of a representation? Where are the invariants that allow us to abstract essential features of solutions and to generalize broadly from one domain to another?

Our hypothesis is that the critical invariants are not local attributes of cognitive agents (e.g., fixed capacity limits on awareness), but abstract properties of dynamical work domains (e.g., constraints on stability). For example, an expert does not attack every situation in the same way. The expert adapts to nuances of specific situations. Similarly, an effective representation must be robust in that it can support multiple strategies that reflect the particular demands of different situations. Thus, much of the focus of measurement is on characterizing the situational constraints. In the terms of Flach and Dominguez (1995), the emphasis is on "use," not on "users." As we attempt to understand and measure work situations, the focus shifts from characterizing behavioral trajectories to characterizing the boundary conditions of situations (i.e., the stability constraints on the distributed information management system).

The emphasis shifts from absolute measures of performance to relative measures. For example, there is less interest in the information-processing rate (bits/s) as a fixed characteristic of a human information-processing system and more interest in the synchronization demands associated with a work domain (e.g., windows of opportunity) and the semantic basis for integrating information into a manageable number of chunks. Note that instability can arise from either not going fast enough or from going too fast! Also, it is clear that the capacity of working memory can be extended indefinitely by integrating information in meaningful chunks. Therefore, knowing minimum reaction time, or that the capacity of working memory is seven plus or minus two, is of limited value to a designer, unless the temporal demands of the task or the semantic bases for chunking information are understood.

This does not mean that the information-processing limits that have been discovered through basic research are unimportant to building a theory of human performance. However, it is beginning to appear that these limits on human information processing are rarely limiting constraints on the performance of experts in their domains of expertise. Experts use their experience with the natural constraints in their domain and the tools available (from advanced computers to post-it notes) to adapt and shape the situation so that tasks that would overwhelm novices can be accomplished with little effort. The fact that experts use the natural constraints in a domain is nicely illustrated in Vicente and Wang's (1998) review of expert/novice differences in memory recall. When natural constraints were broken (e.g., chess pieces randomly placed on the board), experts and novices performed at comparable levels and at levels that reflected general information-processing limitations. However, when the natural constraints were maintained (e.g., an actual game situation), then the expert performance greatly exceeded that of novices. Hutchins' (1995b) discussion of "how a cockpit remembers" nicely illustrates how people manipulate situations to distribute the cognitive load over the work space.

Thus, knowing the fundamental memory constraint of seven plus or minus two is not of much use in the design of work until we can answer the question: Seven plus or minus two what? The answer to this question requires an understanding of the work ecology (e.g., what is a meaningful pattern of chess pieces). Vicente and Wang (1998) illustrated how Rasmussen's abstraction hierarchy might provide a useful framework for characterizing the semantic constraints that allow integration of complex information into manageable chunks.

In conclusion, we appear to be in a phase transition in how we view humans and work. We are shifting from a reductionistic, bottom-up approach to the problem, where the focus was on fundamental "elements" (e.g., therbligs, stages of information processing, resources) that could be manipulated in controlled laboratory investigations. We are moving toward more holistic approaches where abstract properties of the work system (e.g., task semantics) are of particular interest. There is an increased appreciation for the fact that stability in a complex, distributed information system depends on a delicate balance between efficiency and reliability. This balance is an integral property of the whole system, not a local property of any particular element.

At this stage in the transition, there is a great value in generative approaches to research. That is, field observations of the natural work phenomena are critical, so that we can begin to frame the right questions to ask under more controlled conditions (likely to be complex simulations). This is a turbulent stage because, in the context of the field observations, the classical tasks used in most laboratory studies appear to be hopelessly naive. This results in tension between basic and applied practitioners. This is a bit ironic because many of these laboratory tasks were motivated by earlier field observations of Fitts and others. However, the nature of work and the technologies of work have become more complex. Humans today are more likely to be managing (supervising) systems than controlling them (e.g., Sheridan, 1996). The boundary conditions for the management task are very different than those on the control task. We need to emulate Paul Fitts and other early researchers, not by mimicking their paradigms, but by developing new paradigms that are representative of today's work ecologies and that allow strong tests of hypotheses about the nature of the human's role in these systems. A key challenge that must be addressed to make the transition from phenomenological observations to hypothesis testing is to discover general measures that tap into the essence of dynamical work ecologies.

ACKNOWLEDGMENTS

This chapter is adapted from chapter 2 of the following report: J. M. Flach & G. Kuperman (1998), *Victory by design: War, information, and cognitive systems engineering* (AFRHL/HE-WP-TR-1998-00), Wright-Patterson Air Force Base, OH: Air Force Research Laboratory.

REFERENCES

Bennett, K. B., & Flach, J. M. (1992). Graphical displays: Implications for divided attention, focused attention, and problem solving. *Human Factors, 34,* 513–533.

Bennett, K. B., Nagy, A. & Flach, J. M. (1997). Visual displays. In G. Salvendy (Ed.), *Handbook of Human Factors* (659–696). Mahwah, NJ: Lawrence Erlbaum Associates.

Christensen, J. (1947). Psychological factors involved in the design of air navigation plotters. In P. M. Fitts (Ed.), *Psychological research on equipment design* (pp. 73–90). (Report No. 19). Washington, DC: U.S. Government Printing Office.

Christensen, J. (1949). A method for the analysis of complex activities and its application to the job of the Arctic aerial navigator. *Mechanical Engineering, 71.*

Edwards, E. (1988). Introductory overview. In E. L. Wiener & D. C. Nagel (Eds.), *Human factors in aviation* (pp. 3–25). New York: Academic Press.

Endsley, M. R. (1995). Toward a theory of situation awareness in dynamic systems. *Human Factors, 37,* 32–64.

Fitts, P. M., & Jones, R. E. (1947a). *Analysis of factors contributing to 460 "pilot-error" experiences in operating aircraft controls.* (AMC Memorandum Report TSEAA-694-12). Dayton, OH: Wright-Patterson Air Force Base.

Fitts, P. M., & Jones, R. E. (1947b). *Psychological aspects of instrument display, I: Analysis of 270 "pilot-error" experiences in reading and interpreting instruments.* (USAF Air Materiel Command Mem. Report TSEAA-694–12A). Dayton, OH: Wright-Patterson Air Force Base.

Flach, J. M. (1995). Situation awareness: Proceed with caution. *Human Factors, 37,* 149–157.

Flach, J. M. (1999). Ready, fire, aim: Toward a theory of meaning processing systems. In D. Gopher & A. Koriat (Eds.), *Attention and Performance XVII* (pp. 197–221). Mahwah, NJ: Lawrence Erlbaum Associates.

Flach, J. M., & Bennett, K. B. (1996). A theoretical framework for representational design. In R. Parasuraman & M. Mouloua (Eds.), *Automation and human performance: Theory and applications* (pp. 65–87). Mahwah, NJ: Lawrence Erlbaum Associates.

Flach, J. M., & Dominguez, C. O. (1995, July). Use-centered design. *Ergonomics in Design,* 19–24.

Flach, J. M., Hancock, P. A., Caird, J. K., & Vicente, K. J. (1995). *Global prespectives on the ecology of human-machine systems.* Mahwah, NJ: Lawrence Erlbaum Associates.

Flach, J. M., & Rasmussen, J. (in press). Cognitive engineering: Designing for situation awareness. In N. Sarter & R. Amalberti (Eds.), *Cognitive engineering in the aviation domain.* Mahwah, NJ: Lawrence Erlbaum Associates.

Gopher, D., & Donchin, E. (1986). Workload—An examination of the concept. In K. R. Boff, L. Kaufman, & J. P. Thomas (Eds.), *Handbook of perception and human performance* (Vol. 2, pp. 41.1–41.49). New York: Wiley.

Grether, W. F. (1949). Psychological factors in instrument reading. I. The design of long-scale indicators for speed and accuracy of quantitative readings. *Journal of Applied Psychology, 33,* 363–372.

Hancock, P. A., & Meshkati, N. (Eds.). (1988). *Human mental workload.* Amsterdam: North-Holland.

Hammond, K. R. (1993). Naturalistic decision making from a Brunswickian viewpoint: Its past present and future. In G. A. Klein, J. Orsanu, R. Calderwood, & J. Orsanu (Eds.), *Decision making in action: Models and methods* (pp. 205–227). Norwood, NJ: Ablex.

Hutchins, E. (1995). *Cognition in the wild.* Cambridge, MA: MIT Press.

Jenkins, W. O. (1947). The tactual discrimination of shapes for coding aircraft-type controls. In P. M. Fitts (Ed.), *Psychological research on equipment design* (Report No. 19, pp. 189–205). Washington, DC: U.S. Government Printing Office.

Kirlik, A. (1995). Requirements for psychological models to support design: Toward ecological task analysis. In J. M. Flach, P. A. Hancock, J. K. Caird, & K. J. Vicente (Eds.), *Global perspectives on the ecology of human-machine systems* (pp. 68–120). Mahwah, NJ: Lawrence Erlbaum Associates.

Klein, G. (1993). A recognition-primed decision (RPD) model of rapid decision making. In G. A. Klein, J. Orsanu, R. Calderwood, & J. Orsanu (Eds.), *Decision making in action: Models and methods* (pp. 205–227). Norwood, NJ: Ablex.

Klein, G., Calderwood, R., & Clinton-Cirocco, A. (1986). Rapid decision making on the fireground. *Proceedings of the 30th annual meeting of the Human Factors Society* (pp. 576–580). Santa Clara, CA: Human Factors Society.

Klein, G., Orsanu, J., Calderwood, R., & Zsambok, C. (1993). *Decision making in action: Models and methods.* Norwood, NJ: Ablex.

Konz, S. (1987). *Work design: Industrial ergonomics.* Columbus, OH: Publishing Horizons.

Miller, D. P., & Swain, A. D. (1987). Human error and human reliability. In G. Salvendy (Ed.), *Handbook of human factors* (pp. 219–249). New York: Wiley.

O'Donnell, R. D., & Eggemeier, F. T. (1986). Workload assessment methodology. In K. R. Boff, L. Kaufman, & J. P. Thomas (Eds.), *Handbook of perception and human performance* (Vol. 2, pp. 42.1–42.49). New York: Wiley.

Rasmussen, J. (1986). *Information processing and human-machine interaction: An approach to cognitive engineering.* New York: North-Holland.

Rasmussen, J., & Jensen, A. (1973). *A study of mental procedures in electronic troubleshooting.* (Report No. M-1582). Roskilde, Denmark: RISØ National Laboratory.

Rasmussen, J., Pejtersen, A. M., & Goodstein, L. (1994). *Cognitive systems engineering.* New York: Wiley.

Rasmussen, J., & Vicente, K. J. (1989). Coping with human errors though system design: Implications for ecological interface design. *International Journal of Man–Machine Studies, 31,* 517–534.

Rochlin, G. I. (1993). Essential friction: Error-control in organizational behavior. In Akerman (Ed.), *The necessity of friction: Nineteen essays on a vital force* (pp. 196–234). Heidelberg, Germany: Physica-Verlag.

Rochlin, G. I. (1997). *Trapped in the net.* Princeton, NJ: Princeton University Press.

Sanderson, P. M., Flach, J. M., Buttigieg, M. A., & Casey, E. J. (1989). Object displays do not always support better integrated task performance. *Human Factors, 31,* 183–198.

Sheridan, T. B. (1996). Speculations on future relations between humans and automation. In R. Parasuraman & M. Mouloua (Eds.), *Automation and human performance* (pp. 449–460). Mahwah, NJ: Lawrence Erlbaum Associates.

Vicente, K. J. (1997). A history of cognitive engineering: Research at RISØ (1962–1979). In *Proceedings of the Human Factors and Ergonomics Society 41st annual meeting* (pp. 210–214). Santa Monica, CA: Human Factors and Ergonomics Society.

Vicente, K. J., & Wang, J. (1998). An ecological theory of expertise effects in memory recall. *Psychological Review, 105,* 33–57.

Wickens, C. D. (1983). Processing resources in attention. In R. Paraduraman, J. Beatty, & R. Davies (Eds.), *Varieties of attention.* New York: Wiley.

Wickens, C. D. (1984). *Engineering psychology.* Columbus, OH: Merrill.

Woods, D. D. (1991). The cognitive engineering of problem representations. In G. R. S. Weir & J. L. Alty (Eds.), *Human computer interaction and complex systems* (pp. 169–188) London: Academic Press.

2.13　Workload and Situation Awareness

Christopher D. Wickens
University of Illinois

Workload and situation awareness are two constructs that have emerged in applied psychology over the last 25 years, and have generated a tremendous amount of research and academic pondering. They have much in common, but also some important distinctions. In the following, we discuss each concept in turn, then address their commonalties, their critical differences, and finally focus on their interrelation and interplay in complex human performance activities.

MENTAL WORKLOAD

The concept of mental workload has been proposed as an inferred construct that mediates between task difficulty, operator skill, and observed performance (Moray, 1979). Thus, increasing difficulty characteristics of the task (e.g., high time pressure, high working memory demands) or characteristics of the operator (e.g., low skill level), often, but not always degrades performance. However, these variables are said to always increase the mental workload of the task and thereby affect the *potential* for task performance, through the mediating effect of resource demand. The more difficult task or the less skilled operator requires more resources to perform at a constant level, hence leaving fewer resources available to deal with an unexpected increase in task difficulty or the requirement to perform an added "secondary" task.

To the extent that workload can be defined in terms of the resources invested in a task (and therefore inversely related to the reserve resources), one can also speak of workload increasing, as more resources are invested into a task of constant difficulty, by an operator of constant skill level. Thus, the operator who is motivated to "try harder" on a task, performs better and still experiences greater workload (Vidulich & Wickens, 1986; Yeh & Wickens, 1988). This reversal from the usual pattern of correlation (higher workload is more typically associated with poorer performance) is well explained by the resource or "energetics" model of workload (Hockey, Gaillard, & Coles, 1986).

In addition to its status as an inferred construct, two additional characteristics of mental workload provide it with some parallels to situation awareness: the users of the construct, and the nature of measurement. With regard to the first of these, mental workload is an intuitive construct with which operators are quite comfortable; pilots, for example, have no difficulty providing subjective estimates of their workload and understand the nature of the construct. Correspondingly it is a construct for which engineers have desired objective measurement. Such desire is based on known accidents that have been attributable to high workload and the consequent desire to design systems (or train operators) to a level at which workload is below some magic quantifiable "red line." In fact, however, such a scale and the redline level along the scale have remained somewhat elusive. Finally, although users and engineers have generally been comfortable with the construct of mental workload, many cognitive and experimental psychologists have been somewhat less so, in part because of its highly subjective characteristic and the lack of precision of the resource concept that underlies its basis (Navon, 1984).

The study of mental workload, as an inferred construct that must be defined by converging operations, has been associated with one manipulation and three different classes of measurements. As noted, the manipulation is typically one of "task difficulty," that is, some property of the task that is assumed to make it "harder" or "easier." This manipulation in turn is assumed to drive some or all of the three classes of measurable "workload indexes": performance, physiological measures of arousal, and subjective ratings. Performance itself is often dichotomized into performance of the task itself (whose workload is of interest, the primary task) and performance of a secondary or concurrent task (Wickens & Holland, 2000). In each of these three general categories resides a plethora of different candidate measures that have been offered (see Tsang & Wilson, 1997, for a recent review). The range of primary task measures is, of course, as broad as the range of primary tasks for which workload measurement is of interest.

Both the resource model of workload, as described earlier, and empirical experience suggest that the three classes of measures do not always co-vary as difficulty is varied (Yeh & Wickens, 1988). Indeed, as already noted, it is in part the fact that primary task performance does *not* change, even as the other indexes do, that provides validity for the resource construct that underlies much current thinking about mental workload. Further complexities in the modeling of mental workload (e.g., multiple resources, response biases, ceiling effects, and sensitivity ranges), have been used to account for other circumstances in which there is not perfect correlation between workload measures (Yeh & Wickens, 1988).

Finally, although many aspects of mental workload have been extensively researched (some probably have been overresearched), there remain a number of critical workload issues still to be addressed. Most importantly, these concern an understanding, modeling, and prediction of the *consequences* of high workload: If the difficulty of a task or task environment is increased, at what point does task performance begin to degrade (the "red line" problem)? Which tasks are shed or abandoned? When are they resumed? Which ones are still performed (but at degraded levels), and which ones remain "protected" (Chou, Madhaven, & Funk, 1996; Huey & Wickens, 1993; Raby & Wickens, 1994; Schutte & Trujillo, 1997)? This issue of "strategic task management" or workload management has received surprisingly little empirical inquiry, given its importance to safety in high workload environments such as driving, flying, emergency response, or crisis management.

One plausible claim that we revisit later in this chapter is that the maintenance of situation awareness is one of the generic tasks that tends to be abandoned or degraded under conditions of high workload. Before we address this issue, however, we now consider the construct of situation awareness and where the parallels with workload are accurate, and where they break down.

SITUATION AWARENESS

The construct of situation awareness has been defined by Endsley (1995) as: "The perception of elements in the environment within a volume of time and space, the comprehension of their meaning, and the projection of their status in the near future" (p. 36). Although not all researchers in the field are in precise agreement with this definition, there is nevertheless consensus that it appears to capture many of the aspects characteristic of the construct: It is a uniquely cognitive phenomenon, which supports action, but is not part of the action itself; it applies to a dynamic changing environment; and the nature of "the situation" of which awareness is maintained needs to be specified. Thus, where possible, one should refine the definition more specifically to refer to constructs like "geographical hazard awareness," "automation mode awareness," or "task awareness," when this is possible. Finally, the distinction between the product of awareness and the process of maintaining that awareness (facilitated by supporting structures such as long-term memory or attention) is an important part of the understanding (Adams, Tenney, & Pew, 1995).

The cycle of research interest in situation awareness followed that of workload by perhaps 10 to 15 years, with the first few papers emerging in the 1980s (Endsley, 1988; Harwood, Barnett, & Wickens, 1988), the first major conference on the topic in 1985 (Attitude awareness), and a rapid growth in conferences, workshops and publications appearing in the last decade (Garland & Endsley, in press; Gilson, Garland, & Koonce, 1994; NATO AGARD, 1996; *Situation Awareness*, 1995). Furthermore, in parallel with workload, much of the initial interest was spawned in the domain of aviation, but recent efforts have broadened to other domains such as medicine (Gaba, Howard, & Small, 1995) or driving (Gugerty, 1997; see Durso & Gronlund, 1999, for other applications).

Beyond its historical evolution, there are several other parallels between situation awareness and mental workload. Like workload, situation awareness is an inferred construct that must be triangulated with a series of different measurement techniques. Like workload, situation awareness is not the same as performance, but it describes and supports the *potential* for performance (and performance measures are sometimes required for its assessment). Also like workload, it is a concept that operators feel comfortable with, engineers seek to quantify, but with which many psychologists feel a discomfort, because of the lack of precise measurable properties and its association with the illusive construct of "consciousness."

As an inferred construct, which needs triangulation with multiple measures, situation awareness, like workload, has been the recipient of the "three-pronged" approach to measurement: via performance, subjective measures, and physiology (Garland & Endsley, in press). At this point, the parallels break down (and some of the efforts to apply the analogies appear to be misplaced), because of fundamental differences between the two constructs. Mental workload is fundamentally an *energetics* construct, in which the quantitative properties ("how much") are dominant over the qualitative properties ("what kind"), as the most important element. In contrast, situation awareness is fundamentally a *cognitive* concept, in which the critical issue is the operator's accuracy of ongoing understanding of the situation (i.e., a qualitative property). Stated another way, an assessment that a user has a workload rating of 0.80 has more general meaning and value (assuming it to be accurate), than an assessment that a user has 80% situation awareness. For the latter, we must ask Of what is the user aware? The distinction between the two constructs can also be based on the scientist's ability to "score" the observer on the accuracy of his or her subjective rating. If the observer rates subjective workload as a 7 on a 10-point scale, the scientist has little rationale to challenge that rating as incorrect. If the observer rates situation awareness as a 10, it is much easier to assess whether the level of knowledge is accurate or not (and hence to evaluate the appropriateness of the score). For this reason, because self-ratings of awareness may be very wrong (you are not aware of that of which you are not aware), subjective measures of situation awareness do not share the same properties as subjective ratings of workload, and the former must be viewed with some suspicion. (Note that this criticism does not apply to "objective subjective reports" or explicit measures when the operator gives qualitative answers to specific questions, to be discussed later; Endsley, 1995).

As for physiological measures, the energetics properties of workload make it quite amenable to the generally quantitative aspects that many physiological measures can provide (pupil diameter, heart rate variability, electroencephalogram [EEG] amplitude, etc.). The far more qualitative properties of situation awareness (the "what," rather than the "how much"), render it a relatively poor candidate for physiological measures, most of which are fairly impoverished in their abilities to provide qualitative information (compared, for example, to the qualitative richness of language).

It is only in the performance measures that both workload and situation awareness have many common parallels. Here, only the workload measure of primary task performance is paralleled by performance-based measures of situation awareness. This is because one *must* know and specify what the primary task is, to adequately measure its performance. In the same sense, one must know and specify of what information the user should be aware, to assess, by his or her implicit or explicit performance, whether he or she *is* aware of that information. The distinction between implicit and explicit performance measures of situation awareness represents another departure from the analogy with workload for which this distinction is not generally relevant. Explicit measures of situation awareness represent "tests" that are often answered orally (i.e., using language) by the observer (e.g., Where is the nearest aircraft to you?), but they may also be answered by spatial graphics techniques that allow a lot of qualitative information to be conveyed (for example, drawing a map of the territory just traversed or the location of aircraft; Gronlund, Ohrt, Dougherty, Perry, & Manning, 1998). These do not really have a parallel in the mental workload measurement. Implicit measures, on the other hand, represent measures of task performance (i.e., depending on responses, which are not inherently part of the situation awareness construct) that differ depending on whether the user had correct or incorrect situation awareness. Typically, these performance measures involve responses to unexpected or unanticipated events (Durso & Gronlund, 1999; Wickens, in press).

THE INTERACTION:
WORKLOAD AND SITUATION AWARENESS

Although the many parallels between the two constructs break down when measurement is attempted, the two are in many ways tightly linked and intertwined, as we observe cognition and performance when people interact with complex systems. Most importantly, these links are defined by the fact that maintaining a high and accurate level of situation awareness is a resource-intensive cognitive process. Thus, on one hand, one cannot gain accurate situation awareness without expending resources, which may in turn compete with other concurrent cognitive tasks. On the other hand, heavy concurrent task demands may divert resources from the maintenance of situation awareness. (Such a diversion is not necessarily reflected in performance unless an unexpected event occurs, the response to which requires awareness and anticipation of environmental characteristics that were wanting.)

This relation between situation awareness and mental workload (resource cost) is mediated in certain ways by expertise or skill. The skilled operator generally can preserve situation awareness with lower resource cost, because he or she develops a schema in which evolving information can be integrated, develops a mental model that can assist in making predictions, and has calibrated knowledge of the environment that can more effectively guide selective attention to the likely sources of information (Durso & Gronlund, 1999). Thus, situation awareness can often be maintained more efficiently by the skilled operator. However, a dangerous trap is to assume that such an operator necessarily has higher situation awareness. If such an operator relies too heavily on the relatively automatic, resource-free schemata and expectancies of past experience to guide the search for and interpretation of new information, he or she may fail to correctly interpret (or even notice) the unexpected and surprising event. Thus expertise must be coupled with an appropriate allocation of resources to guarantee adequate situation awareness. Importantly, the measures of visual fixation and eye scanning represent a critical contribution to understanding this linkage between situation awareness and resource allocation. The direction of fixation can provide fairly good information about the process of maintaining situation awareness, just as it can often provide useful information about the direction of allocation of visual attention.

Finally, mental workload and situation awareness may be coupled in engineering design remediations. Improved, intuitive and integrated displays are one of the most important solutions to problems of low situation awareness (Sarter & Woods, 1995; Wickens, in press). At the same time, carefully developed displays can often support the cognitive integration of complex dynamic information with a lower resource cost.

In conclusion, Endsley once asked whether workload and situation awareness were "birds of a feather" or "opposite sides of the coin." The arguments made in this chapter are that the answer lies somewhere between. The two share certain characteristics, as do the same feathered birds, but their different associations with the energetics and cognitive characteristics of human performance provide a clear distinction in their birdlike appearances. Finally, their sometimes reciprocal relation, as mediated by the resource concept, is perhaps best captured by the two sides of Endsley's coin.

REFERENCES

Adams, M., Tenney, Y. J., & Pew, R. W. (1995). Situation awareness and the cognitive management of complex systems. *Human Factors, 37*, 85–104.

Chou, C., Madhaven, D., & Funk, K. (1996). Studies of cockpit task management errors. *The International Journal of Aviation Psychology, 6*, 307–320.

Durso, F., & Gronlund, S. (1999). Situation awareness. In F. Durso (Ed.), *Handbook of applied cognition* (pp. 283–314). New York: Wiley.

Endsley, M. R. (1995). Toward a theory of situation awareness in dynamic systems. *Human Factors, 37*, 85–104.

Endsley, M. R. (1988). Design and evaluation of situation awareness enhancement. In *Proceedings, 32nd Annual Meeting of the Human Factors Society* (pp. 97–101). Santa Monica, CA: Human Factors Society.

Gaba, D. M., Howard, S. K., & Small, S. D. (1995). Situation awareness in anesthesiology. *Human Factors, 37*, 20–31.

Garland, D. J., & Endsley, M. R. (in press). *Situation awarenss analysis and measurement.* Mahwah, NJ: Lawrence Erlbaum Associates.

Gilson, R., Garland, D. J., & Koonce, J. M. (1994). *Situation awareness in complex systems.* Daytona Beach, FL: Embry Riddle University Press.

Gronlund, S., Ohrt, D. D., Dougherty, M. R. P., Perry, J. L., & Manning, C. A. (1998). Role of memory in air traffic control. *Journal of Experimental Psychology: Applied, 4,* 263–280.

Gugerty, L. J. (1997). Situation awareness during driving. *Journal of Experimental Psychology: Applied, 3,* 42–66.

Harwood, K., Barnett, B., & Wickens, C. D. (1988). Situation awareness: A conceptual and methodological framework. In R. C. Ginnett (Ed.), *Proceedings of the 11th Psychology in the Department of Defense Symposium* (pp. 316–320). Colorado Springs: United States Air Force Academy.

Hockey, R., Gaillard, A., & Coles, M. (1986). *Energetics and Human Information Processing.* (Dordrecht, Netherlands: Martinus Nijhoff.

Huey, M. B., & Wickens, C. D. (1993). *Workload transition: Implications for individual and team performance.* Washington, DC: National Academy of Sciences.

Moray, N. (1979). *Mental workload: Its theory and measurement.* New York: Plenum.

NATO AGARD. (1996). *Situation awareness: Limitations and enhancements in aviation environments.* Neuilly-Sur-Seine, France: NATO Advisory Group for Aerospace Research and Development.

Navon, D. (1984). Resources: A theoretical soupstone *Psychological Review, 91,* 216–234.

Raby, M., & Wickens, C. D. (1994). Strategic workload management and decision biases in aviation. *International Journal of Aviation Psychology, 4,* 211–240.

Sarter, N. B., & Woods, D. D. (1995). How in the world did we ever get into that mode? Mode error and awareness in supervisory control. *Human Factors, 37,* 5–19.

Schutte, P. C., & Trujillo, A. C. (1997). Flight crew task management in non-normal situations. *Proceedings of the 40th Annual conference of the Human Factors and Ergonomics Society* (pp. 244–248). Santa Monica, CA: Human Factors Society.

Situation Awareness. (1995). In R. D. Gilson (Ed.), *Human Factors, 37*(1).

Tsang, P., & Wilson, G. (1997). Mental workload. In G. Salvendy (Ed.), *Handbook of human factors and ergonomics*. New York: Wiley.

Vidulich, M. A., & Wickens, C. D. (1986). Causes of dissociation between subjective workload measures and performance. *Applied Ergonomics, 17,* 291–296.

Wickens, C. D. (in press). The tradeoff in design for routine and unexpected performance: Implciations of situation awareness. In D. J. Garland & M. R. Endsley (Eds.), *Situation awareness analysis and measurement*. Mahwah, NJ: Lawrence Erlbaum Associates.

Wickens, C. D., & Hollands, J. (2000). *Engineering psychology, and human performance*. Saddle River, NJ: Prentice–Hall.

Yeh, Y. Y., & Wickens, C. D. (1988). Dissociation of performance and subjective measures of workload. *Human Factors, 30,* 111–120.

PART III

Fatigue

THEORY

3.1 Active and Passive Fatigue States

Paula A. Desmond
Texas Tech University

Peter A. Hancock
University of Minnesota

Transportation systems are expected to work around the clock. From transcontinental flight and interstate trucking to container ships and railroads, contemporary society demands nonstop operations. However, there is one component of all transport systems that is not made for continuous activity —the human operator. The conflict of inherent human limitations with our contemporary "24-hour" society leads to a pervasive problem—fatigue. Fatigue represents a most insidious condition because, being personal, it is often hard to identify unequivocally and therefore difficult to measure and thus regulate. This chapter outlines a new model of fatigue that makes a critical distinction between two forms of fatigue state: "passive" and "active" fatigue. In the experimental record on fatigue and common task performance, most studies have examined perceptual-motor capability involved for prolonged periods. *Active fatigue* is then derived from continuous and prolonged, task-related perceptual-motor adjustment. In contrast is a second form of fatigue, *passive fatigue,* which requires system monitoring with either rare or even no overt perceptual-motor response requirements. Closely allied with vigilance, this form of fatigue develops over a number of hours of doing what appears to be nothing at all.

In the present work, we consider the commonalties and differences between these fatigue states. In particular, our concern is a very practical one. Research on fatigue has often been applied to vehicle control. Loss of control, associated with fatigue, is an immediate precursor to accidents in all realms of transportation. In ground transportation in particular, active fatigue is a frequent occurrence. For motorcycles, cars, vans, trucks, and the like, the rider or driver must make adjustments on a momentary basis. As a result, active fatigue is the state experienced by and is familiar to drivers. However, in the light of current "intelligent" transportation developments such as those proposed by intelligent transportation system (ITS) developers, it appears possible and even probable that many aspects of vehicle operation become subject to either full or semiautomated control. Our contention is that under such forms of control, the fatigue experienced by drivers changes radically from active to passive fatigue. In reviewing evidence associated with these two fatigue states, we endeavor to show how such a transition affects transportation operations, errors, and crashes associated with these respective forms of performance stress.

FATIGUE AND ITS IMPORTANCE IN TRANSPORTATION

We have all experienced fatigue, "pulled" all-nighters, forced ourselves to the edge of our limits, or tried to drive that one last hour. Phenomenologically, we can all identify with that "fatigued" feeling. Drooping eyelids, nodding head, startle "wakeups," and microsleeps provide the symptoms of a state of extreme lassitude, the disinclination to continue, and an effortful conscious strategy to fight against the forces conspiring to defeat ongoing performance. Yet despite this shared experience, we still have no completely valid and reliable measure of fatigue level. This is most unfortunate because fatigue plays such an important role in the majority of round-the-clock human–machine operations. In particular, fatigue is a critical problem in transportation. Data from the National Highway Traffic Safety Administration (NHTSA) suggest that of the 6.3 million crashes reported by police from 1989 to 1993, 1.6% involved drowsiness (Knipling, Wang, & Kanianthra, 1995). Moreover, those crashes that involved drowsiness resulted in 1,544 fatalities. The effects of fatigue on operator performance in other modes of transportation such as aviation, air traffic control, marine, and rail operations is also well documented. The following report is from a captain of a commercial jet to the National Aeronautics and Space Administration's (NASA) Aviation Safety Reporting System (ASRS) and illustrates the problem of fatigue in flight operations:

> Landed at (a large midwestern) airport without landing clearance, discovered when switching radios to ground control. Had very early final contact with approach control (20 mi. out at 7000 ft.). [Causal factors included] my involvement in monitoring an inexperienced first officer, crew fatigue, cockpit duties and procedures at requested changeover point. However, the primary factor was crew fatigue caused by late night departure (0230 local time), inability to rest prior to departure (both pilots attempted afternoon naps without success), long duty period (11.5 hrs duty time with en-route stop at

JFK), also, my primary concentration on flying by first officer with low, heavy aircraft experience. Also very clear weather made traffic conditions readily apparent, reducing my awareness of ATC needs. . . . Both pilots were having trouble throughout the let down and approach phases of this last leg of the trip remembering altitudes cleared to, frequencies, and even runway cleared to." (cited in Graeber, 1988)

Lyman and Orlady (1980) examined a database complied by ASRS for fatigue-related errors in air transport crew operations. Their findings showed that from 1976–1980, fatigue was implicated in 77 (3.8%) out of 2,006 reported incidents. However, Lyman and Orlady emphasized that this finding probably underestimates the frequency of errors in which fatigue is involved. When all reports that involved factors that could be directly or indirectly linked to fatigue were included in the analysis, the number of fatigue-related incidents increased to 426 (21.2%). The frequency of fatigue-related errors was higher between midnight and 6 AM. In addition, specific phases of the flight, such as the decent, approach, and landing, were associated with a greater frequency of fatigue-related performance impairments.

The operational characteristics of shipping and air traffic control environments are also vulnerable to fatigue. The typical working pattern for watch standers employed on merchant vessels involves 4 hours of active duty, 8 hours of rest, and 4 hours of duty again. The temporal pattern of marine accidents is similar to that of accidents in other modes of transportation: Most accidents occur late at night and during the early morning. In a recent study, McCallum, Raby, and Rothblum (1996) analyzed 279 marine casualties. The authors derived a "Fatigue Index" score for casualties that was based on three factors: the number of fatigue-related symptoms recorded by the mariner; the number of working hours completed during a 24-hour period before the casualty; and the number of hours slept during a 24-hour period before the casualty. When the authors applied their index of fatigue to the accident data, the findings showed that fatigue was implicated in 16% of critical vessel casualties and 33% of personnel injury casualties.

Luna (1997) highlighted the problem of fatigue in air traffic control and noted: "Safety concerns have been raised because the air traffic controllers (ATCs) often carry an acute debt onto the night-shift where they have little active work to do as they sit in the dark at the nadir of their circadian rhythms" (p. 69).

Recent research has shown that ATCs experience fatigue and even fall asleep during night-shift schedules. Billings and Reynard (1984) examined 22,226 ASRS reports during 1976–1983 and found that ATC errors were implicated in approximately 36% of these reports. Schroeder (1982) examined ATC operational errors from 1969 to 1980 recorded in the System Effectiveness Information System. The majority of errors occurred when workload was moderate or light. Redding (1992) reported that most ATC errors were the result of an inability to maintain situational awareness, lack of attention, incorrect use or misinterpretation of data from the radar, and poor vigilance. Moreover, most errors occurred when the ATCs were operating in conditions of low rather than high workload. Given that the night shift is characterized by low workload conditions and because periods of low workload are associated with the highest percentage of errors, it is highly probable that fatigue is implicated in ATC performance degradation. Luna, French, and Mitcha (1997) found that ATCs operating on night shifts reported greater subjectively measured symptoms of sleep, fatigue, and confusion. In addition, the night-shift ATCs reported, unsurprisingly, lower levels of vigor and general activity.

Fatigue also presents a problem for the rail industry. Hildebrandt, Rohmert, and Rutenfrantz (1975) reported that the monotonous task conditions facing the train driver can result in a reduction in alertness. Smiley's (1990) appraisal of the Hinton train disaster revealed that inadequate vigilance was an important factor in the accident. Edkins and Pollock's (1996) analysis of 112 train accidents and near accidents over a 3-year period in Australia showed that failure of sustained attention was the most important precursory factor in such accidents. Although we have here emphasized fatigue effects on transportation systems, there is no doubt that operator fatigue has been a contributory factor in many of the world's most prominent disasters. Fatigue has been associated with the Chernobyl and Three-Mile Island nuclear reactor disasters (Ehret, 1981) and has been thought to play a role in catastrophes such as Bhopal and Herald of Free Enterprise. Given the pervasive and deleterious nature of fatigue in operations including the various modes of transportation, a case is clearly established for a vigorous research effort to explore the dynamics of this complex behavioral state. However, a critical first issue for researchers investigating fatigue is how to define it, and it is to this problem that we now turn.

DEFINING FATIGUE

Fatigue has traditionally eluded definition. Attempts to define fatigue are associated with an unresolvable conundrum posed by Muscio (1921). According to Muscio, to define any phenomenon (e.g., fatigue), we need to be able to measure it, and to measure it we need a valid and reliable assessment tool. However, we cannot construct a tool without knowing what it is precisely that we wish to measure. On these grounds, most of science is actually rendered inoperable because this implies, tautologically, that we have to define what it is we propose to define. Clearly, this is not possible. Therefore, like all energetic concepts, such as stress, attention, and mental workload, as we each possess a personal knowledge of their occurrence and consider them "real" entities (Polanyi, 1958), we bootstrap our definitions to our experience of reality. As personal experiences, each of these constructs has proved scientifically elusive and none more so than fatigue.

We suggest that this is especially true for fatigue because it is a multidimensional state, only translated to a unitary perception by the unity of consciousness itself. Thus, like James' (1890) classic definition of attention, it appears to be a singular phenomenon, whereas in actuality the unified perception may be arrived at by a combination of many factors. We take fatigue to be a transition state between alertness and somnolence. Fatigue is derived from a spectrum of factors, principal among which are information rate (the temporal frequency of information assimilation) and information structure (the spatial variation of information presentation). These are, respectively, characteristics of the environmental stimulation but also are inextricably linked to endogenous characteristics of the performer such as rhythmic fluctuation in information assimilation capability associated with circadian state (where attempts to separate the perceiver from what is perceived represent a fundamental philosophical flaw; see Hancock, Flach, Caird, & Vicente, 1995). Consequently, fatigue is often associated with long hours of repetitive activity combined with a depressed state of operator response readiness. That the former bears a causal relation to the latter has not therefore escaped our attention.

There are metalevel questions also about the recognition of fatigue states. To be fatigued, we have (at one level) to recognize and associate our particular affective state with the label *fatigue*. This itself requires a degree of self-perception and effort. Consequently, what we most frequently observe in ourselves is the *onset* of a fatigued state. At this juncture, as can be easily inferred, we are not responding or perceiving at our most efficient level. Therefore, the distinction between fatigue induced primarily by hours of work or thermal conditions or the presence of alcohol, can become easily blurred in a state that, by this definition, is at or toward the edge of our conscious capability. Using the model that we have articulated elsewhere (Hancock & Verwey, 1997; Hancock & Warm, 1989), we attempt to show how this multidimensional approach to fatigue can be welded into a single, unified account.

THEORETICAL ISSUES

Recent conceptualizations of stress are useful in developing a model of fatigue. Hancock and Warm (1989) proposed an adaptation model of stress and performance, and following this work, Desmond and Matthews (1996) confirmed that fatigue represents one form of stress. We propose that the adaptation model can be used to account for a wide variety of performance effects under different forms of stress, including how fatigue affects efficiency in multiple-component tasks such as driving. The adaptation model proposes that stress can be identified at three different levels: the input level, the adaptation level, and the output level, respectively. As with stress, we can consider fatigue as the product of some aspect of the input to the individual from the environment, such as heat, noise, and hours of work. The adaptive component of fatigue is highly complex and is evidenced by the attempts individuals make to adjust their behavior in response to the input factors to fatigue. Finally, fatigue can be examined at an output level. Here, performance decrements such as impairment in vehicle control and trajectory provide indicators of fatigue (see Desmond & Matthews, 1997).

The adaptation model of stress suggests that the stressful effects of fatigue can lead to specific changes in the performance strategies adopted by individuals. This model accounts for effects at both low and high levels of stress and is closely linked to contemporary theories of attention (e.g., Wickens, 1987). An input stress may vary between conditions of underload and overload. A zone of comfort is located in the center of the continuum in which minimal adaptation required. At the behavioral level, stress may deplete psychological adaptability, which Hancock and Warm (1989) equated with attentional resources. In this model, attentional resources are assumed to vary both dynamically and adaptively, as advocated by Kahneman (1973).

Desmond and Matthews (1997) have conducted a program of research that has tested hypotheses derived from Hancock and Warm's (1989) model. In a series of simulated studies of driving performance, Desmond and Matthews examined how fatigue interacts with task demands. Hancock and Warm's model predicted that fatigue may impair performance when task demands are low. In these studies, drivers performed both a fatiguing drive, in which they were required to perform a demanding secondary character-detection task, and a control drive with no secondary task. The fatigue manipulation produced an increase in subjective tiredness and physical and perceptual fatigue symptoms. Drivers' lateral control of the vehicle was also assessed on straight and curved road sections during early and late phases of single-task performance (driving only) and during performance of the fatigue-induction procedure (driving and performance of the secondary task). The findings from these studies showed that fatigued drivers' control of vehicle heading deteriorated progressively on straight sections of the road but not on curved sections during and following fatigue induction. These findings suggested that, paradoxically, the fatigued driver may be at particular risk when task demands are obviously low (i.e., straight road) rather than when task demands are higher (i.e., curved road). The findings fit the adaptation model of fatigue that affirms that fatigue may impair performance when task demands are low.

Results from a recent study by Desmond and Hoyes (1996) of workload variation in a simulated air traffic control environment also supported the adaptive model of stress and performance. In this study, task demands were manipulated by varying the number of aircraft to be controlled. Three levels of task demand were examined: low-, medium-, and high-workload conditions. The main performance measure in the study was the mean number of aircraft successfully landed. The findings indicated that subjects in both high- and medium-workload conditions successfully landed a greater number of aircraft than those subjects in the low-workload condition. This suggests that subjects were failing to mobilize their effort effectively when task demands were low (low-workload condition), but when task demands were increased (medium- and high-workload conditions) subjects were able to maintain a constant level of performance.

TOWARD A THEORY OF FATIGUE

The studies reviewed previously illustrate the adaptive nature of fatigue in the performance of complex tasks. We now show how this adaptive component is at the very heart of our theory of fatigue. We propose that attention oscillates between sampling of the environment and the evaluation of self. The proportion of time that the individual spends sampling the environment versus self-evaluation is an adaptive response to the demands at hand. Consequently, in conditions in which there is rich and varied environmental stimulation together with no perceived threat to the self or disturbance to comfort, the large proportion of sampling time is directed to external sources. However, in minimal stimulation conditions with concern for self, the proportion of external sampling decreases.

The overall efficiency of sampling is contingent on level of attention available. Thus, variation in sampling frequency may have little effect if attention is high. However, when attention is depleted, the question of sampling strategy becomes critical. Fatigue occurs in a state of reduced attentional capability. We suggest that reduction might result from two apparently distinct conditions. In the first, attention is depleted by the constant, unavoidable demand placed on it. So, for example, Fancher (personal communication, 1997) has estimated that a driver has to make over 1,000 accelerator adjustments per freeway driving hour. When this demand is combined with the comparable number of steering adjustments, it can be seen that long-distance driving imposes an obligatory high perceptual-motor demand on the driver. As attention is reduced by this continued demand, the sampling frequency to external sources decreases. A consequence of this reduction is a decrease in the frequency of steering and speed control adjustments (essentially equivalent to an increase in periods of open-loop control). As fatigue exerts its effects, increasingly larger excursion corrections occur as the driver attempts to recover from ever greater perturbations resulting from reduced external sampling. Eventually, if the driver does not cease the activity, then a correction becomes of such a magnitude that the vehicle leaves the road, often after a startle "wakeup" following a prolonged period of open-loop control. This is the form that we call *active fatigue*, because the fatigue results from continuous activity.

The second form of fatigue, we call *passive fatigue*. In this condition, the operator is required to monitor a display, but takes rare if any response actions to control the system at hand. This form is more prevalent in pursuits such as process control or transportation systems such as transoceanic flight. However, with ITS technologies such as automated highway system (AHS) and adaptive intelligent cruise control (AICC), this passive fatigue may well be found in future road transport systems. Passive fatigue results from chronic understimulation. Because chronic understimulation is habitually associated with sleep onset, the symptoms are analogous. With an unchanging display, the sampling of the environment diminishes, and in an adaptive response, attention is reduced. This positive feedback continues so that without additional stimulation overall sampling is reduced. The effects on behavior are similar in many ways to those in active fatigue, because the reduction in sampling results in the same increase in the equivalence of open-loop control. In the case of monitoring a display, the probability of missing a signal increases in proportion to the reduction in sampling frequency to external sources.

It also requires attention to monitor self-state. For example, the perceived flow of time is contingent on the attention directed to it (Block, 1990). This sampling also decreases, but it is not unexpected that fatigue states are systematically associated with distortions of time, both time in passing and time in recall (see Hancock, 1998). Similarly, the assessment of self-state is impaired as overall level of attention is reduced. This leads us to conclude that fatigue in general appears to be a single state because it occurs at a threshold where the individual is rapidly losing the ability to accurately assess his or her own condition (see also Hancock & Verwey, 1997). In this way, we can see fatigue in its different forms results from a reduction in attention and its concomitant effects on sampling strategy and sampling frequency to self and surrounding environments.

Desmond, Hancock, and Monette (1998) set out to examine passive fatigue in a study of simulated driving performance. In this study, drivers' performance under automated and manual driving conditions were compared. In the passive fatigue condition, the trajectory of the vehicle was under automated control, and the driver's task was simply to monitor the automated driving system. Failure of the system occurred on three occasions during the 40-minute drive and caused the vehicle to veer toward the edge of the road. In the manual drive, drivers maintained full manual control over the vehicle throughout the duration of the drive. However, drivers experienced occasional "side winds" that caused the vehicle's trajectory to behave in the same way following automation failures. The findings from the study showed that the automated drive was just as effective as the manual drive in inducing subjective stress and fatigue reactions in drivers. Moreover, drivers in the automated condition took longer to recover from the perturbations in trajectory than drivers in the manual condition. Thus, it appears that the passive state of fatigue that emerges from monitoring a system for a prolonged period is just as stressful and tiring as continued performance of the same task (i.e. an active fatigue state) and may result in more deleterious after effects (see also Parasuraman, Mouloua, & Molloy, 1996).

CONCLUSIONS

Fatigue is not going to go away. We have, and are continuing to build, systems that are designed to operate without cessation. We have truly become a 24-four hour society and we need to operate around the clock. With process modifications such as just-in-time manufacturing and delivery, it has become crucial that efficient transportation is available on a permanent basis. Unfortunately, human beings, adaptive as they are, have not evolved for continuous operations. Inevitably, this has meant breaking the day into different work periods, where some are significantly more aversive than others (Folkard, 1996). Despite the best efforts at amelioration, accident frequency still follows the inherent rhythm that underlies human attentional state. Compounding this problem is the emergence of a human role as supervisor and monitor of a largely automated system. In the more sophisticated forms of transport, co-varying with the cost of the vehicle involved, this monitoring role has clearly been emphasized. In a cruel twist of design, humans are least tolerant to this form of activity, or more realistically inactivity, under the adverse effects of fatigue. Unless there is a fundamental re-evaluation of how we structure work (Hancock, 1997), this antagonism will continue. Here, we have shown that fatigue is a pervasive problem and a factor involved in many accidents. We have offered a theoretical model for understanding fatigue. However, much remains to be accomplished on theoretical, experimental, and practical levels if active and passive fatigue effects are to be reduced and eventually eliminated.

REFERENCES

Billings, C. E., & Reynard, W. D. (1984). Human factors in aircraft accidents: Results of a 7-year study. *Aviation, Space and Environmental Medicine, 55,* 960–965.

Block, R. A. (1990). Models of psychological time. In R. A. Block (Ed.), *Cognitive models of psychological time* (pp. 1–35). Hillsdale, NJ: Lawrence Erlbaum Associates.

Desmond, P. A., Hancock, P. A., & Monette, J. L. (1998). Fatigue and automation-induced impairments in simulated driving performance. *Transportation Research Record, 1628,* 8–14.

Desmond, P. A., & Hoyes, T. W. (1996). Workload variation, intrinsic risk and utility in a simulated air traffic control task: Evidence for compensatory effects. *Safety Science, 22,* 87–101.

Desmond, P. A., & Matthews, G. (1996). Task-induced fatigue effects on simulated driving performance. In A. G. Gale (Ed.), *Vision in vehicles* (Vol. 6). Amsterdam: Elsevier.

Desmond, P. A., & Matthews, G. (1997). Implications of task-induced fatigue effects for in-vehicle countermeasures to driver fatigue. *Accident Analysis and Prevention, 29,* 513–523.

Edkins, G. D., & Pollock, C. M. (1996). The influence of sustained attention on railway accidents. In *Proceedings of the second international conference on Fatigue and Transportation: Engineering, Enforcement and Education Solutions* (pp. 257–269). Applecross, Western Australia: Promaco Conventions.

Ehret, C. (1981). New approaches to chronohygiene for the shift worker in the nuclear power industry. In A. Reinberg, N. Vieux, & P. Andlauer (Eds.), *Night and shift work: biological and social aspects* (pp. 263–270). New York: Pergamon Press.

Folkard, S. (1996). Accident black spots. In *Proceedings of the second international conference on Fatigue and Transportation: Engineering, Enforcement and Education Solutions.* Applecross, Western Australia: Promaco Conventions.

Graeber, R. G. (1988). Aircrew fatigue and circadian rhythmicity. In E. L. Wiener & D. C. Nagel (Eds.), *Human factors in aviation* (pp. 305–344). Academic Press, Inc.

Hancock, P. A. (1997). On the future of work. *Ergonomics in Design, 5,* 25–29.

Hancock, P. A. (1998). *On time distortion under stressful conditions.* Manuscript submitted for publication.

Hancock, P. A., Flach, J., Caird, J. K., & Vicente, K. (Eds.). (1995). *Global perspectives on the ecology of human–machine systems.* Mahwah, NJ: Lawrence Erlbaum Associates.

Hancock, P. A., & Verwey, W. B. (1997). Fatigue, workload and adaptive driver systems. *Accident Analysis and Prevention, 29,* 495–506.

Hancock, P. A., & Warm, J. S. (1989). A dynamic model of stress and sustained attention. *Human Factors, 31,* 519–537.

Hilderbrandt, G., Rohmert, W., & Rutenfrantz, J. (1975). 12 and 24 hour rhythms in error frequency of locomotive drivers and the influence of tiredness. *International Journal of Chronobiology, 2,* 175–180.

James, W. (1890). *The principles of psychology.* Cambridge, MA: Harvard University Press.

Kahneman, D. (1973). *Attention and effort.* Englewood Cliffs, NJ: Prentice Hall.

Knipling, R. R., Wang, J. S. & Kanianthra, J. N. (1995). Current NHTSA drowsy driver R&D. In *Proceedings of the 15th international technical conference on Enhanced Safety of Vehicles (ESV), Melbourne, Australia,* (pp. 366–374). Washington, DC: National Highway Traffic Safety Administration.

Luna, T. D. (1997). Air traffic controller shiftwork: What are the implications for aviation safety? A review. *Aviation, Space, and Environmental Medicine, 68,* 69–79.

Luna, T. D., French, J., & Mitcha, J. L. (1997). A study of USAF air traffic controller shiftwork: Sleep, fatigue, activity, and mood analyses. *Aviation, Space and Environmental Medicine, 68,* 18–23.

Lyman, E. G., & Orlady, H. W. (1980). *Fatigue and associated performance decrements in air transport operations.* (NASA Contract NAS2-100060). Mountain View, CA: Battelle Memorial Laboratories, Aviation Safety Reporting System.

McCallum, M. C., Raby, M., & Rothblum, A. M. (1996). *Procedures for investigating and reporting human factors and fatigue contributions to marine casualties.* Seattle, WA: Battelle Research Center.

Muscio, B. (1921). Is a fatigue test possible? (A report to the Industrial Fatigue Research Board). *British Journal of Psychology, 12,* 31–46.

Parasuraman, R., Mouloua, M., & Molloy, R. (1996). Adaptive task allocation enhances monitoring of automated systems. *Human Factors, 38,* 665–679.

Polanyi, M. (1958). *Personal knowledge: Towards a post-critical philosophy.* Chicago: University of Chicago Press.

Redding, R. E. (1992). Analysis of operational errors and workload in air traffic control. In *Proceedings of Human Factors Society 36th annual meeting. Atlanta, GA* (pp. 1321–1325).

Schroeder, D. J. (1982). The loss of prescribed separation between aircraft: How does it occur? In *Proceedings of the Behavioral Objectives in Aviation Automated Systems Symposium, Society of Automotive Engineers* (pp. 257–269). (P-114/821432).

Smiley, A. M. (1990). The Hinton train disaster. *Accident Analysis and Prevention, 22,* 443–445.

Wickens, C. D. (1987). Attention. In P. A. Hancock (Ed.), *Human factors psychology.* Amsterdam: North-Holland.

<table>
<tr><td>**3.2**</td><td></td></tr>
</table>

3.2   Defining Fatigue as a Condition of the Organism and Distinguishing It From Habituation, Adaptation, and Boredom

R. F. Soames Job
James Dalziel
University of Sydney

Fatigue is well recognized as a problem in road safety. Many studies through-out the world have identified its importance, such as in Canada (Nelson, 1997), France (Hamelin, 1987), the United States (Smiley, 1998), England (Horne & Rayner, 1995; Maycock, 1997), and elsewhere. In Australia, esti-mates of the contribution of fatigue to road trauma vary, and its contribu-tion probably varies from state to state in inverse proportion to the density of the population. In the state of New South Wales, the Roads and Traffic Authority (RTA, 1995) estimated that fatigue contributes to 17% of fatal crashes (for Western Australia, see Cercarelli & Ryan, 1996). The fatigue problem is not limited to truck drivers or to country roads (Dalziel & Job, 1997a, 1997b, 1998; Fell & Black, 1997).

Many countermeasures have been designed to address fatigue problems that arise while driving. These may be divided into several categories (Ha-worth, 1996), such as countermeasures that target drivers (including limita-tions on the driving hours of commercial long-distance heavy-vehicle driv-ers, education about fatigue, "driver reviver" stations to encourage drivers to stop), countermeasures that target vehicles (such as some types of proposed fatigue-monitoring systems, e.g., Stein, Allen, & Parseghian, 1996), and countermeasures that target the road environment (such as "rumble strips" on the edges of highways).

THE CONCEPT OF FATIGUE

Problems with the concept of fatigue have plagued relevant research and analysis for some time (Muscio, 1921; see also Hancock & Verwey, 1997). A major problem with the research and analysis of the fatigue problem is the lack of a coherent definition of fatigue itself. This problem exhibits itself in several ways. First, in the analysis of the extent of the problem, the identification of fatigue is vague at best and invalid at worst. For example, the approach of the Roads and Traffic Authority of New South Wales (RTA) to fatigue-related crashes is typical of the problem (except for the fact that at least *some* criteria are specified and related to crash data, unlike other situations where even this is not available). Fatigue is considered to be implicated in a crash if the vehicle travelled to the incorrect side of the road, caused a head-on collision, and was not overtaking another vehicle, and no other relevant factor was identified; or the vehicle ran off the road to the outside of a curve or from a straight road, and excessive speed was not apparent and no other relevant factor was identified (RTA, 1995).

The problem here is that fatigue is not identified via any particular positive feature. Rather, it has become the catchbag of the unsolved cases, rather like past diagnoses of schizophrenia. "There is no apparent involvement of alcohol, vehicle failure, speeding, and so on, and thus call it fatigue" has the same logic as "There is no clear depression or eating disorder or obsession, and so on, and thus call it schizophrenia." Our understanding of schizophrenia did not advance significantly until research took account of the fact that there was no one type called schizophrenia nor one single causal factor (Bentall, 1993; Kavanagh, 1992). Past diagnoses of schizophrenia had thrown together many quite different mental problems for the convenience of having the single category. Similarly, the use of fatigue as the catchbag gives the appearance of providing an explanation of the causes of crashes, yet provides no real understanding.

Indeed, our understanding may be impeded by this process until we look carefully at the differences in the cases currently classified as fatigue. An apparent but incorrect explanation may avert us from the search for a better understanding and may block the path to such understanding through the inclusion of disparate causal factors contributing to the relevant crashes. In the case of crashes identified as owing to fatigue by criteria such as those mentioned earlier, additional possible causes may include being run off the road by other drivers who then failed to stop or being blinded by oncoming headlights or by other undetected causes of driver impairment (drugs, distracting life events, etc.).

Evaluations of the impact of fatigue countermeasures are similarly plagued by the lack of a clear definition or independent variable for fatigue. There is, as yet, no clear measure of fatigue (see Nelson, 1997). The assessment of the impact of a countermeasure requires that fatigue crashes be identified or that some measure of the level of fatigue of targeted road users be measured. Such studies are understandably rare, not only because of inadequate political commitment of funding to the road safety problem in general, but because the measurement issues remain to be resolved.

PROBLEMS WITH CURRENT DEFINITIONS OF FATIGUE

In this chapter, it is argued that currently accepted definitions of fatigue remain problematic. Here, a number of definitions are considered. First, the essential features for a definition of fatigue are asserted and defended. Second, typical definitions and understandings of fatigue are identified and evaluated against these essential features.

The *essential features* of a definition of fatigue are:

1. The definition should identify fatigue as a hypothetical construct, not a performance outcome per se. This feature is advanced on the grounds that fatigue is a state of the person, not simply a feature of his or her behavior.

2. The definition should not identify performance decrement as fatigue. Fatigue may cause performance decrements, but these decrements are, of themselves, not fatigue.

3. The definition should identify the cause of the state of the person. All performance-impairing factors are not the same and for practical and logical reasons should be distinguished on the basis of their causes. Alcohol consumption, illness, and hostility may impair driving, but they are not fatigue.

4. The definition should reflect, as far as possible in logical limits, the meaning ascribed to the term by the general population.

5. On the grounds of following conventional use (Point 4), fatigue should include states arising in the central nervous system (CNS) and the muscles, but not in the sensing neurons (such as in the retina and associated neurons). The consequences of "fatigue" in these locations is quite unlike the conventional usage of the term *fatigue*. These effects are considered in more detail later.

6. In consequence of these features of an appropriate definition, the definition should allow a distinction between fatigue and related phenomena.

There are several *common types* of definitions of fatigue. One of the most popular definitions is that offered by Brown: *"Psychological fatigue is defined as a subjectively experienced disinclination to continue the task"* (Brown, 1994, p. 298). This definition fails to identify the cause of the feeling. One could be disinclined to continue a task because of having succeeded in meeting some criterion performance and thus having no need to go further, or because the task is complete, or because the extrinsic reward for the task has disappeared or changed. None of these constitutes fatigue in the sense that the term is generally employed by people and as the term is related to road safety.

According to Cercarelli and Ryan (1996), *fatigue involves a diminished capacity for work and possibly decrements in attention, perception, decision making, and skill performance.* Again, this definition fails to identify the cause of the problem. Illness, alcohol, or drug consumption could fit this definition.

Fatigue is commonly thought of and measured simply in terms of exposure to the task on which the fatigue may have its effects (e.g., through the impact of hours of driving on driving ability: see Stein et al., 1996). This common usage reasonably describes one common cause of fatigue, but may be misleading. Fatigue may occur on one task after prolonged performance of another, related, task. In short, fatigue is not purely task specific.

Fatigue may refer to *feeling tired, sleepy, or exhausted* (NASA, 1996). This definition identifies the outcome, but not the existence of a hypothetical construct or the causes of the state. It may be best, on practical grounds as well as to allow greater precision, to require that the cause of the outcome form part of the basis on which the hypothetical construct is named.

Fatigue is *"an individual's multi-dimensional physiological-cognitive state associated with stimulus repetition which results in prolonged residence beyond a zone of performance comfort"* (Hancock & Verwey, 1997, p. 497). This definition is closer to meeting the criteria set out earlier. It identifies a hypothetical construct (a state of the organism), a cause (a stimulus repetition), and outcomes (impaired performance). The primary problem with this definition is that the cause is focused only on stimulus repetition. The demands of constant attention and decision making or the repetition of a behavior (rather than stimulus repetition) may also produce consequences that in the typical use of the term would, perhaps even more clearly than the effects of stimulus repetition, be considered fatigue.

A NEW DEFINITION OF FATIGUE

We offer the following definition, in the hope of avoiding the shortcomings previously identified. Fatigue refers to the state of an organism's muscles, viscera, or central nervous system, in which prior physical activity and/or mental processing, in the absence of sufficient rest, results in insufficient cellular capacity or systemwide energy to maintain the original level of activity and/or processing by using normal resources.

The phrase "prior physical activity and/or mental processing" identifies the cause of fatigue and allows for this cause to take several different forms, such as muscular exertion, prolonged attention, attention to a repetitive stimulus, prolonged performance of a complex or repetitive task, and combinations of these kinds of activities.

"In the absence of sufficient rest" has three useful connotations. First, it refers to rest during the ongoing performance of a task. This allows for the possibility that brief (or long) periods of rest during the performance of a task may counteract the effects of activity and/or mental processing. Second, "sufficient rest" is relevant to the state of the organism at the commencement of a task, as it is conceivable that the initial state may already be fatigued in certain circumstances (because of prior sleep loss or performance of some other task immediately before the current task). Third, the phrase acknowledges the many factors that may lead to quicker-than-usual fatigue onset following commencement of a task in a nonfatigued state (these factors may include sleep loss, sleep debt, circadian rhythm disruption, troughs in the circadian cycle, and many other contextual factors that may predispose the organism to faster fatigue onset).

"Insufficient cellular or systemwide energy" identifies the physical state of the organism identified as fatigue. Fatigue may occur at a specific, identifiable muscular or neural cellular level or more commonly as a state of a larger system of cells required for the given task (such as the neural systems required for attention). This systemwide fatigue should, in theory, be reducible to inter- and intracellular processes. These systems could be identified via basic research not related to fatigue, and they will have structural and functional components that it is possible to describe fully given sufficiently advanced techniques in anatomy and/or neuroscience. This distinction is important if fatigue is to be identified as a material phenomenon and in a way that is not ultimately circular.

The phrase "to maintain the original level of activity and/or processing using normal resources" identifies that fatigue involves a change in the organism regarding its activity or mental processing in relation to the systems required for a task, but notes that this is not synonymous with performance decrement. Our definition allows for an individual to continue to perform an activity at the original (or close to equivalent) level of performance by using other resources to offset poorer performance. These may be external (such as chemical stimulants) or internal (such as redirecting energy from another unrelated system[s] to maintain the activity of the now fatigued system[s]). However, the use of alternative sources of energy does not remove fatigue; rather it compensates for existing system fatigue by providing nonstandard resources to allow continued activity or processing.

There are three other issues that this definition helps to address: the subjective experience of fatigue, the progression of fatigue, and identifying the "cutoff" points required for practically distinguishing fatigued from normal systems. First, fatigue has not been identified with a subjective state. However, this does not forestall fatigue's *producing* a subjective experience. Indeed, we suspect that when fatigue of a system first arises, it is typical for a person to experience a subjective awareness of tiredness that results from detection of the drop in system energy. However, if the decision of the organism is not to rest, then fatigue can be expected to progress. Either this involves increasingly poorer levels of activity or mental processing or increasing use of nonstandard resources to maintain previous performance. Although subjective assessment of fatigue effects by an individual may be accurate at lower levels of objective fatigue, it appears poor at higher levels of fatigue (Yabuta, Iizuka, Yanagishima, Kataoka, & Seto, 1985, cited in Haworth, 1996). One byproduct of the progression of fatigue may be impairment of the subjective fatigue detection system, especially in cases where mental resources from other systems are being used to offset existing fatigue. Indeed, this process may lead to more serious fatigue outcomes, in that more and more nonstandard mental resources may be needed to attempt to continue a task because of fatigue of normal systems, and this may also result in extensive impairment of the systems responsible for detection and decision making about fatigue itself. Such an account may be able to explain involuntary sleep onset because of fatigue, perhaps the most devastating result of fatigue (especially on the road).

Second, fatigue progression may be affected by different factors from fatigue onset, and future fatigue research may benefit from a clearer distinction between these processes. For example, if fatigue arises and an organism becomes subjectively aware of this, then a variety of outcomes may result: The organism may choose to rest (either briefly or for a long period, including sleep); the organism may seek external stimulation to offset fatigue (such as chemical stimulants); or the organism may continue to perform the task regardless of the experience of fatigue. How one organism copes with the progression of fatigue (compared to another) is mediated by many factors, such as prior learning and individual differences in the physiology of relevant systems. It may also be mediated by other psychological factors, such as a propensity for risk taking. For example, following the initial subjective awareness of fatigue, a driver may then either choose to stop and rest or to ignore fatigue and continue driving regardless. This decision may be influenced by more general psychological traits not related to fatigue, such as a general propensity for risk taking while driving. Research on taxi drivers provides some support for this view (Dalziel & Job, 1998).

Third, the definition does not identify any specific "cutoff point" for fatigue. Although it acknowledges fatigue as the inability to maintain the original level of activity or processing by using normal resources, it does not attempt to delineate the boundaries of normal and fatigued systems. Ultimately, the question of boundaries is an empirical and practical question, the answer to which depends on both the nature of performance for specific systems (so that normal levels of variability in performance are not mistakenly identified with fatigue) and on practical concerns about what constitutes satisfactory system performance (as opposed to fatigued performance) for given contexts.

In summary, our definition appears to encapsulate the essential features of an adequate definition of fatigue presented earlier. It identifies fatigue as a state of the organism not as the outcome of reduced performance itself; it identifies the cause of this state; and it limits the term to the central nervous system (CNS) and muscles. One consequence of these features is that this definition should distinguish fatigue from related phenomena, such as changes in motivation, habituation, adaptation (and "fatigue") of the senses, and boredom. Confusion between fatigue and these related phenomena is considered next.

ISSUES CONFUSED WITH FATIGUE

First, fatigue may be confused with motivation, in that changes in either may result in reduced performance. However, fatigue is distinguished from motivation in that motivation is not simply related to the energy and cellular capacity to perform a task and in that motivation is not determined solely by previous activity and rest (see Beck, 1978). For example, compared with the general public, taxi drivers have a different motivation for driving, that is, to make money by finding and conveying passengers (Dalziel & Job, 1997b). A taxi driver who has earned a desired set amount for a shift may then lose motivation to continue driving. However, the behavioral outcomes of this loss of motivation could be misinterpreted as fatigue.

Fatigue may also be confused with habituation, in that habituation arises from repeated exposure to a stimulus such that the organism reacts less and less to the repeated occurrences of the stimulus. This may be seen erroneously as a form of fatigue, but in fact is a form of learning, which may be interpreted as involving learning that the repeated stimulus is safe and familiar and need not be attended to as much as before (see Mazur, 1990). For example, a teacher instructing a person learning to drive might note that initially the student was quite anxious before lane changing (showing repeated checking of lanes and cars, sitting forward in the seat, a facial expression of intense concentration, etc.), but that later in the session following many lane changes, the student's behavior before lane changing was limited to a single check of lanes and cars (without sitting forward in the seat or intense facial expressions). This change could be interpreted as fatigue from repeated activity, but equally could be the result of habituation. The present definition distinguishes fatigue from habituation through the mechanism of insufficient cellular capacity or systemwide energy (which is not the cause of habituation). At the more practical level, habituation to a stimulus persists over time during rest, whereas fatigue does not.

Adaptation (and "fatigue") of the senses may be attributed to fatigue of the sense organs or to more adaptive changes to the sensory environment. The former cases are of most relevance here. For example, when the eye is focused for a time on a single color, the neural pathways that detect and respond to that color are highly activated and in consequence are for a brief time less able to respond to stimulation by that color (Murch, 1973). This may be viewed as fatigue but is deliberately excluded by the present definition. The mechanism of this exclusion is the restriction of fatigue to effects on the CNS and the muscles. The rationale for this exclusion is that the effects of such "fatigue" are not effects that would be seen as fatigue in the general population's use of the term. The most obvious effects of this temporary sensory change are the changed perception of color and color aftereffects. That is, if we focus on a red light for some time, our perception of its redness is diminished, and on looking away from it we may see a complementary color (green) as an afterimage. Most of us do not consider this as an example of fatigue, and its effects on driving safety may be quite separate from the effects we normally view as fatigue. In addition, this sensory effect occurs more rapidly than usual fatigue effects, and any countermeasures for its impact are probably different. For these reasons it is not included in the definition of fatigue given here.

Boredom may arise from repetition, as may fatigue. However, boredom refers to the subjective feeling that the task or the stimulation is not interesting. Although boredom and fatigue may be related, they are conceptually distinct. Unlike fatigue, boredom is influenced by previous exposure to the task even though sufficient rest may have occurred in the interim. For example, a commercial driver, who has just enjoyed a holiday free of driving may initially return to commercial driving with a "new lease of life." However, within some time, the driver may (once again) become bored with the driving task, and this boredom may lead to changes in driving behaviors (such as taking less care than initially after return from the holiday). These changes, however, are not the result of fatigue.

CONCLUSION

The final requirement for the definition, to accurately represent the meaning generally given to it is that it accommodates the effects from factors that are well recognized as related to fatigue. These include continued activity on a task, the degree of effort required for the task (Desmond & Matthews, 1997), available energy, rest, circadian rhythm (Folkard, 1997), sleep loss, and sleep debt. The present definition does not exclude any of these. Continued activity is directly accommodated; greater effort presumably corresponds to greater energy expenditure and greater demand on relevant cells or systems, such that energy is more rapidly diminished; available energy is directly accommodated; rest (including its most effective form—sleep) is directly accommodated; and circadian rhythm, sleep loss, and sleep debt are all accommodated via the broader understanding of "sufficient rest." In addition, circadian rhythm may influence system fatigue by allowing less energy availability or reducing cellular capacity at certain times of the day relative to others. The exact mechanisms by which circadian rhythm exerts its influence remain to be determined.

The definition of fatigue presented here allows us to more clearly distinguish between the core phenomena of fatigue and related, although conceptual distinct, phenomena (such as habituation, adaptation, and boredom). The previous analysis also encourages consideration of the possible role of these other phenomena in safety. For example, adaptation to specific visual stimulation may impair vision immediately after the relevant stimulation. This includes detection of movement and speed suppressed by extended exposure to the visual cues of movement (so that it may appear that speed is lower than it actually is). Habituation to certain stimuli, such as other vehicles approaching on side streets (so that we react less and less to these occurrences because they tend to stop), may leave us ill prepared for the occasion when one such vehicle does not stop. Boredom may affect the care with which people drive. These factors deserve research consideration distinct from fatigue.

REFERENCES

Beck, R. C. (1978). *Motivation theories and principles.* Englewood Cliffs, NJ: Prentice Hall.
Bentall, R. P. (1993). Deconstructing the concept of "schizophrenia." *Journal of Mental Health, 2,* 223–238.
Brown, I. D. (1994). Driver fatigue. *Human Factors, 36,* 298–314.
Cercarelli, L. R., & Ryan, G. A. (1996). Long distance driving behaviour of Western Australian drivers. In L. R. Hartley (Ed.), *Proceedings of the second international conference on Fatigue and Transportation: Engineering, enforcement and education solutions* (pp. 35–45). Canning Bridge, Australia: Promaco.
Dalziel, J. R., & Job, R. F. S. (1997a). Motor vehicle accidents, fatigue, and optimism bias in taxi drivers. *Accident Analysis and Prevention, 29,* 489–494.
Dalziel, J. R., & Job, R. F. S. (1997b). *Taxi drivers and road safety.* (Report to the Federal Office of Road Safety). Canberra, Australia: Department of Transport and Regional Development.
Dalziel, J. R., & Job, R. F. S. (1998). Risk-taking and fatigue in taxi drivers. In L. R. Hartley (Ed.), *Managing fatigue in transportation* (pp. 287–299). Oxford: Pergamon.
Desmond, P. A., & Matthews, G. (1997). Implications of task-induced fatigue effects for in-vehicle countermeasures to lower fatigue. *Accident Analysis and Prevention, 29,* 515–523.

Fell, D. L., & Black, B. (1997). Driver fatigue in the city. *Accident Analysis and Prevention, 29*, 463–469.

Folkard, S. (1997). Black times: Temporal determinants of transport safety. *Accident Analysis and Prevention, 29*, 417–430.

Hamelin, P. (1987). Lorry drivers' time habits in work and their involvement in traffic accidents. *Ergonomics, 30*, 1323–1333.

Hancock, P. A., & Verwey, W. B. (1997). Fatigue, workload and adaptive driver systems. *Accident Analysis and Prevention, 29*, 495–506.

Haworth, N. (1996). Factors affecting the success of educational programs to reduce driver fatigue. In L. R. Hartley (Ed.), *Proceedings of the second international conference on Fatigue and Transportation: Engineering, enforcement and education solutions* (pp. 561–572). Canning Bridge, Australia: Promaco.

Horne, J. A., & Rayner, L. A. (1995). Sleep related vehicle accidents. *British Medical Journal, 310*, 565–567.

Kavanagh, D. J. (1992). *Schizophrenia: An overview and practical handbook*. London: Chapman & Hall.

Maycock, G. (1997). Sleepiness and driving: The experiences of U. K. car drivers. *Accident Analysis and Prevention, 29*, 453–462.

Mazur, J. E. (1990). *Learning and behavior* (2nd. ed.). Englewood Cliffs, NJ: Prentice-Hall.

Murch, G. M. (1973). *Visual and auditory perception*. New York: Bobbs-Merrill.

Muscio, B. (1921). Is a test of fatigue possible? *British Journal of Psychology, 12*, 31–46.

NASA. (1996). Fatigue resource directory. In L. R. Hartley (Ed.), *Proceedings of the second international conference on Fatigue and Transportation: Engineering, enforcement and education solutions* (pp. 67–135). Canning Bridge, Australia: Promaco.

Nelson, T. M. (1997). Fatigue, mindset and ecology in the hazard dominant environment. *Accident Analysis and Prevention, 29*, 409–415.

RTA. (1995). *Behavioural issues in road safety*. Sydney: Roads and Traffic Authority of New South Wales.

Smiley, A. (1998, February). *Fatigue management: Lessons from research*. Paper presented to the third international conference on Fatigue and Transportation, Fremantle, Australia.

Stein, A. C., Allen, R. W., & Parseghian, Z. (1996). Monitoring driver alertness using unobtrusive psychophysiological measures. In L. R. Hartley (Ed.), *Proceedings of the second international conference on Fatigue and Transportation: Engineering, enforcement and education solutions* (pp. 149–164). Canning Bridge, Australia: Promaco.

RESEARCH

3.3 Mental Effort Regulation and the Functional Impairment of the Driver

Stephen H. Fairclough
The HUSAT Research Institute, Loughborough University, U.K.

It is significant that the dangers of driver fatigue have not found an equivalent expression in the public arena as related concerns, such as drunk driving. The reasons fatigue has traditionally played a secondary role in road transport safety research and policy were outlined by Brown (1994), namely: (a) that evidence of a causal influence of fatigue in accidents is often circumstantial, and (b) a lack of political will to research the limitation of the hours of work for professional drivers, but underpinning both is the fact that (c) fatigue is a complex and ambiguous concept with no standard measurement index. The latter absence of definition lies at the crux of the fatigue problem, both in terms of research and policy.

In the case of impairment due to alcohol, the Blood Alcohol Content (BAC) measure comprises a broadly linear scale (based on physiological data) with a demonstrable correspondence to road accident involvement (Walls & Brownlie, 1985). Despite a long and varied research history (e.g., Bartley, 1965; Broadbent, 1979; Brown, 1994; Cameron, 1973; Hancock & Verwey, 1997; Muscio, 1921; Thorndike, 1900), no equivalent index of fatigue has been forthcoming. This failure may be a fundamental limitation associated with the concept. Fatigue, like any affective construct such as happiness or anger, is a state that is spontaneously perceived, in the absence of any explicit verbal or numeric framework.

Therefore, all endeavors to measure fatigue represent an attempt to superimpose a continuum upon a psychophysiological entity and, as such, are capable of providing only a better or worse level of symbolic description (see footnote on p. 505 of Hancock and Verwey, 1997). In the absence of an anchor scale (such as BAC in the case of alcohol, or body temperature in the

479

case of impairment due to thermal stressors), the multidimensional measurement of fatigue illustrates the indeterminate nature of the concept. Inevitably, this stems from ambiguity at the conceptual level, which creates related problems of data interpretation. This limitation is particularly striking during attempts to measure the impact of indeterminate fatigue on a complex, real-world activity such as driving.

A recent survey of drivers in the U.K. conducted by Maycock (1995) revealed that 29% of his survey sample (N = 4,600) had experienced feeling close to falling asleep at the wheel during the previous 12 months. Only 7% of the same sample reported fatigue as a contributory factor in accident involvement (adjusted by Maycock to implicate fatigue in between 9% and 10% of accidents for the sample). The general point is made that experiences of extreme fatigue do not have a one-to-one correspondence with accident involvement.

Cameron (1973) was among the earliest theorists to suggest that fatigue be viewed as a generalized stress response extended over a period of time. This stress-based conceptualization of fatigue concerned the moderation of forces acting on the organism (such as increased boredom and sleepiness), and the point at which the organism lost resiliency in the face of rising fatigue. This perspective was advanced by Hancock and Warm (1989), who claimed that the act of sustaining attention constituted a source of stress in its own right. Furthermore, they hypothesized that influences on sustained performance (including fatigue) hinged on three iterative foci: the input stress (i.e., task demand, sleep deprivation, noise), the adaptive or compensatory response to input stress, and the resultant influence on the goal-directed outputs of the individual. Within this conceptualization, the input stress of a 3- to 4-hour drive would be identical for all drivers undertaking such a journey. Differences between individuals are apparent when a range of more or less successful compensatory strategies are employed. A successful compensation strategy is defined by two related goals: to sustain an adequate level of performance and to minimize stress or discomfort. The success or failure of compensatory strategies with respect to both goals is crucial to the discussion that follows. It is argued that the proficiency of compensatory responses is the variable that mediates the relationship between operational fatigue and accidents.

This chapter is concerned with a description and operationalisation of compensatory strategies to cope with fatigue and impairment in more general terms. It is hypothesized that such compensatory mechanisms are based on the investment and conservation of mental effort (Hockey, 1986, 1993, 1997). Furthermore, the measurement of these compensatory mechanisms may constitute an appropriate basis on which to assess the influence of fatigue on behavior.

MENTAL EFFORT AS A COGNITIVE MECHANISM

In his review of mental effort, Mulder (1986) described the development of the concept since the early part of the century. Effort has evolved from a secondary component of general motivation to a central concept in the study of mental workload. In correspondence with this transition, Mulder (1986) is at pains to make the link between effort and attention explicit, hence the conception of effort as a mental activity.

The concept of mental effort arose by inference amidst early models of human cognition almost 30 years ago. Within the Baddeley and Hitch (1974) model of working memory, a finite and effortful central executive was proposed to bolster performance if the memory system was overloaded with information. When Broadbent (1971) conducted his review into the relationship between cognitive performance and the influence of various stressors, such as sleep deprivation, noise, he emphasized an interactive influence of stressors on performance. On the basis of his analysis, Broadbent postulated an upper and a lower mechanism for coping with the effects of stress. The function of the upper mechanism was identical in principle to the central executive proposed by Baddeley and Hitch (1974): to buttress and stabilise the performance level of the lower mechanism as stress or information load increased.

Within both frameworks, mental effort was implicit as a compensatory mechanism within the cognitive system as a whole. Kahneman (1970, 1973) was the first to define effort as an attentional control mechanism within an information-processing framework. His model (Kahneman, 1973) rendered effort synonymous with attention and attentional regulation. Kahneman's (1973) principle of effort regulation was based on a simple feedback loop; task demand was defined by a standard amount of effort and failure to invest effort to an appropriate degree resulted in a degradation of performance. In addition, his model encompassed an explicit economic metaphor that was elaborated several years later (Navon & Gopher, 1979).

The problem of these early theories was the postulation of superordinate controller (i.e., central executive, upper mechanism, allocation policy), which was practically indistinguishable from the processes of its subordinate mechanisms. Therefore, empirical evidence of the effort mechanism or superordinate controller was notoriously difficult to obtain. Also, the popularity of the inverted-U model between arousal and performance (at that time) may have overshadowed mental effort during this formative period. The latter was somewhat inevitable, given that both constructs were operationalized via psychophysiology and were very closely associated at the conceptual level within Kahneman's (1973) model.

The economic character of Kahneman's (1973) model was possibly the crucial feature that rescued mental effort from the conceptual graveyard. Data from a series of studies on visual search and memory search (Schneider & Shiffrin, 1977; Shiffrin & Schneider, 1977) provided evidence for the existence of two modes of processing with an explicit cost metaphor. According to these authors, subjects were capable of either automatic detection (when the task was highly predictable and deterministic) or controlled search (when the task was less predictable and more probabilistic). The latter search activity was characterized as effortful, involving serial processing and being cost-intensive in terms of the cognitive system (i.e., demanding an increased breadth and/or depth of information processing). By contrast, automatic detection ran via effortless, parallel processing and consequently incurred a lower cost on the cognitive system.

A drawback with Kahneman's (1973) model was a lack of specificity concerning the mechanisms by which mental effort enhanced cognition and therefore, improved performance. The notion of hierarchical control has infused all the theories of mental effort discussed so far and should now be made explicit. A prominent example of hierarchical cognitive control is the tripartite framework initially proposed by Rasmussen and Jensen (1974). Within this scheme, the operator may function on a skill-based, automatic level involving a minimal level of mental effort. If problems are encountered outside of the skill base, the operator is forced to search for solutions from a familiar, rule-based repertoire of problem-solving strategies. This process of searching and matching appropriate solutions may incur a higher proportion of controlled processing. If the problem defies solution at the rule-based level, control ascends to a knowledge-based level. This uppermost level is the exclusive domain of controlled processing and entails substantial investment of mental effort (Hockey, 1997).

The mechanism by which the control of performance passes up and down the cognitive hierarchy has been described by Reason (1987, 1990) in his Generic Error Modelling System (GEMS). According to the GEMS, the information processing system operates on the basis of a negative feedback loop as described within control theory (Carver & Scheier, 1982; Miller, Galanter, & Pribram, 1960; Moray, 1981). A control theory perspective contains the implicit assumption that activities are goal-driven and associated with a "plan." The latter corresponds to a hierarchical structure of goals and subgoals (Anderson, 1983; Schank & Abelson, 1977) supplying either explicit or implicit standards of performance (Carver & Scheier, 1981). Performance standards are stored in memory and retrieved when required, and they consequently function as a reference signal for performance. Standards may describe specific subgoals and goals (i.e., discrete accomplishments) or the expected rate of progress towards those goals (i.e., a temporal schedule). Deviation from the plan at the skill-based level automatically evokes rule-based control and a higher level of mental effort and controlled processing (Kluger & DeNisi, 1996). If a corrective action is not found within the rule-based repertoire of solutions (and the deviation from plan is assessed to be sufficiently crucial), a knowledge-based analysis is launched to effectively decompose and reformulate the task plan. This latter strategy demands the highest level of mental effort. If a solution is found at the knowledge-based level and the deviation corrected, cognitive control descends the task hierarchy with an associated decline of mental effort. The implicit assumption within the GEMS is that all learned behaviors are associated with behavioral standards. A continuous process of discrepancy reduction in accordance with those behavioral standards is the tacit principle governing the allocation of mental effort within the cognitive system.

There are two important points regarding mental effort regulation within the GEMS conceptualization. In the first instance, an adaptive hierarchy where mental effort is invested on the basis of the frequency and magnitude of deviations from a plan is effectively a "headless" control system. There is no supervisory controller or central executive system. Consequently, the amount of mental effort invested at any given time is an emergent property based on the interaction between task demand, psychological state and the skill level of the individual. Secondly, it should be apparent that investment of mental effort differs on a qualitative basis, as well as in quantitative terms at each level of the hierarchy. At the skill-based level, mental effort is responsible for standard checks (Allwood, 1984) on task progress. These checks correspond to a negative feedback loop as the standard check is the means of testing for discrepancy. Standard checks may be carried out on a regular or irregular schedule, at more or less frequent intervals. Feedback from checks is characterized as either goal-checking (i.e., has Subgoal A been accomplished?) or scheduling (i.e., how soon should Subgoal A be accomplished?). The frequency of standard checks provides the desired fidelity of performance feedback to the operator. It is hypothesized that an increased frequency of standard checks results in a rise of mental effort.

This analysis does not assume that the amount of mental effort invested is synonymous with the level of the cognitive control hierarchy. An operator may invest an equivalent amount of mental effort at the skill-based level, as he or she would during task analysis at the knowledge-based level. However, an ascent up the cognitive control hierarchy raises the minimum amount of mental effort required to function at each level. Therefore, the minimum amount of mental effort may be invested in a standard check, whereas a higher level is required to search and detect a solution from the rule-based repertoire. At the top of the tree, a knowledge-based task analysis requires the highest minimum investment of mental effort within the hierarchy.

Mulder (1986) hypothesized that mental effort investment may correspond to a dichotomy, namely: (a) in response to cognitive demands or task difficulty or (b) to sustain an adequate level of performance under non-optimal operating conditions, such as excess noise or sleep deprivation. The former is termed *task-effort*, whereas the latter is referred to as *compensatory effort*. Within the current conception, no such distinction is made. It is argued that both (a) and (b) serve to increase input stress (Hancock & Warm, 1989) by rendering adequate performance more "costly" in terms of mental effort. The mechanism of adaptation (i.e., a hierarchy of strategies from an increased frequency of standard checks and positive feedback monitoring, to rule-based corrective action and eventually to knowledge-based analytical decomposition) is assumed to be identical, regardless of the specific source of input stress.

MENTAL EFFORT AND GOAL REGULATION

The allocation of mental effort within a cognitive control hierarchy was described in the previous section. Elements of cognitive control concern error-monitoring and problem-solving in response to higher-level goals (i.e., finish Task A, complete Task B as quickly as possible). It is proposed that the regulation of goals occurs within a higher, hierarchical structure, also operating on the principles of control theory. This conceptualization is taken from the work of Powers (1973, 1980) as elaborated subsequently by Carver and Scheier (1981, 1982). It is important to distinguish between this goal-setting hierarchy that regulates the purposive aspects of behavior and the hierarchy of cognitive control described in the previous section. The former hierarchy sets the level of desirable output to be achieved by the latter control system.

At the topmost level, the goal of behavior may be described at a conceptual principle level. For example, for an operator subjected to a period of task activity, the principle goal may be expressed in terms of a desire to "perform well" or, more formally, to sustain performance at a high level. This superordinate goal provides a behavioral standard in practical terms for the subordinate program level. The program level sets the goal providing impetus for a script (Schank & Abelson, 1977) at the subordinate sequence level. In this case, the program level will set a standard approximating high performance (i.e., reach Goals A, B, and C within the next 60 minutes). The program initiates a sequence of specific actions best-suited to fulfil the superordinate program goal (i.e., check X and Y within Subtask A, repeat for Subtask B, etc.). Within the principle–program–sequence hierarchy,[1] the behavioral standard at each subordinate level are being set and reset by the

---

[1] The original seven-level hierarchy described by Powers (1973, 1980) and by Carver and Scheier (1981, 1982) have been abbreviated for the purpose of the current discussion.

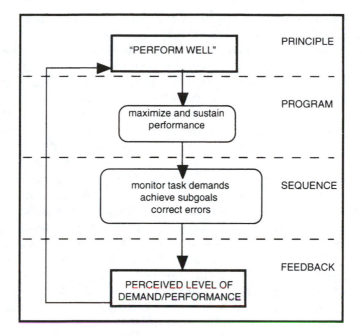

FIG. 3.3.1.   A goal regulation system based on the requirement to perform well.

standard at the superordinate level. Therefore, control is top-down and the
principal goal is the primary regulator for all behavioral standards down to
the sequence level. This goal regulation system is represented in Fig. 3.3.1.

The interaction between cognitive control and the goal-setting hierarchy
may also be subsumed within a control theory metaphor. If the prescribed
sequence of actions fails to achieve a program goal, a new sequence is selected or the existing sequence may be modified. This process corresponds
to troubleshooting at the rule-based level of the cognitive control hierarchy.
If no existing sequences in the rule-based repertoire are adequate, the program level invests effort into the formulation of a new sequence at the
knowledge-based level. If this strategy fails, the operator has no choice but
to lower the behavioral standard at the uppermost principal level (see Fig.
3.3.1), for example, to downshift from a principal goal that maximizes performance to one that simply sustains adequate performance (i.e., "perform
well" is altered to a value-free directive to "sustain performance"). These alterations were termed "aspiration shifts" by Schonpflug (1983). If performance continues to decline or if the goal of sustaining performance becomes
progressively difficult (as in the case of rising fatigue), the principal directive enters a vicious cycle of declining standards. The inevitable conclusion

of this chain of events is the "rising disinclination to continue performance" that characterizes operational fatigue (Brown, 1994).

The preceding description exemplifies performance breakdown with reference to a single, superordinate goal to "perform well." However, it is likely that an operator will bring more than one principal goal to the task environment. As stated previously, an operator facing a prolonged or arduous task may wish to maintain adequate performance (Principal Goal 1) while minimizing stress and personal discomfort (Principal Goal 2). If both principles occupy the same goal-setting hierarchy (Fig. 3.3.2), there may be a problem of complementarity. Obviously, if good performance can be achieved with a minimum of discomfort, there is no conflict. However, if discomfort is a prerequisite for performance (e.g., working in very hot conditions or for a prolonged period), the operator is forced to regulate goals on the basis of a utility judgment at the principal level (Hockey, 1993, 1997; Schonpflug, 1983). In Fig. 3.3.2, this utility judgment is assessed at the uppermost level (indicated by the two-way arrow) on basis of feedback from both task demand and self-rated discomfort. In the incompatible case, the perceived costs and benefits of achieving each principal are assessed, and a directive is formulated as an amalgamation of both principals. This directive may emphasize one principal over the other, which, in turn, is expressed at the program and sequence level. For example, if the task were safety-critical, a responsible operator should emphasize performance over discomfort. On the other hand,

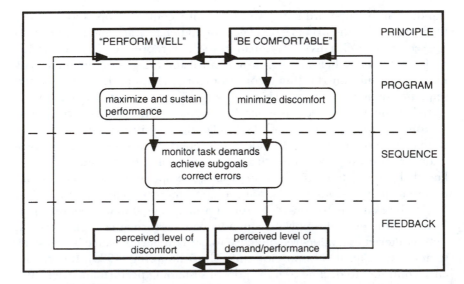

FIG. 3.3.2. The Principle–Program–Sequence hierarchy for co-existing principles of good performance and minimal discomfort.

a task perceived as unimportant may be omitted or performed relatively poorly by a tired operator approaching the end of a long working session.

The authors who subscribed to the hierarchical framework of behavioral regulation (Carver & Scheier, 1981, 1982; Powers, 1973, 1980) emphasized the top-down nature of control. However, the topmost level receives feedback based on output at the sequence level. The subjective acts of appraisal at the feedback level drive the utility process at the principal level. There is evidence that high levels of task demand may increase discomfort (Matthews, Jones, & Chamberlain, 1990) and vice versa, that is, high levels of discomfort may interfere with the ability to perform (Sarason, Sarason, Keefe, Hayes, & Shearin, 1986). Therefore, a degree of confounding is present at the feedback level (indicated by the double-arrow at the feedback level in Fig. 3.3.2).

In the case of prolonged performance, it is possible for the relative primacy of two principal goals to alternate over a period of time. The operator may begin the working session emphasizing the principal of good performance. As time-on-task and fatigue increase, the operator may be inclined to adopt a principle of minimal discomfort. These utility judgments may be performed either at a conscious or an unconscious level. This system of goal regulation provides a basis for the cost–benefit decisions involving mental effort investment described by Kahneman (1973), Schonpflug (1983), and Hockey (1993, 1997). This assessment of utility is the balance sheet by which we self-regulate our principal goals and therefore adapt to changing operational demands.

MENTAL EFFORT AS COGNITIVE-ENERGETICAL REGULATION

The adjustment of task goals at the principle level provides top-down regulatory control over mental effort investment. Within this scheme, mental effort approximates to the exchange system of cognitive system. Effort may be invested at low rate (i.e., controlled processing) invested at high cost per unit time or at a low rate (i.e., automatic processing) invested at low cost per unit time. However, the total effort reserve is always finite. This bounded feature of mental effort determines the necessity for utility and regulation.

The model of effort regulation described by Hockey (1993, 1997) emphasizes utility and hierarchical control. Within his model, an input stress provokes one of two broad responses from the operator, either (a) to invest mental effort in order to reduce a discrepancy or (b) to conserve effort in order to limit the rate of effort investment. This distinction represents a strategic decision to invest and not to invest on a strategic basis.

The modes of mental effort regulation (i.e., investment or conservation), are described in detail by Hockey (1993, 1997). These papers make the crucial point that many studies of performance have revealed only a modest decrement due to impairment, such as fatigue or noise. Hockey cited the investment of mental effort as the mechanism that sustains primary task performance, even when the operator is tired or when operational conditions are difficult (i.e., high task demands, extreme temperatures). However, he made the point that effort investment is associated with a number of costs for the operator. Besides a decline of primary task performance, mental effort regulation makes its presence felt by virtue of latent decrements. These are described as follows (Hockey, 1993, 1997):

1. Subsidiary Task Failure: A selective impairment of peripheral task components, both in the sense of physical localization (attentional narrowing) or low priority within the task hierarchy. This decrement may result from either the investment or conservation of mental effort.

2. Strategic Adjustment: A shift to task strategies where mental effort is minimized. These may include: reducing the number of standard checks, persevering with an inappropriate yet familiar problem-solving solution, reduced reliance on working memory to store information. This within-task shift to routine and familiar performance strategies is usually characteristic of effort conservation.

3. Compensatory Costs: When mental effort is invested, a number of changes may occur in psychophysiology (such as elevated sympathetic nervous activity and suppression of heart rate variability) and subjective appraisal (such as increased mental workload, stress and negative affect). These changes are side effects associated with the investment of mental effort.

4. Fatigue After-effects: Once mental effort has been invested, the regulatory system will attempt to conserve mental effort as soon as task demands or task priority are reduced. In effect, this decrement corresponds to a Strategic Adjustment (b) occurring in the wake of a period of mental effort investment.

Both investment and conservation of mental effort function as coping strategies, which may operate with differing degrees of success. The extent to which either is successful is governed by the influence of latent decrements on task performance. For a highly predictable, simple task, both attentional narrowing (a) and resorting to routinized responses (b) may improve efficiency without degrading performance effectiveness. However, the same latent decrements could easily degrade a complex and highly unpredictable task.

The goal of investment is to maintain an adequate or high level of performance. In terms of Fig. 3.3.3, investment is closely associated with a principal directive to "perform well," whereas the conservation of effort is a tac-

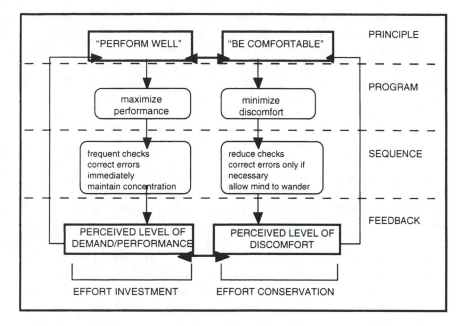

FIG. 3.3.3.   The goal setting hierarchy with two modes of effort regulation: investment and conservation.

tic to reduce the stress associated with continued periods of mental effort investment. The conservation strategy functions to fulfill a goal of minimizing discomfort (Fig. 3.3.3). There is a basic antagonism at work between both principal goals, which finds expression in mutually incompatible directives at the sequence level. As Hockey (1993, 1997) pointed out, periods of sustained investment are associated with a number of compensatory costs, including negative affect (i.e., aversiveness), stress, and discomfort. The conflict between the need to perform and the desire to minimize discomfort is represented in Fig. 3.3.3. This disharmony is the foundation underlying Hancock and Warm's (1989) assertion that the act of sustained attention functions as a stressor. Under conditions of high discomfort, increased pressure to relax performance standards creates cognitive interference, that is, an increased distractibility to non-task-relevant thoughts (Sarason et al., 1986), and consequently even higher levels of mental effort investment are required. The logical result of cognitive interference is that the regulatory system must perform "at a loss," investing a higher level of mental effort than is necessitated on the basis on task demand alone.

A successful investment strategy would sustain or improve the effectiveness or the efficiency of performance, while minimizing the number of latent decrements. For example, under optimal conditions, an investment of mental effort would improve performance without incurring additional stress or a significant fatigue after-effect. In similar terms, the perfect conservation strategy would actively reduce stress and discomfort without any detriment to performance. These successful strategies represent peaks of regulatory control over the cognitive-energetical system. In this sense, optimal regulation shares a number of features with the comfort zone postulated within the Hancock and Warm (1989) model of sustained performance, such as an area of maximal stability for both comfort and performance.

An ineffective investment strategy would fail to impact on performance while maximizing stress. This unfortunate scenario would arise when a fatigued operator who performs poorly seeks to mobilize mental effort to improve performance. If the demands of sustaining task performance outstrip his or her ability to invest mental effort, no improvement is forthcoming and discomfort is increased. Ineffective conservation corresponds to the same outcomes (i.e., poor performance, high stress), but as the result of a different strategy. In this case, the operator has effectively surrendered the possibility of adequate performance in an attempt to minimize stress. However, if the task is safety-critical or if the self-esteem of the individual is jeopardized, stress may not be relieved and could be acerbated.

Within the current conceptualization, it should be emphasized once again that the regulation of mental effort at the principle level is performed on the basis of subjective appraisals of: (a) the task demand and current performance level and (b) a self-appraisal of stress or comfort. This mechanism of discrepancy reduction places a huge reliance on the accuracy of subjective assessment. The ability of the individual to accurately monitor the task and the self places an upper limit on the fidelity of mental effort regulation. In addition, there is evidence that self-assessment may become increasingly inaccurate under conditions of high task demand and fatigue. For example, a number of operating conditions that produce dissociation between subjective estimates of workload and performance have been described by Yeh and Wickens (1988). This dissociation does not hold across all task characteristics (e.g., Warm, Dember, & Hancock, 1996) and may be a function of both the time-scale of measurement and memory loads associated with any given task (Hancock, 1996). However, those tasks involving high time demands (or which load working memory) may reduce effort regulation to an implicit strategy. Brown (1994) and McDonald (1987) made similar points concerning the influence of fatigue on self-monitoring. They each claimed that the detrimental effect of fatigue may extend beyond the impairment of performance and degrade the quality of self-monitoring per se. Therefore, an individual is less able to accurately compensate for rising fatigue or to respond to changes in task demand.

The appropriate regulation of mental effort hinges on the reliability of feedback provided by self-monitoring and appraisal. If an individual is unable to self-monitor with a degree of accuracy, the self-regulating system of mental effort is displaced and may break down completely.

Summary

The points made so far may be summarized as follows:

1. There is no quantifiable anchor scale for fatigue, and this has led to problems of conceptual definition.

2. Fatigue functions as an input stress, which provokes an adaptive response from the individual wishing to maintain performance and task goals.

3. The success of an adaptive response in terms of sustaining performance while minimizing stress accounts for the variable impact of fatigue on performance.

4. An increase of mental effort is associated with an increased fidelity of self-monitoring, effective memory retrieval, and analytical processing of a problem space.

5. Mental effort is a finite resource and must be regulated with respect to principal goals. There is the potential for incompatible principal goals to be activated simultaneously within the goal-setting hierarchy. In this case, mental effort is regulated on a basis of a utility assessment between performance and discomfort.

6. Mental effort may be invested or conserved on the basis of a utility assessment at the principal level. The investment of mental effort is associated with a number of latent decrements, including compensatory costs and fatigue after-effects. A conservation of mental effort causes strategic adjustments within the task to minimize effort investment and compensatory costs. The failure of subsidiary tasks is a decrement common to both investment and conservation.

7. Strategies of mental effort regulation may be effective or ineffective, dependent on how they interact with task characteristics and the extent to which latent decrements impinge on performance and task goals.

8. The regulation of attention during driving behavior may be conceptualized with respect to mental effort. The investment of mental effort takes place in response to increased driving task demands. Effort conservation occurs when demands are low or when the driver is extremely tired.

9. The regulation of mental effort is governed on the basis of subjective feedback regarding performance and discomfort. These subjective appraisals may be rendered inaccurate under certain conditions,

such as when the individual is tired, when task feedback is limited, when the temporal demand of the task is high, or when working memory is loaded.

FUNCTIONAL IMPAIRMENT OF THE DRIVER: AN EXPERIMENTAL STUDY

This section aims to place the system of mental effort regulation described previously within an applied context of driver behavior. *Functional impairment* is a generic term used to describe those occasions when an operator is unable to fulfill the functional requirement of the task. In this particular study, functional impairment was induced during a simulated drive via sleep deprivation and the introduction of an alcoholic beverage. It is argued that the mechanisms of effort regulation are activated to buttress primary task performance under conditions of functional impairment.

The Study

Sixty-four male subjects performed a 120-min. journey in a fixed-base driving simulator. The subjects were matched for age, driving experience, and mean weekly intake of alcohol, and designated into one of four treatment groups. The *control group* subjects received a full night of sleep before the trial and received a placebo drink. The *partial sleep deprivation (PartSD) group* subjects were instructed to sleep for 4 hours between midnight and 4:00 AM on the night before the trial. This group received a placebo drink. The *full sleep deprivation (FullSD) group* subjects were instructed to remain awake throughout the night before the trial and received a placebo drink. The *alcohol-impaired (Alcohol) group* subjects had a full night of sleep on the night before the trial and received an alcoholic drink (a mixture of vodka and lemonade) prior to the experimental session. The amount of alcohol administered was calculated, based on the subject's body weight, to induce a peak Blood Alcohol Content (BAC) level in the range of 0.08 to 0.1% (mg/g).

The simulated 2-hour journey took place in either the early afternoon or mid-afternoon. The journey was divided into three 40-min. sections of continuous driving, separated by a 5-min. break, during which subjects were breathalyzed and completed subjective rating scales. Each section of the journey contained six driving scenarios, which differed with respect to the task demand imposed on the subject. At one extreme, the two-lane road was empty (i.e., an Open Road scenario). The demanding portions of journey involved following a lead vehicle that varied its speed sinusoidally, varying between 50 mph to 70 mph with a cycle time of 30 sec. (i.e., a Sinusoidal Following scenario). Data was captured from three sources: (a) primary task performance (i.e., changes in vehicle control induced by alcohol/sleep deprivation), (b) subjective assessment (i.e., changes in perceptions of subjective workload, sleepiness, sobriety, and related scales), and (c) electrocardiogram (ECG) measures recorded from chest electrodes. For a full report on experimental methodology, the reader is referred to Fairclough and Graham (1999).

Results and Discussion

The subjective assessment of sleepiness provided by the Karolinska scale (Kecklund & Akerstedt, 1993) produced the highest scores for the two sleep-deprived groups. There was no difference in the magnitude of sleepiness between the PartSD and FullSD groups. The alcoholic beverage produced a mean BAC level of 0.081% approximately 15 min. after ingestion, which descended throughout the 120-min. journey to a mean level of 0.055%. According to a subjective rating of alcohol impairment, the mean response of the Alcohol group was "some minor impairment of judgment and reactions" (i.e., the mid-interval of an 8-point scale, compared to an average response of 1.5 from the other three groups). These data are shown in Fig. 3.3.4.

Having established that both impairment manipulations produced expected effects on subjective self-appraisal, the influences of sleep deprivation and alcohol on performance were considered. The analysis of the primary task data revealed a number of latent decrements associated with mental effort regulation. A number of decrements are described in the following sections.

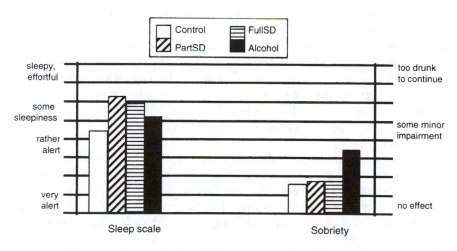

FIG. 3.3.4.   Mean subjective ratings of sleepiness and sobriety for all four treatment groups.

Subsidiary Task Failure. The types of driver errors induced by functional impairment were designated with respect to criticality. At the uppermost level were off-road accidents, that is, when the vehicle left the road with all four wheels at either the right or the left road boundary. Secondary errors, such as lane crossings, occurred when two of the vehicle wheels crossed either the left-side road boundary or the right-side lane boundary.[2] The calculation of minimum Time-To-Line-Crossing (TLC; Godthelp, Milgram, & Blaauw, 1984) was intended to capture those instances of near-lane crossings. For example, the frequency of events where the minimum TLC was 2 sec. (minTLC < 2 sec) refers to those occasions when the subject corrected a potential lane crossing and the lateral velocity of vehicle was 2 sec. from an actual lane crossing.

The analysis of lane-keeping behavior is illustrated in Fig. 3.3.5. The alcohol and full sleep deprivation manipulations produced the highest number of accidents and lane crossings. The PartSD group committed no accidents and were indistinguishable from the Control group in terms of the number of lane crossings observed. However, the analysis of minTLC revealed a higher number of near-crossings for the PartSD group. This pattern revealed a subsidiary degradation of lane-keeping with respect to near-crossings for the PartSD subjects, which did not produce either a higher number of lane crossings or accidents. In this case, it is hypothesized that mental effort was conserved and lane-keeping performance was allowed to degrade to a pre-critical point before any potential error was corrected. This pattern is similar to the widening of the indifference range described by Bartlett (1943). It is proposed that an increase of near-errors, combined with the absence of any primary task decrement, comprises a variety of subsidiary task failure.

A similar effect was observed with respect to lane crossing errors. Given that the right lane was devoid of vehicles for the majority of the journey, the risk of an off-road accident was greater for crossings the left-side road boundary (as opposed to a right-side lane boundary). An analysis of lane crossing frequency revealed that crossing at the right-side boundary occurred more frequently than the left-side road boundary (respective means = 6.6 and 2.2 per 5 min.). In addition, crossings on the right-side boundary were, on average, over twice as long as those on left-side road edge (respective means = 11.5 and 5.4 sec.). Therefore, the degradation of simulated driving occurred in a manner that minimized the risk of accident. This finding illustrated how risk compensation plays a role in the utility assessment underlying effort regulation. The driver prioritizes subgoals and performance standards in a rational manner. This process determines which error forms are less critical and therefore comprise "acceptable" degradation.

---

[2]This simulated environment was modeled on UK roads, therefore the vehicle drives in the left lane of the dual-carriageway.

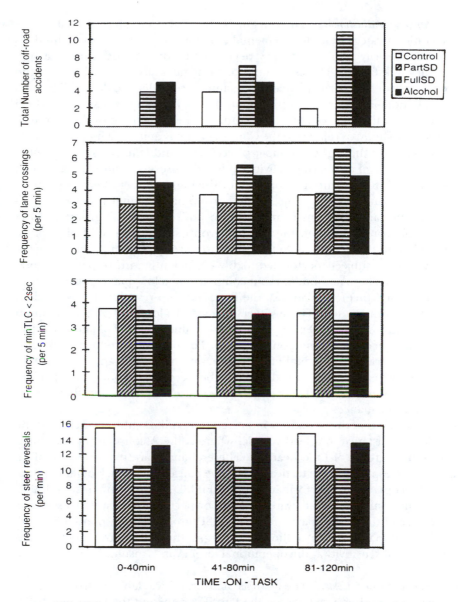

FIG. 3.3.5.   An analysis of lane-keeping behaviour, as represented by multiple variables, displayed by experimental group and time-on-task. TLC = Time-To-Line Crossing; Steer Reversals = frequency of steering wheel zero-crossings.

Strategic Adjustment.   This category of latent decrement represents a conservation strategy to minimize mental effort investment. However, this does not necessarily imply a primary performance decrement or even a subsidiary task failure. The assessment of strategic adjustment focuses on the efficiency of task performance. For example, consider the data in Fig. 3.3.5 as a bottom-up chain of control input, that is, from input as steering wheel activity to various errors of lateral control. The patterning of data (considered as a whole) as relative to the Control group is quite striking. The FullSD subjects have the highest number of accidents and lane crossings, coupled with a low level of steering activity (indexed by reversal rate). Therefore, a low rate of steering input accounted for the high number of errors committed by the FullSD group. The Alcohol group had a higher number of accidents and lane crossings comparative to both the PartSD group and the Control group. However, the Alcohol group sustained a level of steering wheel input that is equivalent to the Control group. The implication here is that Alcohol subjects performed highly inefficiently with respect to lateral control (i.e., high amount of input coupled with a low quality of lateral control performance). By contrast, the PartSD group revealed a very efficient level of lateral control. These subjects provided a low level of steering input equivalent to the FullSD group, but were capable of a level of lateral control that was as accurate as the Control group. The pattern of reduced task activity may characterize mental effort conservation and was apparent in both sleep-deprived groups. The difference between the groups was that the PartSD group were capable of an effective conservation strategy that did not harm performance.

An analysis of time headway indicated that FullSD subjects increased the following distance to a lead vehicle, whereas the Alcohol group exhibited a slight decrease of time headway. These changes are characteristic of an adaptive response at the tactical level in response to feedback of falling standards of operational performance for the FullSD group (Van Winsum, 1996). The implication was that awareness of poor performance caused a strategic increase of the safety margin for the FullSD group. The response of the Alcohol group was to reduce the safety margin, thereby illustrating characteristic lack of sensitivity at the operational level (Van Winsum, 1996).

Compensatory Costs.   The analysis of safety-critical lane control measures revealed no difference between the PartSD group and the Control subjects (Fig. 3.3.5). However, the evidence from the subjective self-assessment questionnaires showed a number of compensatory costs associated with sleep deprivation. A selection of these data are shown in Fig. 3.3.6.

Both sleep-deprived groups registered an equivalent amount of subjective sleepiness and rated the quality of performance as relatively poor (Fig. 3.3.6). There were a number of other compensatory costs that were specific

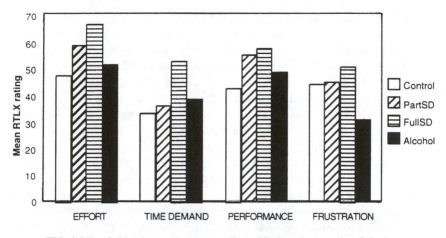

FIG. 3.3.6.   Subjective measures as collected before (pre-test) and during the 120-min. journey by treatment groups. Ratings of *Effort, Time Demands, Frustration,* and *Quality of Performance* originate from the raw ratings on the NASA Task Load Index (TLX) questionnaire (Hart & Staveland, 1988).

to the FullSD group. The detrimental effect of full sleep deprivation increased the level of frustration associated with the simulated driving task. In addition, FullSD subjects experienced a higher level of temporal task demand (Fig. 3.3.6). These findings illustrate how increased discomfort due to fatigue embellished the perception of task demands (at the feedback level, as illustrated in Figs. 3.3.2 and 3.3.3). Both sleep-deprived groups responded by increasing the mobilization of subjective mental effort. By contrast, the net effect of alcohol was to sustain the perception of normative performance and to reduce the level of frustration associated with the task (Fig. 3.3.6). Despite a degraded level of performance, the Alcohol group experienced little conflict at the principal level, due to predictable biases at the level of feedback (Figs. 3.3.2 and 3.3.3).

These data indicate that the sleep-deprived subjects were equally aware of degraded performance and subjective sleepiness. In the case of the FullSD group, these compensatory costs were accompanied by increased frustration and subjective mental workload. Given the obvious performance decrements associated with a full night without sleep (Fig. 3.3.5), it was surprising that the level of subjective sleepiness (Fig. 3.3.4) was equivalent for both the PartSD and the FullSD group. As both groups subjectively perceived an equivalent level of sleepiness (the mean for both groups being seven on a 10-point scale), despite several significant differences with respect to performance and compensatory costs; it is postulated that the FullSD group may have underestimated the magnitude of subjective sleepiness.

A similar bias occurred with respect to the PartSD group's perception that performance was relatively poor. This assessment was not corroborated by a higher number of accidents or lane-crossings. Two alternative explanations are possible, either (a) PartSD subjects responded to sleepiness by raising the standard of performance at the principal level, and therefore perceived poorer performance relative to that standard, or (b) the discomfort associated with sleepiness created a bias in PartSD subjects' assessment of performance, causing a negative perception of actual performance.

Both sleep-deprived groups responded to impairment with a subjective investment of mental effort. This strategy had mixed results. The PartSD group were capable of sustaining performance despite some compensatory costs, whereas the FullSD group experienced a double-decrement (i.e., maximum discomfort and minimum performance). This result may be explained by the analysis of psychophysiological data from the ECG record. The raw ECG data was analyzed to produce the 0.1Hz component of heart rate variability, which has been identified with mental effort and controlled processing (Aasman, Mulder, & Mulder, 1987). It is known that alcohol increases heart rate and distorts the 0.1Hz component (De Waard & Brookhuis, 1991; Gonzalez-Gonzalaz, Llorens, Novoa, & Valeriano, 1992), therefore data from the Alcohol group were not included in this analysis. The raw data from the power spectrum analysis were subjected to a natural log transform (for the purpose of statistical testing) and subtracted from baseline data (obtained during a simulated journey on the day prior to testing). The transformed data are shown in Fig. 3.3.7, note that suppression of the 0.1Hz component is indicative of increased mental effort.

The 0.1Hz component of sinus arrhythmia showed a significant suppression for the PartSD group comparative to the Control group. These data indicate that mental effort investment was sustained at a higher level by the PartSD group to minimize safety-critical errors. In the case of the PartSD group, the effort investment strategy was effective. Primary task performance was maintained with minimum number of compensatory costs (such as increased mental effort, increased sleepiness). The analysis of 0.1Hz component data from the FullSD group showed a linear increase over time. This pattern was indicative of a dissociation between subjective effort and psychophysiological effort. It is hypothesized that fatigue creates this dissociation by demanding a level of mental effort investment that cannot be sustained. At extreme levels of fatigue, compensatory costs are sufficiently prohibitive to automatically provoke a strategy of effort conservation. This may take place unconsciously or even against the wishes of the individual, who persistently attempts an opposing strategy of effort mobilization. This type of dissociation may be indicative of misregulation due to intense conflict between principal goals. The logical response of an exhausted regulatory system is to conserve effort in order to recuperate.

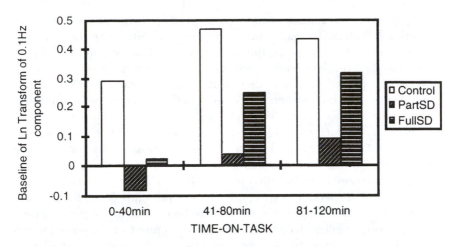

FIG. 3.3.7.   Ln transform of mean power in mid-range frequency band of
HRV analysis (expressed as a baseline) across *condition* and *time-on-task*.

Conclusions

The model of mental effort regulation made a number of predictions concerning the influence of alcohol and sleep deprivation on driving performance. At a basic level, it was assumed that fatigue created conflict between
principle goals, whereas alcohol did not. This pattern was demonstrated in
the analysis of compensatory costs (Fig. 3.3.6). It was shown that latent
decrements in the driving tasks are manifested in the failure of less safety-
critical (i.e., subsidiary) task components. In addition, the driver may adapt
on a strategic basis, either modifying task efficiency (i.e., a reduction of
input without an impact on performance) or adaptively altering the safety
margin as described by Van Winsum (1996).

The contrasting influence of a full night of sleep deprivation compared
with partial sleep deprivation illustrated the difference between ineffective
and effective effort regulation. The former maximized compensatory costs
in conjunction with poor performance, whereas the latter minimized costs
and sustained performance. It was notable that the Control group invested a
lower level of mental effort throughout the task (as indexed by both subjective scales and the 0.1Hz component). This aspect of Control group data
reinforces the view that fatigue increases the minimum requisite level of
mental effort investment. When we are tired, effort investment must increase
proportionately to mitigate the influence of cognitive interference and
other sources of discomfort due to fatigue. However, mental effort is a finite
resource, and a sustained period of investment carries a restrictive burden
of stress and compensatory costs.

Regardless of treatment group, it was predicted that the principle of maintaining high performance would come into conflict with the principle to minimize discomfort during the monotonous, 2-hour journey. It was proposed that this conflict would be intensified in the presence of sleep deprivation. In the case of alcohol, it was postulated that characteristic side-effects of intoxication, including increased confidence and well-being, could introduce predictable biases into feedback loops for both principle goals (Figs. 3.3.2 and 3.3.3). The intoxicated driver would be inclined to make a positive assessment of performance due to increased confidence. In addition, the experience of intoxication may imbue the subject with an increased sense of well-being, thereby reducing discomfort. The net result is that the drunk driver does not experience any conflict at the principle level. By contrast, the sleep-deprived driver is divided between the need to invest mental effort (in order to neutralize sleepiness) and a requirement to conserve mental effort (due to the difficulty of mobilizing effort and sustaining attention in the presence of rising discomfort and stress).

The regulation of mental effort is highly dependent on idiosyncratic performance "standards," changeable levels of self-monitoring with respect to those standards and variable tolerance for stress and discomfort. Therefore, it is highly unlikely that any standard index of fatigue would be workable. However, it is proposed that latent decrements take the stereotypical forms described by Hockey (1993, 1997) and may be operationalized via multi-dimensional data collection. The measurement of latent decrements in conjunction with primary task performance is capable of characterizing the regulatory strategy in operation (i.e., investment or conservation) and the effectiveness of that strategy.

To summarize, the "disinclination to continue performance" during operational fatigue (Brown, 1994) is a rationale process of mental effort regulation, based on the goal to perform being overwhelmed by the goal to reduce stress and discomfort. When humans perform under demanding circumstances or for a prolonged period, the regulation of mental effort is limited by our ability to accurately self-monitor and thereby to sustain task-relevant goals. The subjective feedback loop lies at the crux of the effort regulatory mechanism and forms the basis for all assessments of utility.

ACKNOWLEDGMENTS

The author acknowledges the contribution of Robert Graham to the experimental study described in the Functional Impairment of the Driver section, and Gerry Matthews' comments on an earlier draft of this chapter. This work was performed under the SAVE project (TR 1047), funded by the CEC under the Transport Telematics Programme.

REFERENCES

Aasman, J., Mulder, G., & Mulder, L. J. M. (1987). Operator effort and the measurement of heart rate variability. *Human Factors, 29*(2), 161–170.

Allwood, C. M. (1984). Error detection processes in statistical problem solving. *Cognitive Science, 8,* 413–437.

Anderson, J. R. (1983). *The architecture of cognition.* Cambridge, MA: Harvard University Press.

Baddeley, A. D., & Hitch, G. (1974). Working memory. In G. H. Bower (Ed.), *The psychology of learning and motivation* (Vol. 8, pp. 47–89). New York: Academic.

Bartlett, F. C. (1943). Fatigue following highly skilled work. *Proceedings of the Royal Society (B), 131,* 147–257.

Bartley, S. H. (1965). *Fatigue: Mechanism and management.* Springfield, IL: Charles C. Thomas.

Broadbent, D. E. (1971). *Decision and stress.* London: Academic.

Broadbent, D. E. (1979). Is a fatigue test now possible? *Ergonomics, 22*(12), 1277–1290.

Brown, I. D. (1994). Driver fatigue. *Human Factors, 36*(2), 298–314.

Cameron, C. (1973). A theory of fatigue. *Ergonomics, 16*(5), 633–648.

Carver, C. S., & Scheier, M. F. (1981). *Attention and self-regulation: A control theory approach to human behavior.* New York: Springer-Verlag.

Carver, C. S., & Scheier, M. F. (1982). Control theory: A useful, conceptual framework for personality-social, clinical, and health psychology. *Psychological Bulletin, 92*(1), 111–135.

De Waard, D., & Brookhuis, K. A. (1991). Assessing driver status: A demonstration experiment on the road. *Accident Analysis and Prevention, 23*(4), 297–301.

Fairclough, S. H., & Graham, R. (1999). Impairment of driving performance caused by sleep deprivation or alcohol: A comparative study. *Human Factors, 41*(1), 118–128.

Godthelp, J., Milgram, P., & Blaauw, G. J. (1984). The development of a time-related measure to describe driving strategy. *Human Factors, 26,* 257–268.

Gonzalez-Gonzalaz, J., Llorens, A. M., Novoa, A. M., & Valeriano, J. J. C. (1992). Effect of acute alcohol ingestion on short-term heart rate fluctuations. *Journal of Studies on Alcohol, 53*(1), 86–90.

Hancock, P. A. (1996). Effects of control order, augmented feedback, input device and practice on tracking performance and perceived workload. *Ergonomics, 39*(9), 1146–1162.

Hancock, P. A., & Verwey, W. B. (1997). Fatigue, workload and adaptive driver systems. *Accident Analysis and Prevention, 29*(4), 495–506.

Hancock, P. A., & Warm, J. S. (1989). A dynamic model of stress and sustained attention. *Human Factors, 31*(5), 519–537.

Hart, S. G., & Staveland, L. E. (1988). Development of the NASA-TLX (Task Load Index): Results of empirical and theoretical research. In P. A. Hancock & N. Meshkati (Eds.), *Human mental workload* (pp. 139–183). Amsterdam: North-Holland.

Hockey, G. R. J. (1986). Changes in operator efficiency as a function of environmental stress, fatigue, and circadian rhythms. In K. R. Boff, l. Kaufman, & J. P. Thomas (Eds.), *Handbook of perception and human performance* (Vol. 2, pp. 44.43–44.44). New York: Wiley.

Hockey, G. R. J. (1993). Cognitive-energetical control mechanisms in the management of work demands and psychological health. In A. Baddeley & L. Weiskrantz (Eds.), *Attention: Selection, awareness and control* (pp. 328–345). Oxford: Clarendon.

Hockey, G. R. J. (1997). Compensatory control in the regulation of human performance under stress and high workload: a cognitive-energetical framework. *Biological Psychology, 45,* 73–93.

Kahneman, D. (1970). Remarks on attentional control. In A. F. Sanders (Ed.), *Attention and performance* (Vol. 3, pp. 118–131). Amsterdam: North-Holland.

Kahneman, D. (1973). *Attention and effort.* Englewood Cliffs NJ: Prentice-Hall.

Kecklund, G., & Akerstedt, T. (1993). Sleepiness in long distance truck driving: an ambulatory EEG study of night driving. *Ergonomics, 36*(9), 1007–1018.

Kluger, A. N., & DeNisi, A. (1996). The effects of feedback interventions on performance: A historical review, a meta-analysis, and a preliminary feedback intervention theory. *Psychological Bulletin, 119*(2), 254–284.

Matthews, G., Jones, D. M., & Chamberlain, A. G. (1990). Refining the measurement of mood: The UWIST Mood Adjective Checklist. *British Journal of Psychology, 81*, 17–42.

Maycock, G. (1995). Driver sleepiness as a factor in car and HGV accidents (Rep. No. 169). Crowthorne, UK: Transport Research Laboratory.

McDonald, N. (1987). Fatigue and driving. *Alcohol, Drugs and Driving, 5*(3), 185–192.

Miller, G. A., Galanter, E., & Pribram, K. H. (1960). *Plans and the structure of behaviour.* New York: Holt, Rinehart & Winston.

Moray, N. (1981). Feedback and the control of skilled behaviour. In D. Holding (Ed.), *Human skills* (pp. 15–39). Chichester: Wiley.

Mulder, G. (1986). The concept and measurement of mental effort. In G. R. J. Hockey, A. W. K. Gaillard & M. G. H. Coles (Eds.), *Energetical issues in research on human information processing* (pp. 175–198). Dordrecht, The Netherlands: Martinus Nijhoff.

Muscio, B. (1921). Is a fatigue test possible? *British Journal of Psychology, 12,* 31–46.

Navon, D., & Gopher, D. (1979). On the economy of the human-processing system. *Psychological Review, 86*(3), 214–225.

Powers, W. T. (1973). Feedback: Beyond behaviour. *Science, 179,* 351–356.

Powers, W. T. (1980). A systems approach to consciousness. In J. M. Davidson & R. J. Davidson (Eds.), *Psychobiology of consciousness* (pp. 217–242). New York: Plenum.

Rasmussen, J., & Jensen, A. (1974). Mental procedures in real-life tasks: A case study of electronic troubleshooting. *Ergonomics, 17,* 293–307.

Reason, J. (1987). Generic Error-Modelling System (GEMS): A cognitive framework for locating common human error forms. In J. Rasmussen, K. Duncan, & J. Leplat (Eds.), *New technology and human error* (pp. 63–83). Chichester: Wiley.

Reason, J. (1990). *Human error.* Cambridge: Cambridge University Press.

Sarason, I. G., Sarason, B. R., Keefe, D. E., Hayes, B. E., & Shearin, E. N. (1986). Cognitive interference: Situational determinants and traitlike characteristics. *Journal of Personality and Social Psychology, 57,* 691–706.

Schank, R. C., & Abelson, R. P. (1977). *Scripts, plans, goals and planning.* Hillsdale, NJ: Lawrence Erlbaum Associates.

Schneider, W., & Shiffrin, R. M. (1977). Controlled and automatic human information processing I. Detection, search and attention. *Psychological Review, 84*(1), 1–66.

Schonpflug, W. (1983). Coping efficiency and situational demands. In G. R. J. Hockey (Ed.), *Stress and fatigue in human performance* (pp. 299–330). Chichester: Wiley.

Shiffrin, R. M., & Schneider, W. (1977). Controlled and automatic human information processing II. Perceptual learning, automatic attending and a general theory. *Psychological Review, 84*(2), 127–190.

Thorndike, E. (1900). Mental fatigue. *Psychological Review, 7,* 466–482.

Van Winsum, W. (1996). *From adaptive control to adaptive driver behaviour.* Unpublished doctoral dissertation, Rijksuniversiteit Groningen, The Netherlands.

Walls, H. J., & Brownlie, A. R. (1985). *Drink, drugs & driving.* London: Sweet & Maxwell.

Warm, J. S., Dember, W. N., & Hancock, P. A. (1996). Vigilance and workload in automated systems. In R. Parasuraman & M. Mouloua (Eds.), *Automation and human peformance: Theory and application* (pp. 183–200). Mahwah, NJ: Lawrence Erlbaum Associates.

Yeh, Y. Y., & Wickens, C. D. (1988). Dissociation of performance and subjective measures of workload. *Human Factors, 30*(1), 111–120.

A Heavy Vehicle Drowsy Driver Detection and Warning System: Scientific Issues and Technical Challenges

Paul S. Rau

National Highway Traffic Safety Administration

Even though loss of alertness has been detected in laboratory driving simulators with impressive accuracy, there are numerous scientific issues and technical challenges associated with developing a field-operational drowsiness detection and warning system. The key *scientific issues* are related to the development of fieldable detection models and warning systems. Issues include model validation, individualized versus generalized monitoring, and detection and warning versus activity-based maintenance. The key *technical challenges* are related to system operability and acceptance. Challenges include system upkeep and calibration, driver and vehicle compatibility, risk compensation and migration, alertness restoration, and operational reliability. This chapter provides an overview of the drowsy driver problem in the United States, a description of the National Highway Traffic Safety Administration's drowsy driver technology program, and an introduction to some of the scientific issues and technical challenges that confront system deployment.

Research is underway at the U.S. National Highway Traffic Safety Administration (NHTSA) to develop, test, and evaluate a prototype drowsy driver detection and warning system for commercial vehicle drivers (1996–1998). Even though the loss of alertness in drivers has been detected in laboratory driving simulators with impressive accuracy (Wierwille, Lewin, Fairbanks, & Neal, 1996), there remain numerous scientific issues and technical challenges associated with field deployment. This chapter provides an overview of the drowsy driver problem in the United States, a description of NHTSA's

drowsy driver technology program, and an introduction to some of the scientific issues and technical challenges that confront system deployment.

The *scientific issues* discussed are related to the development of detection
models and warning systems. The discussion includes the issues of model
validation, individualized versus generalized monitoring, and detection and
warning versus activity-based maintenance. *Technical challenges* relate to system operability and acceptance, including system upkeep and calibration,
driver and vehicle compatibility, risk compensation and migration, alertness
restoration, and operational reliability.

Although the list of issues and challenges is not exhaustive, it provides an
initial framework suggestive of deployment alternatives as the detection and
warning system is developed. The prototype development team is presently
charged to fully understand these concerns and to complete the initial prototype. Ultimately, the final system seeks to reduce the annual numbers of
injuries and deaths associated with drowsiness.

PROBLEM SIZE

Currently, our understanding of the drowsy driver problem in the United
States is based on NHTSA's revised estimates for the 5-year period between
1989 and 1993 (Knipling, Wang, & Kanianthra, 1995). An average annual
total of 6.3 million police-reported crashes occurred during this period. Of
these, approximately 100,000 crashes per year (1.6% of 6.3 million) were
identified on police crash reports (PCR) where drowsiness was indicated
and from a review of "drift-out-of-lane" crashes *not* specifically indicated but
that had drowsiness characteristics. Approximately 71,000 of all drowsy-
related crashes involved nonfatal injuries, whereas 1,357 drowsy-related fatal
crashes resulted in 1,544 fatalities (3.6% of all fatal crashes), as reported by
the Fatality Analysis Reporting System (FARS). Nevertheless, many run-off-
roadway crashes are not reported or cannot be verified by police, suggesting
that the problem is much larger than previously estimated.

As for differences between cars and trucks, approximately 96% of annual
drowsy driver crashes (96,000 total including 1,429 fatalities) involved drivers of passenger vehicles, whereas only 3.3% (3,300 total including 84 fatalities) involved drivers of combination-unit trucks. Nevertheless, drowsiness
was cited in more truck crash involvements (.82%) than passenger vehicle
crashes (.52%). In addition, the risk of a drowsiness-related crash in a combination-unit truck's operational life is 4.5 times greater than that of passenger vehicles, because of greater exposure (60K versus 11K miles/year),
longer operational life (15 versus 13 yrs), and more night driving (Knipling
& Wang, 1994). There is also a greater likelihood of injury in heavy vehicle
crashes. Approximately 37% of the truck-related drowsy driver fatalities and

20% of the nonfatal injuries occurred to individuals *outside* the truck, compared with 12% of the fatalities and 13% of the nonfatal injuries from drowsy passenger vehicle drivers.

DROWSY DRIVER TECHNOLOGY PROGRAM

The objective of NHTSA's Drowsy Driver Technology Program is to develop, test, and evaluate a prototype drowsy driver detection and warning system for commercial motor vehicle drivers. One of the key tasks of the program is to develop drowsiness detection models and algorithms based on field data. However, laboratory-based experiments will also be conducted to suggest sensors and algorithms for further validation in the context of over-the-road driving. There are a variety of university, industry, and government partners associated with the laboratory and field study elements of the program.

First, in partnership with the University of Pennsylvania (funded by the Federal Highway Administration's Office of Motor Carriers), candidate sensors are being validated by monitoring sleep-deprived subjects in a controlled laboratory setting. Subjects undergo vigilance and cognitive tests while deprived of sleep. Specifically, polysomnographic and performance measures are collected continuously; subjects are either "alerted" or "not-alerted" about their drowsiness as they become drowsy over a 20-hour period. Alerted and unalerted conditions are experimentally comparable because the presence or absence of an alerting stimuli could alter the response characteristic of certain devices. As another part of the validation process, "blind" data from the experiments are provided to the vendors of each device to determine when the drowsiness episodes occurred (prospective phase). Successful device vendors from the prospective phase receive algorithms from each of the other device vendors, as an opportunity to improve the detectability of their respective methods (retrospective phase).

Second, NHTSA's principal industry partner for building the prototype system is Carnegie Mellon Research Institute (CMRI), in Pittsburgh, PA. CMRI is the technical lead on the project and has outfitted several commercial trucks (courtesy of Pitt-Ohio Express, Inc.) with numerous sensors and an automated data collection system. Field studies are designed to unobtrusively monitor commercial truck drivers over 10-hour overnight express runs. In the procedure, numerous performance and behavioral measures are collected as the foundation for developing detection models. This field work is guided by drowsiness detection procedures, which were developed under NHTSA sponsorship over a 5-year period, based on driving studies in simulators (Wierwille et al., 1996). However, a number of new detection model and algorithm approaches are also being developed and tested from the new field data, including the measures from sensors validated in the laboratory phase.

Finally, another government agency partner is the Naval Health Research Center (NHRC) in San Diego, CA. NHRC provides special expertise in monitoring drowsiness from a recently developed method of processing electroencephalograph (EEG) signals. NHRC's role on the team is to assist in the development of field-based drowsiness detection models and to provide a psychophysiological index of drowsiness previously developed under contract with the Office of Naval Research. The validity of the NHRC drowsiness detection metric are also examined under the prospective and retrospective phases of the laboratory study.

SCIENTIFIC ISSUES—DETECTION MODELS AND WARNING SYSTEMS

Model Validation

Model validation is the principal activity of the program. These models derive their ability to detect changes in alertness from relations among factors, the correlations between which are built up from data collected during observed levels of alertness. As a result, models represent relations among the conditions required for drowsiness to be detected. For example, conditions may include drifting out of lane, excessive lane deviations, drift and jerk steering, percentage of eye closure, and so on. Thus, a model might specify that a certain magnitude of deviation in a lane can be expected from a certain percentage of eye closure. In addition to prediction, models also specify the relative importance of relations among the measures such that we might also gain an improved understanding of the important behavioral and performance components of driving.

Performance and Physiology. As the program goal is to develop a prototype system, one of our most important considerations is implementation. For example, we do not expect that commercial drivers will accept a system that requires a driver to don a cap wired with electrodes. Nevertheless, a model could be based on a measure like EEG if shown to be valid. Such a "gold" standard or yardstick by which drowsiness can be measured is important for building models that relate specific changes in physiology to driving performance. Thus, one option is to detect drowsiness based on performance inputs alone, once a strong relationship between driving performance and physiology has been established.

Another modeling option is to base the detection on a valid psychophysical index alone, if it could be measured unobtrusively. Specifically, ocular movement will soon be measured unobtrusively from inside the vehicle (a 1998 NHTSA Small Business Innovative Research program initiative). This capability might provide direct access to an ocular index of drowsiness. Thus, the validation of an ocular index of drowsiness might result in models that relate driving performance to ocular measures, models that relate ocular measures to other previously validated psychophysical indices of drowsiness, and/or models that relate driving performance to ocular and/or other valid indices.

Normative Weighting and Event-Driven Models. It is possible that quantitative models alone can*not* be produced from the measures obtained in the field study. Therefore, it is an option to explore improving the quantitative models with various qualitative data related to normative trends in drowsy driving. For example, according to data from NHTSA's General Estimates System (GES), police reported that drowsy-related crashes occur most frequently between 1:00 and 5:00 AM, and again in late afternoon between 3:00 and 6:00 PM. Information is also available about the number of drowsy related crashes, based on the number of hours driven. Therefore, in a normatively weighted model, a qualitative rule could be used to mediate the alarm/warning threshold of a data-driven model according to population trends.

Similarly, model validity might also be improved by using knowledge about events that occur during the particular time line of travel. For example, information about the frequency of stops, duration of stops, regularity of speed, number of passengers, changes in air flow and temperature, and noise levels could be measured and used to modify the detection capability of the model. Such an algorithm would detect departures from previously determined normative levels. For example, the alarm/warning threshold of a detection system could be lowered when there is an absence of an environmental change; a monotonous environment might indicate a precondition for drowsiness. Therefore, the validity of a quantitative detection model might be improved by using qualitative information about the population of drivers and/or about the experience of a particular driver. It is also possible that the most useful detection model might be based on the qualitative information alone.

In sum, there are numerous modeling opportunities, all of which offer promise in producing an operational system. As a result, the prototype system could be based on some combination of driving performance, ocular behavior, and/or the inclusion of normative and event-based heuristics.

Individualized Versus Generalized Models

Individualized versus generalized models are distinguished as those that either detect loss of alertness in a single driver or among all drivers, respectively. The issue is that quantitative models use the response data from only a small sample of drivers. Therefore, predictions about a larger population of drivers must be derived statistically. Nevertheless, any large differences among individual drivers could overwhelm any otherwise significant effect related to the group. It is, therefore, likely that a detection model could be improved by using specific knowledge about an individual driver. Moreover, individualized models could include normative or event information, as previously described. Last, some technologies have been shown to detect an individual drowsy "signature." For example, certain classes of neural networks can learn baseline driver behavior and then warn the driver about departures from normal "alert" patterns. Both group-based and individualized models are potential outcomes of the research.

Detection and Warning Versus Activity-Based Maintenance

Detection and warning versus activity-based maintenance is an issue that contrasts the detection-modeling approach of our program, with an "activity-based" approach that requires continuous driver interaction. For example, there are several devices that could alert the driver when a specific behavior fails. One device sounds an alarm when any change in steering wheel motion stops. Presumably, moments of motionless steering may indicate that the driver has fallen asleep. However, there are numerous differences in driving style and roadway conditions that result in motionless steering. Thus, when avoiding frequent alarms during normal periods of motionless steering, steering could become erratic and unsafe.

Another device measures a driver's reaction time to a small light, which is illuminated following a *random* elapsed period of time. A button must be pressed within 3 seconds after the small light is illuminated, or else a buzzer sounds. This secondary task forces the driver to monitor a specific location inside the vehicle for the random occurrence of a single light. Monitoring a random event, not related to vehicle operation, could dangerously divide attention away from the roadway and mirrors. Still another device includes two alarms: one, if a button is not pressed before an *adjustable* time interval, and two, if a second button is not pressed within another adjustable time period following the occurrence of the first alarm.

There are many compromises to driver safety in using activity-based alertness maintenance devices. Nevertheless, some form of activity-based device, which does not interfere with safe driving, might provide a useful countermeasure to drowsiness. Perhaps a future system might offer some combination of passive detection and alarm/warning methods, with an activity-based system.

TECHNICAL CHALLENGES— OPERABILITY AND ACCEPTANCE

System Upkeep and Calibration

System upkeep and calibration is perhaps the most important technical challenge in designing a generally useful system. The system must be easy to learn, easy to use, and easy to maintain. However, certain sensors might be more difficult for passenger vehicle owners to maintain and calibrate on a regular basis. For commercial carriers, upkeep and calibration might be achieved during regular periods of maintenance.

The difficulty of this challenge depends on which sensors are required to support a valid detection model. For example, a camera-based lane-keeping system might require regular lens cleaning and alignment checking. Thus, for a nontechnically oriented consumer, the system might also require a performance monitoring and fault localization (PMFL) device to automatically inform drivers if their system performance degrades. Our challenge, then, is that regardless of how valid and reliable in detecting drowsiness, the fielded system must be easily maintained and calibrated.

Driver–Vehicle Compatibility

Driver–vehicle compatibility is presented not so much as a challenge of system design, but as an activity for building engineering models of driver–vehicle interaction. There exist various guidelines on vehicle *interfaces,* but there are no known models that specifically address driver–vehicle *interaction.* Such models would constitute computational methods for predicting human performance in vehicles. As a start, cognitive models of driver–vehicle interaction could arise from, as well as contribute to, the existing wealth of knowledge from cognitive science and cognitive psychology. The challenge is to focus that knowledge as an organized framework of methods, whereby quantitative models of driver–vehicle interaction may be developed. Other specialty areas in human factors have previously begun this process. For example, in the area of user–computer interaction, there exist a number of models that characterize the interaction (not necessarily the interface) between users and computer systems. Many of the basic components of previous models of human–machine dialog could also be applied to develop predictive models of driver–vehicle interaction. The present research contributes to this knowledge base, specifically with regard to the models developed that specify relations between physiology and performance.

Risk Compensation and Migration

Risk compensation and migration relate to diminished operational effectiveness because of the misuse of a countermeasure by drivers, as well as because of external sources of probability *not* associated with the detectability of a device. First, risk compensation refers to the undesired use of a countermeasure that reduces a driver's awareness of the actual risks associated with certain risky driving behaviors. For example, depending on how the system reports loss of alertness, drivers may use the information to continue driving. It is well known that drivers are often motivated to keep driving, even under impaired levels of drowsiness. Drivers persist in driving drowsy for many reasons including proximity to their destination, safety concerns about sleeping at rest areas, lodging alternatives, and delays in schedule. The technical challenge is to minimize risk compensation through the design of the user interface. For example, a continuous "fuel gauge" display of alertness might encourage drivers to continue driving, whereas a single threshold alarm would communicate that falling asleep at the wheel is imminent.

Second, risk migration refers to externally determined probabilities, which can affect the overall performance of the system. For example, there have been informal reports suggesting that with roadside rumble strips, there are fewer run-off-road crashes for those road segments that contain rumble strips. However, overall, the same number of crashes occur on that particular highway. It is as though the incidences "migrate" to subsequent road segments without the rumble strips. There are no models to predict this phenomenon, but it suggests that there are other probabilities involved that could influence the effectiveness of the system. Therefore, part of the challenge is confronting problems that are unexpected.

Alertness Restoration

Alertness restoration relates to the response characteristic of a particular detection system. Specifically, the output of one detector may result from a direct measurement of some parameter related to drowsiness, such as heart rate, temperature, or EEG. Another detector may infer drowsiness from a quantitative and/or qualitative model. Both methods might detect drowsiness. However, the strategy for restoring alertness may be different depending on the speed and accuracy with which detection occurs in each. Whereas a direct measure of drowsiness might keep up with a rapid detection-warning cycle, an inferential model based on the accumulation of evidence about drowsiness may not. Therefore, in the latter case, a relatively stronger alerting stimulus may be required to sustain alertness during the period required for the model to integrate its data. This scenario suggests the need to quantify the alerting effectiveness of various warning systems as a function of specific detection methods. The challenge is compounded by the fact that the increasing homeostatic pressure to sleep may eventually overwhelm the effectiveness of any warning system to sustain alertness. Therefore, the relation between the detection and warning components of a system could be further mediated by the deterioration in alertness over time. Quantifying these relations would provide an index for selecting among alternative detection and warning subsystems during sustained driving.

Operational Reliability

Operational reliability relates to interactions among selected sensors and to the accumulation of detection errors, which adversely affect detection. Eventually, the sensors in the final system will be optimized to achieve desired performance. However, each sensor contributes some probability of detection error. Therefore, system development work must account for the potential additivity of error among sensors, which could overwhelm the detectability of any single sensor. In sum, the effectiveness of any sensor is constrained by its reliability, both in relation to the sensitivity of any particular sensor and in combination with the detectability of other sensors in the system.

The combined performance of sensors is also a concern about system cost. By understanding how sensors operate collectively, we can better estimate the relative costs associated with levels of benefit. For example, a 5% improvement in detectability may cost hundreds of dollars in purchase, maintenance, and calibration. Therefore, some consumers may elect not to purchase certain small gains in effectiveness. The challenge, then, is to quantify these cost and benefit tradeoffs.

CONCLUSIONS

This chapter introduced a few of the scientific issues and technical challenges confronting system deployment. While the framework is not exhaustive, it does suggest several deployment alternatives as the detection and warning system is developed. As development continues, it is likely that the issues and challenges will not be fully reconciled in the final system design. Nevertheless, as we proceed through the system development and validation process, we hope to better understand those constraints.

REFERENCES

Knipling, R. R., & Wang, J. S. (1994, November). *Crashes and fatalities related to driver drowsiness/ fatigue* (Research Note). Washington, DC: National Highway Traffic Safety Administration.

Knipling, R. R., Wang, J. S., & Kanianthra, J. N. (1995, May). Current NHTSA Drowsy Driver R&D. In *Proceedings of the 15th international technical Conference on Enhanced Safety of Vehicles (ESV)*, Melbourne, Australia (pp. 366–374). Washington, DC: National Highway Traffic Safety Administration.

Wierwille, W. W., Lewin, M. G., Fairbanks, RJ, & Neal, V. L. (1996, September). *Research on vehicle-based driver status/performance monitoring.* (Virginia Polytechnic Institute and State University Report No. ISE 96-06 through 96-08, Contract Number DTNH 22-91-Y-07266, NHTSA Report No. DOT HS 808 640). Washington, DC: National Highway Traffic Safety Administration.

Examining Work Schedules for Fatigue: It's Not Just Hours of Work

3.5

Roger R. Rosa
National Institute for Occupational Safety and Health

One of the primary sources of fatigue for most adults is the job they perform for their livelihood. The degree and quality of fatigue that can be attributed to work varies with the type of work performed, ranging from profound muscular fatigue associated with heavy physical labor to mental fatigue associated with sedentary cognitive or perceptual tasks. These qualitative differences in fatigue are situational, depending on the types and varieties of tasks performed in a particular job. Fatigue can also be affected, however, by more general aspects of work organization, which can be examined across different sorts of jobs, such as the degree of autonomy workers have in choosing the sequence of tasks or the amount of control they have over the pacing of work. A fundamental aspect of work organization is the manner in which on-duty and rest time are scheduled. The work schedule, perhaps more than any other activity, dictates how employed individuals plan their weekly routine and organize other important responsibilities in their lives. The purpose of the present chapter is explore ways to examine work schedules to identify potential sources of fatigue. For practical purposes, we define fatigue very broadly as any reduction in the capacity or motivation to perform, which may decrease efficiency, increase operational errors or accidents, or otherwise compromise the safety or health of workers or those who might depend on them.

Our basis of comparison for evaluating work schedules is the standard daytime 40-hour week of five consecutive 8-hour days, occurring sometime between 7:00 AM and 6:00 PM, with weekends off. Although other work schedules are being instituted with increasing frequency, an 8-hour/five-shift daytime work week is the schedule experienced by the large majority of nonfarm workers in the United States and other industrialized countries (Rones, Ilg, & Gardner, 1997). As such, this schedule can be considered to produce the modal level of fatigue for a full-time worker.

Concerns about excessive fatigue arise from schedules that add to the average demand of the daytime 40-hour/5-day week by some combination of increased hours of work, reduced opportunities for rest, or work at times of reduced capacity. It may be intuitively obvious that fatigue increases with the amount of time that work is performed, but the rate at which fatigue accrues with hours worked is modulated by other aspects of the work schedule, such as the time at which work occurs or the degree and quality of rest obtained prior to work. In general, the schedule-related fatigue experienced by a worker will be a function of the number of hours worked, the timing of work in the 24-hour day (i.e., what shift is worked), how many work shifts occur before a rest day, how many rest days are taken before a return to work, how much rest is taken between work shifts, how much rest is taken during the shift, and how variable the timing of the shift is.

WORK SCHEDULE FEATURES

Shift Timing

We start by examining when work occurs within the 24-hour day. Prior to actually performing work, humans carry a capacity to perform that varies in a regular pattern around the clock. As diurnal animals, human capacity is highest during the daytime and reaches a low point in the middle of the night. Several studies have demonstrated oscillations in physiological and behavioral functions over a 24-hour period (Wever, 1986). Activity in most of the physiological functions is reduced at night and, because many of these functions are involved in the production of energy required for work, it can be expected that the capacity to perform work is also reduced at night. These oscillations in the 24-hour day are termed *circadian rhythms,* derived from the Latin meaning "about or around a day."

Important physiological circadian rhythms for energy output include sympathetic nervous system activity and production of glycocorticoids, catecholamines, and pituitary hormones (reviewed by Wojtczak-Jaroszowa, 1977). Reduction of activity at night in these functions limits the availability of glycogen and lipid stores for production of energy, which may limit the ability to perform. Time-of-day studies of physiological responses to a normal work shift, or to acute (3–30 min) bouts of exercise generally support the suggestion that overall capacity for physical work is reduced at night. Cohen and Muehl (1977), for example, reported that both resting pulse rate and pulse rate following 3 minutes of rowing were lowest in the middle of the night. Wojtczak-Jaroszowa and Banaskiewicz (1974) reported decreased values for maximum oxygen uptake (VO_2 max) following maximal exercise at night. Ilmarinen, Ilmarinen, Korhonen, and Nurminen (1980) reported that heart rate, muscle strength, and neuromuscular coordination (balance) were lowest, perceived exertion was highest, and recovery after exercise was slowest, when these tests occurred at night. Cabri, De Witte, and Clarys (1988) reported that blood pressure was most perturbed at night following maximal isokinetic muscle contractions.

In parallel with circadian variations in physiological functions and exercise ability, several other behavioral tasks and indexes of subjective state indicate reduced night-time capacity (see Folkard, 1996; Monk, 1990, for reviews). With some minor variations, measures of laboratory types of performance tasks such as simple and choice reaction time, visual search, perceptual-motor tracking, or short-term memory generally indicate maximal performance in the afternoon (2:00 to 5:00 PM) and minimal performance in the early morning (2:00 to 6:00 AM). Measures of real-world tasks such as frequency of meter-reading errors or missed railroad signals, minor accidents by hospital workers, or speed of switchboard-answering responses show similar patterns (Monk, Folkard, & Wedderburn, 1996), as do subjective reports of reduced alertness, feelings of tiredness and fatigue, or irritable and depressed mood (Monk, 1990). Observations of highly experienced female thread spinners, for example, indicated that they produced 10% fewer units during night shift compared to day shift (Wojtczak-Jaroszowa, 1977).

The influence of circadian arousal depends on whether, and to what extent, the circadian system has adapted to waking activity during the night-time hours. Such adaptation occurs slowly, over several nights, to a routine of night-time activity and daytime sleep. It is even hypothetically possible to completely invert the high and low points of the circadian rhythm if a worker strictly adheres to an active night-time schedule with sleep during the daytime (Åkerstedt, 1985). Such complete inversion rarely occurs, however, because societal pressures make it attractive for the worker to revert to a daytime routine on nonwork days. These rest days usually occur at 3- to 7-day intervals, which are frequent enough to interrupt the slow progression of circadian adaptation to the night-time routine.

Night-time reductions in performance, alertness, or mood are associated not only with the endogenous circadian reductions in physiological activity/arousal, but also with the fact that humans normally are asleep at night. The sleep–wake cycle is itself a circadian rhythm because it varies in a regular cycle during the 24-hour day. To a greater extent than that observed in other physiological rhythms, however, sleep and waking are under voluntary control by humans. Such control allows humans to remain awake at night despite appreciable physiological pressure by the sleep–wake system to be asleep at that time and also allows the possibility for sleep in the daytime despite physiological pressure, from circadian physiological arousal systems, to be awake at that time.

The degree of adaptation of the circadian arousal system to a night-time orientation affects the ability to sleep during the daytime. The lowest quantity and the poorest quality of daytime sleep are obtained when there is no adaptation of the circadian system to night-time work. The partial sleep deprivation that results from difficulties with daytime sleep then contributes to the experience of fatigue and sleepiness during the subsequent night shift. The first or second of a series of night shifts is usually the most fatiguing because the unadapted circadian rhythm (i.e., a rhythm in the daytime peak orientation) enhances night-time fatigue directly and also compromises rest and recovery through inadequate daytime sleep. Social or domestic incentives to remain awake during the daytime and evening hours can compound fatigue levels during night work because sleep could be skipped altogether (Gadbois, 1981). A strong daytime orientation of circadian rhythms could promote this tactic by increasing the likelihood that a person would feel active and awake and thus downplay the need for sleep. Studies of shift-worker sleep support these generalizations. Both retrospective cross-sectional surveys and prospective sleep-diary types of studies have indicated that night-shift workers consistently obtain less sleep than day- or evening-shift workers (Knauth et al., 1980). Furthermore, night-shift workers often report their daytime sleep to be lighter, more fragmented, and less restful than sleep at night (Lavie et al., 1989; Walsh, Tepas, & Moss, 1981). Increasing experience with shift work apparently does not result in adaptation of sleep patterns, because older shift workers still show decreased daytime sleep (Tepas, Duchon, & Gersten, 1993).

Night shift is not the only shift that can reduce the amount of sleep. Very early starts for day (or morning shift) can truncate sleep taken before that shift and also increase self-perceived sleepiness and fatigue during subsequent waking hours (Kecklund & Åkerstedt, 1995; Knauth et al., 1980). Worksite surveys suggest that reduced sleep prior to an early-morning shift is probably a result of difficulty in advancing the evening retiring time to try to obtain an adequate amount of sleep (Folkard & Barton, 1993). This observation is consistent with laboratory studies of circadian sleep propensity, which have demonstrated very long sleep latencies when sleep is attempted in the early evening (Lavie, 1991). Social incentives could also play a role in delaying retiring as the early evening is a typical time to meet with family and friends. Delaying the morning start time can result in increased total sleep and reductions of fatigue during the day shift (Rosa, Härmä, Pulli, Mulder, & Näsman, 1996).

In summary, the degree of fatigue experienced by the worker as function of time of day is a combination of effects from oscillations in circadian arousal and the adequacy of sleep. Sleep and subsequent waking alertness are adversely affected to the degree to which working hours intrude on the normal night-time sleeping hours. This tendency is most obvious in night shift workers who must sleep during the daytime but also may be apparent in those individuals who start work very early in the morning. The relative impact of sleep and circadian rhythms depends on the extent of adaptation of these factors to a particular shift. Complete adaptation to night shift may be possible but occurs infrequently because of social and physiological impetuses to be active during the daytime.

Shift Rotation

The apparent tension between the physiological advantages of adaptation to nocturnal activity and the social advantages of diurnal activity has stimulated a debate over the most appropriate system for staffing the night or early morning shifts. Short of banning night-time work altogether, which is not possible in critical public health, safety, and utility services and is not deemed efficient or cost effective in several other industries, it becomes a question of whether night work should be covered by a permanent or a rotating staff of workers. Permanent staffing would presumably maximize the physiological adaptation of those workers to a particular shift. Rotational staffing, on the other hand, would allow more social and domestic contact with family and friends because the isolation of night work would be experienced intermittently.

Speed of Rotation.   A more general way to characterize the question of whether to institute permanent or fixed shifts or rotating shifts is to ask how fast should shifts rotate or how many consecutive shifts should be worked before changing to a different work time. Rotation could range from none (fixed shifts) or very slow (e.g., every 3–4 wks) to very rapid (e.g., every 2–3 days). In his review, Wilkinson (1992) argued that reasonable physiological adaptation could be achieved by workers on permanent night or very slow rotating shifts and that fatigue and performance could be maintained at levels comparable to day workers. He cautioned, however, that there are no worksite studies adequately comparing fixed and rotating shifts, perhaps because, as Knauth (1995) suggested, there are too many differences between the settings and the workforces using these types of shifts. Folkard (1992) has argued that the decision about whether or not to use permanent shifts should depend on the work situation and the relative desirability of adaptation to night work. Rapidly rotating systems (e.g., 2 days, 2 evenings, 2 nights, 2 rest days) permit no appreciable adaptation to night work as too few night shifts are experienced before rotating to a rest day or another shift. Under those circumstances, all nights shifts are fatiguing, from the perspective of circadian rhythms, but the night-time discomfort and potential compromises in performance disappear quickly for a given worker. Across groups of workers, however, rapid rotations raise the possibility that all night shifts are staffed with unadapted individuals who are quite fatigued from reduced circadian arousal and partial sleep loss. Consequently, if rapidly rotating systems are to be used, the work tasks would need to be structured so that maximal demands do not occur during the night shifts. Steps should also be taken by the workers themselves to maximize their opportunities for sleep before night shifts so that at least one potential source of fatigue is reduced. If nighttime tasks are too demanding, a slower rotation might be more effective from the standpoint of fatigue because some of the workforce could be partially adapted and could potentially compensate for co-workers who most recently rotated to night shift. It has also been suggested that shifting to and from night work so frequently over the years, under rapid rotations, might be too stressful and lead to health problems. Empirical evidence to support this suggestion, however, is not available. An alternative viewpoint asserts that the least stress is experienced by maintaining circadian rhythms in their normal daytime peak orientation (i.e., no adaptation to night work), which would be expected from rapidly rotating systems. Supportive empirical evidence for this hypothesis, however, is not available.

The reality in the United States appears to be that neither very slow nor very rapid systems (more common in Europe) are in common use. Although reliable statistics are not available, it appears that weekly rotations are used most frequently. Such systems are poor to mediocre in terms of physiological adaptation (Knauth, 1995) and require sacrificing access to day and evening activities for an entire week (every third week) while working night shifts.

Direction of Rotation.   The sequence of changes of shift times or the direction of change with respect to the clock also might influence fatigue. It has been suggested that schedules that rotate in a counterclockwise direction (i.e., morning following evening following night) can produce more fatigue than those that rotate in a clockwise direction (i.e., morning to evening to night). Three related factors favor rotation in a clockwise direction. First, laboratory studies of circadian rhythms suggest that the cyclic period of the endogenous circadian pacemaker (the internal clock that controls rhythmic timing) runs slightly longer than 24 hours if allowed to "free-run," or oscillate without respect to external time cues (Czeisler, Moore-Ede, & Coleman, 1982). This internal circadian "push" would favor a delaying or clockwise type of schedule. In a related manner, it is easier through voluntary activity to delay sleep onset than to advance it with respect to the clock. Thus it has been argued that more sleep is obtained when the work schedule forces a worker to delay sleep rather than try to sleep at earlier and earlier times. A third factor is more organizational, having to do with the so-called "quick change," or "quick return," which is a brief rest interval often scheduled between shifts in counterclockwise rotation systems. Schedules with quick changes require a worker to rotate to a new shift approximately 8 hours after completing the previous shift (Barton & Folkard, 1993). For example, a rotation to night shift at 9:00 PM might be required after completion of the day shift at 3:00 PM. Such a brief rest interval between shifts at that time of day makes it both impractical and physiologically difficult to obtain adequate sleep to help reduce sleepiness on the night shift. Delaying systems, by contrast, often allow at least 24 hours of free time before rotating to the new shift time, which would provide greater opportunities for rest and sleep. Studies directly comparing advancing and delaying systems are rare. Epstein et al. (cited in Knauth, 1995) reported improved sleep quality after changing from an advancing to a delaying shift rotation, whereas Landen et al. (cited in Knauth, 1995) reported fewer health complaints on a delaying system. In a cross-sectional analysis, Barton and Folkard (1993) reported that advancing systems, especially those with quick changes, produced the most health complaints, including chronic fatigue.

Daily Work–Rest Ratios

Fatigue as a function of accumulated hours of work must be examined both on a daily basis in terms of time working and time resting and on a weekly basis in terms of consecutive work days and free days. The interaction of these factors with shift timing and shift rotation must also be considered. On a daily basis, questions of excessive fatigue often arise when the daily ratio exceeds the standard 8 hours of work to 16 hours of rest. This occurs when overtime hours are required or a compressed work week has been instituted whereby daily hours are increased, (e.g., to 10 or 12 hours) and consecutive work days are decreased (e.g., to 3 or 4 days).

Overtime effects on health, performance, and fatigue have been reviewed in detail by Spurgeon, Harrington, and Cooper (1997) and Rosa (1995). Most studies of overtime have been concerned with cumulative effects on health and well-being, and few have examined fatigue directly. In general, overtime has been associated with stress-related health complaints, where fatigue, from long hours combined with high workloads, has been implicated only indirectly. Fatigue from long hours has also been implicated in studies of manufacturing early in this century when it was demonstrated that reductions in the 50- to 60-hour workweek had little negative effect on productivity (Spurgeon et al., 1997). Case studies of hospital infectious outbreaks also pointed to fatigue from overtime hours as a factor contributing to failure to comply with aseptic practices (Rosa, 1995). That is, fatigued hospital workers often skipped cleanliness steps. Such behaviors are consistent with the speed–accuracy trade-offs we observed in standard laboratory tests after a week of 12-hour workdays (Rosa & Colligan, 1988).

Scheduled work shifts longer than 8 hours, as in compressed workweeks, have been instituted with increasing frequency in industrialized countries in the past 20 years (Rones et al., 1997), and much review and debate have occurred about their relative advantages and disadvantages. The costs and benefits of extended work shifts have been discussed in several recent reviews, especially with reference to the more popular 12-hour shifts (Rosa, 1995; Smith, Folkard, Tucker, & Macdonald, 1998). A principle disadvantage of extended shifts recognized by all reviewers is the potential risks from the additional daily fatigue, especially during night shift when the circadian downturn in arousal could add to fatigue from additional hours of work.

Self-ratings of sleepiness from one of our work-site comparisons of 8-hour and 12-hour shifts (Rosa, Colligan, & Lewis, 1989), presented in Fig. 3.5.1, illustrate the interaction of time of day and accumulated fatigue from hours of work. The values in the figure were derived from a multiple regression equation, which modeled sleepiness as a linear function of consecutive hours worked and a sinusoidal circadian function with peak sleepiness in the early morning and minimal sleepiness in the late afternoon. The combined effects of both functions are apparent in the steeper rise in sleepiness

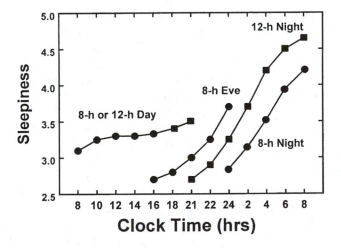

FIG. 3.5.1.   Self-rated sleepiness of control room operators during 8-hour and 12-hour shifts at a power plant. Values in the figure were derived from a multiple regression equation modeling sleepiness as a linear function of consecutive hours worked interacting with a sinusoidal circadian function with maximum sleepiness in the early morning and minimum sleepiness in the late afternoon.

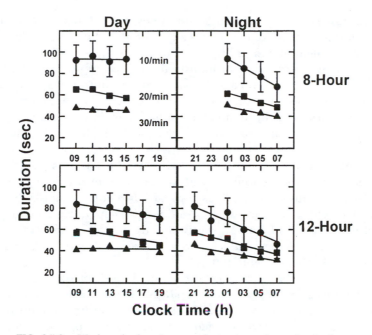

FIG. 3.5.2.   Work cycle durations producing a moderate level of perceived upper extremity fatigue during a simulated manual assembly task
performed at three repetition rates during 8-hour and 12-hour day and
night shifts.

across night shifts as compared to day shifts and the greater overall sleepiness during 12-hour as compared to 8-hour shifts. Figure 3.5.2 illustrates
how physical workload could interact with hours worked and time of day. In
that laboratory study (Rosa, Bonnet, & Cole, 1998), subjects were instructed
to perform a manual work simulation at three repetition rates and to stop
the work trial when they reached a given level of perceived muscular fatigue.
As shown in the figure, the duration of the work trials decreased to lower levels (i.e., muscular fatigue increased) during 12-hour shifts. In addition,
fatigue increased more quickly across the night shifts compared with day
shifts. It also is apparent that the subjects were willing to work longer at
slower repetition rates early in the shift but that this difference among workloads was appreciably smaller by the end of the shift, especially night shift.
Both Fig. 3.5.1 and Fig. 3.5.2 serve as reminders that work schedule and
workload factors need to be examined in combination to obtain a realistic
picture of fatigue effects. The figures reinforce the notion that extra hours
of work added to the early-morning downturn in circadian arousal is likely
to produce the greatest fatigue.

Daily Rest Breaks. Daily work–rest ratios are a function not only of length of shift but also of how rest is allocated in a shift. Our impression is that a 30- to 60-minute meal break and one or two 10- to 15-minute coffee breaks remain the standard for industrialized countries despite the lack of quotable statistics on rest breaks. Thus, actual work during a nominal 8-hour shift could be reduced by an hour or so. In certain jobs, additional breaks might be distributed throughout the shift under intense working conditions, such as in air traffic control, which requires a high degree of sustained attention at the monitoring station. Much recent research has been devoted to demonstrating the value of more frequent brief rest breaks (e.g., 3–5 minutes) distributed throughout the shift as a countermeasure to physical fatigue from repetitive tasks (such as keyboard data entry; Galinsky, Swanson, Sauter, Hurrell, & Schleifer, 2000). Additional recuperative value might be added to rest breaks by brief exercise or stretching to counteract fatigue from static postures or to induce transient arousal (Henning, Jacques, Kissel, Sullivan, & Alteras-Webb, 1997).

For night workers, "enriching" the value of a rest break might take the form of a brief nap to counteract sleepiness. Several recent demonstrations of napping before or during night work have demonstrated improvements in performance and alertness if naps are taken as a supplement to, rather than as a replacement for, a normal main period of sleep (see Rosa, 1993, for review). Data in Fig. 3.5.3, taken from actual shift workers performing a laboratory work simulation, illustrate the improvements in alertness after napping (Sallinen, Härmä, Åkerstedt, Rosa, & Lillqvist, 1998). As shown, choice reaction time at the end of the night shift was substantially reduced after 30- and 50-minute naps, taken sometime during the shift, when compared to a no-nap control condition.

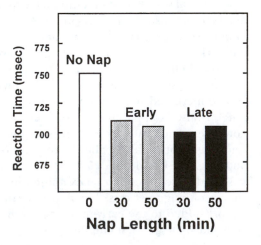

FIG. 3.5.3. Choice reaction time at 7:00 AM at the end of simulated night shifts following no nap or 30-minute and 50-minute naps taken during the first (early) or second (late) half of the shifts.

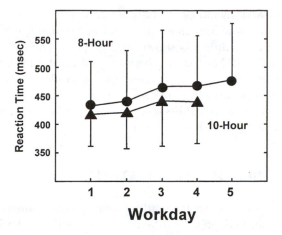

FIG. 3.5.4.   Choice reaction time during a 5-day week of
8-hour shifts and a 4-day week of 10-hour shifts worked
by air traffic control specialists. Daily values are the average of three test sessions per day.

Weekly Work–Rest Ratios

Fatigue also might be expected to accumulate over consecutive days worked,
and recovery might be expected over consecutive rest days between bouts of
work. Against a standard of 5 days of work and 2 days of rest, there are examples of weekly 8-hour shift rotations of up to 7 consecutive days followed by
2 rest days (Rosa et al., 1989), compressed workweeks of 3 or 4 consecutive
extended workdays followed by 3 or 4 rest days, and schedules at distant
mines or offshore oil rigs where 14 consecutive work days at the remote site
might be followed by 14 days of home rest (Duchon, Keran, & Smith, 1994;
Parkes, 1994). Added to scheduled workdays is the possibility for unscheduled overtime days as demand requires. Accumulated fatigue across consecutive workdays is illustrated in Fig. 3.5.4 where similar progressive increases
in choice reaction time are apparent across a 5-day week of 8-hour shifts and
a 4-day week of 10-hour shifts in air traffic control specialists (Schroeder,
Rosa, & Witt, 1998).

Fatigue from night work can be expected to compound accumulated fatigue from consecutive shifts. As mentioned previously, circadian adaptation over consecutive night shifts could reduce some of the circadian effect.
Knauth (1995), however, has pointed to several studies demonstrating progressive increases in accidents and errors over consecutive night shifts, which
suggests that circadian adaptation may be too slow to counteract fatigue or,
alternatively, that sufficient sleep between night shifts is not obtained often.
In our own comparisons of 6- or 7-day workweeks of 8-hour shifts to three- or

four-day compressed workweeks of 12-hour shifts (Rosa et al., 1989; Rosa & Bonnet, 1993), an apparent sleep dept did accumulate over consecutive shifts under both schedules, as shown in Fig. 3.5.5. The debt was greater, however, under the 12-hour shift schedule, perhaps because there were 4 fewer hours of free time per day under that schedule. Indeed, reports of sacrificing sleep because of the work schedule were much more frequent at the end of a week of 12-hour shifts compared with a week of 8-hour shifts, as shown in Fig. 3.5.6 (Rosa & Bonnet, 1993).

SUMMARY OF WORK SCHEDULE FEATURES

Schedule features that potentially affect fatigue are summarized in Table 3.5.1. In the present chapter, we have reviewed briefly the basic features of shift timing, rotation, and ratios of work to rest, which have the potential to interact to determine the basal level of fatigue that a worker brings to the job. In addition to the basic features, there are two other major factors mentioned in Table 3.5.1 that could interact with other features of the work schedule. One factor is the degree of predictability of the schedule, which influences the worker's ability to plan ahead for rest. Although there is little direct empirical research on this factor, there are job situations requiring emergency responses or "on call" availability that may require a rapid change from offduty to onduty status. Consequently, there may be occasions when

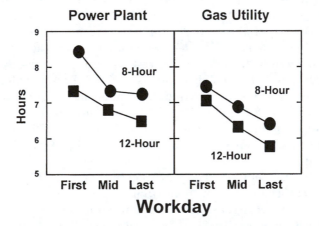

FIG. 3.5.5.   Total sleep time across a 7-day week of 8-hour shifts and a 4-day week of 12-hour shifts at two worksites. Values were calculated from sleep diaries kept by participating workers. "Mid" represents the average of the third and fourth workdays of each schedule.

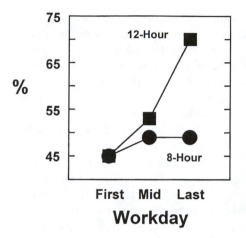

FIG. 3.5.6. Percentage of workers at a natural gas utility who reported sacrificing sleep because of their work schedule across a 7-day week of 8-hour shifts and a 4-day week of 12-hour shifts. "Mid" represents the average of the third and fourth workdays of each schedule.

TABLE 3.5.1
Work Schedule Features Potentially Affecting Fatigue

Feature	Variable	Example
Time of shift	Day, evening, or night	
Shift rotation		
Permanent	Fixed shift times (no rotation)	
Rotating	Changing shift times	
Speed	No. workdays before shift change	Rapid: 2 days/shift
		Slow: 21 days/shift
Direction	Clockwise or counterclockwise change	
Work–rest ratios		
Daily	Work hours/rest hours	8 h work/16 h rest
		12 h work/12 h rest
	Overtime work hours	
	Rest Breaks	
Weekly	No. workdays/no. rest days	5 workdays/2 rest days
		7 workdays/3 rest days
	Overtime workdays	
Predictability	Can affect any other variable	Emergency "on-call"
		Unplanned overtime
		Demand-based scheduling
Flexibility	Selectable start–end times	

the worker must respond without having time for proper rest or sleep. Such occasions are not restricted to public safety or health care situations but may also occur in jobs where products or services are delivered on short notice in the face of fluctuating market demands, as in "just in time" manufacturing. Unscheduled overtime hours are often accrued at those times, occasionally as an addendum to a scheduled extended work shift, which creates the possibility for 16 to 20 hours of consecutive work.

A second factor related to schedule predictability is the degree of flexibility available to the worker for setting starting and ending times for work. In such "flex-time" situations, there is some degree of latitude in setting work hours as long as a minimum number of work hours are achieved over a week (or some other period of days). Flexible scheduling can potentially be used to control fatigue as it allows the worker to adjust the timing of work and rest so that fatigue can be maintained at acceptable levels. On the negative side, flexibility could allow the individual to accumulate many consecutive work hours in a short period as a strategy to gain extra free days at the end of a workweek. Thus, the worker may risk excessive fatigue on a given day to have the luxury of one or more additional free days. Some limits on flex time, such as restricting work-time variations to 1–3 hours, could be instituted to safeguard against such a strategy.

CONCLUSION

The message to be taken from this chapter is that one must examine combinations of work schedule features and look for interactions among them to determine sources of fatigue and to design ways to reduce excessive fatigue. One must resist the temptation to focus on a single dimension of the schedule as the source of all problems or as the answer to all prayers. If a work schedule needs change, it is recommended that any new schedule should be instituted provisionally and all dimensions should be evaluated carefully for fatigue, both at work and during free time when workplace safety sanctions may be absent.

REFERENCES

Åkerstedt, T. (1985). Adjustment of physiological and circadian rhythms and the sleep-wake cycle to shiftwork. In S. Folkard & T. H. Monk (Eds.), *Hours of work: Temporal factors in work scheduling* (pp. 185–198). Chichester, England: Wiley.

Barton, J., & Folkard, S. (1993). Advancing versus delaying shift systems. *Ergonomics, 35*, 59–64.

Cabri, J., De Witte, B., & Clarys, J. P. (1988). Circadian variation in blood pressure responses to muscular exercise. *Ergonomics, 31*, 1559–1565.

Cohen, C. J., & Muehl, G. E. (1977). Human circadian rhythms in resting and exercise pulse rates. *Ergonomics, 20,* 475–479.

Czeisler, C. A., Moore-Ede, M. C., & Coleman, R. M. (1982). Rotating shift work schedules that disrupt sleep are improved by applying circadian principles. *Science, 217,* 460–463.

Duchon, J. C., Keran, C. M., & Smith, T. J. (1994). Extended workdays in an underground mine: A work performance analysis. *Human Factors, 36,* 258–269.

Folkard, S. (1992). Is there a "best compromise" shift system? *Ergonomics, 35,* 1425–1446.

Folkard, S. (1996). Effects on performance efficiency. In W. P. Colquhoun, G. Costa, S. Folkard, & P. Knauth (Eds.), *Shiftwork: Problems and solutions* (pp. 65–87). Frankfort am Main: Peter Lang.

Folkard S., & Barton J. (1993). Does the "forbidden zone" for sleep onset influence morning sleep duration? *Ergonomics, 36,* 85–92.

Gadbois, C. (1981). Women on night shift: Interdependence of sleep and off-the-job activities. In A. Reinberg, N. Vieux, & P. Andlauer (Eds.), *Night and shift work: Biological and social aspects* (pp. 223–227). Oxford: Pergamon Press.

Galinsky, T., Swanson, N., Sauter, S., Hurrell, J., & Schleifer, L. (2000). A field study of supplementary rest breaks for data entry operators. *Ergonomics, 43,* 622–638.

Henning, R. A., Jacques P., Kissel G. V., Sullivan A. B., & Alteras-Webb S. M. (1997). Frequent short rest breaks from computer work: Effects on productivity and well-being at two field sites. *Ergonomics, 40,* 78–91.

Ilmarinen, J., Ilmarinen, R., Korhonen, O., & Nurminen, M. (1980). Circadian variation of physiological functions related to physical work capacity. *Scandinavian Journal of Work, Environment, and Health, 6,* 112–122.

Kecklund, G., & Åkerstedt, T. (1995). Effects of timing of shifts on sleepiness and sleep duration. *Journal of Sleep Research, 4*(Suppl. 2), 47–50.

Knauth P. (1995). Speed and direction of shift rotation. *Journal of Sleep Research, 4*(Suppl. 2), 41–46.

Knauth, P., Landau, K., Droge, C., Schwitteck, M., Widynski, M., & Rutenfranz, J. (1980). Duration of sleep depending on the type of shift work. *International Archives of Occupational and Environmental Health, 46,* 167–177.

Lavie, P. (1991). The 24-hour sleep propensity function (SPF): Practical and theoretical implications. In T. H. Monk (Ed.), *Sleep, sleepiness and performance* (pp. 65–93). Chichester, England: Wiley.

Lavie, P., Chillag N., Epstein, R., Tzischinsky, O., Givon, R., Fuchs, S., & Shahal, B. (1989). Sleep disturbances in shift workers: A marker for maladaptation syndrome. *Work Stress, 3,* 33–40.

Monk, T. H. (1990). Shiftworker performance. In A. Scott (Ed.), *Occupational medicine state of the art reviews, vol. 5: Shiftwork* (pp. 183–198). Philadelphia: Hanley & Belfus.

Monk, T. H., Folkard, S., & Wedderburn A. I. (1996). Maintaining safety and high performance on shiftwork. *Applied Ergonomics, 2,* 17–23.

Parkes, K. R. (1994). Sleep patterns, shiftwork, and individual differences: A comparison of onshore and offshore control-room operators. *Ergonomics, 3,* 827–844.

Rones, P. L., Ilg, R. E., & Gardner, J. M. (1997, April). Trends in hours of work since the mid-1970s. *Monthly Labor Review,* 3–14.

Rosa, R. R. (1993). Napping at home and alertness on the job in rotating shiftworkers. *Sleep, 16,* 727–735.

Rosa, R. R. (1995). Extended workshifts and excessive fatigue. *Journal of Sleep Research, 4,* 51–56.

Rosa, R. R., & Bonnet, M. H. (1993). Performance and alertness on 8-hour and 12-hour rotating shifts at a natural gas utility. *Ergonomics, 36,* 1177–1193.

Rosa, R. R., Bonnet, M. H., & Cole, L. L. (1998). Work schedule and task factors in upper extremity fatigue. *Human Factors, 40,* 150–158.

Rosa, R. R., & Colligan, M. J. (1988). Long workdays versus restdays: Assessing fatigue and alertness with a portable performance battery. *Human Factors, 30,* 305–317.

Rosa, R. R., Colligan, M. J., & Lewis, P. (1989). Extended work days: Effects of 8-hour and 12-hour rotating shift schedules on performance, subjective alertness, sleep patterns, and psychosocial variables. *Work Stress, 3,* 21–32.

Rosa, R. R., Härmä, M., Pulli, K., Mulder, M., & Näsman, O. (1996). Rescheduling a three-shift system at a steel rolling mill: Effects of a 1-hour delay of shift starting times on sleep and alertness in younger and older workers. *Occupational and Environmental Medicine, 53,* 677–685.

Sallinen, M., Härmä, M., Åkerstedt, T., Rosa, R. R., & Lillqvist, O. (1998). Promoting alertness with a short nap during a night shift. *Journal of Sleep Research, 7,* 240–247.

Schroeder, D., Rosa, R. R., & Witt, A. (1998). Effects of 8- versus 10-hour work schedules on the performance of air traffic control specialists. *International Journal of Industrial Ergonomics, 21,* 307–321.

Smith, L., Folkard, S., Tucker, P., & Macdonald, I. (1998). Work shift duration: A review comparing eight hour and 12 hour shift systems. *Occupational and Environmental Medicine, 55,* 217–229.

Spurgeon, A., Harrington, J. M., & Cooper, C. L. (1997). Health and safety problems associated with long working hours: A review of the current position. *Occupational and Environmental Medicine, 54,* 367–375.

Tepas, D. I., Duchon, J. C., & Gersten, A. H. (1993). Shift work and the older worker. *Experimental Aging Research, 19,* 295–320.

Walsh, J. K., Tepas, D. I., & Moss, P. D. (1981). The EEG sleep of night and rotating shift workers. In L. C. Johnson, D. I. Tepas, W. P. Colquhoun, & M. J. Colligan (Eds.), *Biological rhythms, sleep, and shift work* (pp. 371–381). New York: Spectrum.

Wever, R. A. (1986). Characteristics of circadian rhythms in human functions. *Journal of Neural Transmission, 21,* 323–373.

Wilkinson, R. T. (1992). How fast should the night shift rotate? *Ergonomics, 35,* 1425–1446.

Wojtczak-Jaroszowa, J. (1977). *Physiological and psychological aspects of night and shift work.* (NIOSH Technical Report [DHEW (NIOSH)] No. 78-113). Cincinnati: National Institute for Occupational Safety and Health.

Wojtczak-Jaroszowa, J., & Banaszkiewicz, A. (1974). Physical work capacity during the day and at night. *Ergonomics, 17,* 193–198.

PRACTICE

Managing Fatigue in the Road Transport Industry: An Occupational Safety and Health Solution

Laurence R. Hartley
Pauline Arnold
Murdoch University

RESEARCH IN WESTERN AUSTRALIA ON FATIGUE IN THE TRANSPORT INDUSTRY

As elsewhere, there is concern in Australia about the part played by fatigue in truck crashes. Several Australian states restrict a trucker's hours of service to 11 or 12 per 24 hour-period in an attempt to limit fatigue as a causal factor in road crashes involving heavy goods vehicles. These regulations are enforced by requiring drivers to maintain a log book of their hours of service, which must be produced on demand from a police officer. While fatigue has been found to be a significant contributor to road crashes (e.g., Hartley, Arnold, Penna, Hichstadt, Corry, & Feyer, 1996a; Haworth, Heffernan, & Horne, 1989: Ryan & Spittle, 1995; Sweatman, Ogden, Haworth, Pearson, & Vulcan, 1990; Tyson, 1992; Williamson, Feyer, Coumarelos, & Jenkins, 1992), the effectiveness of specific hours of service restrictions for the control of fatigue and resultant reduction of crashes is not clear.

There is little readily available information about truck drivers' work practices in Australia and how these might contribute to the development of fatigue. Although some states in Australia have legislation restricting hours of service and others do not, the only information available prior to this study came from a national survey of drivers working mainly in those states that have prescriptive driving and working hours (Williamson et al., 1992). As it is likely that work practices are different when hours of service are not

531

restricted, the present research obtained information from truck drivers and transport companies in an Australian state, Western Australia (WA), which does not enforce restrictions on hours of service. The aim of the study reported here was to obtain information about hours of work and sleep from drivers operating in a state without restrictions on hours of service. (These are referred to later as unregulated drivers.) It was also intended to have both drivers and companies identify the causes of fatigue and the effective countermeasures they use to manage it. Where possible, these findings are compared with those of Williamson et al.'s drivers, 90% of whom were from eastern states of Australia with restrictions on hours of service. (These are referred to as regulated drivers in later sections). A full report of the findings of the study are presented in Hartley, Arnold, Penna, Hochstadt, Corry, and Feyer (1996a, 1996b, 1996c).

Methodology

Driver Survey.   During a 7-day period, 1,249 truck drivers were invited to participate in the survey. Two hundred and one of the drivers had already been interviewed at other locations. Of the remainder, 638 (60.9%) agreed to be interviewed using a standardized driver's questionnaire. The most common reasons given for not participating were having insufficient time and being uninterested.

Data were collected by teams of research assistants at six roadhouses (road-side businesses that sell gasoline, serve refreshments, and provide restroom facilities) in Western Australia. Because these drivers were not working prescribed hours, they are referred to as unregulated drivers. Roadhouses were chosen so as to provide a sample of drivers from each major long-distance transport route in the state. Research assistants approached drivers entering the roadhouses to invite them to participate in the study at all times that the roadhouses were open for business, in most cases 24 hours per day.

Drivers were asked to provide details about their driving and nondriving work schedules and the amount of sleep they had obtained in the past week. Drivers were asked about the frequency of fatigue-related events such as nodding off while driving, crashes, and near misses. These data were collected using a 5-point scale, anchored from "very frequently" to "never." Drivers were also asked how frequently they experienced fatigue and how frequently it was experienced by, or posed a danger for, others. Drivers were not provided with any description of fatigue. These data were also collected using a 5-point scale, anchored from "always" to "never."

They were also asked to indicate those things that caused them to feel fatigue while driving and what measures they took to manage it.

Company Survey.   Management representatives of transport companies operating in Western Australia were also interviewed using a second standardized company questionnaire. Eighty-eight companies were contacted by telephone, eighty-four agreed to participate, and each nominated someone knowledgeable about transport operations for the interview. As there is no reliable count of the number of transport companies operating in the state, it is impossible to report what proportion these 84 companies represent. Rather, interviews were solicited from a spread of companies representing large, medium, and small, city and country operations with a variety of freight types identified in a traffic count of trucks on major highways.

Company representatives were asked to indicate how frequently their drivers experienced fatigue and how great a problem it was for the transport industry. A definition of fatigue was not provided. These data were collected using a 5-point scale, anchored from "always" to "never." They were asked what caused drivers to experience fatigue. Data were analysed using chi-square analyses. Only comparisons in which differences exceed the 1% probability level (i.e., $p < 0.01$) are reported. This stringent level of confidence was adopted because of the large number of comparisons made

Results

Hours of Driving and Sleep. Because new regulations currently being discussed in Australia propose to limit hours of service to 14 hours in any 24-hour period, the number of unregulated drivers exceeding 14 hours was examined. Two methods were used to estimate the proportion of drivers exceeding 14 hours of service per 24-hour period. The first method was based on drivers' retrospective reports of their activities over the 24 hours preceding being interviewed. The second method relied on their prospective estimates of activities until arrival at their ultimate trip destination. As Table 3.6.1 shows, these estimates suggest about 38% of drivers exceed 14 hours of driving in the 24-hour period. When other nondriving work was taken into account, the proportion exceeding 14 hours of work per 24-hour period increased by about 13%.

TABLE 3.6.1
Percentage of Unregulated Drivers Exceeding 14 Hours Driving and
Nondriving Work in 24 Hours

	Estimates		
	Retrospective Over Previous 24 Hours	*Prospective Until End of Current Trip*	*Average Estimate*
Drivers exceeding 14 hours driving in 24 hours	33%	43%	38%
Drivers exceeding 14 hours of driving & nondriving work in 24 hours	47%	55%	51%

Some discussion has also proposed restrictions on hours of service over the week, with 72 hours suggested as the upper limit. The present data show that 17.5% of unregulated drivers exceeded 72 hours of driving in the week. When nondriving work is added, 30% worked in excess of 72 hours. Eleven percent of drivers had worked more than 90 hours in the previous week. The national survey conducted by Williamson et al. (1992) reported that about 35% of their regulated drivers exceeded 72 hours of work per week. Insofar as the present data from an unregulated state and Williamson et al.'s data are comparable, a greater proportion of truck drivers work longer hours each week in the regulated states. (As Williamson et al. did not report daily hours of service, we are unable to provide a similar comparison for these hours.)

Reports of hours of sleep were also collected both for the week preceding the interview and for the night prior to commencing the current journey. About 20% of unregulated drivers had less than 6 hours of sleep before their current journey. The mean was 8.25 hours of sleep. This figure is similar to the 7.5 hours of sleep before departure obtained by Williamson et al.'s (1992) study of mainly regulated drivers.

One third of unregulated drivers reported they had obtained in excess of 8 hours of sleep on typical workdays in the preceding week. Over half reported they slept between 4 and 8 hours on average. Of great concern, however, is that about 12.5% of unregulated drivers reported having had less than 4 hours of sleep on one or more of their working days in the week preceding the interview. Thus these drivers are likely to be operating their vehicles while having a significant sleep debt, and this may have serious implications for road safety.

Hazardous Events.   Five percent of the unregulated drivers reported having experienced a hazardous, fatigue-related event, such as nodding off, on their current journey. Significantly more of these events were reported by drivers who also reported having slept less than 6 hours before commencing the trip (Table 3.6.2). That is, the 20% of drivers who reported having had less than 6 hours of sleep reported 40% of the hazardous events.

Drivers reported the frequency with which they nodded off or had near misses over the previous 9 months on a 5-point scale ranging from never to very frequently. Fourteen percent of drivers admitted to nodding off at least occasionally while driving. About 16% reported having near misses at least occasionally. Responses about the frequency of nodding off were cross-tabulated with hours of sleep obtained prior to the current journey (see Table 3.6.2). A greater than expected proportion of drivers who said they had had less than 6 hours of sleep prior to departure also said they had nodded off while driving occasionally or more often in the last 9 months.

A further 5% of drivers reported having had a crash in the previous 9 months. When asked if they thought their crash was related to fatigue, 12% thought it was.

TABLE 3.6.2
Number of Unregulated Drivers Reporting
Near Misses and Nodding Off While Driving
by Hours Slept Before Current Journey
(Percentages in Parentheses)

| | Hours Slept Prior to Current Journey | | | |
	Less Than 6 Hours	6–10 Hours	More Than 10 Hours	Total
Near Misses				
Occasionally or more	28 (22.0)	61 (15.9)	13 (13.8)	102 (16.9)
Rarely	28 (22.0)	71 (18.5)	14 (14.9)	113 (18.7)
Never	71 (56.0)	251 (65.5)	67 (71.3)	389 (64.4)
No. and percentage of drivers	127 (100)	383 (100)	94 (100)	604 (100)
Nodding Off				
Occasionally or more	32 (25.2)	46 (12.1)	9 (9.5)	87 (14.4)
Rarely	29 (22.8)	62 (16.3)	17 (17.9)	108 (17.9)
Never	66 (52.0)	273 (71.6)	69 (72.6)	408 (67.7)
No. and percentage of drivers	127 (100)	381 (100)	95 (100)	603 (100)

TABLE 3.6.3
Perceptions of Fatigue as a Driver and Company Problem in an
Unregulated State (Percentages in Parentheses)

| | Perceptions of Fatigue as a Problem for Themselves | | Perceptions of Fatigue as a Problem for Others | |
	Drivers	Companies	Drivers	Companies
Always	28 (4.4)	0 (0)	75 (12.3)	9 (10.7)
Often	37 (5.9)	1 (1.2)	160 (26.3)	26 (31.0)
Sometimes	191 (30.2)	9 (10.7)	323 (53.1)	38 (45.2)
Rarely	143 (22.6)	37 (44.0)	30 (4.9)	7 (8.3)
Never	233 (36.8)	37 (44.0)	20 (3.3)	4 (4.8)
No. of drivers/companies	632 (100)	84 (100)	608 (100)	84 (100)

Perceptions of Fatigue as a Problem for Truck Drivers. The unregulated drivers rated the frequency with which fatigue was a problem for themselves on a 5-point scale ranging from never to always. Company representatives were also asked how often their drivers had a problem with fatigue. Table 3.6.3 shows that about 10% of drivers reported fatigue to be a problem they always or often experienced. Only 1% of company representatives thought their drivers always or often had problems with fatigue. Conversely, about 60% of drivers and 88% of managers considered it was rarely or never a problem for themselves or their drivers.

The data in Table 3.6.3 can also be usefully compared with data shown from Williamson et al's. (1992) study of mainly regulated drivers. Twenty-nine percent of Williamson et al.'s drivers considered fatigue to be a problem on most or every trip, and 15.3% considered it to be rarely or never a problem. Thirty-five percent of these drivers considered it to be at least a major or substantial problem, and 15% considered it to be no problem. It is probably most useful to compare the top and bottom of the scales in the two studies. To the extent to which the present unregulated drivers and Williamson et al.'s data are comparable, fewer unregulated drivers (4.5%) consider fatigue to be always a problem than do their regulated counterparts consider it to be a major problem (8.6–10.7%). Rather more unregulated drivers consider it to be never a problem (37%) as compared with regulated drivers of whom only 15% feel fatigued very rarely or consider it is no problem.

In sharp contrast to ratings of the frequency of fatigue as a personal problem, drivers and company representatives considered that it is a much more common problem for others. Thus, Table 3.6.3 also shows that about 12% of WA drivers thought that other truck drivers always experienced fatigue-related problems, and 26% thought they often did. Similarly, 11% of company representatives thought that fatigue was always a problem for other companies, and 31% thought it often was. Conversely, only 8% of drivers and 13% of company representatives considered it was rarely or never a problem for other drivers or other companies. These data from unregulated drivers can again be usefully compared with those drawn from Williamson et al. (1992), where 38% of predominantly regulated drivers thought fatigue was a major problem. Fatigue appears to be regarded at least as significant a problem, if not a greater one, for the regulated as compared with unregulated industry drivers.

Contributors to Fatigue and Strategies Used to Manage Fatigue.   Drivers and company representatives were asked to name three main factors causing fatigue. The principal causes identified were driving long hours, loading, delays in loading, lack of sleep, tight schedules, and dawn driving. Williamson et al. (1992) also found driving long hours, lack of sleep, loading, and dawn driving were principal causes of fatigue by regulated rather than unregulated drivers. Prescriptive hours of service appears to have done little to diminish the principle causes of fatigue in states with this regulation.

There were some differences between the causes identified by drivers and company representatives. As Table 3.6.4 shows nearly 70% of company managers thought that long hours of driving per se are a main contributor to fatigue whereas less than 40% of drivers named long hours. About half the company representatives thought lack of sleep contributed to fatigue whereas about one third of drivers thought so. Companies also identified inexperi-

TABLE 3.6.4
Contributors to Fatigue
Reported by Unregulated Drivers ($N = 638$)
and Company Representatives ($N = 84$)

	Drivers	Companies
Driving long hours	244 (38.2)	58 (69.0)
Loading/unloading	213 (33.4)	18 (21.4)
Delays in loading	206 (32.4)	6 (7.1)
Lack of sleep	206 (32.4)	41 (48.8)
Over-tight delivery schedules	135 (21.2)	30 (35.7)
Driving between 2–5 AM	135 (21.2)	16 (19.0)
Breakdowns	97 (15.2)	8 (9.5)
Irregular trip schedules	63 (9.9)	11 (13.1)
Poor rest in truck	57 (8.9)	9 (10.7)
Irregular rest hours on road	52 (8.2)	6 (7.1)
Inexperience	36 (5.6)	16 (19.0)
Other	75 (11.8)	16 (19.0)
Nothing	59 (9.2)	1 (1.2)

Note. More than one strategy could be reported.
Percentages in parentheses.

ence as a cause of fatigue more often than did drivers. In contrast, more drivers blamed both loading the truck and delays in loading for their fatigue while fewer company representatives identified these two causes.

Drivers were asked what strategies they used to manage their driving fatigue, and Table 3.6.5 shows their responses. Large proportions of drivers said they obtained a good night's sleep prior to departure or could pull over to the side of the road to rest when they felt tired. Drinking caffeine beverages was reported by more than two thirds of unregulated drivers, and almost half reported use of nicotine products. One in six reported using drugs to manage their fatigue. These data from unregulated drivers can be compared with the predominantly regulated drivers' responses in Williamson et al. (1992). About the same proportion of their drivers report pulling over when tired. Prescriptive hours of service do not seem to have reduced the need to rest and sleep as compared to self-regulations in the industry. The proportion of drivers reporting caffeine and taking nicotine as useful was comparable in the two studies. However, the regulated drivers reported taking drugs about twice as frequently as unregulated drivers, and those taking drugs reported it to be the most effective countermeasure.

Unregulated drivers who were waged employees ($N = 348$) were asked what their companies did to help them manage their fatigue. Ninety percent of drivers reported that their companies engaged or provided opportunities to use appropriate countermeasures. These include things like giving

TABLE 3.6.5
Strategies Used by Unregulated Drivers to Deal With Fatigue

	Frequency	Percentage
Pull over when tired	521	81.7
Drink caffeine beverages	437	68.5
Good night's sleep before departure	398	62.4
Smoke cigarettes/chew nicotine gum	279	43.7
Keep fit and healthy	269	42.2
Sleep regular hours	205	32.1
Eat lollies/chocolates	199	31.2
Take pills/drugs	104	16.3
Drink alcohol	29	4.5
Other	3	0.5
Nothing	32	5.0

Note. More than one strategy could be reported.
($N = 638$).

drivers ample time to do trips (60%), allowing adequate time off between trips (51%), and providing help with loading/unloading (50%). Similarly, when asked what more their companies should be doing, nearly 40% said there was nothing more to be done. Importantly, 10% of drivers said their company did nothing to alleviate fatigue. More than expected of these drivers also reported working in excess of 14 hours in a day or 72 hours in the week prior to being interviewed.

Discussion

The findings about hours worked per day show that a large proportion of unregulated truck drivers in this study work in excess of 14 hours in typical working days. Conversely, this suggests that many drivers obtain less than 10 hours off between work shifts. Concern about these conditions of work is reinforced by the finding that one eighth of drivers get less than 4 hours sleep on one or more of their working days during a week. However, the hours these drivers spend working and sleeping are comparable with those of drivers in states that regulate hours of service (Hartley, Arnold, Penna, Hochstadt, Corry, & Feyer, 1996a; Williamson et al., 1992). Thus, attempting to limit hours of service does not appear to reduce the time spent at work in the road transport industry.

The concern that one eighth of drivers obtained 4 hours of sleep or less on one or more nights and thus may experience sleep debt and an increased risk of an accident was reinforced by the relation between hours slept and nodding off and near misses reported by drivers. Drivers who had less than 6 hours sleep before the trip were over-represented among those nodding off and reporting near misses. A plausible explanation is that the current trip with few hours of prejourney sleep is typical of these drivers. That is, they regularly have few hours of sleep prior to commencing their journeys.

Twelve percent of drivers who reported having had a crash in the previous 9 months identified fatigue as a contributory factor. This is comparable to Maycock's (1995) data for English car drivers, of whom 15% cited tiredness as a contributor to their accidents on major motorways.

Many unregulated drivers and company representatives reported fatigue to often be a problem for other drivers (37%), but considered themselves (10%) or their companies' drivers to be relatively infrequently affected by fatigue. This pattern of responses is consistent with many findings of drivers' overconfidence in their own capabilities and low levels of risk. While being aware of the problems faced by others, drivers commonly consider themselves to be immune or more able to cope with problems than are others.

The data reported about unregulated drivers' perceptions about fatigue was compared with those reported by Williamson et al. (1992) for mainly regulated drivers. About 28% of the regulated drivers reported feeling fatigue on most or all trips they made, and only 15% said it was a very rare event. Thirty-five percent of regulated drivers considered fatigue to be a major or substantial problem for themselves (Hartley et al., 1996a). The results suggest unregulated drivers perceive that fatigue is a problem for themselves less frequently than do regulated drivers (10% versus 28–35%). Similarly, fewer unregulated drivers considered fatigue to be a general industry problem than did regulated drivers (37% versus 78%). These differences in frequency ratings may be due to differences in the attention paid to fatigue as a safety problem in regulated and unregulated states. The enforcement of hours of service regulations and large-scale public awareness campaigns in the regulated states may have made the problem more salient for Williamson et al.'s drivers.

These findings also suggest that, as has been found for road user risk in general, drivers appear to be overconfident about their own resilience to fatigue even though they recognize that others are at risk of experiencing fatigue while driving. Both drivers and company managers rated themselves or their drivers as having a problem with fatigue much less frequently than other drivers. This finding is similar for drivers operating in states where discussion about fatigue has been greater. Although individual awareness of the problem was greater in those drivers, they still rated fatigue to be a far greater problem for other drivers. This suggests that intiatives to raise awareness of the fatigue problem will have difficulty changing many drivers' views about their own vulnerability even though they may have increased awareness of the problem in general.

The perceptions about contributors to fatigue obtained from unregulated drivers and company managers can be compared with Williamson et al.'s results. Poor roads (58%), poor weather (48%), and heavy city traffic (25%) were identified as major causes of fatigue in Williamson et al.'s sample, but these things were not mentioned as causes by unregulated drivers. Compared with regulated drivers, fewer unregulated drivers blamed loading and unloading activities (47% versus 33%) and driving at dawn as causing fatigue (56% versus 21%). Importantly, regulated drivers were more likely than unregulated drivers to report that long hours of service (49% versus 38%) and lack of sleep during the trip (40% versus 32%) caused them fatigue, despite operating under restricted hours of driving.

Drivers in Williamson et al.'s study were also asked what strategies they used to manage their driving fatigue. Like the unregulated drivers (32%), a large proportion of the drivers from the national survey of mainly regulated drivers mentioned regular sleeping habits (78%). Drinking caffeine beverages was also reported by a large proportion of both regulated and unregulated drivers (78% and 69%). Similar proportions of both groups also reported use of nicotine products (47% and 44%). However, twice as many regulated drivers reported taking drugs as did unregulated drivers (32% and 16%). Fifty-three percent of the regulated drivers rated drugs as "among the most helpful" strategy available to them. The only other strategy rated as "among the most helpful" by a large proportion (45%) of the regulated drivers was sleeping. If drug use by drivers is taken as an indicator of fatigue, then fatigue is a greater problem in states with prescriptive hours of service than in the state with self-regulation (Hartley et al., 1996a).

The breadth of causes of fatigue identified by drivers highlights the necessity for multifaceted fatigue management system in the road transport industry. Effective fatigue management requires attention to all causes of fatigue, not just to hours of service. The disparity between drivers' and companies' views about causes and effective countermeasures needs to be resolved; unless there is agreement on what factors need controlling and how it should be done, effective management is hard to achieve.

DEVELOPMENT OF THE WESTERN AUSTRALIAN STRATEGY FOR MANAGING FATIGUE IN THE ROAD TRANSPORT INDUSTRY

The research described previously (Hartley et al., 1996a, 1996b, 1996c), while certainly not suggesting that the problem of driver fatigue under the self-regulating policy in WA is more serious than in other Australian states with hours of service regulations, nevertheless recognized the problem does exist and therefore needs to be addressed in Western Australia as in other states. The WA Department of Transport determined that what was required was a system that was flexible enough to accommodate the very widespread population centers of WA; that required companies and drivers to demonstrate adequate fatigue management practices; and that could be enforced by reference to standards and guidelines. An approach that can provide this flexibility is by way of the Duty of Care provision as required under the Occupational Safety and Health Act.

An Overview of the Occupational Safety and Health Act in Western Australia

The Occupational Safety and Health Act sets objectives to promote and improve occupational safety and health standards in Western Australia. The broad provisions of the act are supported by the Occupational Safety and Health Regulations that detail minimum requirements for specific hazards and work practices and guidance material such as approved codes of practices as described later. The act contains a general Duty of Care, which describes the responsibilities of people in relation to safety and health at work. Employers must, so far as is practicable:

- Provide a workplace and safe system of work so that, as far as practicable, employees are not exposed to hazards.
- Provide employees with information, instruction, training, and supervision to enable them to work in a safe manner.
- Consult and cooperate with safety and health representatives in matters related to safety and health at work.

Responsibility also extends to workers:

- Employees are required to take reasonable care to ensure their own safety and health at work and the safety and health of others affected by their work.
- Self-employed persons must take reasonable care to ensure their own safety and health at work and, as far as practicable, ensure their work does not affect the safety and health of others.

For a commercial driver, a vehicle is a workplace, and others affected by their work include other road users.

An important element of the act is that safer systems of work are required where they are practicable. *Practicable* means that the cost of introducing an intervention to reduce risk is less than the cost of the injury resulting from the hazard. Because fatigue-related crashes are typically severe and costly for road users, a variety of measures to reduce fatigue can be justified. The single most important defense to a charge of negligence is for employers to have in written form the practices and procedures that lead to safe and efficient operations. This is a fatigue management system. To comply with the Duty of Care provision, a fatigue management system should be in place; if the company does not have a documented fatigue management system, it could not demonstrate it had a safe system of work and would be in breach of the act.

The Company Fatigue Management System

Transport operators usually have a risk management program. A driver fatigue management system is another component within the risk management program. Documentation of policies and procedures associated with the fatigue management system provides practical evidence that a system is in place and is actively working to manage driver fatigue. Record keeping is also important. Records are an essential part of an overall risk management program as they provide a history of a particular driver or management activity. This information may be of vital importance in any legal action.

Code of Practice

A company's fatigue management system should address a number of key areas. To assist industry in meeting its duty in these areas, guidance is required on appropriate standards. The provision of a Code of Practice, developed in conjunction with industry, is intended to offer this guidance. A Code of Practice also provides guidance to the authorities. In the event of a government agency or another competent authority investigating an incident involving driver fatigue, a comparison is made between the system of work and the recognized acceptable standard. The Code of Practice developed in WA is intended to provide a standard for transport industry operations based on research, comprehensive in its coverage of the causes of fatigue, and including appropriate measures for excluding them from schedules and rosters and providing countermeasures.

Standards and Guidelines in the Code of Practice

The code contains detailed standards and guidelines covering incident reporting, record keeping, scheduling, rostering, time working, rest periods, fitness for duty, health, management practices, workplace conditions, vehicle safety and standards, training, policy and procedures, responsibilities, management of noncompliance with the fatigue management system, record keeping, and documentation and review processes. For each standard, a variety of fatigue control measures is also proposed for occasions when it is not practicable to adhere to a standard for reasons of obtaining better sleep; delays resulting from accidents, traffic, or weather; or to allow for provision of improved rest facilities or environments. These are not detailed here but include using two-up or shared driving, calling on relief drivers, amending the trip schedule or driver roster. The following standards are proposed.

Scheduling. A key factor in managing driver fatigue is how a company schedules or plans individual trips to meet freight tasks. Where practicable and reasonable, scheduling practices should include appropriate pre-trip or forward planning to minimize fatigue. A driver should not be required to drive unreasonable distances in insufficient time and without sufficient notice and adequate rest. Scheduling practices should not put the delivery of a load before a driver's safety, health, or welfare.

To meet the standards, scheduling should ensure that:

- A driver is given at least 24 hours' notice to prepare for *time working* of 14 hours or more.
- A driver is not required to exceed 168 hours of *active work* in 14 days.
- *Active work* does not average more than 14 hours per 24 hours over 14 days.
- Total time *not working* in any 24 hours is at least 8 hours.
- A solo driver has the opportunity for at least 6 hours of continuous sleep in a 24-hour period.
- Continuous periods of *active work* do not exceed 5 hours.
- Minimum *short breaks* total 30 minutes in 5½ hours.
- Flexible schedules permit *short breaks* or discretionary sleep.
- Driving is minimized if solo driver does not have the opportunity for at least 6 hours of continuous sleep in 24 hours.
- There are no consecutive periods of *active work* time exceeding 14 hours in 24 hours.
- Where night shift operations occur, active hours of work are reduced to reflect the higher crash rate from fatigue between 10:00 PM and 6:00 AM.
- The opportunity for sleep is maximized by minimizing very early departures.

Rostering. Rosters are the driver's planned pattern of work and rest for a week or more. A driver's roster and workload should be arranged to maximize the opportunity for a driver to recover from the effects or onset of fatigue.

To meet the standards, rostering should ensure that:

- A driver does not exceed 168 hours *active work* in 14 days.
- A driver has a least one day of time *not working* in 7 days or two in 14 days.

- Irregular or unfamiliar work rosters are minimized.
- Schedules and rosters that depart from daytime operations are minimized when drivers return from leave.
- Total *time not working* is at least 8 hours per 24 hours.
- Minimum *short breaks* total 30 minutes in 5½ hours.
- A solo driver has opportunity for at least 6 hours of continuous sleep in 24 hours.

Readiness for Duty.   Drivers should be aware of the impact of activities such as second jobs, recreational activities, sports, insufficient sleep, consumption of alcohol and drugs, prescribed or otherwise, and situations stressful of their well-being and capacity to work effectively. These activities may affect their state of fatigue, especially cumulative fatigue, and capacity to drive safely.

To meet standards, readiness for duty means that a driver must be in a fit state for work when presenting for duty.

Health.   Poor health and fitness of drivers is a contributing factor to fatigue, and its effective management is critical to the safe operation of a vehicle. A health management system should be developed and implemented to identify and assist those drivers who are at risk. The system should include medical history, sleep disorders, diet, alcohol, substance abuse or dependency, and lifestyle. The system should also promote better health management.

To meet the standards, health management systems should ensure:

- Medical examinations, at least every 3 years until 49 years of age and every year thereafter, in accordance with the National Road Transport Commission Medical Examination of Commercial Drivers or the Road Transport Forum (RTF) Driver Health Program.
- Assessment of sleep disorders, other fatigue related conditions, and health problems such as diabetes.
- Provision of appropriate employee assistance programs where practicable.
- The provision of information and assistance to promote management of driver health.
- Training of drivers for risk factors for poor health and control measures.
- Drivers are informed of benefits of good dietary intake and necessity for exercise to combat obesity, which can result in risk factors such as obstructive sleep apnea, a common sleep disorder causing day time sleepiness.

- Encourage a healthy lifestyle program in the workplace.
- Encourage drivers to take healthy foods in vehicle to eat on trip and avoid excessive consumption of high calorie food, especially at one sitting, which may cause sleepiness.

Workplace Conditions. Unsafe and unsuitable workplace conditions contribute to fatigue. The ergonomic design standards of a vehicle cabin are important if a driver is to operate a vehicle safely on a road. Unsuitable depot facilities may prevent drivers from obtaining adequate sleep to reduce the effects of fatigue. Operators should ensure workplaces comply with the Occupational Safety and Health Act and relevant Australian design rule specifications.

To meet the standards, workplace conditions should ensure as far as practicable:

- Meet appropriate Australian standards for seating and sleeping accommodation.
- Vehicles that are used for sleep during periods of *time not working* should be fitted with, as a minimum standard:
  In a truck—a sleeper berth that meets ADR42 (Sleeper berths).
  In a tour bus/coach—adequate sleeping accommodation as prescribed by legislation.
- Vehicle cabin meets the requirement of the Occupational Safety and Health Act and includes, as a minimum, ventilation in accordance with ADR 42.17 and seating suspension that is adjustable to driver's weight and height.
- Depots provide safe and suitable fatigue management facilities for rest.
- Truck cabins are air conditioned where practicable, are comfortable and checked before trip.

Training and Education. Training and education must ensure that all employees, contractors, and managers understand the meaning of fatigue and have the knowledge and skills to practice effective fatigue management and comply with the fatigue management system. Training should be structured and programmed to meet the training needs of the participants. All training and education provided should be documented and participation recorded.

To meet the standard, training and education must include:

- Duties imposed by the Occupational Safety and Health Act.
- The penalties for failure to comply with the Occupational Safety and Health Act.

- Introduce training before commencing work.
- Management of driver fatigue and strategies for making lifestyle changes.
- Training and education programs are documented and employee attendance is recorded.

Responsibilities.   The success of a fatigue management system is dependent on the operator, clients, and drivers knowing and practicing their responsibilities and authorities to ensure policies, procedures, and contingency actions are performed as required by the fatigue management system. Responsibilities included in the fatigue management system should be defined and encompassed in position and job descriptions, which should be kept current.

To meet the standards, responsibilities should include:

- The operator should develop the fatigue management system in consultation with drivers and suppliers.
- Duties of the operator and drivers under the Occupational Safety and Health Act.
- Where appropriate the operator should delegate staff to implement the fatigue management system.
- Maintaining records of trip schedules, rosters, time working, and other information to show that the company is conforming with its fatigue management system.

Documentation and Records.   Employers should give consideration to keeping records of all regular and irregular trips, driver's schedules, and rosters. These could be based on trip sheets, pay records, and delivery dockets. They must show sufficient information for an auditor to determine that the company and its drivers have conformed to the fatigue management system.

To meet the standards, documentation and records should include:

- A fatigue management system that documents how the company and its drivers address the agreed operating standards.
- Documents that record all actual regular and irregular trip time schedules, driver's schedules, and rosters.
- In the event that an agreed standard is not met, the control measure(s) that have been adopted should be recorded.
- These should include all trips performed, including details of any trip alterations.
- Personnel records, kept on a confidential basis, include copies of current medical certificates and details of any work restrictions imposed and applicable rehabilitation programs.

Management of Incidents. A fatigue management program should require all unsafe incidents to be recorded. This information should be used to target unsafe practices and prevent injuries and damage. Comprehensive and thorough reporting of all unsafe incidents at work is required in a fatigue management system.

To meet the standards, management of incidents should ensure:

- All unsafe incidents that may cause a hazard or potential injury or harm are reported.
- Sufficient information is collected for action to be taken to prevent a future occurrence of the cause of the unsafe incident.
- Procedures to prevent any further harm or injuries due to this cause.
- Policies that promote and encourage all employees, subcontractors, and relief staff to report all unsafe incidents, including those where there has been no injury or damage.
- Procedures are in place to monitor, record, and investigate all incidents and to take corrective action.
- A review of the fatigue management system after each unsafe incident.

Enforcement of the Code

The code is enforceable under the Occupational Safety and Health act, which provides for a *work improvement* or a *work prohibition* notice to be served on an individual or company when an inspector from the enforcement agency, the Department of Occupational Safety and Health, believes the duty of care principle has been breached. Inspection of the workplace can be at the instigation of a complaint or at any time. The improvement notice states a time by which the breach must be rectified. The prohibition notice states that the work must cease. The notices may stipulate how the breach in duty of care is to be rectified, such as by adherence to the industry Code of Practice or the establishment of a fatigue management system in conformity with the code. The individual or company may appeal to the Commissioner for Health and Safety and then to a Health and Safety Magistrate. In the event that a corporate body is found guilty of the breach, members of the body may also be accountable, such as company managers. Fines for an employer found guilty of a breach of the act are $100,000, and $200,000 if death or injury is caused. Fines for employees are $10,000 and $20,000 for these offenses.

Criminal sanctions can also be applied to companies for safety breaches.

Training, Implementation, and Evaluation

The industry Code of Practices was submitted for comment to the industry and public in late 1997, and comments were considered for incorporation.

The code was field tested with transport companies during early 1998 for full implementation in late 1998. A training program has been developed to assist implementation. The evaluation strategy draws on a number of potential indicators of the impact of the code, including deaths and injuries, lost work day due to injuries, working hours and work practices in the industry, and the costs and benefits to the industry and community.

The incentives for companies to introduce fatigue management conforming to the Code of Practice are that they can demonstrate a safer system of work and avoid prosecution under the Occupational Safety and Health Act; they obtain discounted insurance premiums; some customers require fatigue management systems in their transport companies; and adherence to the code is a defense in the event of companies being prosecuted for unsafe practices. The code can also be used by drivers to demand improved scheduling and rostering from their supervisors; and it can be used by companies to insist that their customers cease demanding impossible delivery schedules or face prosecution for doing so under the Act.

ACKNOWLEDGMENTS

This research was commissioned by the Western Australian Road Transport Industry Advisory Council and funded by the Traffic Board of Western Australia. The industry Code of Practice was developed by the authors in conjunction with the WA Road Transport Industry Advisory Council.

REFERENCES

Hartley, L. R. (Ed.). (1996). *Recommendations of the second international conference on Fatigue.* Murdoch University, Western Australia: Institute for Research in Safety and Transport.

Hartley, L. R., Arnold, P. K., Penna, F., Hichstadt, D., Corry, A. & Feyer, A-M. (1996a). *Fatigue in the WA transport industry: The principle and comparative findings.* (Institute for Research in Safety and Transport Report No. 117 ISBN 0 86905-534-8. Perth: Western Australia Department of Transport.

Hartley, L. R., Arnold, P. K., Penna, F., Hochstadt, D., Corry, A. & Feyer, A-M. (1996b). *Fatigue in the WA transport industry: The drivers' perspective.* (Institute for Research in Safety and Transport Report No. 118 ISBN 0 86905-536-4). Perth: Western Australia Department of Transport.

Hartley, L. R., Arnold, P. K., Penna, F., Hochstadt, D., Corry, A. & Feyer, A-M. (1996c). *Fatigue in the WA transport industry: The company perspective.* (Institute for Research in Safety and Transport Report No. 119 ISBN 0 86905-536-4). Perth: Western Australia Department of Transport.

Haworth, N. L., Heffernan, C. J., & Horne, E. J. (1989). *Fatigue in truck crashes.* Melbourne: Monash University Accident Research Centre.

Maycock, G. (1995). *Driver sleepiness as a factor in car and HGV accidents.* Crowthorne, England: Transport Research Laoratories TRL 169.

Ryan, G. A., & Spittle, J. (1995). The frequency of fatigue in truck crashes. In *Proceedings of the National Road Safety Research and Enforcement Conference* (pp. 97–110). Perth, WA: Promaco Conventions.

Sweatman, P., Ogden, K., Haworth, N. L., Pearson, R., & Vulcan, P. (1990). *Heavy vehicle safety on major NSW highways; A study of crashes and countermeasures.* Sydney: Roads and Traffic Authority of New South Wales, Department of Transport and Communications.

Tyson, A. H. (1992). *Articulated truck crashes in South Australia 1978–1987: A discussion of countermeasures.* Adelaide: Office of Road Safety, South Australian Department of Road Transport.

Williamson, A. M., Feyer, A. M., Coumarelos, C., & Jenkins, T. (1992). *Strategies to combat fatigue in the long distance road transport industry.* Canberra: Federal Office of Road Safety (CR 108).

Broadening Our View of Effective Solutions to Commercial Driver Fatigue

Anne-Marie Feyer
*New Zealand Environmental and Occupational
Health Research Centre*

Ann M. Williamson
University of New South Wales

Fatigue has emerged as a major occupational hazard for commercial long-distance drivers. For example, there is increasing recognition that it is a major risk factor for crashes involving heavy vehicles in Australia (Howarth, Hefferman & Horne, 1989; Sweatman, Ogden, Haworth, Vulcan & Pearson, 1990) and elsewhere (Hamelin, 1987; Mitler et al., 1988; National Transportation Safety Board, 1995; van Ouwerkerk, 1987). Howarth, Triggs, and Grey (1988) estimated that, for articulated vehicles in Australia, between 5 and 10% of all crashes, 20 to 30% of casualty crashes, and 25 to 35% of fatal crashes are probably caused by fatigue. For particular types of crashes, the involvement of fatigue may be much higher, for example, 40% to 50% of fatal single vehicle semitrailer crashes are probably fatigue related (Howarth et al., 1988). If accidents where fatigue is a contributory rather than primary cause are considered, as many as 60% of heavy vehicle crashes have been reported as involving fatigue to some extent (Sweatman et al., 1990).

The prevalence of exposure to fatigue underscores the magnitude of the problem. A substantial proportion of professional drivers report that they drive while fatigued on a regular basis. Approximately one half of drivers surveyed in a large national sample in Australia reported experiencing fatigue on their last trip and indeed on at least one half of their trips overall (Williamson, Feyer, Coumarelos, & Jenkins, 1992). Other surveys have found that heavy vehicle drivers commonly report driving near to the edge of sleep or even having fallen asleep at the wheel recently (Howarth, Vulcan, Schulze, & Foddy, 1990; van Owerkerk, 1987). The prevalence of fatigue among professional drivers, particularly given the level of exposure to driving, is of serious concern. There is clear evidence both from experimental studies (Brookhuis & De Waard, 1993; Mackie & Miller, 1978; Ranney & Gawron, 1987) as well as from drivers themselves (Williamson et al., 1992) that fatigue impairs driving skills and is thus likely to play a role in crashes.

The search for effective ways of dealing with commercial driver fatigue under operational conditions has been slow to advance, however. In most parts of the world, the focus has been on setting parameters for regulating working/driving hours that manage driver fatigue. This regulatory approach has not been particularly successful, however. Despite quite different parameters to be found in regulations around the world, commercial driver fatigue remains a serious problem (U.S. Department of Transportation, Federal Highway Administration, 1990). In Australia, for example, those states with unregulated working hours regimes do not have very different crash profiles from those states that are regulated (Hartley et al., 1996). One possible explanation for this situation is that the focus of regulatory frameworks, largely duration of work, is too narrow. Evidence from other industrial settings with round-the-clock operation suggests that the pattern of work and rest, rather than merely consecutive hours of operation, is important for managing safe operation. Night work, timing of work periods, number of work periods in succession, and time off between periods of work are just some of the characteristics that together have been implicated in determining safety in 24-hour operations (Folkard, 1996; Knauth, 1996). Given that none of these factors is reflected in current regulatory approaches, it is hardly surprising that, as they stand, the regulatory regimes have been of limited effectiveness. Yet, there has been reluctance on the part of regulators to relinquish the traditional regulatory approach. To a large extent, this reluctance reflects the absence of alternatives that have demonstrated effectiveness for managing fatigue.

Recent data suggest that aspects of operational practice other than the length of time-on-task have a major influence on fatigue among long-distance drivers. This chapter provides an overview of this work on the relation between work practices and fatigue and the implications of these data for more effective management of fatigue among commercial long-distance drivers.

WORK PRACTICES IN THE LONG-DISTANCE ROAD TRANSPORT INDUSTRY IN AUSTRALIA

Three main work practices operate in long-distance road transport in Australia. The standard practice is single or solo operation, where one driver takes the load from point of origin to point of destination. Two alternatives currently operate in Australia, both reflecting the industry's attempts to keep the truck on the road for 24 hours a day and to cover the necessary distances while managing driver performance. In staged operations, drivers from different points of origin renedezvous en route and exchange trucks or loads. Both drivers then return to their respective starting point with their new cargo. In the two-up regime, two drivers complete the trip from point of origin to point of destination, sharing the driving.

Over a series of studies, Feyer and Williamson have examined the impact of these work practices on fatigue (Feyer & Williamson, 1996; Feyer, Williamson, & Friswell, 1995, 1997; Feyer, Williamson, Jenkin, & Higgins, 1993; Williamson et al., 1992; Williamson, Feyer, & Friswell, 1996; Williamson, Feyer, Friswell, & Leslie, 1994). These studies have revealed that it is not the work practices in principle that are critical. Rather, different work practices, as they are currently operationalized, vary in terms of the extent to which they incorporate factors that fundamentally contribute to fatigue and/or the extent to which they provide opportunities for the driver to take remedial action. Three main factors have emerged as key determinants of the extent to which fatigue in commercial long-distance driving is managed: acute fatigue from a period of work, chronic or accumulated fatigue over several periods of work, and the nature of effective work/rest balance to address both acute and chronic fatigue.

Acute Fatigue

Two main factors are likely to contribute to acute fatigue over a driving shift: duration of the work period and the time of day at which it occurs. Research on time-on-task suggests that, in general, the longer the duration the greater the fatigue and the greater the loss of performance capacity (Krueger, 1989). Driving is no different. Accident risk increases with time on the road (Hamelin, 1987; Jones & Stein, 1987; Mackie & Miller, 1978; van Ouwerkerk, 1987). The occurrence of fatigue also increases with time on the road (Harris & Mackie; 1972; Mackie & Miller, 1978). Acute fatigue is likely to be a problem for long-distance drivers because the hours that are worked in the long-distance road transport industry are far outside the norm for most working populations (Williamson et al., 1992). Duration of work periods has, therefore, been the most readily acknowledged cause of commercial driver fatigue and has been the cornerstone of most regulatory frameworks.

Research on the impact of time of day on performance and accident risk is also vast (Åkerstedt, 1994; Folkard & Monk, 1979; Smith, Folkard & Poole, 1994). Clear evidence documents its importance in road transport safety. Not only are night-time operations more prone to fatigue, but the impact of time of day on truck crashes is also well recognized (Hamelin, 1987; Kecklund & Åkerstedt, 1995; Mackie & Miller, 1978; van Ouwerkerk, 1987). Yet, virtually no attention has been paid to the impact of time of day in regulatory frameworks for driving hours.

ACUTE FATIGUE IN THE LONG-DISTANCE
ROAD TRANSPORT INDUSTRY IN AUSTRALIA

The distances that need to be covered in Australia are vast, and work periods are accordingly lengthy. As discussed earlier, operational practices used in the industry aim to manage driver fatigue associated with the need to cover these long distances by providing a relief driver as part of a driving team or keeping trips shorter through use of staged operations. Results from our early work suggested that these operational attempts to manage fatigue may be undermined by the way in which the practices are actually implemented. First, it is clear that the relation between fatigue experience and time on the job is not an entirely linear one. In a survey-based study, long-distance drivers responded to a questionnaire that obtained information about work, rest, and fatigue experience (for further details, see Williamson et al., 1992). Drivers were classified in each of two ways, according to the operation they drive on the last trip and according to their employment status. Table 3.7.1 shows work hours on the last trip and fatigue experience for each type of

TABLE 3.7.1

Work Duration, Perception of Fatigue, and Experience of Fatigue
Reported by Drivers Driving Each Operation (% of Drivers in Each
Category) and by Employment Status (% Drivers in Each Category)

	Single (N = 844)	Two-up (N = 45)	Staged (N = 60)	Employee (N = 731)	Owner (N = 215)
Duration of last trip					
1–11 hrs	16.6	5.7	60.3	21.8	9.5
12–17 hrs	29.0	8.6	31.0	32.2	16.5
18–29 hrs	27.2	34.3	5.2	23.3	35.0
>29 hrs	27.1	51.4	3.4	22.7	39.0
Perception of fatigue as a personal problem					
A problem	38.2	27.3	16.7	34.6	36.6
Not a problem	61.8	72.7	83.3	65.4	63.4
Experience of fatigue on the last trip					
Yes	57.8	50.0	39.7	55.2	60.2
No	42.2	50.0	60.3	44.8	39.8
Typical onset of fatigue					
Mean hours	12.58	18.63	9.33	12.12	14.22

Data are from A.-M. Feyer and A. M. Williamson (1995), Work and rest in the long distance road transport industry in Australia, *Work and Stress*, 9(2/3), 198–205. Copyright 1995 by Taylor & Francis Ltd., *http://www.tandf.co.uk/journals*.

driving operation and by employment status. The survey revealed that the role of duration of work period was, in part, unexpected. Overall, shorter hours were associated with less fatigue. However, the longest hours were not necessarily associated with the worst fatigue experience, and shorter hours only partially offset fatigue occurrence. Fewer two-up drivers and owner drivers reported fatigue as a problem in general or on the last trip, even though they reported the longest working hours. In contrast, single and employee groups worked considerably shorter hours but were higher reporters of fatigue. In addition, shorter hours in this industry do not appear to prevent the onset of fatigue. Staged drivers, while working the shortest hours of any group, reported experiencing fatigue much earlier in a work period. Other aspects of operational practice besides time on the job are important determinants of acute fatigue.

Further examination of the survey data revealed several factors accounting for fatigue experience. Figure 3.7.1 shows the distribution of work, rest, and fatigue on their last trip for all drivers surveyed and for each operational subgroup. The distribution of fatigue occurrence followed the pattern that would be expected based on circadian principles. For all groups of drivers, fatigue occurrence peaked in the early hours of the morning with a second smaller peak in the early hours of the afternoon. From Fig. 3.7.1, it is also clear that the pattern of work for staged drivers was different from that of all other groups in that they worked mainly overnight. The timing of their work period, which consistently occurred at the low point of the circadian rhythm, is likely to have undermined the benefits of shorter trips for staged drivers.

The other outstanding feature of the data obvious from Fig. 3.7.1 is the relation between the distribution of fatigue and of rest breaks. Concordance between the distributions was least for staged drivers and greatest for two-up drivers and owner drivers. In other words, for most drivers, the distribution of breaks did not peak at time of peak fatigue. This finding may account for the fact that duration of work period did not seem to account for all the variance in fatigue experience. When the timing of work and rest was arranged to coincide with periods of fatigue, drivers fared better, despite working longer hours. In contrast, when fatigue and rest did not coincide well, as was the case for stage drivers, shorter trips did not entirely offset the cost of working when fatigued.

From these data, it seems that the duration of work and its timing are interactive as influences of fatigue. This interpretation is corroborated by data about the role of these factors in crashes. Accident probability increased with both hours of driving and time of day, but the combined effects of the two risk factors are multiplicative rather than additive, so that the midnight-to-dawn period amplifies the effect of duration (Moore-Ede, Campbell, & Baker, 1988). Even relatively shorter work periods can be fatiguing if they occur at night.

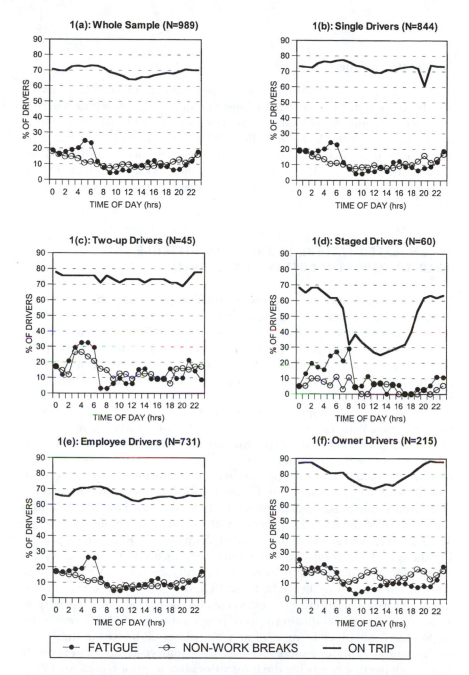

FIG. 3.7.1. The proportion of drivers on a trip each hour of the day, and, of those, the proportion reporting taking a break and the proportion reporting fatigue at each hour of the day (expressed as a percentage of total sample, of each type of operational and each employment classification). From "Work and Rest in the Long Distance Road Transport Industry in Australia," by A.-M. Feyer and A. M. Williamson, 1995, *Work and Stress, 9*(2/3), 198–205. Copyright 1995 by Taylor & Francis Ltd., http://www.tandf.co.uk/journals. Reprinted with permission.

Accumulated Fatigue

Acute fatigue, from both duration of any spell of work and circadian factors, is only one part of the problem for long-distance drivers, however. There is evidence indicating that fatigue accumulates across days of schedules (Jovanis & Kaneko, 1990; Mackie & Miller, 1978). This problem is particularly acute where there is accumulated sleep debt because much of the work occurs at night (Bonnet, 1994; Mitler et al., 1988). With accumulated sleep debt, performance deteriorates over successive night shifts (Tilley, Wilkinson, Warren, Watson, & Drud, 1982).

ACCUMULATED FATIGUE IN THE LONG-DISTANCE ROAD TRANSPORT INDUSTRY IN AUSTRALIA

Several findings from two onroad studies underpin the importance of accumulated fatigue as part of the problem to be addressed if driver fatigue is to be managed effectively. For both these studies, drivers were assessed via subjective, physiological, and performance measures during regular operational trips to examine the relaion between characteristics of operational practice and the development of fatigue (Williamson et al., 1994, 1996; Feyer et al., 1995, 1997).

For many drivers, fatigue is accumulated even before any given driving period begins. Williamson et al. (1994, 1996) found that the potential benefits of staged operations to keep trips shorter and enable more rest to be taken at home were undermined by the overall patterns of work and rest that are the hallmark of such operations. The pattern for these drivers was typically one of high workload in the previous 7 days. The high workload was not so much due to the total hours worked, but was based on their relatively high degree of night work combined with relatively low amounts of night sleep. Drivers averaged only 6.3 hours of sleep per day in the week preceding the staged trip compared with community norms of 7.5 hours (Hyppa, Kronholm, & Mattlar, 1991). In addition, only one half (51.8%) of their sleep occurred during the night hours. These figures suggest that drivers may well have been suffering the effects of chronic partial sleep deprivations (CPSD). Such work/rest patterns were proposed to be the likely reason for higher levels of fatigue being evident at the outset of the first of three trips in the study. In all, three successive trips were undertaken by drivers over a week, with the first one being the most vulnerable to prior typical work/rest patterns. The second and third trips drivers undertook were mostly preceded by a break of at least 24 hours, that is, the experimental regime was not as demanding of drivers as their normal workweek. Most important for the present discussion, pretrip levels of fatigue appeared to be an important determinant of later fatigue. When drivers started the trip more tired, they remained at higher levels of fatigue throughout the trip (Williamson et al., 1996).

The relation between pretrip work/rest patterns and fatigue was seen even more clearly in the results of another onroad study. The impact of driving on a very long trip typical of remote zone driving in Australia was examined, and the fatigue experience of two-up and single drivers was compared (Feyer et al., 1995, 1997). Operational differences were clear in pretrip fatigue levels. Fatigue was found to be significantly higher at pretrip for two-up drivers than for single drivers in the study and remained so for most of the trip (Feyer et al., 1995). In fact, although fatigue increased for all drivers over the trip, two-up drivers were also quicker to reach higher levels of fatigue on the trip.

When possible reasons for greater fatigue at the outset among two-up drivers was examined, recent work history again provided some insights. Two-up and single drivers obtained equivalent amounts of sleep in the night prior to the studied trip (7 to 8 hours) and had similar lags between waking and starting work on the day of departure. Once at work, however, there were clear differences between the groups in terms of the duration and type of pretrip work activities undertaken. Two-up drivers spent nearly twice as much time at work before commencing their trips and were more than twice as likely as single drivers to be involved in loading activities, with an average of 7 to 8 hours spent involved in such activities before departure. It is not surprising that the impact of what was essentially a full day of physical work before driving commenced is reflected in higher pretrip fatigue levels, reported just before departure by two-up drivers.

The relation between fatigue and immediate pretrip work was directly examined in this study. Analysis revealed that pretrip work was an important determinant of the level of fatigue reported by drivers before commencing their trips. Figure 3.7.2 shows the relation between self-reported fatigue before the trip and time spent in pretrip work in general, and loading specifically, on the day of the measured trip for all 37 drivers in the study. Those drivers who did more loading and yard work were significantly higher reporters of fatigue before the trip commenced.

Relatively little work has examined the impact of nondriving work as part of the overall workload of truck drivers. Yet, it is clear that when drivers do most of their own loading, they report that they are often tired before setting out on a trip (Feyer et al., 1997; Howarth et al., 1990). Moreover, when combined with subsequent long periods of driving, physical work can have adverse effects on fatigue and performance (Mackie & Miller, 1978). Importantly, it has not been usual for regulations to clearly recognize the impact of such additional workload. Enforcement has tended to concentrate on driving hours, possibly because they provide a more enforceable target.

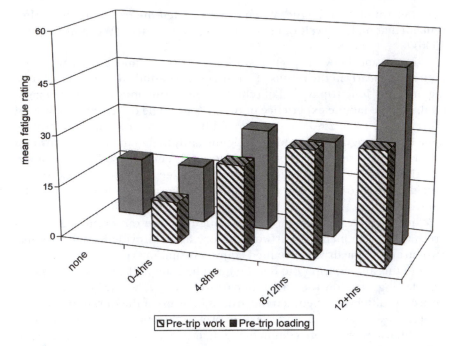

FIG. 3.7.2. The influence of duration of pre-trip work in general, and pre-trip loading in particular, on self-reported fatigue at the outset of the trip.

Besides work on the day of departure, work history in the previous week is also likely to have contributed to pretrip fatigue for two-up drivers. Single and two-up drivers spent statistically equivalent amounts of time working, although single drivers reported spending more time driving. However, work and rest were differently distributed, so that two-up drivers obtained less of their rest and recovery at night. Thus, despite single drivers having driven more and rested less in the previous week, they appeared to have been better rested at the beginning of the trip than two-up drivers because a greater proportion of their rest was taken at night.

Pretrip fatigue, accumulated through pretrip work and/or through CPSD, appears to be a powerful determinant of later fatigue and performance. Although it is acknowledged that sleep deprivation in general has a profound effect on human functioning, it has been suggested that partial sleep deprivation may in fact have a much stronger effect on subsequent performance than either short-term or long-term sleep deprivation (Pilcher & Huffcut, 1996). This raised important questions about ongoing work and rest schedules under which long-distance drivers operate and suggests the importance of managing recovery from accumulated fatigue in those schedules.

EFFECTIVE RECOVERY
IN THE MANAGEMENT OF FATIGUE

The major path to recovery from fatigue is, of course, sleep. Considerable evidence documents the importance of time of day in the occurrence of fatigue and therefore, the importance of night rest for restoring alertness (Åkerstedt, 1995; Bonnet, 1994). Research examining the advantages to be gained from naps suggests that their benefits depend on their duration, on the length of the preceding sleep loss, and on the time of the day during which sleep is taken (Angus, Pigeau, & Heslefrave, 1992; Dinges, Orne, Whitehouse, & Orne, 1987; Gillberg, 1984). When naps are effective, their effects on performance have been reported to be considerable, both in terms of extent of improvement in performance and longevity of the improvement (Dinges, 1992). On the other hand, taking rest in split periods, rather than in consolidated overnight blocks, has been reported as a risk factor for truck crashes (Hertz, 1988; National Transportation Safety Board, 1995).

From our data, there is clear evidence of a consistent distribution of self-reported fatigue across the day and night for drivers. Fatigue occurrence followed the pattern expected on the basis of circadian principles, and better fatigue management was reported when drivers reported taking rest at these vulnerable times of day (Feyer & Williamson, 1995). Moreover, the finding that those drivers who undertook the longest trips, owner drivers and two-up drivers, were able to stave off the effects of fatigue for longer than any other group, may reflect that these drivers were able to enhance the impact of rest by its timing. These findings raise the issue that the ability to achieve effective rest may be a critical feature of fatigue management for long-distance drivers.

We examined the role of rest in maintenance and recovery of alertness directly, in the onroad evaluation of two-up driving and single driving in remote zones of Australia (Feyer et al., 1995, 1997). The method for the study, described in detail elsewhere (Feyer et al., 1995), included performance and subjective and physiological measures of fatigue. The trip selected for the study was the 4,500 kilometer round trip between Perth and Broome, on Australia's western coastline, and was completed by 22 (11 pairs) two-up drivers and 15 single drivers. These trips extended over several days, taking between 80 and 119 hours, depending on the operation used. Although involving only one trip, the sequence of work and rest activities over the trip in fact provided an opportunity to study what amounts to work for a week. The study also provided an unusual opportunity to examine accumulated fatigue. During two-up operations, drivers are likely to experience a gradual buildup of accumulated fatigue across the trip. On the one hand, drivers have access to regular breaks, with evidence suggesting that shorter sleeps taken regularly and early in a period of work can be most effective (Dinges, White-house, Orne, & Orne, 1988; Hartley, 1974). On the other hand, despite reg-ular access to breaks throughout the day and night, drivers doing two-up operations forego regular overnight rest because of the duration of the very long trips, so that chronic partial sleep deprivation is likely to be a problem. As discussed previously, overall, the study showed, not surprisingly, that fa-tigue increased for all drivers over the trip. Although overall, two-up drivers showed greater fatigue compared with single drivers, some ways of doing two-up were less fatiguing than single driving. Important differences were seen among two-up drivers in terms of trip length and distribution of rest obtained across the 4- to 5-day trip. Three different trip subgroups were identified. The first group undertook the trip in typical two-up fashion, with no stationary rest. The second group also took regular in-vehicle rest breaks, like the first group, but its trips were around one third longer. This group did an additional leg of the trip beyond the anticipated mid-trip point, do-ing the longest trip of any of the groups. The third group covered the same distance as the first group, but in addition to in-vehicle rest it also had an overnight stationary rest at trip midpoint.

Striking differences were seen in recovery and maintenance of alertness associated with these operational differences. Figure 3.7.3 shows self-re-ported fatigue at the four major milestones of the trip, at pretrip, before and mid-trip point, after the mid-trip point, and at post-trip. Overnight station-ary test for two-up drivers at the time of peak fatigue, at mid-trip, was associ-ated with significant reductions in fatigue levels after the break and allowed these drivers to finish with the lowest levels of fatigue of any group, includ-ing single drivers. Two-up drivers who had no stationary rest but had the shortest trip duration of any group showed a significant increase in alertness on the last leg of the trip and also fared better than single drivers. Although single drivers showed substantial recovery of alertness after stationary overnight rest at mid-trip, this recovery was not maintained over the home-ward leg of the trip with significant decrease in alertness evident at the end of the trip. In contrast, two-up drivers who did much longer trips and did these trips without the benefit of overnight stationary rest at mid-trip, showed little sign of recovery at any point in the trip. By including an addi-tional sector to the trip at midpoint, increasing the overall trip duration by around one third, these drivers accrued significantly greater fatigue than the other groups. They continued to deteriorate after the midpoint, ending the trip more tired than any other group. It is noteworthy that increases in self-reported fatigue were associated with poorer performance and reduced arousal (Feyer et al., 1995, 1997).

The pattern of fatigue development in the two-up groups was not related to general differences in break activity of the groups. The groups were re-markably similar in all respects, including the nature, duration, and distri-

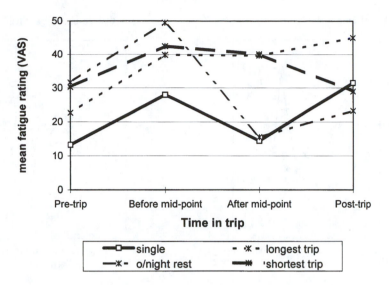

FIG. 3.7.3.   Mean self-reported fatigue at milestones in the trip for single and two-up subgroups (higher ratings indicate higher fatigue). From "Balancing Work and Rest to Combat Driver Fatigue: An Investigation of Two-Up Driving in Australia," by A.-M. Feyer, A. M. Williamson, and R. Friswell, 1997, *Accident Analysis and Prevention, 29*(4), 541–553. Copyright 1997 by Elsevier Science. Reprinted with permission.

bution of breaks. Two-up drivers shared the driving in approximately equal proportions, changing over at around 4- to 5-hour intervals. There was one notable exception: Only one two-up group obtained stationary overnight rest. Rather, the patterns of fatigue development reflected changes in the effectiveness of breaks over the trip. Figure 3.7.4 shows the pre- and post-break alertness for each break in the trip, for each type of two-up trip. As two-up trips became longer, breaks became increasingly ineffective in the latter part of the trip and totally lost their effectiveness at the end of the longest trip. Where two-up drivers had a stationary rest, breaks continued to be restorative, and in fact, these drivers appeared to gain more from the rest breaks on the homeward leg of the trip than any other group. Shorter trips also allowed drivers to benefit from breaks throughout the trip, although prebreak fatigue levels remained constant over the trip. Thus, where work practices kept fatigue under control, such as on shorter trips and where trips incorporated overnight stationary rest, rest breaks were more likely to be helpful. Where fatigue continued to build up, such as on longer two-up trips without the benefit of overnight stationary rest, rest breaks provided poor recovery once fatigue had accumulated.

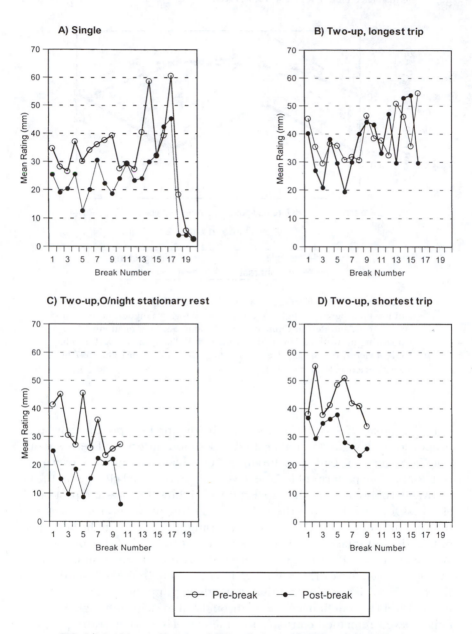

FIG. 3.7.4. Mean self-reported fatigue before and after each break in the trip for single and two-up subgroups (higher ratings indicate higher fatigue). From "Balancing Work and Rest to Combat Driver Fatigue: An Investigation of Two-Up Driving in Australia," by A.-M. Feyer, A. M. Williamson, and R. Friswell (1997). *Accident Analysis and Prevention, 29*(4), 541–553. Copyright 1997 by Elsevier Science. Reprinted with permission.

CONCLUSIONS

Taken overall, the present findings clearly suggest that commercial driver fatigue must be viewed as part of the whole pattern of work and rest. The most effective improvements in managing fatigue must take account of overall work practices, including activities in the past week, activities before driving begins, as well as the way the trip is structured. Chronic fatigue as well as acute fatigue appears to be a major hazard of the job of long-distance drivers. Neither current operational and rest practices nor hours of service regulations recognize these hazards. Preoccupation with limitations of total hours of driving in a day or a week has dominated the search for solutions. In contrast, there is now growing evidence that patterns of work and rest are far more important determinants of fatigue. When drivers do very long trips, these could be offset by judicious use of team driving and overnight rest, while much shorter trips were clearly not immune to the effects of extensive night work. Fatigue accumulated before trips adds to the accumulation of fatigue once trips begin. All rest breaks are not of equal value to recuperation, their benefit depending critically on aspects of their timing. If management of fatigue is to be improved, many parameters that are part of the overall balance between work and rest need to be part of the equation, not just time on the job.

REFERENCES

Åkerstedt, T. (1994). Work injuries and time of day—national data. *Work hours, sleepiness, and accidents. Stress Research Reports, 248,* 106.

Åkerstedt, T. (1995). Work hours, sleepiness and accidents. Introduction and summary. *Journal of Sleep Research, 4*(Suppl. 2), 1–3.

Angus, R. G., Pigeau, R. A., & Heselgrave, R. J. (1992). Sustained operations studies: From the field. In C. Stampi (Ed.), *Why we nap: Evolution, chronobiology and functions of polyphasic and ultrashort sleep* (pp. 217–241). Boston: Birkhauser.

Bonnet, M. H. (1994). Sleep deprivation. In M. H. Kryger, T. C. Roth, & W. C. Dement (Eds.), *Principles and practice of sleep medicine* (2nd ed., pp. 50–67). Philadelphia: W. B. Saunders.

Brookuis, K. A., & De Waard, D. (1993). The use of psychophysiology to assess driver status. *Ergonomics, 36,* 1099–1110.

Dinges, D. F. (1992). Adult napping and its effects on the ability to function. In C. Stampi (Ed.), *Why we nap: Evolution, chronobiology and functions of polyphasic and ultrashort sleep* (pp. 119–134). Boston: Birkhauser.

Dinges, D. F., Orne, M. T., Whitehouse, W. G., & Orne E. C. (1987). Temporal placement of a nap for alertness: Contributions of circadian phase and prior wakefulness. *Sleep, 10,* 313–329.

Dinges, D. F., Whitehouse, W. G., Orne, E. C., & Orne, M. T. (1988). The benefits of a nap during prolonged work and wakefulness. *Work and Stress, 2,* 139–153.

Feyer, A.-M., & Williamson, A. M. (1995). Work and rest in the long distance road transport industry. *Work and Stress, 9,* 199–205.

Feyer, A.-M., Williamson, A. M., & Friswell R. (1995). *Strategies to combat fatigue in the long distance road transport industry, Stage 2: Evaluation of two-up operations.* (Report No. CR 158). Canberra: Federal Office of Road Safety.

Feyer, A.-M., Williamson, A. M., & Friswell, R. (1997). Balancing work and rest to combat driver fatigue: An investigation of two-up driving in Australia. *Accident Analysis and Prevention, 29,* 541–553.

Feyer, A.-M., Williamson, A. M., Jenkin, R. A., & Higgins, T. (1993). *Strategies to combat fatigue in the long distance road transport industry: The bus and coach perspective.* (Report No. CR 122). Canberra: Federal Office of Road Safety.

Folkard, S. (1996). Effects of performance efficiency. In W. P. Colquhoun, G. Cost, S. Folkard, & P. Knauth (Eds.), *Shiftwork: Problems and solutions* (pp. 67–87). Frankfurt am Main: Peter Lang.

Folkard, S., & Monk, T. H. (1979). Shiftwork and performance. *Human Factors, 21,* 483–492.

Gilberg, M. (1984). The effects of two alternative timings of a one hour nap on early morning performance. *Biological Psychology, 19,* 45–54.

Hamelin, P. (1987). Lorry drivers' time habits in work and their involvement in traffic accidents. *Ergonomics, 30,* 1323–1333.

Harris, W., & Mackie, R. R. (1972). *A study of the relationship among fatigue, hours of service and safety operations of truck and bus operators.* (Report No. BMCS-RD-71-2). Washington, DC: Bureau of Motor Carrier Safety.

Hartley, L. R. (1974). A comparison of continuous and distributed reduced sleep schedules. *Quarterly Journal of Experimental Psychology, 26,* 8–14.

Hartley, L. R., Arnold, P. K., Penna, F., Hochstadt, D., Corry, A., & Feyer, A.-M. (1997). *Fatigue in the Western Australian road transport industry, Part 1: The principle and comparative findings.* (Technical report).

Hertz, R. (1988). Tractor trailer driver fatality: The role of nonconsecutive rest in a sleeper berth. *Accident Analysis and Prevention, 20,* 431–439.

Howarth, N. L., Heffernan, C. L., & Horne, E. J. (1989). *Fatigue in truck crashes.* (Report No. 3). Melbourne: Monash University Accident Research Centre.

Howarth, N. L., Triggs, T. J., & Grey, E. M. (1988). *Driver fatigue: Concepts, measurement and crash countermeasures.* (Report No. CR 72). Canberra: Federal Office of Road Safety.

Howarth, N. L., Vulcan, A. P., Schulze, M. T., & Foddy, B. (1990). *Truck driver behaviour and perceptions study.* (Report No. 18). Melbourne: VicRoads.

Hyppa, M. T., Kronholm, E., & Mattlar, C. E. (1991). Mental well being of sleepers in a random population sample. *British Journal of Medical Phsychology, 64,* 25–34.

Jones, I. S., & Stein, H. S. (1987). *Effect of driver hours of service on tractor trailer crash involvement.* Arlington, VA: Insurance Institute for Highway Safety.

Jovanis, P. P., & Kaneko, T. (1990). *Exploratory analysis of motor carrier accident risk and daily patterns.* (Research Report UCD-TRG-RR-90-10). Davis: Transportation Research Group, University of California at Davis.

Kecklund, G., & Åkerstedt, T. (1995). Time of day and Swedish road accidents. *Shiftwork International Newsletter, 12,* 31.

Knauth, P. (1996). Design of shiftwork systems. In W. P. Colquhoun, G. Costa, S. Folkard, & P. Knauth (Eds.), *Shiftwork: Problems and solutions* (pp. 155–173). Frankfurt am Main: Peter Lang.

Krueger, G. P. (1989). Sustained work, fatigue, sleep loss and performance: A review of the issues. *Work and Stress, 3,* 129–141.

Mackie, R. R., & Miller, J. C. (1978). *Effects of hours of service of schedules and cargo loading on truck and bus driver fatigue.* (Contract No. DOT-HS-5-01142). Washington, DC: Federal Highway Administration.

Mitler, M. M., Carskadon, M. A., Czeisler, C. A., Dement, W. C., Dinges, D. F., & Graeber, R. C.

(1988). Catastrophes, sleep, and public policy: Consensus report. *Accident Analysis and Prevention, 17,* 67–73.

Moore-Ede, M., Campbell, S., & Baker, T. (1988). Falling asleep behind the wheel: Research priorities to improve driver alertness and highway safety. In *Proceedings of Federal Highway Administration Symposium on Truck and Bus Driver Fatigue.* Washington, DC: Federal Highway Administration.

National Transportation Safety Board (1995). *Factors that affect fatigue in heavy truck accidents.* (PB95-917001, NTSB/SS-95/01). Washington, DC: Author.

Pilcher, J. J., & Huffcut, A. I. (1996). Effects of sleep deprivation of performance: A meta-analysis. *Sleep, 19,* 318–326.

Ranney, T., & Gawron, V. J. (1987). Driving performance as a function of time on the road. In *Proceedings of the Human Factors Society 31st Annual Meeting.* Santa Monica, CA: Human Factors Society.

Smith, L., Folkard, S., & Poole, C. J. M. (1994). Increased injuries on night shift. *Lancet, 344,* 1137–1139.

Sweatman, P. F., Ogden, K. J., Howarth, N. L., Vulcan, A. P., & Pearson, R. A. (1990). *NSW heavy vehicle crash study final technical report.* (Report No. CR 92). Canberra: Federal Office of Road Safety.

Tilley, A. J., Wilkinson, R. T., Warren, P. S. G., Watson, W. B., & Drud, M. (1982). The sleep and performance of shiftworkers. *Human Factors, 24,* 624–641.

U.S. Department of Transportation, Federal Highway Administration. (1990). *Hours of service study: Report to Congress.* Washington, DC.

van Ouwerkerk, F. (1987). *Relationship between road and transport working conditions, fatigue, health and traffic safety.* (Report No. Vk87-01). Haren: University of Groningen, Haren, Netherlands, Traffic Research Centre.

Williamson, A. M., Feyer, A.-M., Coumarelos, C., & Jenkins, T. (1992). *Strategies to combat fatigue in the long distance road transport industry, Stage 1: The industry perspective.* (Report No. CR 108). Canberra: Federal Office of Road Safety.

Williamson, A. M., Feyer, A.-M., & Friswell, R. (1996). Impact of work practices on fatigue in long distance truck drivers. *Accident Analysis and Prevention, 28,* 709–719.

Williamson, A. M., Feyer, A.-M., Friswell, R., & Leslie, D. (1994). *Strategies to combat fatigue in the long distance road transport industry, Stage 2: Evaluation of alternative work practices.* (Report No. CR 144). Canberra: Federal Office of Road Safety.

3.8 Fatigue and Workload in the Maritime Industry

Mireille Raby
John D. Lee
University of Iowa

Fatigue is perceived to be a common occurrence in the maritime industry; however, there is little quantitative information about both the nature and extent of fatigue in this industry. Several recent research projects sponsored by the U.S. Coast Guard (USCG) sought to address this deficiency (McCallum, Raby, & Rothblum, 1996; Sanquist, Raby, Maloney, & Carvalhais, 1996). This chapter examines some general contributors to fatigue in the maritime industry through an accident analysis to define the role of fatigue in maritime safety and field studies to describe sleep patterns and workload fluctuations that may contribute to fatigue.

In the maritime industry, estimates of the proportion of accidents attributable to human factors range between 50 and 90% (Perrow, 1984; Wagenaar & Groeneweg, 1987). Although many sources have provided estimates of the role of human error in contributing to transportation accidents, the incidence of fatigue as a causal or contributing factor in transportation accidents is not clear. Lauber and Kayten (1988) noted this difficulty in quantifying the role of fatigue, then cited several case studies where fatigue, sleepiness, and disruption of circadian rhythm have played a role in major transportation accidents, including several maritime casualties. As in other domains, maritime accidents linked to fatigue seem to be associated with sleep disruptions caused by the need to work throughout the day. Accident rates vary through the day with a peak in the early morning hours and a slight rise in the early afternoon. Figure 3.8.1 shows an example of this circadian pattern. This graph may reflect circadian variation in alertness and the 24-hour work demands common to most commercial maritime vessels.

The standard work schedule of watch standers on ocean-going vessels consists of two watches of 4 hours separated by an 8-hour rest period. For

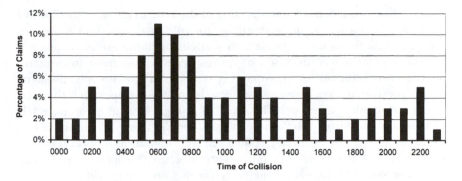

FIG. 3.8.1. Percent of shipping collisions as a function of time of collision (from Sanquist, Raby, Maloney, and Carvalhais, 1996).

example, one mate stands watch from 4:00 AM to 8:00 AM and 4:00 PM to 8:00 PM, another mate stands watch from midnight to 4:00 AM and noon to 4:00 PM, and a third mate stands watch from 8:00 AM to noon and 8:00 PM to midnight. In addition to the watch standers, crews often contain several members who work a more standard daytime shift. Beyond the watch schedule, mariners often work overtime during their off-watch hours to complete repair, maintenance, and administrative tasks. The combination of watch standing and overtime work may result in more than 10 hours of work each day, with a 12-hour day being common. Furthermore, this schedule is often maintained continually, without any days off, for many weeks. This work schedule makes it impossible to sleep more than 8 hours at a time and forces mariners to sleep in two periods of 2 to 6 hours. Vibration, noise, light intrusion, and ship motion can make sleep during these short periods difficult. During storms, ship motion can be so extreme that it may be difficult to remain in the bunk. Reviewing previous fatigue research suggests these environmental characteristics undermine alertness (Sanquist et al., 1996). Beyond environmental factors, task demands associated with cargo loading, navigating into ports, and attending to unanticipated repairs can lead to long periods of intense work that can interrupt and disrupt the routine sleep schedules.

ACCIDENT ANALYSIS

The pattern of results in Fig. 3.8.1 suggests circadian variations in mariner alertness and implicates fatigue as a contributing factor to maritime accidents. Obtaining a better understanding of the human factors contributions to accidents is a key to understanding the role of fatigue and improving safety. Emphasis on investigation and analysis of human-related causes of marine accidents has increased in recent years within most investigative agencies across the world. A comparative review of several investigative agencies suggests that most investigative agencies adopt the approach that accidents and incidents are rarely the result of a single human error, especially in complex industries (McCallum & Raby, 1995). These accidents or incidents are usually the result of a coincidental combination of several failures or deficiencies already present in the system before the accident (Norman, 1980; Perrow, 1984; Reason, 1990; Wagenaar & Groeneweg, 1987).

Although most investigative agencies adopt a similar conceptual approach, their human factors taxonomies and investigative tools seem to vary. For example, in the mid-1990s, the National Transportation Safety Board (NTSB) in the United States was using a taxonomy that it developed and that had been proved successful since 1982 (NTSB, 1990). Its taxonomy classifies the human involvement in all transportation accidents in terms of behavioral, medical, operational, task-related, equipment design, and environmental factors. The International Civil Aviation Organization (ICAO, 1993) has adopted an approach to human factors investigation that is both based on the SHEL model (Hawkins, 1987) and an accident causation model introduced by Reason (1990). The Marine Accident Investigation Branch (MAIB) in England was adopting a new approach that addresses six main human factors: external bodies liaison, company and organization, crew factors, equipment, working environment, and individuals (McCallum & Raby, 1995). However, in 1998, the International Maritime Organization (IMO) presented a set of Guidelines for the Investigation of Human Factors in Marine Casualties and Incidents that also make use of the SHEL and GEMS model (Reason, 1990).

The investigating, analyzing, and reporting tools depend on the resources and purpose of the investigation. The purpose behind the investigation (e.g., to promote safety, regulate, or assign blame) guides the nature and amount of information gathered. An investigative agency that has a dual role of improving safety and regulating has more difficulty than others in obtaining human factors information, especially if that information may incriminate the individual involved. Furthermore, an investigative agency that spends several weeks or months investigating a few major cases can investigate the contributory and causal factors in more depth than one with only a few hours or days for each one of the numerous minor cases to investigate. Depending on their aims and resources, their analysis tools are more or less complex. For example, the Nuclear Regulatory Commission uses an extensive methodology that includes, event sequencing, barrier analysis, root cause analysis, and programmatic cause analysis (Paradies, Unger, Haas, &

Terranova, 1993). Finally, their databases also vary. A database whose primary function is administrative contains minimal human factors information and may not have any indications of fatigue contributions, whereas a database aimed at improving safety might have a specific human factors domain that describes the accident's unsafe conditions or actions (McCallum & Raby, 1995).

ANALYSIS OF FATIGUE AS A CONTRIBUTOR TO MARITIME ACCIDENTS

The U.S. Coast Guard sponsored the accident analysis presented in this chapter. Some of the project's goals were to develop and implement procedures for investigating and reporting human factors contributions to marine accidents and to determine the applicability of these procedures and explore their potential as a general approach for broader use (McCallum et al., 1996). These procedures had to consider the objectives and constraints of the USCG Casualty Investigation Process. Some of the constraints included a large body (approximately 160) of investigating officers who have spent an average of 12 months serving as an investigator (Byers, Hill, & Rothblum, 1994). These investigating officers typically open approximately three new cases per week and have a limited amount of time and resources to investigate each accident or incident. A large number of the cases are investigated by using telephone interviews or with limited site visits. Therefore, an efficient investigation methodology was needed to leverage the limited experience and resources available for investigation.

This project used a two-step procedure consisting of an initial human factors screening followed by a fatigue investigation of human factors cases. Separate reporting forms were developed for each step to be used in the field by investigating officers. The initial step of the investigation was conducted to identify cases with a direct human factors contribution to the casualty. Investigating officers determined if there were any individuals who, through their decisions, actions, or inactions, contributed directly to the outcome or severity of the casualty. The individuals' decisions, actions, or inactions had to be directly linked to the *immediate* series of events leading to the casualty. *Latent* errors were not considered in this investigation process (Reason, 1990). If no individuals were identified, it was assumed that the casualty did not have any direct human factors contribution, and the investigating officers completed only the first form. However, if one or more contributing individuals were identified, the second step was to investigate the potential contribution of fatigue.

The information gathered in the second step focused on the factors contributing to fatigue and the performance consequences of fatigue. Table

TABLE 3.8.1
Fatigue Information Gathered During the Investigations

1. Mariner's experience and job position
2. Mariner's schedule and activities on the casualty day
3. A 72-hour work/rest schedule
4. Number of days off in the last 30 days
5. Symptoms of fatigue and contributing factors to fatigue
6. Mariner's decision or action
7. Mariner's opinion on the contribution of fatigue to the casualty

3.8.1 summarizes some of the fatigue-related information collected by the investigating officers. In addition, they made judgments on the fatigue contribution, type of human error involvement, and accuracy of the information obtained. The choice to gather the following fatigue information was based on a review of the fatigue literature, a comparative study of other investigative agencies, and their approach to the investigation of fatigue-related factors, as well as on the resources and limitations imposed on the USCG investigating officers (McCallum & Raby, 1995). In a more recent publication, the Transportation Safety Board (TSB) of Canada (1997) suggests a similar set of fatigue information to guide the investigation of fatigue in transportation crashes.

Four Marine Safety Offices (MSOs) and a total of 42 investigating officers supported this project. Each investigating officer received 1 day of training that focused on basic human factors and human error concepts and some general understanding of fatigue. Data collection spanned a period of 6 months, collecting data from all vessel and personnel injury cases from the targeted MSOs. The analysis of these data focused on identifying and characterizing accidents with *direct human factors* contribution and those with a *fatigue* contribution.

Direct Human Factors Contributions

Of the 397 cases received, 118 vessel accidents were classified as "minor" because they involved limited property damage with no risk to the loss of the vessel or personnel injury. These cases were excluded from the analyses. The remaining accidents were organized into "critical vessel" accidents (135 cases) and "personnel injury" accidents (135 cases). In addition, 9 cases involved both a critical vessel casualty and a personnel injury. These 9 cases were included in both the analyses of "critical vessel" and "personnel injury" accidents, resulting in a total of 144 accidents in each group. Of the 144 critical vessel accidents, 76 (53%) were classified as having a direct human contribution (i.e., an individual's decisions, actions, or inactions contributed directly to the casualty). A much higher rate 131 of cases (91%) was found for the 144 personnel injuries. Across all personnel injury and critical vessel accidents, the rate of direct human factors involvement was 71%. This relatively low percentage of human factors involvement is due, in part, to a strict definition of human factors involvement, in which only immediate and direct actions were considered, and latent failures were discarded.

An analysis of vessel casualty types indicates that direct human factors links were most prevalent in collisions (81%), allisions (69%), and groundings (56%). In general, these are accidents in which actions or inactions are often directly linked to the accident, such as inadequate vessel navigation. In contrast, human factors were less likely to be direct contributors to founderings and sinkings (39%), fires (17%), and floodings (7%). In these cases, equipment failure or poor maintenance practices were frequently cited as the cause. Although these latter causes are also a general class of human factors, they were not considered *direct immediate* human factors contributions, because the human error may have occurred days or weeks prior to the casualty. An additional analysis of industry segments involved in critical vessel accidents suggested that direct human contributions were markedly higher than the average (53%) for the 52 accidents involving tugs and barges (67%).

Similar analyses of the personnel injury accidents indicated a uniformly high rate of direct human factors contributions across personnel injury type, the most prevalent being with sprains/strains (97%) and slips/falls (94%). With regard to the industry segments, the highest level of direct human factors contributions were found aboard fishing vessels (92%) and tugs and barges (92%).

Fatigue Contributions

The study used two strategies to identify accidents in which fatigue was a contributing factor. First, the investigating officers' and mariners' judgment was used to identify fatigue contributions in each case. The advantage of this approach was the intimate knowledge that the investigating officer and/or mariner may have had about the specific conditions of the casualty. The disadvantages include the lack of expertise among mariners and investigating officers about the diagnosis of fatigue, as well as the potential biases of over- or under-reporting fatigue cases. The second strategy was to define fatigue using factors identified in the literature and gathered during the investigation, thereby applying an externally developed and validated fatigue criterion.

Using the first strategy, the analysis suggests that, when asked about the possibility of fatigue contribution to the casualty, mariners indicated that fatigue was a contributor in 17 of 98 cases (17%). Investigating officers judged that fatigue was a contributor in 23% of the cases. Overall, investigating officers and mariners agreed on 74 of 86 cases (86%), with mariners being less likely to classify an accident as having a fatigue contribution.

In the second approach, the relation between the investigating officers' and mariners' subjective judgments and the objective indicators of fatigue contribution was investigated. First, an accident was judged to have a fatigue contribution if *either* the investigating officer or the mariner judged fatigue to be a contributor. Second, several factors were included in a multiple regression analysis of the objective fatigue indicators. Only three factors were found to significantly contribute to the multiple R. Equation 1 provides the resulting *Fatigue Index Equation* for computing a *Fatigue Index Score* for an accident case ($F_{(3,57)} = 16.6$, $p < .0001$; $R^2 = 0.466$):

Fatigue Index Score = [4.39 × (number of fatigue symptoms)] +
$$[1.25 \times \text{(hours worked in last 24 hours)}] + \qquad (1)$$
[0.93 × (hours slept in last 24 hours)] + 39.75.

Fatigue index scores were computed for the 93 cases with data available for the three factors included in the equation. The distribution of the Fatigue Index scores, along with the judgment about fatigue, was reviewed, and a cutoff score of 50 was selected. Any case with a Fatigue Index score greater than 50 was classified as an accident with a fatigue contribution. This cutoff score resulted in a level of agreement on 80% of the cases between investigating officers and mariners and the Fatigue Index classification.

The Fatigue Index cutoff score was next used to classify all critical vessel and personnel injury casualty cases to determine the contribution of fatigue. Of the 199 critical vessel and personnel injury cases judged to have a direct human factors contribution, 88 cases had sufficient data to evaluate the contribution of fatigue by using the Fatigue Index procedure. Figure 3.8.2 sug-

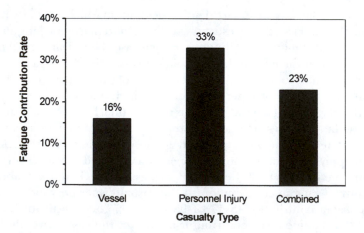

FIG. 3.8.2.   Estimated rates of fatigue contributions using the Fatigue Index Score approach.

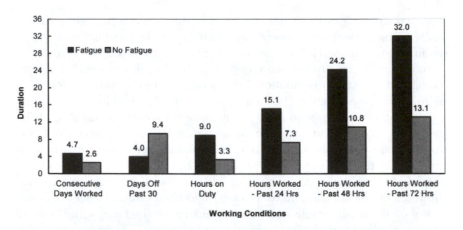

FIG. 3.8.3.   Comparison of working conditions found to significantly contribute to fatigue-related critical vessel accidents.

gests that with this Fatigue Index procedure, the estimated fatigue contribution rate was 16% for critical vessel accidents, 33% for personnel injuries, and 23% for the combined set of critical vessel and personnel injury accidents.

Although the number of casualty cases classified according to their fatigue contribution by using this second methodology was limited, the cases were further analyzed to assess the general value of this approach in characterizing accidents with a fatigue contribution. Of interest was the analysis of the working conditions contributing to fatigue-related accidents. Eighteen factors were analyzed to identify their contribution to fatigue in critical vessel accidents. Figure 3.8.3 illustrates six of the eight separate factors that were found to differ significantly with the fatigue classification of critical vessel accidents. The factors with the greatest level of significance were hours on duty prior to the casualty, $F_{(1,39)} = 20.68$, $p < .001$; hours worked in the 24, 48, and 72 hours preceding the casualty, with $F_{(1,39)}$ values of respectively 31.74, 24.84, and 23.88 with $p < .001$; the mariner's report that the work schedule during the casualty was different from the normal schedule, $\chi^2 = 7.43$, $p < .01$; and the existence of company or union policies regarding work hours, $\chi^2 = 4.71$, $p < .05$.

A similar set of analyses conducted for the personnel injuries showed that the highest frequencies of fatigue contributions were also found in injuries involving slips/falls and sprains/strains. The analyses of working conditions showed clear differences between personnel injury cases classified as fatigue or not. Fatigue-related cases had an average of 7.7 hours on duty prior to the casualty compared with 3.2 hours ($F_{(1,52)} = 13.46$, $p < .001$) and mariners reported an average of 14.3 hours worked in the 24 hours preceding the casualty compared with 8.4 hours ($F_{(1,52)} = 34.07$, $p < .001$). Similar p-values were obtained for hours worked in the 48 and 72 hours preceding the casualty.

In summary, this project was aimed at developing procedures for investigating and reporting information directly relevant to the contribution of fatigue in marine accidents. The analysis procedure resulted in estimates that fatigue was a contributor to 16% of critical vessel accidents and 33% of the personnel injury accidents. The main indicators of fatigue were hours worked and slept in the last 24 hours and the number of fatigue symptoms reported by the mariner. Hours on duty at the time of casualty, hours reported working in the previous 24 to 72 hours, and a change in schedule showed a significant contribution to fatigue-related critical vessel accidents. Although the small sample limits the application of the Fatigue Index score to other situations, the global approach and methodology developed in this project could transfer to the investigation of fatigue in other domains or to other human factors contributions to marine accidents.

FIELD STUDIES OF MARITIME OPERATIONS AND THE CONTRIBUTORS TO FATIGUE

An accident usually represents the unusual or abnormal situation, the exception. To develop a better understanding of fatigue on-board vessels, there is a need to learn about normal operating conditions. A very limited number of studies have examined sleeping and working on-board vessels (Fletcher, Colquhoun, Knauth, DeVol, & Plett, 1988; Rutenfranz et al., 1988). To expand on these studies and to complement the accident analysis described earlier, the U.S. Coast Guard sponsored two field studies to explore working and sleeping conditions on-board commercial vessels. One field study focused on measuring mariner fatigue and the factors that influence fatigue (Sanquist et al., 1996). The second study focused on understanding maritime operations and the associated task demands (Lee et al., 1997). Both studies examined ocean-going vessels that included tankers and container ships.

In each study, one or two experimenters boarded the vessels for several days to introduce the data collection protocol, observe operations, and conduct interviews. Approximately 140 mariners participated in each study. The primary data collection tool of the first study was a logbook. Each day, for a period of 10 to 30 days, mariners recorded their sleep and work times, sleep quality, ratings of alertness, and symptoms of fatigue. In addition to the logbook, each study included questionnaires and individual interviews. The primary data collection tool in the second study was a pair of structured interviews that generated detailed information concerning the specific tasks performed during different parts of the voyage, the priority of the tasks, the personnel needed to perform the tasks, and the duration of the tasks. The data from these interviews were complemented with observations of all phases of vessel operation (e.g., port calls, cargo loading, cargo unloading, and open-ocean transit).

Data from these studies provide a comprehensive description of mariner alertness and the factors that influence its fluctuation. To explore the prevalence of fatigue, Sanquist, Raby, Forsythe, and Carvalhais (1997) identified critical fatigue factors based on prior research. These factors were associated with impaired performance and/or imminent sleep onset that occur with subjective ratings of low levels of alertness (Akerstedt & Folkard, 1995) and sleep reduction (Gillberg, 1995). Analysis of the logbook data revealed that 11% of the data showed critical levels of subjective alertness (rated alertness levels below 3) and 7.4% of the data showed critical levels of sleep reduction (4 or fewer hours of sleep in the past 24 hours). The watch standers on the 0400 to 0800 watch schedule seemed most affected, and were found to sleep less than 4 hours per 24-hour period 22% of the time. These data were confirmed by the average amount of sleep the mariners received. In particular, the watch standers on the 4-on, 8-off schedule experienced considerable disruptions to their sleep. The average total sleep duration per day for the watch standers was 6.6 hours, which was obtained in fragmented periods of less than 5 hours in duration (Sanquist et al., 1997). This compares to a reported average of 7.9 hours of sleep while at home and an ideal sleep duration of 8.1 hours per day. Compared with mariners' ideal or at-home sleep, mariners suffer a sleep debt of approximately 1.3 hours. The data suggest that an important contributor to the fatigue experienced by mariners is the fractured sleep periods. These results suggest that in addition to the total time to sleep, sleep continuity is also important. Thus, sleep hygiene, the ability to control the sleep environment, and work-related sleep disruptions are important contributors to fatigue.

The second field study examined the task demands and how they may modulate time available to sleep, particularly in the context of crew size and the introduction of labor saving technology. This study collected detailed information on over 100 tasks that occur during a typical voyage. Many of these tasks have been influenced by recent developments of labor saving technology. This technology has made it possible to reduce crew sizes from 30 to 40 people to 12 to 18 crew members. Although much of this technology has greatly enhanced operating efficiency and safety, there is a clear danger for unanticipated negative consequences. Specifically, just as clumsy automation in other domains has reduced workload during routine situations and elevated it during abnormal situations (Bainbridge, 1983; Woods, Potter, Johannesen, & Holloway, 1991), advanced maritime technology may also introduce unanticipated workload peaks. A discrete event simulation was developed based on the task data collected on the vessels. This model simulated the tasks and activities that occur during a voyage for each person in the crew. The output of the model included time available to sleep, hours worked each day, and instances where sleep is interrupted or curtailed due to task demands. The output of the model corresponded very closely to the logbook data collected in the first field study. The mean number of hours worked per person per day predicted by the model was 11.49 compared to a mean of 11.45 for the logbook data. The correlation between the predicted and actual number of hours worked per day was 0.82 (Lee et al., 1997).

The interviews with crew members and analysis of model output demonstrated the potential for clumsy automation in the maritime industry. In addition to clumsy automation at the micro level (individual mariners working for minutes or hours with a single piece of automation), this study identified clumsy automation at the macro level (multiple mariners working for days and weeks with multiple pieces of automation). This macro-level clumsy automation leads to periods where the crew, which has been reduced because of the introduction of new technology, is forced to work long hours and endure significant sleep disruptions (Lee & Morgan, 1994). For example, sophisticated navigation technology and advanced engine control systems have greatly reduced the workload during ocean transits, making it possible to operate with fewer crew during these periods. However, this automation does little to reduce the workload associated with cargo loading, docking, and response to unexpected emergencies. Thus, much maritime automation is clumsy at a macro level, particularly when its introduction is paired with crew reduction: the automation makes low workload periods less demanding, but increases the workload of demanding periods. Careful consideration of technological innovations and how they contribute to sleep disruption is a critical consideration in the maritime industry where the personnel available to perform a task are limited to those on board and assistance cannot be requested from other locations.

CONCLUSION

This chapter presented some recent findings on fatigue and touched on the role of workload in the maritime environment. Fatigue is a difficult phenomenon to quantify and its impact on safety even more difficult to assess. Two field studies and an accident analysis help reveal the causes of fatigue and its role in maritime safety. Field study data described routine ship operations and showed how these operations affect sleep and workload, and their impact on alertness and performance on board vessels. The results described here suggest that, like many 24-hour industries, fatigue poses a considerable risk to the maritime industry. The field study data also describe the task demands that can disrupt sleep. Some of these disruptions result from what might be termed macro-level clumsy automation, where technology makes it possible to reduce crew sizes during normal operations, but may lead to workload peaks that affect the remaining crew members and can extend for several days, disrupting sleep and inducing fatigue. Evidence during normal operating conditions complements the information gathered from critical situations, which are represented in the accident analyses. The accident analysis suggested that fatigue is an important contributing factor to accidents, where it was identified as a contributing factor in 16% of vessel accidents and 33% of personnel injury cases. The field studies show that the situations that result in fatigue-related accidents are not isolated incidents in the maritime industry. Disrupted sleep and chronically low levels of alertness are characteristics of the industry that need to be carefully considered by the regulating agencies, the operating companies, and the employees that work onboard these vessels.

ACKNOWLEDGMENTS

The views expressed in this chapter are those of the authors and are not official U.S. Coast Guard policy. We wish to acknowledge gratefully the contributions of our sponsor at the USCG Research and Development Center, Dr. Anita M. Rothblum, and our colleagues, Dr. Marvin C. McCallum, Dr. Thomas F. Sanquist, Dr. Alvah Bittner Jr., and Mrs. Alice M. Forsythe. This work was supported by the USCG R&D Center, under Contract DTCG39-94-D-E00777.

REFERENCES

Akerstedt, T., & Folkard, S. (1995). Validation of the S and C components of the three-process model of alertness regulation. *Sleep, 18,* 1–6.

Bainbridge, L. (1983). Ironies of automation. *Automatica, 19*(6), 775–779.

Byers, J. C., Hill, S. G., and Rothblum, A. M. (1994). *U.S. Coast Guard marine casualty investigation and reporting: Analysis and recommendations for improvement.* (Report No. CG-D-13-95; AD-A298380). Groton, CT: U.S. Coast Guard Research and Development Center.

Fletcher, N., Colquhoun, W. P., Knauth, P., DeVol, D., & Plett, R. (1988). Work at sea: A study of sleep, and of circadian rhythms in physiological and psychological functions, in Watch-keepers on merchant vessels, VI: A sea trial of an alternative watchkeeping system for the merchant marine. *International Archives of Occupational and Environmental Health, 61,* 51–57.

Gillberg, M. (1995). Sleepiness and its relation to the length, content, and continuity of sleep. *Journal of Sleep Research, 4*(Suppl. 2), 37–40.

Hawkins, F. H. (1987). *Human factors in flight.* Aldershot, UK: Gower Technical Press.

ICAO. (1993). *Human factors digest no. 7—Investigation of human factors in accidents and incidents.* (Circular 240-AN/144). Montreal, Canada: International Civil Aviation Organization.

International Marine Organization. (1998). *Role of the human element in maritime casualties: Joint ILO/IMO ad hoc working group on investigation of human factors in maritime casualties.* (Maritime Safety Committee MSC 69/13/1). London: Author.

Lauber, J. K., & Kayten, P. J. (1988). Sleepiness, circadian dysrhythmia, and fatigue in transportation system accidents. *Sleep, 11*(6), 503–512.

Lee, J. D., McCallum, M. C., Maloney, A. L., & Rothblum, A. M. (1997). *Validation and sensitivity analysis of a crew size evaluation method.* (Report No. CG-D-25–97). Groton, CT: U.S. Coast Guard Research and Development Center.

Lee, J. D., & Morgan, J. (1994). Identifying clumsy automation at the macro level: Development of a tool to estimate ship staffing requirements. *Proceedings of the Human Factors and Ergonomics Society 38th Annual Meeting, 2,* 878–882.

McCallum, M. C., & Raby, M. (1995, April). *Human factors in U.S. Coast Guard marine casualty investigation and reporting.* (Letter Report). Seattle, WA: Battelle Seattle Research Center.

McCallum, M. C., Raby, M., & Rothblum, A. M. (1996). *Procedures for investigating and reporting human factors and fatigue contributions to marine accidents.* (Report No. CG-D-09-97; AD-A-323392). Groton, CT: U.S. Coast Guard Research and Development Center.

Norman, D. A. (1980). *Errors in human performance.* (Technical Report). La Jolla, San Diego: University of California, San Diego, Center for Human Information Processing.

NTSB. (1990). Outline of the human performance investigation. In *The marine accident investigation manual.* Washington, DC: National Transportation Safety Board.

Paradies, M., Unger, L., Haas, P., & Terranova, M. (1993). *Development of the NRC's human performance investigation process,* Vols. 1–3. (Report NUREG/CR-5455). Washington, DC: Nuclear Regulatory Commission.

Perrow, C. (1984). *Normal accidents: Living with high-risk technologies.* New York: Basic Books.

Reason, J. (1990). *Human error.* Cambridge: Cambridge University Press.

Rutenfranz, J., Plett, R., Knauth, P., Condon, R., DeVol, D., Fletcher, N., Eickhoff, S., Schmidt, K. H., Donis, R., & Colquhoun, W. P. (1988). Work at sea: A study of sleep, and of circadian rhythms in physiological and psychological functions, in watchkeepers on merchant vessels, II: Sleep duration, and subjective ratings of sleep quality. *International Archives of Occupational and Environmental Health, 60,* 331–339.

Sanquist, T. F., Raby, M., Forsythe, A., & Carvalhais, A. B. (1997). Work hours, sleep patterns, and fatigue among merchant marine personnel. *Journal of Sleep Research, 6,* 245–251.

Sanquist, T. F., Raby, M., Maloney, A. L., & Carvalhais, A. B. (1996). *Fatigue and alertness in merchant marine personnel: A field study of work and sleep patterns.* (Report No. CG-D-06-97; AD-A322126). Groton, CT: U.S. Coast Guard Research and Development Center.

Transportation Safety Board. (1997). *A guide for investigating for fatigue.* (Internal Report). Ottawa, Ontario: Transportation Safety Board of Canada.

Wagenaar, W. A., & Groeneweg, J. (1987). Accidents at sea: Multiple causes and impossible consequences. *International Journal of Man–Machine Studies, 27,* 587–598.

Woods, D. D., Potter, S. S., Johannesen, L., & Holloway, M. (1991). *Human interaction with intelligent systems: Trends, problems, new directions.* (CSEL Report 1991-001). Columbus: The Ohio State University.

COMMENTARY

3.9 An Overview of Fatigue

Valerie J. Gawron
Veridian Engineering

Jonathan French
United States Air Force Research Laboratory

Doug Funke
Veridian Engineering

Fatigue is multifaceted and complex. The effects of fatigue not only overlap the areas of performance, physiology, cognition, and emotion, but also combine with other states, such as boredom or drowsiness (McDonald, 1989). The two general types of fatigue are peripheral (physical) fatigue and central (mental) fatigue. Physical fatigue is generally defined as a reduction in capacity to perform physical work as a function of preceding physical effort. Mental fatigue is inferred from decrements in performance on tasks requiring alertness and the manipulation and retrieval of information stored in memory (Stern, Boyer, & Schroeder, 1994). Since World War II (Reid, 1948), both types of fatigue are known to have influenced performance.

Fatigue was originally conceived of as measurable decrements in performance of an activity caused by the extended time for performing the activity (Bartlett, 1953). This definition led to the traditional understanding that as a subject's time on task increases, his or her performance decreases in a linear fashion (McDonald, 1989). However, a linear relation between driving time and accident rates has not been found. Instead, researchers have found a variety of high and low performance rates with different driving durations (Blom & Pokorny, 1985; Harris, 1977; Harris et al., 1972). From these studies evolved the question of whether the task or some other associated factors create the fatigue.

A 1983 study naval of aviation accidents suggested that the major contributors to fatigue may be the total time spent at work (inclusive of the task)

581

and the time of day at which the work occurs (Borowsky & Wall, 1983). Other researchers have also studied this relation (Blom & Pokorny, 1985; Hertz, 1987). In 1996, Stoner used anonymous questionnaires, physiological measurements (mean arterial pressure, pulse, and pulse pressure), the Armed Forces Vision Tester, and hematologic measurements (complete blood count and sedimentation rate) to try to predict early fatigue in 42 Navy EP-3E aircrews. However, none of the physiological or hematologic measurements varied between fatigue and normal states, although 14% of personnel did show increased tendencies for visual phorias. In a study of sleep duration, subjects limited to mean sleep durations of 5.2 hours per night for 4 weeks, 4.3 hours per night for 4 nights, and 5.3 hours per night for 18 nights showed no performance decrements in logical reasoning or auditory vigilance but did show performance decrements in a ability to ignore distracting information (Blagrove, Alexander, & Horne, 1995).

Landing or takeoff under emergency conditions, dealing with inflight emergencies, and ingress under fire are difficult circumstances for the most vigilant crews, and burdening crews with fatigue from sleep deprivation can cripple their abilities to deal with emergencies. In fact, the Air Force Safety Center relates that there were 92 Class A incidents—those that involve over $1 million worth of damage or loss of life—between 1972 and 1995 (Palmer, Gentner, Schopper, & Sottile, 1996). Sixty per cent of these incidents were related to sleep deprivation. Long transmeridian flights have almost three times the incident rate of shorter flights (Graeber, 1987). The fact that human error or loss of cockpit coordination causes as many as 75% of accidents strongly implicates fatigue and circadian disruption (Graeber, 1988). Dramatic deficiencies in operational performance are also easily observed in simulator tests (in commercial airline aircraft simulators, e.g., Moore-Ede, 1993).

Repetitive missions may lead to chronic fatigue and increase the human cost of long endurance missions. Frequently, these operations involve multiple days of sustained operations with little sleep or poor sleep between missions (Neville, Bisson, French, Boll, & Storm, 1994). The impairment associated with repetitive, long-duration flights was documented during Operations Desert Shield and Desert Storm (Bisson et al., 1992; French et al., 1992) and in a simulator study (French, Bisson, Neville, Mitcha, & Storm, 1994). Much can be done to prepare for long duration activities (Ferrer, Bisson, & French, 1995).

FATIGUE FACTORS

Age

People of different ages incur different problems with fatigue. For example, younger drivers have a tendency to resist fatigue, which is one factor leading them to drive more at night than do older drivers. These younger drivers cause proportionately more accidents by falling asleep at the wheel than do older drivers, who tend to avoid prolonged driving at night and take rest breaks (Harris & Mackie, 1972). However, older drivers suffer more attention lapses attributable to fatigue during the afternoon hours, especially after lunch (Smith, 1989).

Langlois, Smolensky, Hsi, and Weir (1985) performed a similar analysis of single-vehicle accidents and found a major diurnal peak from 1:00 AM to 6:00 AM and a secondary peak, especially among older drivers, from 1:00 PM to 5:00 PM (Langlois et al., 1985). These peaks appear to follow a general pattern associated with circadian rhythms (discussed in more detail hereafter) and the same hourly distribution of fatigue-related problems in many other activities (Mitler et al., 1988).

Work–Rest Schedules and Sleep Deprivation

Irregular work and rest scheduling, like that experienced by many professional drivers, disrupts normal circadian rhythms. This disruption has been associated with fatigue and performance decrements (Tilley, Wilkinson, Warren, Watson, & Drud, 1982). For example, professional drivers who use sleeper berths to accumulate rest are three times more likely to be involved in a fatal crash than their counterparts who do not use sleeper berths (Hertz, 1987). This effect has been attributed to federal regulations that allow drivers to accumulate the required 8 hours of off-duty time in two nonconsecutive periods. Drivers with sleeper berths also reported more subjective fatigue than other drivers, and studies of steering, lane tracking, and lane drifting have shown a performance decrement for drivers with sleeper berths (Mackie & Miller, 1978).

The effects of shift work on sleep schedules are critical, because sleep during the day is not equivalent to sleep during the night. A study of train engineers on irregular work schedules showed that subjects who went to sleep after midnight slept for shorter periods of time and that subjects who slept during the day had a qualitatively different type of sleep from that obtained at night (Foret & Lantin, 1972). For professional drivers, shift work poses two problems: maintaining concentration and attention at a time when the circadian rhythm is at a low level, and obtaining enough quality sleep when circadian rhythms and social factors are not favorable for this (McDonald, 1981). Performance measures, obtained in both lab and operational settings, show circadian fluctuations that range from 10% to 190% of mean performance, depending primarily on the task. The greatest decrements occur during the circadian trough, which is typically between 2:00 AM and 7:00 AM (Kelly, 1996), so that working at these hours can be a particular challenge. Lack of sleep causes more concentration and attentional problems while driving and probably leads to cumulative sleep loss and fatigue. In interviews, nearly all shift-working truck drivers said they suffered from drowsiness at some time while working, and 60% suffered from drowsiness very often while working (Edmondson & Oldman, 1974). Finally, night work can lead to health problems, including sleep disorders, gastrointestinal disorders (colitis, gastroduodenitis, peptic ulcers), neuropsychic disorders (chronic fatigue, depression), and cardiovascular problems (hypertension, ischemic heart disease; Costa, 1996).

Sleep deprivation can have serious effects on performance and safety, and circadian effects on cognition compound the degradation in performance caused by extended workdays (Tilley et al., 1982). When sleep debt accumulates, performance degrades linearly—especially on tasks that are low in motivating qualities—while continuing to follow a circadian rhythm. The decline is steeper during the circadian trough than at other times, and this decline seems unaffected by motivation (Elsmore et al., 1995). In subjects with restricted night sleep, Carskadon and Dement (1981) found a striking difference in objective measures of sleepiness between 4 and 5 hours of night sleep. Two nights with 4 hours of sleep each produced "pathological" levels of sleepiness (Carsakadon & Dement, 1981). However, some evidence suggests that people can be conditioned to need less sleep. In a limited sample of young adult couples, gradual reduction of sleep to 4.4 to 5.5 hours per night over 6 to 8 months did not result in cognitive performance decrements (Friedmann et al., 1977).

In a study of the effects of schedule on mood, fatigue, and vigor, it was found that once C-141 crews surpassed the usual 125-hour limit on cumulative flight hours per month, both flight hours and sleep hours obtained in the past 24 and 48 hours affected subjective measures of vigor (French et al., 1992). A study of two-crew, long haul, night flight operations (Samel et al., 1997) concluded that two consecutive night flights and a short layover resulted in ratings of critically high fatigue and brain activity indicating low vigilance. Naps can be helpful (Dingus, Hardee, & Wierwille, 1987). The benefits of naps are greatest if they are taken in the first 24 hours of operation. Nevertheless, lack of sleep affects military operations. For example, infantry soldiers who were totally deprived of sleep for 3 days showed little deterioration in their riflery skills, but their cognitive performance dropped precipitously each day. Then, 4 hours of night sleep on Day 4 resulted in large performance improvements *except during the circadian trough*. After obtaining 4 hours of sleep on each of Days 4 through 6, performance recovered from approximately 50% to 88% of control levels (Haslem, 1982).

Recovery sleep periods are essential for returning performance to baseline levels. After seven days with only 5 hours of sleep each, one full night of sleep followed by a daytime nap returned measures of sleepiness and fatigue to prerestriction levels (Carskadon & Dement, 1981). After 40 hours awake, cognitive performance returned to its baseline with 4 hours of night sleep and with 8 hours of night sleep following 64 hours awake (Rosa, Bonnet, & Warm, 1983). Two nights of sleep may be sufficient for full recovery from 3 days of total sleep deprivation and 6 days of partial sleep deprivation (4 hours of sleep per night; Haslem, 1982). However, performance during the circadian trough may be slower to return to baseline levels.

Motivational Factors and Coping

The results of laboratory tests cannot necessarily be generalized to real-world performance because personnel may be more motivated in their real-world experiences, which can offset many of the effects of fatigue (Boff & Lincoln, 1988). Humans also possess a large reservoir of performance capacity that, when called on, can overcome the effects of fatigue (O'Hanlon, 1981). In an excellent review of fatigue associated with long military engagements, Johnson and Naitoh reported that it is difficult to predict the effect of prolonged sleep loss of less than 60 hours on performance in highly motivated military forces (Johnson & Naitoh, 1974). Individuals have many ways of coping with fatigue to allow capable performance. Some of these techniques are increasing the following distance between vehicles (Fuller, 1981), lowering risk thresholds (Riemersma, Sanders, Wildervanck, & Gaillard, 1997), and increasing effort (Mackie & Miller, 1978).

Individuals choose and modify coping strategies and motivational factors based on their fatigue levels and the changing demands of the tasks. This self-monitoring accounts for the fact that critical incidents are rare, even when demanding work requirements stretch an individual's resources to the maximum (McDonald, Fuller, & White, 1989). For example, professional truck drivers with experience in night driving learn how to cope with fatigue. Drivers who make these compensations eventually move to safer, better run companies, and drivers who find it hard to cope leave the occupation (Hamelin, 1987; McDonald, 1989; Wyckoff, 1979). Also, many studies have shown that prolonged time at a task tends to produce changes in attitude toward the task. Eventually, participants favor speed at the expense of accuracy (McFarland, 1953; Welford, 1968). In driving situations, the frequency of risky overtaking maneuvers by subjects increased with hours driven (Brown, Tickner, & Simmonds, 1970).

Total Time Spent Working

The total time spent working is a more important cause of fatigue than total time on a particular task (McDonald, Fuller, & White, 1991). In 1985, Transportation Research and Marketing found that 41% of heavy truck crashes involved drivers whose total onduty time was 16 or more consecutive hours.

Onduty time includes time spent on any assigned task: driving, waiting for dispatch, loading, and unloading (Transportation Research and Marketing, 1985). One study showed that drivers with shifts exceeding 14 hours have nearly three times the accident rate of their counterparts on 10-hour shifts. Drivers on 11-hour shifts have twice the accidents of those on shorter shifts (Hamelin, 1981, 1987). In addition, the accident data for one trucking company showed that the ratio of expected to actual accidents increased between the seventh and tenth hours of driving (Harris & Mackie, 1972).

Circadian Disruption

The effects of sleep deprivation on reaction time and accuracy are well known. In particular, circadian phase disruption is a debilitating consequence of rapid, transmeridian travel. Physiological functions critical to maintaining vigilance on duty (e.g., sleep inertia, neuroendocrine levels, immune response, gastrointestinal function, physical strength) are severely disrupted by long duration travel (Fevre-Montange et al., 1981). Studies of reaction time, logical reasoning, and simulated flight have showed that complex cognitive abilities are similarly affected (Bruener et al., 1970).

Circadian rhythms have a marked effect on performance. For example, the postlunch and predawn hours are periods of decreased driving ability (Langlois et al., 1985). Accident data have suggested a marked contribution of fatigue to accidents during the night hours (McDonald, 1978). The time of day when sleep is taken also affects fatigue—sleep taken during the day is typically shorter in duration and less restful than night sleep (Tepas & Carvalhis, 1990). These factors are discussed in more detail later in this chapter.

The deterioration of physical and mental performance associated with circadian desynchrony is well documented in aviation human factors research (Klein, Wegmann, Athanassenas, Hohlweck, & Kuklinski, 1976; Moore-Ede, 1993; Ribak et al., 1983; Wright, Vogel, Sampson, Knapik, & Patton, 1983). Aircrews are at particular risk because they often must take off at night; consequently experiencing altered light and dark cycles. Light exposure can disrupt normal circadian cycles (Reiter, 1991), so that recovery from mission-induced fatigue may take longer for crews exposed to unusual light cycles.

Flight has profound effects on sleep (Sasaki, Kurosaki, Spinweber, Graeber, & Takahashi, 1993). In most individuals, traveling east is more disruptive than traveling west (Winget, DeRoshia, Markley, & Holley, 1984). For example, commercial airline pilots suffer more fragmentation of sleep after eastbound travel, and older aircrew are more sensitive to rapid travel than are younger members. The likely cause of this phenomenon is the fact that the period of the circadian rhythm is longer than 24 hours. Rapid eastbound travel shortens the geophysical experience (e.g., solar cues; Sack, Lewy, Blood, Keith, & Nakagawa, 1992).

Circadian rhythm can take longer than 1 week to fully recover from trans-meridian travel. The realignment of rhythm is compounded by the fact that different circadian patterns adjust at different rates (Graeber, Dement, Nicholson, Sasaki, & Wegmann, 1986). Therefore, the severest symptoms of "jet lag" might not be experienced until the second or third day after arrival, when differential adjustment causes different rhythms to be farther out of phase.

FATIGUE COUNTERMEASURES

Nonpharmacological techniques can improve rest during operations, and powerful pharmaceutical agents are available to promote sleep and maintain alertness. Also, a new class of compounds and techniques, chronobiotics, can be used to adjust circadian phase prior to deployment.

Nonpharmaceutical Techniques

Nonpharmaceutical techniques provide limited but effective fatigue management with careful adherence to crew rest guidelines. Aircrews typically are not trained to be sensitive to their own fatigue levels and instead are expected to overcome fatigue. Thus, fatigue awareness training may be useful. Being more alert to fatigue in themselves and in others may enable crews to better gauge and respond to fatigue-related problems. Also, physical activity and dietary stimulants can temporarily mask fatigue (Mitler et al., 1988).

Sleep training may be an effective means to reduce stress and promote sleep (by rapidly relaxing large muscle groups; Naitoh, Kelly, & Babkoff, 1992). Careful attention to sleep environment and duration may permit a few minutes of sleep to provide hours of vigilance. Because sleep latency is reduced at different times of the day, sleep schedules can be designed to maximize crew rest periods, minimize recovery time, and allow safer, extended work cycles.

Aircrew members report many personal coping strategies: (a) performing tasks one step at a time, (b) conversations with other crew members, (c) cold drinks, (d) walking around, (e) being as organized as possible, (f) staying busy and active, (g) keeping the mind busy and active, (h) making extra effort to fight fatigue, (i) eating, and (j) discussing fatigue symptoms with other crew members (Petrie & Dawson, 1997). Recommended strategies include preplanned cockpit rests, physical exercise at night, dietary control (eating carbohydrates to induce sleep, eating protein to induce wakefulness), and light therapy (Ferrer et al., 1995). Institutional support (e.g., for protecting sleep periods) is critical to effective transition from daytime to nighttime flying (Comperatore & Allan, 1997). Other organizational actions include increasing personnel staffing, modifying tasks, dividing workload, cross-training personnel, decreasing aircrew physical load, enhancing physical fitness, realistic training, creating night teams, planning rest and sleep discipline, and providing recovery sleep between combat episodes (Krueger, 1991). In addition, the Air Force Research Laboratory has developed procedures for countering fatigue effects of shift work and long-distance travel (Ferrer et al., 1995).

Pharmaceutical Agents

Pharmaceutical techniques have always been an effective means to promote vigilance or induce sleep, and new compounds are available that are safer (better tolerated) and more efficacious. However, the chronopharmacological and interaction effects with other compounds are important considerations. New, safer "Go" and "No-Go" agents have a great deal of promise for maximizing alertness and enhancing performance.

"Go" Agents. Stimulants ranging from amphetamines to caffeine can relieve fatigue. Unlike hypnotic compounds, there are no clear guidelines with regard to substance, dosage, or timing for stimulants. New agents (ergogenic and eugregaric compounds) like Modafinil may be better tolerated than familiar stimulants, and procedures to use them should be made available (Lyons & French, 1992). Past experience has shown that carefully regulated stimulant administration can be very successful at maintaining vigilance during prolonged operations. Recently, stimulant use by aircrews was banned. However, some research indicates that "the elimination of amphetamine use has put aircrews at increased actual risk for the sake of eliminating theoretical use" (Cornum, Caldwell, & Cornum, 1997, p. 57).

Amphetamines were discovered in the 1930s and they have been used in every major conflict since then. There are still many questions about stimulant use. For example, the number of days that individuals can function effectively on stimulants is not well established, and the most effective dosing regimens over multiple days are not known. Evidence from laboratory studies suggests a continual, but gradual, degradation in performance past 60 hours awake, even with repeated administration of stimulants (Pigeau et al., 1995). The Air Force has used dextroamphetamine in the past (Emonson & Vanderbeek, 1995; Senechal, 1988). Simulator studies (Caldwell, 1996; Caldwell, Caldwell, & Crowley, 1997) and flight evaluations of UH-60 pilots (Caldwell, Caldwell, Crowley, & Jones, 1995; Caldwell et al., 1996) support the use of dextroamphetamine. In one study, dextroamphetamine caused no clinically detectable adverse effects on six male and six female UH-60 pilots (Caldwell, 1996). Nevertheless, methamphetamine might be better used for the few days of activity. Methamphetamine is more centrally active, has a shorter plasma half-life, and causes fewer side effects than Dexedrine (Medical Economics, 1997).

A number of nonamphetamine stimulants have potential for use. Far less is known about Modafinil than about amphetamines, but it does seem to have considerable promise for blunting fatigue without the euphoria, hyperactivity, or abuse potential of traditional stimulants. A study of 41 military personnel showed that Modafinil might be a useful alternative to amphetamines in sustained operations (Pigeau et al., 1995). However, Modafinil is not commercially available in the US at present. Pemoline is a newer stimulant that has a fairly long plasma half-life (as much as 12 hours). However, long-duration effects may not be an advantage because they prevent users from getting rest if rest opportunities become available soon after the stimulant is administered. Some evidence suggests that Pemoline may be associated with acute liver toxicity. Methylphenidate is a short-acting stimulant, but it lacks the central activity of the amphetamines.

Finally, caffeine is perhaps the oldest and most reliable of the stimulants. Relatively small doses of caffeine (200 mg) improve response times, vigilance, and alertness during periods of sleep deprivation (Bonnet & Arand, 1994). However, the effects of caffeine on fatigue are not as profound as those of amphetamines. A sustained-release caffeine capsule may be available in the future to prolong the plasma activity, but, again, long-duration effects may not be an advantage if rest becomes available.

"No-Go" Agents.   Sedatives are available to help overcome circadian rhythm problems. Melatonin is a widely available, nonaddictive means to promote sleep. Restoril is currently recommended because its pharmacodynamics in Air Force operations is well known. Zolpidem (Ambien) is a newer, shorter acting sedative compound that is clinically well known. It is an alternative for Restoril and may help overcome sleeplessness antagonized by the wrong circadian phase. Zolpidem has received much acclaim, and it is currently the world's best-selling sedative. More significantly, though, Zolpidem has been approved for use by Army aviators with command approval, based on laboratory evaluations using 18 Army aviators. The Italian Air Force has found termazepan effective for inducing and maintaining diurnal sleep (Porcu, 1997).

Sedatives do have drawbacks. Idiosyncratic reactions, such as hallucinations, can occur in some people (Caldwell et al., 1997). Beyond 1 week of use, many sedatives (particularly the older benzodiazepines like Restoril) begin to interfere with sleep. In fact, with most of the suggested sedatives, "rebound insomnia" can occur after only a few days of use. Short-acting benzodiazepines, such as triazolam (Halcyon), can induce rebound insomnia even on subsequent attempts at sleep, and personality changes have been reported.

Chronobiotics

The unique properties of the pineal hormone Melatonin make it extremely appealing as a nonaddictive means to promote sleep and perhaps to rapidly shift biological rhythms (Arendt, 1994). The use of Melatonin to counteract desynchronosis was demonstrated during an army-training mission to the Middle East (Compertore, Liberman, Kirby, Adams, & Crowley, 1996). Combinations of Melatonin and bright light schedules may provide the most potent means yet to preshift the circadian timing system and quickly prepare crews to function at peak performance in a new temporal environment.

Environmental light improves performance degraded by nocturnal circadian effects. The use of bright illumination is possible with existing lighting facilities in most planning, control, and communication environments. Bright light also may be useful in rapidly preadjusting circadian phase to night operations. Exposure to bright light at critical times in the early morning seems to quickly regulate circadian phase within a day or two (Dijak, Beersma, Daan, & Lewy, 1989; Honma, Honma, & Wada, 1987). Laboratory studies have shown that carefully controlled light exposures of a few hours can reverse circadian phase by as much as 12 hours (Jewett, Kronauer, & Czeisler, 1991). The timing of light exposure is critical to these effects. Light in the early morning hours before the temperature minimum (usually between 1:00 AM and 4:00 AM) delays the circadian phase, whereas light after the temperature minimum (between 4:00 AM and 7:00 AM) advances the phase. Exposure to light at other times is less likely to affect circadian phase (Pittendrigh, 1981). These results are relatively new, however, and should be tested in field conditions because there are some occasions when reverse effects are possible (Eastman, 1992; Jewett, Kronauer, & Czeisler, 1994). Social cues (zeitgebers) also have small but demonstrable effects on altering circadian phase, although not as profoundly as bright light or Melatonin (Weaver, 1985).

REFERENCES

Arendt, J. (1994). Clinical perspectives for Melatonin and its agonists. *Biological Psychiatry, 35,* 1–2.

Bartlett, F. C. (1953). Psychological criteria for fatigue. In W. F. Floyd & A. T. Welford (Eds.), *Symposium on fatigue.* London: H. K. Lewis.

Bisson, R. U., Neville, K. J., Boll, P. A., French, J., Ercoline, W. R., McDaniel, R. L., & Storm, W. F. (1992). Digital flight data as a measure of pilot performance associated with fatigue from continuous operations during Desert Persian Gulf conflict. In *Nutrition metabolic disorders and lifestyle of aircrew* (pp. 12–54). Neuilly-sur-Seine, France: NATO.

Blagrove, M., Alexander, C., & Horne, J. A. (1995). The effects of chronic sleep reduction on the performance of tasks sensitive to sleep deprivation. *Applied Cognitive Psychology, 9,* 21–40.

Blom, D. H. J., & Pokorny, M. L. I. (1985). *Accidents of bus drivers: An epidemiological approach.* Leiden: Nederlands Instituut voor Praeventieve Gesondheidszorg.

Boff, K. R., & Lincoln, J. E. (Eds.). (1988). *Engineering data compendium: Human perception and performance.* Dayton, OH: Wright-Patterson Air Force Base.

Bonnet, M. H., & Arand, D. L. (1994). The use of prophylactic naps and caffeine to maintain performance during a continuous operation. *Ergonomics, 37,* 1009–1020.

Borowsky, M. S., & Wall, R. (1983). Naval aviation mishaps and fatigue. *Aviation, Space, and Environmental Medicine, 54,* 6, 535–538.

Brown, I. D., Tickner, A. H., & Simmonds, C. V. (1970). Effect of prolonged driving on overtaking criteria. *Ergonomics, 13,* 239–242.

Bruener, H., Holtmann, H., Klein, K., Rehme, H., Steinhoff, W., Stoltz, J., & Wegmann, H. (1970). Circadian rhythm of pilots' efficiency and effects of multiple time zone travel. *Aerospace Medicine, 41,* 125–132.

Caldwell, J. A. (1996). Effects of operationally effective doses of dextroamphetamine on heart rates and blood pressures of Army aviators. *Military Medicine, 161,* 673–678.

Caldwell, J. A., Caldwell, J. L., & Crowley, J. S. (1997). Sustaining female helicopter pilot performance with Dexedrine during sleep deprivation. *International Journal of Aviation Psychology, 7,* 15–36.

Caldwell, J. A., Caldwell, J. L., Crowley, J. S., & Jones, H. D. (1995). Sustaining helicopter pilot performance with Dexedrine during periods of sleep deprivation. *Aviation, Space, and Environmental Medicine, 66,* 930–937.

Caldwell, J. A., Caldwell, J. L., Lewis, J. A., Jones, H. D., Reardon, M. J., Jones, R., Colon, J., Pegues, A., Nillard, R., Johnson, P., Woodrum, J., & Hegdon, A. (1996, December). *Sustainment of Helicopter Pilot Performance* (Rep. No. USAARL 97-05). Fort Rucker, AL: U.S. Army Aeromedical Research Laboratory.

Caldwell, J. A., Jones, R. W., Caldwell, J. L., Colon, J. A., Pegues, A., Iverson, L., Roberts, K. A., Ramspott, S., Springer, W. D., & Gardner, S. J. (1997, February). *The efficacy of hypnotic-induced prophylactic naps for the maintenance of alertness and performance in sustained operations* (Rep. No. USARRL 97-10). Fort Rucker, AL: U.S. Army Aeromedical Research Laboratory.

Carskadon, M. A., & Dement, W. C. (1981). Nocturnal determinants of daytime sleepiness. *Sleep, 5,* S73–S81.

Comperatore, C. A., & Allan, L. W. (1997). Effectiveness of institutionalized countermeasures on crew rest during rapid transitions from daytime to nighttime flying. *International Journal of Aviation Psychology, 7,* 139–148.

Comperatore, C. A., Liberman, H., Kirby, A. W., Adams, B., & Crowley, J. S. (1996, October). *Melatonin efficacy in aviation missions requiring rapid deployment and night operations.* (USAARL 97-03). Fort Rucker, AL: U.S. Army Aeromedical Research Laboratory.

Cornum, R., Caldwell, J., & Cornum, K. (1997). Stimulant use in extended flight operations. *Airpower Journal, 11*(1), 53–58.

Costa, G. (1996). The impact of shift and night work on health. *Applied Ergonomics, 27,* 9–16.

Dijak, D. J., Beersma, D. G., Daan, S., & Lewy, A. J. (1989). Bright morning light advances the human circadian system without affecting NREM sleep homeostasis. *American Journal of Physiology, 256,* R106–R111.

Dingus, T. A., Hardee, H. L., & Wierwille, W. W. (1987). Development of models for on-board detection of driver impairment. *Accident Analysis Prevention, 19,* 271–283.

Eastman, J. (1992). High intensity light for circadian adaptation to a 12 h shift of the sleep schedule. *American Journal of Physiology, 263,* R428–436.

Edmondson, J. L., & Oldman, M. (1974). *An interview study of heavy goods vehicle drivers.* (Contract Rep. No. 74/14). Southampton, England: Institute of Sound and Vibration Research, University of Southampton.

Elsmore, T. F., Hegge, F. W., Naitoh, P., Kelly, T., Schlangen, K., & Gomez, S. (1995). *A com-*

parison of the effects of sleep deprivation on synthetic work performance and a conventional perform-ance assessment battery (Rep. No. 95-6). San Diego, CA: Naval Health Research Center.

Emonson, D. L., & Vanderbeek, R. D. (1995). The use of amphetamines in US Air Force tacti-cal operations during Desert Shield and Storm. *Aviation, Space, and Environmental Medi-cine, 66,* 260–263.

Ferrer, C. F., Bisson, R. U., & French, J. (1995). Circadian rhythm desynchronosis in military deployments: A review of current strategies. *Aviation, Space, and Environmental Medicine, 66,* 571–578.

Fevre-Montange, M., Van Cauter, E., Refetoff, S., Desir, D., Tourniaire, J., & Copinschi, G. (1981). Effects of "jet lag" on hormonal patterns: Adaptation of melatonin circadian rhythms. *Journal of Clinical Endocrinology and Metabolism, 52,* 642 -649.

Foret, J., & Lantin, G. (1972). The sleep of train drivers: an example of the effects of irregular work schedules on sleep. In W. P. Colquhoun (Ed.), *Aspects of human efficiency* (pp. 273–282). London: English Universities Press.

French, J., Bisson, R., Neville, K. J., Mitcha, J., & Storm, W. F. (1994). Crew fatigue during sim-ulated, long duration B-1B bomber missions. *Aviation, Space, and Environmental Medicine, 65,* A1–6.

French, J., Neville, K., Boll, P., Bisson, R., Slater T., Armstrong, S., Storm, W., Ercoline, W., & McDaniel, R. (1992). *Subjective mood and fatigue of C-141 crew during Desert Storm* (Nu-trition Metabolic Disorders and Lifestyle of Aircrew). Neuilly-sur-Seine, France: NATO-AGARD.

Friedmann, J., Globus, G., Huntley, A., Mullaney, D., Naitoh, P., & Johnson, L. (1977). Perfor-mance and mood during and after gradual sleep reduction. *Psychophysiology, 14,* 245–250.

Fuller, R. G. C. (1981). Determinants of time headway adopted by truck drivers. *Ergonomics, 24,* 463 -474.

Graeber, R. C. (1987). Sleep in space. In *Proceedings of the 28th NATO DRG seminar: Sleep and its implications for the military* (pp. 59–69). Neuilly-sur-Seine: NATO.

Graeber, R. C. (1988). Aircrew fatigue and circadian rhythmicity. In E. L. Wiener & D. C. Nagel (Eds.), *Human factors in aviation,* (pp. 305–344). New York: Academic.

Graeber, R., Dement W., Nicholson, A., Sasaki, M., & Wegmann, H. (1986). International co-operative study of aircrew layover sleep: Operational summary. *Aviation, Space, and Envi-ronmental Medicine, 57,* B10–B13.

Hamelin, P. (1981). Les conditions temporelles de travail des conducteurs routièrs et la sécu-rité routière [Work conditions for truck drivers and their security]. *Travail Humain, 44,* 5–21.

Hamelin, P. (1987). Lorry drivers' time habits in work and their involvement in traffic acci-dents. *Ergonomics, 30,* 1323–1333.

Harris, W. (1977). Fatigue, circadian rhythm and truck accidents. In R. R. Mackie (Ed.), *Vigi-lance: Theory, operational performance and physiological correlates* (pp. 230–248). New York: Plenum.

Harris, W., & Mackie, R. R. (1972). *A study on the relationships among fatigue, hours of service, and safety of operations of truck and bus drivers* (Rep. No. 1727-2). Goleta, CA: Human Factors Research.

Harris W., Mackie, R. R., Abrams, C., Buckner, D. N., Harabedin, A., O'Hanlon, J. F., & Starks, J. R. (1972). *A study of the relationships among fatigue, hours of service and safety of operations of truck and bus drivers* (Rep. No. 1727-2). Goleta, CA: Human Factors Research.

Haslem, D. R. (1982). Sleep loss, recovery sleep, and military performance. *Ergonomics, 25,* 163–178.

Hertz, R. (1987). Sleeper berth use as a risk factor for tractor-trailer driver fatality. In *31st An-nual Proceedings for Automotive Medicine* (pp. 215–228). New Orleans, LA: Association for the Advancement of Automotive Medicine.

Honma, K., Honma, S., & Wada T. (1987). Phase dependent shift of free-running human circadian rhythms in response to a single bright light pulse. *Experientia, 43,* 1205–1207.

Jewett, M., Kronauer, R., & Czeisler, C. (1991). Light induced suppression of endogenous circadian amplitude in humans [Letter]. *Nature, 350,* 59–62.

Jewett, M., Kronauer, R., & Czeisler, C. (1994). Phase amplitude resetting of the human circadian pacemaker via bright light: a further analysis. *Journal of Biological Rhythms, 9,* 295–314.

Johnson L., & Naitoh, P. (1974). *The operational consequences of sleep deprivation and sleep deficit.*
(AGARD AG-193). Neuilly-sur-Seine: NATO AGARD.

Kelly, T. L. (1996). *Circadian rhythms: Importance of models of cognitive performance* (Rep. No. 96-
1). San Diego, CA: Naval Health Research Center.

Klein, K., Wegmann, H., Athanassenas, G., Hohlweck, H., & Kuklinski, P. (1976). Air operations and circadian performance rhythms. *Aviation, Space, and Environmental Medicine, 47,*
221–229.

Krueger, G. P. (1991). Sustained military performance in continuous operations: Combatant
fatigue, rest and sleep needs. In R. Gal & A. D. Mangelsdorff (Eds.), *Handbook of military
psychology* (pp. 103–137). New York: Wiley.

Langlois, P. H., Smolensky, M. H., Hsi, B. P., & Weir, F. W. (1985). Temporal patterns of reported single-vehicle car and truck accidents in Texas, U.S.A., during 1980–1983. *Chronobiology International, 2,* 131–146.

Lyons, T., & French, J. (1992). Modafinil, the unusual characteristics of a new stimulant. *Aviation, Space, and Environmental Medicine, 62,* 432–435.

Mackie, R. R., & Miller, J. C. (1978). *Effects of hours of service, regularity of schedules and cargo
loading on truck and bus driver fatigue* (Rep. No. 1765-F). Goleta, CA: Human Factors Research.

McDonald, N. (1978). *Fatigue, safety and the working conditions of heavy goods vehicle drivers.* Unpublished doctoral dissertation, University of Dublin.

McDonald, N. (1981). Safety and regulations restricting the hours of driving of goods vehicle
drivers. *Ergonomics, 24,* 475–485.

McDonald, N. (1989). Fatigue and driving. *Alcohol, Drugs and Driving, 5,* 185–191.

McDonald, N., Fuller, R., & White, G. (1989). Fatigue and safety: A reassessment. *Proceedings
of the 5th international symposium on Aviation Psychology* (Vol. 2, pp. 551–556). Columbus:
The Ohio State University.

McDonald, N., Fuller, R., & White, G. (1991). Fatigue and accidents: A comparison across
modes of transport. In E. Farmer (Ed.), *Stress and error in aviation* (Proceedings of the 17th
WEAAP conference, Vol. 2, pp. 125–133). Avebury, England: Avebury Technical.

McFarland, R. A. (1953). *Human factors in air transportation.* New York: McGraw-Hill.

Medical Economics. (1997). *Physicians' desk reference guide to drug interactions, side effects, indications, and contraindications.* Montvale, NJ: Author.

Mitler, C. C., Carskadon, M. A., Czeisler, C. A., Dement, W. C., Dinges, D. F., & Graeber, R. C.
(1988). Catastrophes, sleep, and public policy: Consensus report. *Sleep, 11,* 100–109.

Moore-Ede, M. (1993). Aviation safety and pilot error. In M. Moore-Ede (Ed.), *The twenty-four
hour society* (pp. 81–95). New York: Addison-Wesley.

Naitoh, P., Kelly, T., & Babkoff, H. (1992). *Sleep inertia: Is there a worst time to wake up?* (NHRC
Rep. No. 91–145). Washington, DC: U.S. Printing Office.

Neville, K. J., Bisson, R. U., French, J., Boll, P. A., & Storm, W. F. (1994). Subjective fatigue of
C-141 aircrews during Operation Desert Storm. *Human Factors, 36,* 339–349.

O'Hanlon, J. F. (1981). Critical Tracking Task (CTT) sensitivity to fatigue in truck drivers. In
B. J. Oborne & J. A. Levis (Eds.), *Human factors in transport research* (pp. 78–106). London:
Academic.

Palmer, B., Gentner, F., Schopper, A., & Sottile, A. (1996). Review and analysis: Scientific review of air mobility command and crew rest policy and fatigue issues. *Fatigue Issues,* 1–2.

Petrie, K. J., & Dawson, A. G. (1997). Symptoms of fatigue and coping strategies in international pilots. *International Journal of Aviation Psychology, 7,* 251–258.

Pigeau, R., Naitoh, P., Buguet, A., McCann, C., Baranski, J., Taylor, M., Thompson, M., & Mack, I. (1995). Modafinil, d-amphetamine and placebo during 64 hours of sustained mental work, I. Effects on mood, fatigue, cognitive performance and body temperature. *Journal of Sleep Research, 4,* 212–228.

Pittendrigh, C. (1981). Circadian systems: entrainment. In J. Aschoff (Ed.), *Biological rhythms: Handbook of behavioral neurobiology* (Vol. 4, pp. 95–124). New York: Plenum.

Porcu, S. (1997). Acutely shifting the sleep–wake cycle: Nighttime sleepiness after diurnal administration of temazepam or placebo. *Aviation, Space, and Environmental Medicine, 68,* 688–694.

Reid, D. D. (1948). Fluctuations in navigator performance during operational sorties. In E. J. Earnaley & P. B. Warr (Eds.), *Aircrew stress in wartime operations.* New York: Academic.

Reiter, R. J. (1991). Pineal melatonin: Cell biology of its synthesis and of its physiological interactions. *Cell Biology, 12,* 151–180.

Ribak, J., Ashkenazi, I. E., Klepfish, A., Avgar, D., Tall, J., Kallner, B., & Noyman, Y. (1983). Diurnal rhythmicity and air force flight accidents due to pilot error. *Aviation, Space, and Environmental Medicine,* 1096–1099.

Riemersma, J. B. J., Sanders, A. F., Wildervanck, C., & Gaillard, A. W. (1997). Performance decrement during prolonged night driving. In R. R. Mackie (Ed.), *Vigilance* (pp. 237–263). New York: Plenum.

Rosa, R., Bonnet, M. H., & Warm, J. S. (1983). Recovery of performance during sleep following sleep deprivation. *Psychophysiology, 20,* 152–159.

Sack, R. L., Lewy, A. J., Blood, M. L., Keith, L. D., & Nakagawa, H. (1992). Circadian rhythm abnormalities in totally blind people-Incidence and clinical significance. *Journal of Clinical Endocrinological Metabolism, 75,* 127–134.

Samel, A., Wegmann, H. M., Vejoda, M., Drescher, J., Gundel, A., Manzey, D., & Wenzel, J. (1997). Two-crew operations: Stress and fatigue during long-haul night flights. *Aviation, Space, and Environmental Medicine, 68,* 679–687.

Sasaki, M., Kurosaki, Y., Spinweber, C. L., Graeber, R. C., & Takahashi, T. (1993). Flight crew sleep during multiple layover polar flights. *Aviation, Space, and Environmental Medicine, 64,* 641–647.

Senechal, P. K. (1988). Flight surgeon support of combat operations at RAF Upper Heyford. *Aviation, Space, and Environmental Medicine, 59,* 776–777.

Smith, A. (1989). Diurnal variations in performance. In A. M. Colley & J. R. Beech (Eds.), *Acquisition and performance in cognitive skills* (pp. 301–325). New York: Wiley.

Stern, J. A., Boyer, D., & Schroeder, D. (1994). Blink rate: A possible measure of fatigue. *Human Factors, 36,* 285–297.

Stoner, J. D. (1996). Aircrew fatigue monitoring during sustained flight operations from Souda Bay, Crete, Greece. *Aviation, Space, and Environmental Medicine, 67,* 863–866.

Tepas, D. I., & Carvalhis, A. B. (1990). Sleep patterns of shiftworkers. In A. Scott (Ed.), *Occupational medicine: State of the art reviews* (pp. 199–208). Philadelphia: Hanley & Belfus.

Tilley, A. J., Wilkinson, R. T., Warren, P. S. G., Watson, B., & Drud, M. (1982). The sleep and performance of shift workers. *Human Factors, 24,* 629–641.

Transportation Research and Marketing (1985). *A report on the determination and evaluation of the role of fatigue in heavy truck accidents.* Falls Church, VA: AAA Foundation for Traffic Safety.

Weaver, R. (1985). Use of light to treat jet lag: Differential effects of normal and bright artificial light on human circadian rhythms. *Annual Report of the NY Academy of Science, 453,* 282–304.

Welford, A. T. (1968). *Fundamentals of skill.* London: Methuen.

Winget, C. M., DeRoshia, C. W., Markley, C. L., & Holley, D. C. (1984). A review of human physiological and performance changes associated with desynchronosis of biological rhythms. *Aviation, Space, and Environmental Medicine, 55*, 1085–1096.

Wright, J. E., Vogel, J. A., Sampson, J. B., Knapik, J. J., & Patton, J. F. (1983). Effects of travel across time zones (jet lag) on exercise capacity and performance. *Aviation, Space, and Environmental Medicine, 54*, 132–137.

Wyckoff, D. (1979). *Truck drivers in America.* New York: Free Press.

3.10 Coping With Driver Fatigue: Is the Long Journey Nearly Over?

Ivan Brown
Ivan Brown Associates, Cambridge, England

Researching fatigue in transport systems has been a long haul, and it remains debatable that valid, practicable, and socially acceptable solutions to the variety of potential fatigue problems that still plague transport operators are currently available. This unsatisfactory situation results not so much from disagreement on the definition of fatigue, or its causation, or its possible countermeasures, as from the conflicts that are seen to exist between the effects of those countermeasures and the economic viability of those systems in which the problems largely exist. This commentary thus presents the view that fatigue as it is commonly experienced today, and especially in transport systems, is essentially of society's own making, although it is suffered mainly by individuals working in those systems.

DEFINITION

In human terms the word *fatigue* has been used to imply a diminished capacity for work. In other words, is has usually described a state recognized by its consequences rather than its nature or its causes. Hence it is not surprising that many earlier writers on the subject confessed that fatigue was a complex phenomenon that was difficult to define precisely, because the consequences of fatigue are many and various. A broad distinction has long been made between its psychological and physiological effects, but it was probably the detailed study by Bartley and Chute (1947) that confirmed the current view that the term *fatigue* should be used to identify only the psychological aspects of the phenomenon. Furthermore, these authors subdivided the psychological aspects of fatigue into performance decrements on the one hand and cognition of negative personal states on the other. However, performance decrements like physiological impairments are almost certainly a specific function of the task being performed and of its physical and social environmental conditions. They therefore have insufficient generality as contributors to a broadly useful definition of human fatigue. Decrements in performance would, in any case, simply index the *effectiveness* with which a task was being carried out, telling us nothing about the psychological state of the individual performing the task. If we want to theorize about the state of fatigue, predict its consequences for individuals in specific circumstances, and hope to counteract its adverse effects, we must know something about the way in which fatigue affects operator *efficiency*.

Considerations of this kind appear to have been behind Nelson's (1989) development of the concept of fatigue as "a declarative state based upon self-perceptions of body distress, mood and performance." In operational terms, Nelson defined fatigue as "the condition the person reaches as a result of sustained activity wherein he or she *declares* being unable to continue the activity further." This may well be analogous to the way in which the term fatigue is defined in engineering terms, but it fails to match the common experience of fatigue as a graded state in which individuals experiencing these self-perceptions are nevertheless able to continue performing for considerable periods and in which actually being unable to continue with the task in hand is but the end point. For this reason, I have tended to define the state of human fatigue as "*a subjectively experienced disinclination to continue performing the task in hand because of perceived reductions in personal efficiency*" (see Brown, 1993).

CAUSATION

Fatigue, as defined in this way and as commonly recognized, is clearly part of the human condition. Biological systems can only expend so much energy per unit of time before a certain amount of muscular recuperation becomes necessary. This limitation is unlikely to prove life threatening, in the absence of predators, so long as the individual in question remains self-paced or is under the control of other individuals who recognize the importance of rest pauses for efficient prolonged activity. With this in mind, is seems clear that our present problem of fatigue in transport system has its origins in the industrial revolution, when the need to maximize returns from capital investments was seen to require maximal operation of new technological systems. Fatigue in those times was still largely muscular in origin and caused mainly by the prolonged expenditure of energy. As technology advanced, it may reasonably be assumed that the required expenditure of energy was reduced for many tasks and that prolonged working hours became a major cause of fatigue, whatever the nature of the task's demands.

There appears to have been little general interest in the consequences of such fatigue for the individual until World War I, when it became essential to increase the production and use of armaments while mininizing the consequences of human performance errors, and this period may be seen as the beginning of a scientific interest in the problem of fatigue. For example, in England an Industrial Fatigue Research Board (the forerunner of the Medical Research Council) was set up to examine the problem and explore ways of dealing with fatigue-related human errors. The board's success in reducing production errors by reducing the number of working hours per shift no doubt focussed attention on the "time-on-task" aspect of fatigue causation, and this became the major target during subsequent attempts to minimize effects of fatigue in transport systems. Hence most legislative countermeasures against driver fatigue have imposed constraints on the hours spent in direct vehicle operation, although more enlightened legislators also constrained the total number of working hours per shift, because they recognized that fatigue can be transferred from one task to another. (It is relevant here to point out that, for this reason, the title of this chapter refers to "driver" fatigue rather than "driving" fatigue).

It was not until the early 1960s that research on maritime watch keeping and industrial shift systems established the causal relation between biological circadian rhythms and performance efficiency. The results of such comprehensive research over the past 35 years or so now confirm that fatigue is not simply (perhaps not mainly with many types of work) a function of the time spent in performing a task. The times of day in which the task is carried out are now seen to be crucial determinants of performance efficiency, with the early morning hours and the postlunch period likely to be associated with greater sensations of fatigue and with exacerbated effects of time-on-task for the majority of individuals. Over the same period, research on the quality and quantity of sleep taken by working individuals has identified the essential contributions of these variables to the experience of fatigue, making excessive "time since sleep" a major factor in fatigue causation.

In summary, it is now generally accepted that the principal determinants of fatigue are unduly prolonged working hours, working during troughs in the circadian cycle of biological activation, and insufficient and/or poor quality sleep prior to commencing work. Whether fatigue is actually *experienced* by individuals therefore depends on the ways in which their work is organized and scheduled. If their task demands continuous attention, if the consequences of momentary distraction could be disastrous, and if they have no control over the way in which their work is scheduled over each 24 hours or longer periods, then they are clearly more at risk from fatigue than individuals who are largely self-paced. It may therefore not be going too far to assert that individual fatigue results from society's ill-considered attempts to maximize the exploitation of technological developments by operating human/technological systems in an inhuman way. The implications of this claim for the risk of fatigue experienced by vehicle drivers attempting to meet society's demands for goods and services are obvious.

MEDIATING FACTORS

A number of factors outside the immediate circumstances of task and environment obviously mediate the individual's experience of fatigue. These include alcohol and drugs, age, personality, training and experience with the task in hand, and physical fitness. They are not considered in detail here (although see Brown, 1993), but they have clear implications for subsequent considerations of possible countermeasures against fatigue.

EXPERIENCE OF FATIGUE

On the definition presented earlier, it is clearly essential to understand individual *experiences* of fatigue, as well as its causes, if one is to identify and implement appropriate countermeasures. Some of the very informative early research on such experiences was summarized by Bartlett (1948) in his report on the initial signs of skill fatigue among aircrew during long wartime flights. He identified the principal signs as inaccurate timing of control movements, leading to a loss of smoothness in manipulating the vehicle controls; a tendency for the individual to require larger than normal changes in stimuli before responding; an apparent reduction in the normal span of anticipation; and a heightened sensitivity to bodily changes, manifested largely in more aggressive responses toward other people, but also toward the vehicle and its control equipment. This list of symptoms is probably representative of all the main ways in which an individual vehicle driver experiences fatigue. It includes the decrement in performance of control actions, which the individual perceives as a decline in effectiveness. It includes the need for more information on task and environmental demands before reacting, which may signal a decline in sensory and/or perceptual effectiveness, but which may also be indicative of increasing uncertainty about personal competence. It includes a decline in anticipatory responding, which again may reflect impaired sensory and/or perceptual effectiveness (including states variously termed "highway hypnosis," "driving without awareness," and "loss of situational awareness"), but which may also indicate a strategic change in the way the individual chooses to carry out the task, following a recognition that efficiency of task performance is declining (see, e.g.,

Sperandio, 1978; Welford, 1978). Finally, it includes a cognition of mood changes and a generalized emotional rejection of the current working conditions. All these experiences have something to tell us about the design of valid and practicable countermeasures against fatigue.

It is relevant here to distinguish the experience of fatigue from that of *boredom,* which is also "a subjectively experienced disinclination to continue performing the task in hand." This may conveniently be done by modifying the previous definition of fatigue so that the reference to "perceived reductions in efficiency" is replaced by "perceived monotony and experienced tedium." This definition of boredom accepts that a bored individual is not necessarily less efficient at performing a given task, but lacks the motivation to continue.

It is also relevant here to comment briefly on the relations among fatigue, as defined here, informational workload, and stress, before considering the questions of measurement and countermeasure design. In engineering terms, stress is measured objectively as the relation between imposed workload and the capacity of a given material for supporting that load. Analogously, in human terms, an objective measure of "stress" may be derived from the relation between actual workload and actual information-processing capacity. If individuals are confronted by too many sources of information and/or if their normal capacity for processing information is impaired, they are "stressed," in objective terms. Again, in engineering terms, the effect of stress is strain, measured objectively as a relative adverse change in the stressed material. By analogy, "strain" in human terms may be calculated from measured decrements in the normal efficiency of performance. This strain may be perceived directly by the individual in question, via task feedback, as a decrement in normal levels of performance. Alternatively, it may be perceived indirectly as a requirement for increased effort (e.g., for more frequent or more wide-ranging eye movements) to be certain of coping adequately with excessive workload demands. In either case, if the perceived strain continues unrelieved, it is experienced as "fatigue." In summary, an adverse mismatch between task workload and processing capacity as they are perceived by the individual is experienced as stress and the resulting strain, if it persists, is perceived as entry into a state of fatigue.

There is, however, an additional factor in human fatigue that has no analogy in engineering terms. This is the individual's sense of responsibility for any adverse consequences of continuing to perform a task at perceptibly reduced levels of efficiency. The importance of this factor in fatigue causation was revealed very clearly in a study of hours of work and safety at sea (Brown, 1989), conducted in connection with considerations of compliance with international legislation on seafarers' working arrangements. As part of that study, anecdotal evidence on personal fatigue experiences was collected

from a variety of professional seafarers, who were asked to submit anonymous reports on conflicts and near accidents that could have been related
in some way to their working arrangements. As might be expected, the
detailed experiences they reported included muscular pains, headaches,
eyestrain, mental and physical exhaustion, loss of appetite, drowsiness, insomnia, microsleep incidents, general anxiety, time stress, reduced comprehension, complete disorientation, loss of concentration, distraction, for-
getfulness, "tunnel vision," carelessness, extended reaction and decision
times, impaired co-ordination of control skills, erroneous control operation,
persistence with erroneous interpretation of events, and inability to relax
during rest periods. In spite of these experiences, they invariably continued
performing their tasks to the end of their shift, because they were responsible for some aspect of the command or control of their vessels and because,
in many cases, they were irreplaceable. This response exemplifies the *conflict*
that Bartley and Chute (1947) concluded was an essential component in the
fatigue syndrome. Fatigued individuals who are responsible for operation
and control in transport systems inevitably experience conflict between a
self-imposed need or externally imposed demand to complete their given
tasks and their desire to preserve the safety of themselves and others. Resolving this conflict satisfactorily is the key to the avoidance of adverse fatigue
effects on safety in the running of transport systems, but how is this resolution to be achieved?

MEASUREMENT OF FATIGUE

Measurement of fatigue is clearly as important to the consideration of countermeasures against its adverse effects as it is to the understanding of the nature of that state. Research can be replicated and errors corrected, but the
unwise implementation of ill-considered countermeasures may have disastrous effects before they can be properly identified and remedied. Given the
definition of fatigue presented earlier, the only valid way of measuring it
seems to be via self-reports by the individual at risk. However, this method
may be unreliable as an input to countermeasures, given the conflict between an individual's responsibilities discussed previously. Hence, research
continues into objective methods of measuring fatigue's effects, which must
be both valid and reliable and also readily implemented in transport systems
(e.g., see Hartley, 1995).

Not surprisingly, attention has focused on in-vehicle methods of measuring those changes in control skills among fatigued individuals which have
been well-known for half a century. As Bartlett (1948) reported, one principal early sign of fatigue (at least among aircrew) was a tendency to require

larger than normal changes in stimuli before responding. This "sign," among road vehicle drivers, has been recorded as a delay in steering corrections until a much larger than normal correction becomes necessary to restore the vehicle to its required course (see Iizuka, Yanagishima, Kataoka, & Seno, 1985). In practice, this delay can be identified most clearly and measured via the atypically large response that follows it or, more directly, via continuous anticipatory measurement of the time before which the vehicle leaves the road if no change to its current heading is made (often termed "time to line-crossing"). This observed delay in normal steering corrections may also reflect the "retraction of the normal span of anticipation" reported by Bartlett among fatigued individuals. Iizuka et al. found that it was not just declining frequency with which steering corrections are made that signalled driver fatigue, but also the magnitude and speed of the first two corrective movements made following a long period of steering activity. This appears to reflect Bartlett's finding that fatigue leads to a loss of smoothness in manipulating the vehicle's controls.

"Delayed responding" and "loss of smoothness" can obviously be measured via other vehicle control skills. It has long been known that a generalized measure of such skills and of certain changes in them can easily be obtained by recording the lateral and longitudinal accelerations imposed on the vehicle (see Brown, 1962). However, the accelerations imposed on modern vehicles under typical operating conditions are so low that valid departure from normal behavior by fatigued drivers would be difficult to measure with any acceptable level of reliability.

Reports of "retraction in the normal span of anticipation" among fatigued individuals have served to focus attention on drivers' eye movements as a source of cues to fatigue onset. Affordable technological systems are readily available to record small changes in frequency, duration, and scatter of visual fixations and to implement alternative remedial measures when these changes exceed some valid criterion of fatigue onset.

Such measures of changes in vehicle control skills and visual behavior have received sufficient research and development attention in recent years that we can be fairly certain of their validity as signs of driver fatigue. The crucial problem remains the reliability with which such measures can be obtained in real traffic conditions. How does one distinguish fatigue-related changes in "normal" control usage or visual behavior from those changes that have been necessitated by alterations in road or traffic conditions? This problem of "criterion setting" in order to strike an acceptable balance between "missed signal" and "false alarms" is one that seems likely to plague the reliable measurement of driver fatigue for some time to come. The other problem is to find a way of dealing with the mediating effects on fatigue measurements of individual differences, such as alcohol, drugs, age, personality, training, experience, and physical fitness.

COUNTERMEASURES AGAINST DRIVER FATIGUE

Road safety countermeasures in general are categorized as education, enforcement, or engineering. It follows from our definition of fatigue that education must remain of central importance to its identification and avoidance, because fatigue is suffered by a wide variety of drivers who are likely to be out of reach of enforcement and engineering countermeasures. Given that fatigue is a subjective experience, drivers must have a clear understanding of its causes and symptoms, if they are to be in a position to implement one or more remedial actions before becoming accident involved. This need is becoming increasingly recognized, as evidenced by the road signs warning of the dangers of driver fatigue being installed in certain countries and by the inclusion of questions relating to fatigue in new theory tests for European drivers. It is not so clear that the effectiveness of such educational countermeasures has been sufficiently tested empirically. Clearly there is a need to do so, because the majority of fatigued drivers complete their journeys safely and this experience, in time, devalues the message that "fatigue kills." For drivers who are not so lucky, it is necessary for them to appreciate the causal relation between the symptoms they were experiencing and their accident involvement, if they are to change their fatigue-inducing behavior. However, it is not simply the presence or absence of educational/informational countermeasures against fatigue that need to be tested empirically, but also the *method* by which such information is conveyed to drivers. In addition to off-road and roadside displays, which may well be ignored, new in-vehicle technological systems for driver support provide opportunities to present fatigue-related information as warning outputs from the measurement of departures from normal control activity or visual behavior. This more personal and specific feedback on fatigue symptoms seems likely to have a greater educational impact on driver behavior than do the more general displays of information currently being introduced.

Enforcement of legislation aimed at controlling fatigue effects also seems an essential requirement. Although "time-on-task" is not the sole cause of fatigue, legislation on working hours does, at least, provided some control over muscular fatigue during spells of continuous driving and over the "time since sleep" variable in fatigue causation among professional groups of drivers. The need for an adequate period of continuous sleep between workshifts and for rest breaks between periods of continuous work is well established. The technology is there to record continuous working by heavy goods vehicle and coach drivers (e.g., the tachograph), and this works well if the legislation is enforced. However, fatigue is also a substantial problem among drivers of private vehicles who are not subject to the legislation (e.g., see Maycock, 1995), therefore enforcement can be only a partial, albeit essential, solution to the problem.

Legislative measures against certain mediating factors in fatigue causation are also in place in most developed countries, aiming to limit the effects of, for example, alcohol and drugs, among all classes of drivers. Among professional drivers, legislation additionally aims to control the mediating influences of age, training, and driving experience. Recent research findings associating obesity with the contribution of sleep apnea to driver fatigue suggest that legislation controlling the physical fitness of professional drivers may well be needed to limit the mediating effect of this personal characteristic on susceptibility to fatigue.

Engineering appears to offer the most scope for fatigue countermeasures among all driver groups, both private and professional. In-vehicle systems for driver monitoring, support, and control are becoming increasingly reliable and affordable. Given sound empirical evidence of their effectiveness and reliability and the political will of legislators to enforce their implementation and use where serious problems of driver fatigue remain, there is no reason that the problem should not be brought under control. As indicated earlier, changes in drivers' steering behavior appear to offer the most soundly based, face-valid, reliable, and applicable measure of fatigue onset, with criterion setting remaining as the principal objection to implementation of the appropriate technology.

The other major problem is finding a reliable and socially acceptable output from such a fatigue detection system which would safely limit further driving by dangerously fatigued individuals. Given the obvious conflict that exists between drivers' experiences of fatigue symptoms and their desire, or their employer's demand, to complete the journey, it is clear that simple warnings of fatigue onset are of limited effectiveness, although they may serve an educational function. On the other hand, the enforced termination of a journey would in most cases be socially unacceptable and possibly dangerous. In many countries, the opportunities to park a heavy goods vehicle offroad are seriously limited. What appears to be required is a practical and acceptable way of supporting fatigued drivers in their decision to terminate their journey on genuine safety grounds.

It is not difficult to envisage the way in which such a support system might work among professional drivers. The output from an in-vehicle fatigued driver support system could simply be networked to communicate with employers and to provide a variety of helpful responses, ranging from advice on local rest facilities to the import of a replacement driver. A similar network could be envisaged to link private drivers with their national motoring organizations. In both cases, cost will be a limiting factor, and legislation may be necessary to overcome objections on these grounds among transport operators, but the feasibility of such a system seems clearly established.

JOURNEY'S END?

The question posed by the title of this chapter asks how near are we to coping with driver fatigue. This is obviously a multifaceted question. It seems generally accepted that the principal causes of fatigue are known and that individual characteristics that can mediate susceptibility to fatigue are reasonably well understood. The various symptoms experienced by fatigued individuals are well documented, and technological systems capable of detecting those symptoms are in an advanced stage of development. Thus we seem to be coping well with the valid *identification* of driver fatigue. Implementing reliable and socially acceptable countermeasures is the major hurdle to successfully coping with the complete problem, as it affects road safety. So long as society demands a round-the-clock supply of goods and services, transport operators tend to require round-the-clock drivers. Legislation and enforcement may limit many of the adverse effects of time-on-task, time-of-day, and time-since-sleep on driver performance, but conflicts inevitably persist between personal and market pressures to complete a journey after experiencing fatigue on the one hand and concerns for road safety on the other. The central dilemma is that fatigue is a subjective experience, best known by the individual in question, but that individual is often not the best person to rely on for a safe and effective resolution of his or her fatigued state. Technological systems exist that allow us to transfer responsibility for the individual fatigued driver's problem to others and to communicate their helpful responses back to the driver. However, journey's end will not be reached until cost-effective methods of implementing this type of fatigue countermeasure can be agreed on and widely implemented in our transport systems.

REFERENCES

Bartlett, F. C. (1948). A note on early signs of skill fatigue. *MRC Flying Personnel Research Committee* (Report No. FPC703). London: Medical Research Council.

Bartley, S. H., & Chute, F. L. (1947). *Fatigue and impairment in man.* New York: McGraw-Hill.

Brown, I. D. (1962). Studies of component movements, consistency and spare capacity of car drivers. *Annals of Occupational Hygiene, 5,* 131–143.

Brown, I. D. (1989). *Study into hours of work, fatigue, and safety at sea.* Unpublished report prepared for the Shipping Policy and Registration Division of the Department of Transport, England.

Brown, I. D. (1993). Driver fatigue and road safety. *Alcohol, Drugs and Driving, 9,* 239–252.

Hartley, L. (Ed.). (1995). *Fatigue and driving.* London: Taylor & Francis.

Iizuka, H., Yanagishima, T., Kataoka, Y., & Seno, T. (1985). The development of drowsiness warning devices. Paper presented at the *10th international conference on Experimental Safety Vehicles,* Oxford, UK.

Maycock, G. (1995). *Driver sleepiness as a factor in car and HGV accidents* (Report No. 169). Crowthorne, Berkshire, UK: Transport Research Laboratory.

Nelson, T. M. (1989). Subjective factors relating to fatigue. *Alcohol, Drugs and Driving, 5,* 193–214.

Sperandio, J.-C. (1978). The regulation of working methods as a function of work-load among air traffic controllers. *Ergonomics, 21,* 195–202.

Welford, A. T. (1978). Mental workload as a function of demand, capacity, strategy and skill. *Ergonomics, 21,* 151–167.

$$\underline{3.11}$$

What Is Stress and
What Is Fatigue?

Donald I. Tepas
Jana M. Price
University of Connecticut

Stress and *fatigue* are words with a long history of use by scientist, practitioners, and the public. In both science and the workplace, these words are liberally and frequently employed, often with the assumption that they promote communication, concern, and change. Unfortunately, this is not always true. Research clearly indicates that both of these words refer to multidimensional and interacting constructs. Those who use stress and fatigue as references often fail to recognize this complexity and use these words in confusing ways.

This chapter is not intended to be a comprehensive review of either the stress or fatigue literature. It begins with critical discussions of how the terms *stress* and *fatigue* have been and are being used. This is followed by a consideration of the benefits and problems associated with this terminology. Using shiftwork studies as a paradigm, we propose that fatigue and stress are seductive words but that their scientific use continues to be dated and counterproductive. Rather than use the terms *fatigue* and *stress*, shiftwork researchers focus on the development of better tools to assess workplace health and safety in a quantitative way. Collectively, these tools are termed *Workware*. The features of the Workware approach are outlined by using *Work-Shift Usability Testing* as an example of how and why Workware is evolving. In any case, the primary focus of fatigue and stress research should be on improving methodology, rather than on seeking general measures for these global terms.

WHAT IS STRESS?

Definitions of stress are varied and sometimes conflicting. In everyday conversation, stress may be referred to as an affliction or something that happens to a person. An event can be "stressful" or can leave one "stressed out." In popular magazines, articles instruct readers about ways to reduce stress through changes in diet, exercise, work habits, meditation, and numerous other ways.

Stress and stress-related concepts have had a major impact on the workplace. One study (Grippa & Durbin, 1986) found that workers' compensation claims for stress-related disability rose from 5% in 1980 to 14% in 1986. In Western Australia, there was a sevenfold increase in "work-related stress" between 1988 and 1994 (Worksafe Western Australia, 1996). The cost of these claims is significant, because this study showed that workers lost an average of 50.8 work days per work-related stress claim and women lost an average of 58.5 days. Thus, its not surprising that the concept of stress is regarded with concern even though definitions of stress are often wanting.

The lack of a clear definition of stress does not only occur in popular parlance. In the academic literature, stress has also been defined in a variety of ways. Some view it as a physiological response to an external stimulus (Selye, 1985). It is also said to be the result of "overload," when environmental demands outweigh an individual's capabilities or resources (McGrath, 1970) or when demands exceed control (Karasek & Theorell, 1990). Others define stress as a salient negative discrepancy between the individual's perceived state and desired state (Edwards, 1988).

Even criticizing the nebulous nature of the definition of stress has become passé. Authors often decry the lack of consensus on how to present the issue of stress and then in an attempt to clarify this problematic construct create new terms with similar connotations. This adds to the general confusion in defining the overall concept. Figure 3.11.1 provides a good example of this. This figure lists a few of the many constructs that are associated with stress in the literature. It is not intended as a comprehensive list, and the terms are simply listed in alphabetical order.

SOME CONSTRUCTS IN THE RESEARCH LITERATURE WITH STRESS CONNOTATION

ADAPTATION	ANXIETY	AROUSAL
BURNOUT	COPING	EXERTION
EXHAUSTION	EXPOSURE	FATIGUE
HARDINESS	MENTAL LOAD	REPETITIVENESS
STRAIN	STRESSOR	TENSION

FIG. 3.11.1.   A collection of terms associated with stress in the research literature.

Many researchers have made reference to *stressors* as precursors to stress, dysfunction, or health problems. Selye (1985) defined stressors as any agents that evoke the *general adaptation syndrome,* which includes states of alarm, adaptation, and exhaustion. According to Selye, physical, psychological, and emotional stimuli can all serve as stressors. Diseases such as peptic ulcers, high blood pressure, and heart disease are all said to be influenced by the presence of stressors. In an article by Cohen (1985), four different types of stressors are identified: acute time-limited events, such as a parachute jump; stress-event sequences, like a job loss; chronic intermittent stessors, for example, a conflict with neighbors; and, chronic stress, such as long-term job stress.

Strain is another variable frequently associated with stress. It has been defined as an outcome variable or as the consequence of being in an environment with stressors (Fletcher, 1991). According to Fletcher, strain may be physiological, psychological, behavioral, or medical, and it may also be short term or long term. Fletcher (1988) noted that certain mediators such as personality, satisfaction, or social factors may affect the relation between occupational stressors and strain symptoms.

Another term used to discuss a set of variables that mediate the stressor–strain relationship is *coping.* Coping is referred to as a means to deter or mitigate the effect of stressors. It may include efforts to improve person–environment fit by changing the environment or by using thoughts and actions to relieve the emotional effects of stress. Other forms of coping are said to include avoidance behaviors, denial, and drug taking (Monat & Lazarus, 1985). According to Edwards (1988), coping is a process triggered by stress. In this interpretation, the success or failure of coping is determined by individual resources, the nature of the stress, and other situation factors.

In these studies and others, the measurements used to gauge stress levels varied widely. Physiologically, there have been attempts to measure stress in numerous ways, such as catecholamines in the urine (Frankenhauser, 1986), uric acid, or serum cholesterol level (Rahe, Ryman, & Biersner, 1976). Psychologically, researchers have presumed to examine stress with mood checklists (Mackay, Cox, Burrows, & Lazzerini, 1978), surveys that track recent life events (Holmes & Rahe, 1967), and other measures. There is little reason to conclude that all these variables are measuring the same construct.

A review of the literature on stress does seem to indicate that many of the terms associated with stress have a common theme or family of ideas underlying them. However, the distinctions that have been made in the literature are great enough to cause concern and to confuse the average reader. In their 1982 paper, Sharit and Salvendy noted, "The terms mental load, fatigue and arousal have often been used synonymously with stress" (p. 131). They continued by attempting to give unique definitions of these terms. However, it is difficult at times to see where the distinguishing features lie.

For example, at one point they stated, "If fatigue takes the form of a generalized stress response, the methods used for assessment should presumably be those applied to studies of stress" (p. 132). Here it becomes evident that drawing a distinction among these varied concepts may be difficult, if not impossible to accomplish.

WHAT IS FATIGUE?

Colloquial use of the word *fatigue* by the populace is also common. For most of these users, fatigue refers to symptoms. In general, these symptoms are perceived to be the product of work or play, amid conditions that diminish performance. As a corollary, unlike many symptoms or diseases, fatigue is often assumed to be something that can be easily removed or reduced by rest. Thus, the nonscientist adopts what one might term a *restitutional theory of fatigue*. It is a theory that has evolved by intuitions and experience.

The focus of science on what might be termed the fatigue problem is lengthy and varies with history. Many investigators have recognized that *rest*, like fatigue, is multidimensional. Rest does not always assure a full recovery of performance. However, restoration-by-rest continues to be an explicit or implicit part of the thoughts that many scholars have about fatigue. A manifestation of this is the fact that many investigators and theories of fatigue show a narrow and major interest in how long a person has been working at a task (time-on-task). These studies are perhaps best viewed as conscious or unconscious manifestations of the restitutional theory of fatigue. Sleep research and chronobiology have demonstrated that a simple restitutional theory is no longer tenable. Both time-on-task and time-of-day must be considered (Tepas, Paley, & Popkin, 1997).

Dissatisfaction with the catchall concept of fatigue, as used by the general public, has certainly led to research aimed at obtaining a better understanding of this concept. However, the results of decades of fatigue research are sometimes confusing, often ignored, and certainly not easy to comprehend. For the most part, the scientific study of fatigue has not consolidated our thinking or solved many problems. Figure 3.11.2 lists some of the constructs that have been attributed to fatigue in the fatigue research literature. This is not a comprehensive list, and the constructs are simply listed in alphabetical order. An examination of this list is quite revealing. Some specific examples illustrate these problems.

Activation and *arousal* are used as synonyms by some investigators and not by others. For one investigator, they may suggest a unitary neurophysiological mechanism (Grandjean, 1968). Others proposed that a unitary mechanism is an untenable or a weak approach to complex behavior (Hancock & Warm, 1989; Hancock, Chignell, & Vercruyssen, 1990). *Alertness* and *sleepiness* seem to be opposites. Some investigators (Malmo & Surwillo, 1960) con-

SOME CONSTRUCTS IN THE
RESEARCH LITERATURE WITH
FATIGUE CONNOTATION

ACTIVATION	**ALERTNESS**	**AROUSAL**
BOREDOM	**DISCOMFORT**	**DISORGANIZATION**
EXHAUSTION	**IMPAIRMENT**	**INATTENTION**
MONOTONY	**MOTIVATION**	**REPETITIVENESS**
SLEEPINESS	**STRESS**	**TENSION**

FIG. 3.11.2.   A collection of terms associated with fatigue in the research literature.

cluded that the fatigue produced by sleep deprivation is alerting, whereas others proposed that it is an impairment associated with sleepiness and lowered levels of activation (Williams, Lubin, & Goodnow, 1959). *Boredom* can be seen as an emotional response to *monotony* (Davies, Shackleton, & Parasuraman, 1983), the peril of automation (Thackray, 1981), or a feature that increases the satisfaction of workers (Stagner, 1975).

The classic paper on fatigue by Broadbent (1979) turned to *stress* and *motivation* as the author searched for a new fatigue test. Longitudinal study and *chronic* effects were suggested as appropriate ergonomic approaches. The importance of examining the chronic impact of fatigue is not a new idea. Poore (1875) noted that fatigue comes in both general and local forms, yet both may be *acute* or *chronic*. MacDougall (1899) pointed out that earlier German investigators make a sharp distinction between fatigue (*Ermudung*) and weariness (*Mudigkeit*), suggesting that they may be manifest in quite different ways. Contemporary investigators have often failed to make this acute-chronic distinction clear. This represents a major gap in our understanding of the concept of fatigue. In the ergonomic literature, longitudinal research on fatigue is rare. Empirical evidence, supporting or rejecting the hypothesis that the time course and symptoms of acute fatigue recapitulate those of chronic fatigue, is uncommon.

Perhaps the largest gap in our understanding of fatigue is the one associated with developments in chronopsychology and sleep research. These ideas are not new and are quite well established. In the sleep research area, there is a fairly clear rejection of a general restitutional theory (Horne, 1979). The brain is very active during sleep, and it is no longer reasonable to view the brain as resting during sleep. At times, parts of the brain may be much more active during sleep than during wakefulness. Furthermore, the time required to recover from sleep deprivation is much shorter than the duration of the sleep-deprivation period.

Chronopsychology has repeatedly demonstrated that alertness and performance vary with time-of-day, even when individuals are placed in time isolation (Monk & Tepas, 1985). When living on earth in a day–night environment, people are diurnal. That is, they are active and perform best during the day. Performance at night is generally poorer. Thus, performance is subject to circadian (about-a-day) variations. Because these rhythms persist when an individual is placed in a controlled environment for an extended period, it is reasonable to conclude that these variations are determined by both internal biological clocks (endogenous rhythms) and external social-environmental changes (exogenous variables). Alertness, sleepiness, performance, and various mood measures have all been shown to have circadian variation. All of these variables peak during the day and bottom out at night, but they do not all peak at the same time (Tepas et al., 1997).

Clearly, our understanding of fatigue must incorporate these developments in chronopsychology and sleep research. In assessing fatigue, a narrow focus on the task type and the duration of performance is no longer adequate. Time-of-day becomes just as important as time-on-task, and both time dimensions must be considered if fatigue is to be fully assessed. Just as the findings in chronopsychology indicate that all hours of the day cannot be considered equal with regard to fatigue, sleep research confirms that there is no simple relation between fatigue measures and recovery time.

Although the ergonomic fatigue literature does include some modest recognition that time-of-day must be measured or controlled, systematic studies are missing. It is difficult to fully understand what fatigue is unless both time-on-task and time-of-day are considered.

HOW SCIENCE BENEFITS AND SUFFERS FROM THESE DISTINCTIONS

In some respects, the constructs of *stress* and *fatigue* can be beneficial. As the previous discussions make clear, both the public and the scientific uses of these two constructs are often global and at the same time multidimensional. In practice, these characteristics allow one to use either or both constructs to explain a wide range of perceptions, experiences, and data. They may be explanatory, but more often they are merely explicatory. That is, they are not meant to describe specific phenomena, but rather are catchall terms to communicate a general meaning.

It is appropriate to ask whether these two expletives should be deleted. Simple, unidimensional explanations can, in general, be helpful. They are often the modus operandi for the communication of complex constructs to a less informed populace. They can be useful devices for attracting the interest, concern, and support of the masses. The term *drug* provides a good example. The introduction of a drug into the human body may be harmful, beneficial, or inconsequential. For many drugs, impact is determined by a number of variables, including dosage, how the drug is administered, the time of day the drug is administered, how much time has passed since the drug was administered, the individual's history of previous exposures to the drug, and the health of the recipient. Predicting the impact of a drug is not a unidimensional task.

Accurate prediction of drug impact is possible, but it requires knowledge of all these variables and their interactions. Science has developed a tool box of reliable *methods* that can be used for the study of drug impact. Tool selection varies depending on the drug under study. For new drugs, specification of effect awaits the completion of clinical trials with actual users. Assigning the label "drug" to a substance is beneficial to the degree that it communicates that it should be approached with caution. It does not tell us whether the drug is toxic, curative, addictive, or harmless. Once reliable user testing is completed, then additional labels are attached, instructions are provided, and/or availability is controlled.

Like the term *drug*, the research literature has suggested that both fatigue and stress can also have a harmful, beneficial, or insignificant impact. In some cases, conditions of fatigue and stress that threaten the health or safety of workers can be easily identified and tagged. Unfortunately, many of the methods used for this are qualitative, not quantitative. Perhaps more important for both fatigue and stress, we lack clear terminology to discriminate between risky and beneficial fatigue or stress conditions. This makes communication difficult and promotes confusion.

Muscio (1922) concluded that "the term fatigue [should] be absolutely banished from precise scientific discussion, and consequently that attempts to obtain a fatigue test be abandoned" (p. 31). In the preface of a recent volume on stress measurement, the editors noted that "measures are often chosen because of their availability or visibility rather than because they are the most appropriate for answering investigators' questions" (Cohen, Kessler, & Gordon, 1995, p. vi). As they have evolved in the last 40 years, the constructs of fatigue and stress no longer have simple and intuitive connotations. For the public, and for many scientists as well, fatigue and stress are unidimensional terms invariantly associated with risk. However, a simple and accurate measure of all fatigue and/or stress has not been demonstrated. It is appropriate to again entertain the proposal that the search for *the* fatigue test or *the* stress test should be abandoned. This does not mean that one should no longer entertain the hypothesis that fatigue and stress are different things.

SHIFT WORK, STRESS, AND FATIGUE

Early shift work investigators described night work as stressful and/or fatiguing. Their goal was to identify and eliminate those work schedules that lead

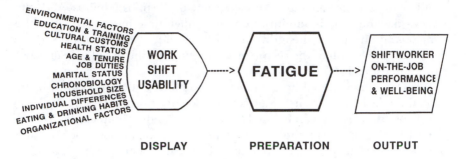

FIG. 3.11.3.   A schematic displaying the wide range of variables that have been shown to interact with work schedules to impact fatigue, performance, and well-being.

to unacceptable stress and/or fatigue. This approach viewed humans as circadian beings with an internal biological clock driving their chronopsychology. In this orientation, the task for the ergonomist was to install a work schedule that was "chronobiologically correct" and to eliminate/minimize worker exposure to nocturnal work hours. This led to a search for the single "best" work schedule design.

Subsequent research, however, clearly demonstrated that work schedules interact with a wide range of variables. These variables include many workplace and nonworkplace factors like social networks, performance, sleep habits, cultural practices, workload, and environmental conditions. The interaction of these variables with a given work schedule is said to set the level of worker *fatigue,* which then modifies the workers' performance and health. A simplified diagram of this is presented in Fig. 3.11.3. In most instances, one can easily talk of *stress* rather than fatigue and adopt the similar diagram presented in Fig. 3.11.4. In either case, studies verifying the impact of each of the variables listed are in the research literature.

For many investigators, an explanation of shift work problems based on the combined effects of multiple *stressors* is attractive. For example, some studies report higher than expected rates of gastrointestinal complaint among shift workers as a symptom of stress. However, it is not at all clear that stressors are responsible for these complaints. They may, in some workplaces, be simply related to food service methods (Rutenfranz, 1982), food selections available (Tepas, 1990), or bacterial infections. Epidemiological research by Swedish and American investigators related shift work to the incidence of coronary heart disease (Knutsson, 1989; Kawachi et al., 1995). For some, these studies, together with the sleep literature, favor a fatigue interpretation. For this interpretation, night shift work is associated with chronic sleep reduction and concomitant sleepiness. These epidemiology

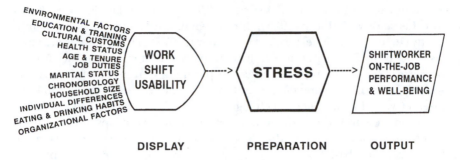

FIG. 3.11.4.   A schematic displaying the wide range of variables that have
been shown to interact with work schedules to impact stress, performance,
and well-being.

studies also demonstrated that the exposure time (in years) is a significant
factor. A subsequent report suggested that job strain may not be related to
work schedule (Knutsson, 1995).

For many investigators, an explanation based on the combined effects of
multiple *fatigue* factors is more attractive than an explanation simply based
on sleep reduction. Shift work investigators often use sleep length as a
benchmark for work schedule effects (Tepas & Mahan, 1989). As a rule,
workers sleep significantly less when they work at night and sleep in the day.
Research on shift workers has also confirmed that these night-shift reductions in sleep length are associated with decrements in performance (Tepas,
Walsh, Moss, & Armstrong, 1981). Longitudinal (Gersten, 1987) and cross-
sectional studies (Tepas, Duchon, & Gersten, 1993) have also demonstrated
that these reductions in sleep do *not*, as a rule, disappear with exposure to
shift work, and they are not as evident on nonworkdays (Tepas & Carvalhais,
1990). This makes it easy to think of chronic fatigue, produced by sleep loss
and manifest with concomitant sleepiness, as a critical factor.

Like the epidemiological studies, sleep research also demonstrated that
both acute and chronic exposure must be considered. In this case, chronic
sleep deprivation is relevant, and some would say that these findings favor
fatigue as the critical condition. Because social and environmental factors
have also been related to shift work tolerance, it is difficult in any case to
adopt any univariate stress or fatigue model of the impact of night shift work.

One might conclude from our discussion that fatigue and stress are the
same thing and that it is reasonable to use the terms interchangeably. However, this is *not* true. Using total sleep deprivation as a method to reveal differences, Froberg and his associates demonstrated that there is circadian
variation in adrenaline excretion, noradrenaline, performance, and fatigue
ratings. Of special interest is their finding that the peak of circadian vari-

ation for each of these variables was at a *different* time-of-day (Froberg, Karls-son, Lennart, & Lidberg, 1975), and adrenaline excretion was *not* significantly correlated with performance (Froberg, 1977). These 20-year-old studies of total sleep deprivation firmly demonstrate that circadian variation in performance, fatigue reports, and stress physiology are not identical. A reasonable conclusion is that fatigue and stress are not the same thing, but both interact with chronobiological change.

Recent research on myocardial infarction appears to upset any notions that fatigue is more relevant than, *or* the same as, stress. Three reports in the literature, each based on a large sample, showed that there is a septadian (day-of-the-week) pattern in the occurrence of cardiac problems (Peters, McQuillan, Resnick, & Gold, 1996; Peters, Brooks, Zoble, Liebseon, & Seals, 1996; Willich et al., 1994). All three studies reported a significant peak in cardiac problems on Mondays. Fig. 3.11.5 is based on one of these studies (Willich et al., 1994). This study suggested that this septadian peak on Monday is most evident in employed workers. Although this relation to work needs verification and further analysis, it suggests that cardiac problems are associated with the beginning rather than the end of the workweek.

A chronic or cognitive stress interpretation of these findings is possible. At best, this explanation is a post hoc interpretation. One must also recognize that almost any fatigue interpretation requires a peak later in the workweek or at the end of the workweek. Thus, it is very difficult to support a simple fatigue interpretation of these results. Yet, one cannot deny that work often increases fatigue. We suggest that contemporary research on the impact of work schedules demonstrates that stress and fatigue are distinct, complex, and interactive constructs. These two constructs are useful when the goal is to communicate the presence of a threatening occupational health or safety risk that might be avoided. Although fatigue and stress appear to be diverse and disunited terms, one must recognize that contemporary use of these terms often has little or no heuristic value when significant risk is not apparent.

WHAT SHOULD BE DONE?

As we pointed out earlier, the complexity of stress and fatigue as concepts was recognized by science more than 100 years ago. Over 75 years ago, Muscio seriously concluded that the term *fatigue* should be abandoned. We suspect he would say the same thing today and would expand his position to include modern concepts of stress. About 20 years ago, Broadbent attempted to argue that science was at last approaching the point where one might finally be able to develop a measure of fatigue and/or stress. Since Broadbent wrote his paper, research has more fully demonstrated that fatigue and

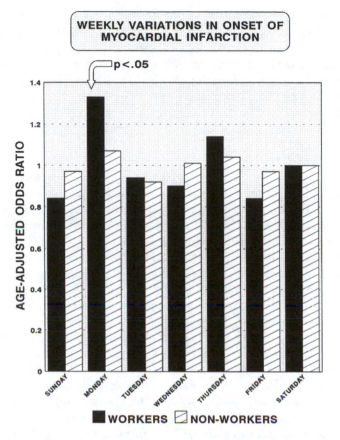

FIG. 3.11.5.   Weekly variations in the onset of myocardial infarction based on numbers reported by Willich et al. (1994) using data
from 2,636 patients with myocardial infarction events.

stress are distinct but multivariable constructs. There is little reason to conclude that his expectation has yet been realized.

The contemporary scientific literature suggests that the search for an invariant measure of fatigue or stress should be abandoned. It is unreasonable
to expect that a complex system can be reliably assessed with a single or simple measure. This does not mean that the terms *fatigue* and *stress* should be
dropped. The problem is that we use these terms indiscriminately, which
makes public risk perception difficult. In addition, for members of the ergonomic profession, the use of these two terms simply perpetuates an impossible methodological search for an ideal simple measure of multiple
complex constructs. This search for a simple conceptual solution distracts
from discovery, innovation, and prompt solution of practical problems.

Developments in the shift work area make the need for a new approach to work and stress apparent. A recent conservative estimate is that over 10,000 different work schedules are in use (Knauth, 1997). This provides a practical demonstration of the multiplicity of variables involved in doing work. Experts are no longer searching for *the* ideal work schedule, one that fits all work equally well. They are designing new work schedule systems using a wide array of methodological tools and approaches. What they need is a toolbox of reliable methods so that they can select the right tool for every application.

The term *workware* is suggested as one overall term that might be used to describe the array of tools and approaches needed in the work-scheduling area. A computer software program is a set of ordered instructions used by a human operator for directing and timing the operation of computer hardware. Workware is a set of instructions that can be used by workers as an aid in directing and timing their performance of a job. As is noted in Fig. 3.11.6, *work shift usability testing* is a Workware methodology similar to *software usability testing*. The term *software* refers to an array of tools used to program computer hardware to perform tasks and/or to evaluate hardware-software systems. Workware tools are used to guide people in scheduling and performing work and/or evaluating work systems.

Usability testing is not a novel approach for the ergonomics professional. It has been used for years with many tasks, including the design and evaluation of tools, devices, displays, toys, and software. This form of testing refers to methods used to ascertain the value of a design, not simply to an effort to assess the quality or ability of an individual. The user aids in the identification of desirable and undesirable designs, but in most cases does not select the final design. Alternative designs are often considered. A fundamental feature of the usability approach to design is the assumption that *use testing with real users is a required and fundamental approach* in most cases (Nielsen, 1997).

Work shift usability testing retains this fundamental assumption that real users should participate in assessing the value of a work schedule design. In addition, it assumes that a *macroergonomic* approach is needed. That is, one must recognize that work shifts, as well as changing work shifts, impact and interact with many of the basic ways in which an organization operates. In this approach, work shift assessment is the first phase of work shift usability testing. The details of this multiple-step process of work shift usability testing have been described in a previous paper (Tepas, 1999).

The application of work shift usability testing methods is not the only form Workware can take. Future Workware methodology might include quantitative ways to make work shift forecasts; improved worker behavior self-assessment methods; Workware training programs that work; Workware shift registration; and, the establishment of Workware data warehouses.

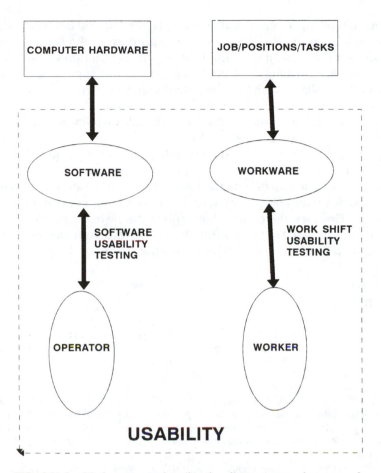

FIG. 3.11.6. Workware can describe the diverse approaches to work scheduling. Just as computer software is manipulated by a user to direct the performance of hardware, Workware may be used by a worker to shape job timing and performance.

These Workware tools are methods that the professional can select from in designing a wide variety of alternative work-scheduling tasks, not just shift-work problems. Common to all these proposed Workware approaches should be these fundamental assumptions: Testing should be done with real users; quantitative measurement is required; reliability, validity, and long-term impact must be assessed; and, the results of applications should be retained and studied.

For over a century, people have endeavored to learn more about the concepts of stress and fatigue. Advances in knowledge have been slow and much too qualitative. Work shift usability testing is not a radical new approach for ergonomic professionals. Workware is an approach to the issues of fatigue and stress, which attempts to avoid the stigma of using these old labels. It communicates methods more than concepts. In this approach, acceptable work shifts are viewed as the product of a design process in which alternative designs are considered, selected, and then evaluated.

Workware and more specifically the work shift usability testing approach emphasize the importance of developing reliable and valid work schedule design methodologies, rather than searching for a simple singular way to measure a complex process. These approaches recognize the value of a systems approach, quantitative methodology, and consideration of alternative solutions. Stress and fatigue are terms that are sometimes helpful in colloquial use. However, their true value as problem-solving constructs remains to be demonstrated. For the present, their use in the study, design, and evaluation of work systems should be avoided whenever possible.

REFERENCES

Broadbent, D. E. (1979). Is a fatigue test now possible? *Ergonomics, 22,* 1277–1290.

Cohen, F. (1985). Stress and bodily illness. In A. Monat & R. S. Lazarus (Eds.), *Stress and coping, an anthology* (pp. 40–54). New York: Columbia University Press.

Cohen, S., Kessler, R. C., & Gordon, L. U. (1995). Measuring stress. New York: Oxford University Press.

Davies, D. R., Shackleton, V. J., & Parasuraman, R. (1983). Monotony and boredom. In G. R. J. Hockey (Ed.), *Stress and fatigue in human performance* (pp. 1–34). New York: Wiley.

Edwards, J. R. (1988). The determinants and consequences of coping with stress. In C. L. Cooper & R. Payne (Eds.), *Causing, coping and consequences of stress at work* (pp. 233–263). New York: Wiley.

Fletcher, B. (1988). The epidemiology of occupational stress. In C. L. Cooper & R. Payne (Eds.), *Causing, coping and consequences of stress at work.* New York: Wiley.

Fletcher, B. (1991). *Work, stress, disease and life expectancy.* New York: Wiley.

Frankenhauser, M. (1986). A psychobiological framework for research on human stress and coping. In M. H. Appley & R. Trumbull (Eds.), *Dynamics of stress: Physiological, psychological and social perspectives* (pp. 101–116). New York: Plenum.

Froberg, J. E. (1997). Twenty-four-hour patterns in human performance, subjective and physiological variables and differences between morning and evening active subjects. *Biological Psychology, 5,* 119–134.

Froberg, J. E., Karlsson, C. G., Lennart, L., and Lidberg, L. (1975). Circadian rhythms of catecholamine excretion, shooting range performance and self-ratings of fatigue during sleep deprivation. *Biological Psychology, 2,* 175–188.

Gersten, A. H. (1987). *Adaptation in rotating shift workers: A six year follow-up study.* Unpublished doctoral dissertation, Illinois Institute of Technology, Chicago.

Grandjean, E. (1968). Fatigue: Its physiological and psychological significance. *Ergonomics, 11,* 427–436.

Grippa, A. J., & Durbin, D. (1986). Workers' compensation occupational disease claims. *National Council on Compensation Insurance Digest, 1*(2), 15–23.

Hancock, P. A., Chignell, M. H., & Vercruyssen, M. (1990). Predicting the effects of stress on performance. *Proceedings of the Human Factors Society 34th Annual Meeting* (pp. 1081–1085).

Hancock, P. A., & Warm, J. S. (1989). A dynamic model of stress and sustained attention. *Human Factors, 31,* 519–537.

Holmes, T. H., & Rahe, R. H. (1967). The social readjustment rating scale. *Journal of Psychosomatic Research, 11,* 213–218.

Horne, J. A. (1979). Resitution and human sleep: A critical review. *Physiological Psychology, 7,* 115–125.

Karasek, R., & Theorell, T. (1990) *Healthy work: Stress, productivity and the reconstruction of working life.* New York: Basic Books.

Kawachi, I., Colditz, G. A., Stampfer, M. J., Willett, W. C., Manson, J. E., Speizer, F. E., & Hennekens, C. H. (1995). Prospective study of shift work and risk of coronary heart disease in women. *Circulation, 92,* 3178–3182.

Knauth, P. (1997). Innovative working times. *Shiftwork International Newsletter, 14,* 2.

Knutsson, A. (1989). Shift work and coronary heart disease. *Scandinavian Journal of Social Medicine* (Suppl. 44).

Knutsson, A. (1995). Prevalence of job strain in workers with odd work schedules. *Shiftwork International Newsletter, 12,* 82.

MacDougall, R. (1899). Fatigue. *Psychological Review, 6,* 203–208.

Mackay, C., Cox, T., Burrows, G., & Lazzerini, T. (1978). An inventory for the measurement of self-reported stress and arousal. *British Journal of Social and Clinical Psychology, 17,* 283–284.

Malmo, R. B., & Surwillo, W. W. (1960). Sleep deprivation: Changes in performance and physiological indicants of activation. *Psychological Monographs: General and Applied, 74* (15, Whole No. 502).

McGrath, J. E. (1970). *Social and psychological factors in stress.* New York: Holt, Reinhart & Winston.

Monat, A., & Lazarus, R. S. (1985). Stress and coping: Some current issues and controversies. In A. Monat & R. S. Lazarus (Eds.), *Stress and coping: An anthology* (pp. 1–12). New York: Columbia University Press.

Monk, T. H., & Tepas, D. I. (1985). Shift work. In: C. L. Cooper & M. J. Smith (Eds.), *Job stress and blue collar work* (pp. 65–84). New York: Wiley.

Muscio, B. (1922). Is a fatigue test possible? *The British Journal of Psychology: General Section, 12,* 31–46.

Nielsen, J. (1998). Usability testing. In G. Salvendy (Ed.), *Handbook of human factors and ergonomics* (2nd ed., pp. 1543–1568). New York: Wiley.

Peters, R. W., Brooks, M. M., Zoble, R. G., Liebson, P. R., & Seals, A. A. (1996). Chronobiology of acute myocardial infraction: Cardiac arrythmia suppression trial (CAST) experience. *American Journal of Cardiology, 78,* 1198–1201.

Peters, R. W., McQuillan, S., Resnick, S. K., & Gold, M. R. (1996). Increased Monday incidence of life-threatening ventricular arrythmias: Experience with third-generation implantable defilibrator. *Circulation, 94,* 1346–1349.

Poore, G. V. (1875, July 31). On Fatigue. *The Lancet,* 163–164.

Rahne, R. H., Ryman, D. H., & Biersner, R. J. (1976). Serum uric acid, cholesterol, and psychological moods throughout stressful naval training. *Aviation Space and Environmental Medicine, 47,* 883–888.

Rutenfranz, J. (1982). Occupational health measures of night- and shift-workers. *Journal of Human Ergology, 11* (Suppl., 67–86).

Selye, H. (1985). History and present status of the stress concept. In A. Monat & R. S. Lazarus (Eds.), *Stress and coping: An anthology* (pp. 13–17). New York: Columbia University Press.

Sharit, J., & Salvendy, G. (1982). Occupational stress: Review and reappraisal. *Human Factors,* *24*(2), 129–162.

Stagner, R. (1975). Boredom on the assembly line: Age and personality variables. *Industrial* *Gerontology, 7,* 23–44.

Tepas, D. I. (1990). Do eating and drinking habits interact with work schedule variables? *Work* *and Stress, 4,* 203–211.

Tepas, D. I. (1999). Work shift usability testing. In W. Karwowski & W. S. Marris (Eds.), *The oc-* *cupational ergonomics handbook* (pp. 1741–1758). Boca Raton, FL: CRC Press.

Tepas, D. I., & Carvalhais, A. B. (1990). Sleep patterns of shiftworkers. In A. J. Scott (Ed.), *Shiftwork* (pp. 199–208). Philadelphia: Henley & Belfus.

Tepas, D. I., & Mahan, R. P. (1989). The many meanings of sleep. *Work & Stress, 3,* 93–102.

Tepas, D. I., Duchon, J. C., & Gersten, A. H. (1993). Shiftwork and the older worker. *Experi-* *mental Aging Research, 19,* 295–320.

Tepas, D. I., Paley, M. J., & Popkin, S. M. (1997). Work schedules and sustained performance. In G. Salvendy (Ed.), *Handbook of human factors and ergonomics* (2nd ed., pp. 1021–1058). New York: Wiley.

Tepas, D. I., Walsh, J. K., Moss, P. D., & Armstrong, D. (1981). Polysomnographic correlates of shift worker performance in the laboratory. In A. Reinberg, N. Vieux, & P. Analauer (Eds.), *Night and shift work: Biological and social aspects* (pp. 179–186). Oxford: Pergamon Press.

Thackray, R. I. (1981). The stress of boredom and monotony. *Psychosomatic Medicine, 43,* 165–176.

Williams, H. L., Lubin, A., & Goodnow, J. J. (1959). Impaired performance with acute sleep loss. *Psychological Monographs, 73,* 1–26.

Willich, S. N., Lowel, H., Lewis, M., Hormann, A., Arntz, H-R., & Kell, U. (1994). Weekly vari-ation of acute myocardial infarction: Increased Monday risk in the working population. *Circulation, 90,* 87–93.

WorkSafe Western Australia (1996). *Safetyline: The Magazine.* [On-line]. http://www.wt.com. au/~dohswa/magazine/sline_31/article5.htm

<table>
<tr><td>3.12</td></tr>
</table>

3.12 Stress, Workload, and Fatigue as Three Biobehavioral States: A General Overview

Anthony W. K. Gaillard
TNO Human Factors

A framework is presented in which mental load, stress, and fatigue are regarded as distinct biobehavioral states generated by different factors in the work environment, involving different mechanisms, and resulting in different cognitive, affective, and bodily reactions. Macroconcepts, such as mental load and stress, are often confounded because they refer to similar phenomena and are used interchangeably in daily life. A high workload, for example, is assumed to result in stress and fatigue reactions, such as psychosomatic and psychological complaints, whereas it is also maintained that working under conditions of stress and fatigue enhances the subjective workload. Moreover, some people are able to work efficiently under high levels of work pressure without experiencing feels of stress, whereas stress reactions may occur when the workload is low or when there is no workload at all (see also Gaillard & Wientjes, 1994).

Although the states of workload, stress, and fatigue may occur simultaneously and sometimes are difficult to distinguish, they should not be regarded as synonymous, and their differences should be specified. A proper distinction among these concepts is not only important for theory building but also for the reconstruction of the work environment. It may lead to recommendations that aim to enhance the work efficiency of employees while reducing the probability that stress responses occur and health risks increase.

The confusion between these concepts originates in their poor definition. Mental load may refer either to the objective demands imposed by the task (e.g., complexity, pacing) or to the subjective judgment of the operator with regard to the task demands. In most theories, workload refers to the limitations in the information-processing capacity of the operator. It may, however, also encompass feelings of work pressure, which has a more emotional connotation. The concept of stress has an even larger variety of meanings:

- An *input* variable referring to either work demands (difficulty, time pressure), emotional threat (accident, potential violence or loss), or adverse environments (noise, sleep loss, drugs, etc.).
- An *output* variable referring to a pattern of behavioral, subjective, and physiological responses, often labeled as strain.
- A *state* in which we feel strained and threatened on the basis of a subjective evaluation of the situation.
- A *process* that gradually results in a dysfunctional state degrading the work capacity and the potential to recover from work.

In the current framework, a high workload is regarded as an important but not a critical factor in the development of stress symptoms. Even under unfavorable conditions, it is possible to work intensively and to be highly activated without feelings of strain or psychosomatic complaints. In contrast, working conditions that provide few possibilities for control and little social support are associated with reduced well-being and increased health risks. As outlined later, mental load and stress are regarded as two states different in energy mobilization, coping style, aftereffects, and mood.

Fatigue is seen as a response of mind and body to the reduction in resources because of the execution of a mental task and a warning for the increasing risk of performance failure. Fatigue is not only determined by the amount of work done, but also by what still has to be done. This work-related fatigue is regarded as a state induced by enduring task performance. Fatigue, however, may also refer to a subjective complaint encompassing a general feeling of lack of energy, which is not necessarily related to the amount of work. Chronic fatigue (and burnout) appears to be a symptom of chronic stress and could be regarded as a process (discussed later). Thus, fatigue may be seen as a state characterized by the aftereffects of preceding states. Fatigue, resulting from a demanding task that has been completed successfully, is quite different from that resulting from a day full of arguments and irritation on the work floor.

Before further discussing the relation between mental load, stress, and fatigue, the relation between cognition and energy or between psychological and physiological processes is discussed because these factors play an essential role in specifying the concepts.

STATE REGULATION AND PERFORMANCE EFFICIENCY

Models in human factors are often based on the computer metaphor: Information coming from the work environment is transformed in sequential stages, resulting in an overt response (Sanders, 1983; Wickens, 1992). This metaphor can also be used to illustrate the distinction between process and state: The software stands for the processing of cognitive information and the hardware for the state of the brain. The main reason for postulating that changes in state affect the efficiency of cognitive processing is that computational models are not able to account for variations in human performance, in particular when human beings have to perform under demanding or threatening conditions. Therefore, models adopted from human factors or human performance research have been extended with concepts such as "state," "resources," or "energetical mechanisms."

The distinction between process and state has along history in psychology (see Hockey, Coles, & Gaillard, 1986, for a review). Hebb maintained as early as 1955 that "arousal is an energizer not a guide," indicating that an optimal state of the brain may facilitate information processing, although it does not guide behavior. The state of mind and body is regulated by a variety of energetic mechanisms. "Energetics" is used as a generic term encompassing all mechanisms that energize and regulate the organism and directly or indirectly influence psychological processing. The term *energetics* is preferred above commonly used labels such as arousal, effort, fatigue, and activation, because it does not have the specific theoretical connotations (Hockey et al., 1986). The state of the organism may be inferred from physiological measures and subjective ratings (e.g., on mental effort, emotions, and fatigue) or indirectly through the effects of environmental stressors (e.g., sleep loss and long-term performance) and aspecific task variables (e.g., feedback, bonus) on performance measures.

Under normal circumstances, our energetic state is in line with the activities we want to undertake. Energetical mechanisms regulate our brain and body in such a way that they are in an optimal state to execute the task and to process information. The majority of bodily processes are regulated automatically and unconsciously. The execution, and even the planning, of a task prompts energetic mechanisms to adapt our body gradually into a state that is optimal to perform the task. An optimal state is a prerequisite for efficient processing and determines the capacity available to execute a task. In most instances, we do not have to pay attention to the continuous adaptation, which gives us the opportunity to concentrate on the more interesting aspects of life. Only when our state is far from optimal because of the fatigue or strong emotions do we realize how much cognitive processing is affected by our state. Because these effects are mostly outside our control, we can only attempt to modulate them and to adapt to the situation. When planning our daily activities, we may take into account possible fluctuations in our state from fatigue or time-of-day effects. For example, someone having problems in getting started in the morning may make appointments only after the coffee break.

Besides the endogenous effects of our organism, such as the circadian rhythm, and the influence of environmental factors, such as noise, sleep loss, and drugs, three types of energy mobilizations may be distinguished that determine performance efficiency: Task-induced activation refers to the stimulating effects of a task or work environment; internally guided mental effort involves voluntary energy mobilization under conditions of a mental load; and emotional arousal refers to energy mobilization in stressful threatening situations.

Task-Induced Activation

When we know that a particular task has to be done in the near future, we can prepare for certain activities on a cognitive level, but also on an energetic level. Only thinking about the task to be done, affects the regulation of our state: Energetic mechanisms are activated to reach an optimal state. The relation between state and efficiency is dependent on the demands of the task on the one hand and the availability of processing resources on the other. When there is abundant processing capacity, a deviation from the optimal state does not result in a reduction of performance efficiency. However, in complex or novel task situations that require all our resources, even a small deviation from the optimal may result in a performance decrement. Thus, for simple, well-trained tasks that do not require many resources, the range in which optimal performance can be obtained is larger when the amount of resources needed for the task approaches the available capacity. We therefore prefer an easy task over a difficult one when we are tired. In the evening, for example, we may be too tired to write a technical paper, but we may read a novel.

The Role of Mental Effort

When the actual state does not deviate too much from the optimal state, we are still able to perform the task, but a slower pace or less accurately. We can also attempt to maintain performance at the same level by mobilizing extra energy through mental effort. This "try harder" response can be maintained only for a relatively short period, because the physiological and psychological costs are high, which induces cognitive strain and mental fatigue. Kahneman (1973) identified effort with the action of maintaining a task activity in focal attention; effort is needed when lapses in attention immediately result in performance deterioration. This may be the case in the following, apparently quite different, situations (see also Gaillard, 1993): (a) the energetic state is not optimal because of sleep loss or fatigue; (b) emotions disrupt the energetic state and continuously attract attention; (c) the task is attention demanding because of inconsistent or varying input–output relations or heavy demands on working memory; (d) the task environment is complex and attention has to be divided between different tasks; and (e) new skills have to be acquired in a learning situation.

These situations have in common that an employee may have problems in maintaining the task set. In the first two, concentration on the task may be the problem. In the third and fourth, the task continuously requires attention, because the task set is changing and because controlled processing is required. In the fifth, the task set is still to be developed.

Emotion: The Third Level

The role of affective processes has been largely neglected in cognitive psychology and human performance research and also in human factors and ergonomics (e.g., Eysenck, 1982). Consequently, most theories distinguish only between state and process, dichotomizing between energy (e.g., arousal, activation, or effort) on the one hand and psychological processes on the other. Emotions may be regarded as a third layer between the processing on cognitive and bodily levels. Emotions play an important role in motivating people to initiate and maintain a task in the first place, but they may also interfere with cognitive processing. In particular, under time pressure or threatening conditions, the regulation of our emotions is critical for efficient task performance. When we are uncertain about the goals of the task, our success to meet the criteria, or the rewards to be gained, in combination with fatigue or threat, there is competition from other goals than the execution of a task. Because sustaining an effortful state is subjectively aversive and has its costs, it may conflict with other personal goals, such as maintaining well-being and positive affect (see also Hockey, 1997). In particular, intense and negative emotions may reduce performance efficiency in several ways: (a) they may disrupt the state regulation, which makes it less optimal for task performance because of over-reactivity; (b) they may be so distracting that they directly interfere with the processing of the task information; or (c) they may cause psychosomatic complaints that also demand attention. Intense emotions, such as anxiety, have "control precedence"; that is, they continually beg for attention, with the result that less capacity is available for processing task-relevant information.

The three types of energy mobilization described here may be used to characterize the states of mental load and stress. Under *normal conditions,* the energy mobilization is generated by the planning and execution of a task, which is mostly sufficient to perform the task properly. When the task is data limited, the investment of more energy in the task via mental effort does not improve performance. Given a balanced work/rest schedule, the employee does not experience psychosomatic complaints and feelings of fatigue.

Under high levels of *mental load,* when the task is attention demanding, extra energy is mobilized through mental effort. Mental effort is aimed at improving or maintaining performance efficiency. Mental effort is a normal and healthy aspect of an active coping strategy to meet work demands that are experienced as a challenge. This type of mobilization is largely under the control of the employee, which increases well-being and reduces the risk of psychosomatic complaints. Also the recovery is rather fast, and feelings of "healthy" fatigue may be accompanied by satisfaction when the task has been completed successfully.

Under *stress,* the energy mobilization is dominated by negative emotions over which we have limited control. The situation is experienced as threatening, which results in strain and psychosomatic complaints. The enhanced activation is not instrumental to the execution of the task; it may even distract attention, disturb the state, and therefore become dysfunctional and reduce performance efficiency. Because we have limited control over this type of energy mobilization, it can result in maladaptive activation patterns: overreactivity, protracted recovery, and sustained activation, which are associated with performance deterioration and increased health risks.

Other states related to mental load and stress can be defined in a similar way. *Understimulation or underload:* In vigilance situations, both the task and the work environment are not very stimulating and not inherently motivating. Consequently, the task-induced activation may be not sufficient for an efficient task performance. In particular, in combination with fatigue, activation may drop below an acceptable level. Although the workload is rather low, the employee may experience negative feelings. Via mental effort, extra energy may be mobilized, which is only possible when the employee is well-motivated by incentives outside the task (e.g., salary, social control, etc.).

Adverse environment: The task has to be executed under the influence of environmental stressors, such as heat, noise, vibration, or working at night, which may disrupt the energetic state of the employee physically. Thus, environmental stressors influence our cognitive and affective processes through their negative effects on the body. With "stress" defined previously, and in the next section, the route is just the other way around. We have particular cognitions (e.g., time pressure) or emotions (anxiety) that lead to bodily reactions.

MENTAL LOAD AND STRESS COMPARED

In the research centered around the concepts of "workload" and "stress," two approaches may be distinguished that differ in their background, theoretical framework, and methodology. Research on workload is concerned with the efficient performance of complex or demanding tasks, mostly in a technical environment, whereas stress research concentrates on the work–health relation in a psychosocial working environment.

Workload research uses cognitive-energetic models based on human factors and cognitive psychology (i.e., Hancock, 1986; Hockey et al., 1986; Wickens, 1992), which describe how cognitive processing is affected by state variables (e.g., sleep loss, long-term performance). The research is particularly relevant when the capacity of the operator is just enough or not sufficient to get the requested work done in the time available. Mental load refers to the ratio between the current processing capacity of the operator and the capacity required by the task (e.g., Gopher & Donchin, 1986; Hancock, 1987). By examining the effects of task demands on workload, this research aims at getting a better insight in the factors that determine the workload of an employee and the risk for overload and errors. The ultimate aim of this research is to develop procedures to reduce the workload and to redesign the work environment to improve performance efficiency.

Theories in stress are adopted from industrial, social and personality psychology (Kahn, 1981; Karasek & Theorell, 1990). Stress theories describe the relation between the person and the environment. The evaluation of the situation (Lazarus & Folkman, 1984) or the perceived controllability of the situation (Karasek & Theorell, 1990) is central in these theories. The work–health relation is examined by investigating the influence of the work environment on well-being, psychosomatic complaints, and health risks.

At first sight, the two approaches have the same goal in that their primary focus is on the balance between demands and resources and the uncertainty that the task may be finished successfully. The two areas, however, differ in the breadth of their approach. Research on workload is limited to the processing capacity required for the execution of the task, whereas in stress research the discrepancy refers to the fit between the characteristics of the person and those of the environment.

In stress theories, "demands" and "resources" are conceptualized more broadly. Resources refer not only to processing capacity but also to personality traits, coping strategies, and social skills. Resources can also encompass the availability of means (e.g., supplies, apparatus, information) to perform the task. Similarly "demands" in stress theories refer not only to the requirements of the task, but to the entire work environment, including its social, physical, and organizational aspects. In workload theories, "demands" refer only to the specific requirements of the task and the internal resources needed to execute the task.

In both types of theories, the operator is assumed to resolve the mismatch between demands and resources by either reducing the demands or increasing the resources. The two types of theories differ, however, in the way in which the operator is assumed to do this. In workload theories, the operator can either increase the available resources by mobilizing extra energy through mental effort or change the work strategy either by changing the speed–accuracy trade-off (i.e., making more errors) or by concentrating on the most relevant aspects of the job. In stress theories, control refers to the decision latitude the operator has to adapt to the situation and to the freedom to change or organize the work environment, in terms of supplies, apparatus, information, and manpower.

In workload theories, the level of control is limited to the execution of the task and based on knowledge of results of the task performance. In stress theories, controllability refers also to the physical and social environment. The central executive continuously monitors the environment for potential threats and danger on the one hand, and interesting opportunities on the other. This appraisal is guided by the goals we have and is based on our norms and values. In addition to task information, the operator takes into account the psychosocial aspects of the situation (e.g., social approval or social support).

In summary, mental load and stress refer to two states that are both induced by a perceived discrepancy between the demands required and the resources available. This discrepancy triggers the mobilization of energy and increases the level of activation. The two states differ, however, in the way in which the individual responds to the situation. The two states are distinguished by the following characteristics:

Energy mobilization: The energy mobilization under mental load is attuned to the demands of the task and aimed at improving or maintaining performance efficiency by focussing attention. Mental effort is a normal and healthy coping strategy to meet work demands. Under stress, the enhanced activation is not instrumental to the execution of the task. The increased activation may even be distracting, dysfunctional, and efficiency reducing.

State regulation: Energy mobilization under a high workload is limited to the period in which the task has to be executed. In reasonable time, the increased activation returns to a resting level after the task has been completed. In a stress state, the activation persists outside the task situation and inhibits recovery. Activation and disturbance of the energetic system may even continue when the trigger for the stress response no longer exists.

Mood: Under mental load, the situation is experienced as a challenge; this state is accomplished by positive emotions and results in feelings of accomplishment; feelings of fatigue afterward may even be experienced as positive. Under stress, the situation is experienced as threatening and results in strain and negative emotions.

Coping strategy: Under mental load, coping is oriented toward the execution of the task, whereas stress is oriented toward self-protection. Under mental load, actions are taken aimed at solving the problems. The stress state is characterized by a defensive style and palliative reactions aimed at reducing the negative effects of (potential) threats.

ACTIVATION AND EFFICIENCY

Although most investigators in human performance and psychophysiology agree that a multidimensional framework is necessary to describe the complex relation between the energetic mechanisms and the psychological processes, they still tend to describe their results in a one-dimensional view, which originally stems from the flight/fight reaction described by Cannon (1932). States of high activation are associated with emergency reactions, inefficient performance, and anxiety. In this tradition, negative views on physiological activation are predominant. The activity of the sympathetic nervous system has been coupled with anxiety, depression, and cardiovascular disease, whereas stress management has become almost identical to the reduction of that activation (Holmes & Rahe, 1967; Selye, 1956). Also, more recent investigators such as Karasek and Levi (see Karasek & Theorell, 1990, p. 91), still use a one-dimensional model of physiological activation (inverted U-shaped curve). This is particularly surprising because their stress model (described later) is multidimensional.

In human performance research, increases in task demands, mostly established by making the task more difficult, are assumed to enhance the level of activation, either directly or via enhanced effort or stress. However, studies on the relation between activation and performance efficiency have revealed ambiguous and contradictory results (e.g., Eysenck, 1982). Reliable results appear to be found only at the extremes. Performance is found to be reduced either because of a too low level of activation owing to sleep loss and fatigue or a too high level owing to anxiety or stress. It is quite possible, however, that performance deterioration and high activation are independently affected by the stressor, because high levels of activation are often accompanied by intensive emotions that at the same time can be distracting and decrease performance.

The relation between the energetic state and performance efficiency is often assumed to be an inverted U-shaped curve (see also Hebb, 1955). A reduction in performance efficiency caused by sleep loss or fatigue is assumed to be caused by a low level of activation, whereas the negative influence of anxiety or stress is explained by an excessively high level of activation. So far, the U-curve hypothesis has received scant empirical support, and a number of methodological problems have been raised against this type of research (e.g., Neiss, 1988). The most important one is the lack of agreement among researchers about how to objectively determine the different levels of activation. It has also been questioned whether the inverted-U is a correlational or a causal hypothesis. Events or manipulations, either in the laboratory or in daily life, that enhance the level of activation may at the same time elicit emotions and distractions that reduce processing capacity and performance efficiency directly. A third problem is that (one-dimensional) activation theories do not discriminate between different types of energy mobilization and hardly specify the effects on emotional and cognitive processes and the consequences this may have for the working behavior of the employee. As a result, activation theories are not able to explain why under some conditions efficient performance is possible even with high levels of activation, whereas debilitating states that degrade performance may also occur at medium or low levels of activation. It appears that emotions, in particular when negative (e.g., anxiety, worry, and depression), reduce performance efficiency, both at low and high levels of activation.

Only a few attempts have been made to develop multidimensional models, and it is still questionable how this might be achieved (Hockey, 1997; Hockey et al., 1986). Instead of a one-dimensional model, one could conceptualize the state of our brain and body as determined by the activity of many energetic mechanisms. Every task, whether physical or mental, has its specific position in this multidimensional space where performance is most efficient. This model of state regulation may be regarded as a multidimensional inverted U-shaped curve model (see also Hancock, 1986). The more similar the actual state is to the optimal state, the more efficient the processing in a particular task is. Deviations from the optimal state may be compensated by mobilizing extra energy through mental effort. Task situations with high attentional demands are assumed to be more sensitive to deviations from the optimal state, because the area in which efficient processing can take place is smaller. The function of our energetic system is to continually bring and maintain our mind and body in a condition that is optimal for the processing required by the activities we want to do. The system of energetic mechanisms takes care of the continuous adaptation to the demands of the environment on the one hand and the recovery from demanding and stressful episodes on the other.

This way of thinking is quite different from most resource theories (Kahneman, 1973; Sanders, 1983; Wickens, 1992), which assume that performance decrements are produced by shortages in resources. These theories explain the effects on performance efficiency in terms of the availability of energy supplies, whereas the present theory stresses the regulatory aspects. In the present framework, it is assumed that preparing and executing a task under normal circumstances (task-induced activation) have different effects on the energetic state than when extra energy is mobilized through mental effort (as discussed earlier). In a similar way, Hockey (1997) postulated in his compensatory control model two separate but connected loops, one for routine regulatory activity and one for effort-based control. From this it follows that there is no direct relation between task difficulty and effort. Increases in task difficulty, or decreases in task performance, only result in the mobilization of extra energy through mental effort, when a performance decrement is not acceptable for operators and when they are motivated to prevent it. Moreover, they are able to do this only when they receive appropriate feedback on performance. This is also the reason that nonspecific task variables, such as feedback or bonus, have larger effects on indicators of effort than task difficulty in itself. Thus, whether in a particular task situation effort is mobilized depends not only on the characteristics of the task, but also on the motivation and personality of the subject.

Such a multidimensional "regularity" view of energetic states is better able to accommodate a number of experimental observation than can one-dimensional "volume" theories. A one-dimensional theory cannot explain:

Why some types of enhanced physiological activation (for example, relaxed jogging) are assumed to be healthy, whereas other types are not.

Why people are able to work very hard without experiencing negative effects.

Why stress reactions also occur at medium or low levels of activation; for example, when people perform a monotonous task under conditions of underload or work in physical and social isolation. People may even experience stress reactions when they work half-time or when they are not working at all.

Why there is no direct relation between the level of activation and performance efficiency. For example, the number of working hours per week is not related to health outcomes. Employees in active jobs with high psychological demands that presumably have a higher level of activation do not have increased health risks and psychological complaints compared with employees in passive jobs. Employees in active jobs even appear to have more energy in the evening to engage in active hobbies and social activities (see also Karasek & Theorell, 1990).

Thus, a high level of activation is not always accompanied by a reduction in performance efficiency, well-being, and health, which implies that there are different patterns of reactivity to work demands that have different consequences for performance efficiency and health risks.

WORK AND HEALTH

Ideas about the relation between work and health have changed dramatically over the past 10 years. In the classic view, employees, in particular their "weak" personalities, were held accountable for stress problems, such as psychosomatic complaints, sleeping problems, and unhealthy lifestyles. Not only researchers but also governments and employers have become convinced that some characteristics of the work environment have negative effects on health and well-being. Life expectancy has been estimated to be

shortened by 7 years by unhealthy psychosocial factors in the work environment (Johnson, Hall, & Theorell, 1989).

Little is known, however, about the underlying mechanisms that mediate
between work demands, stress reactions, and increased health risks. Classical epidemiological studies based on correlations between work characteristics and health indexes provide only limited insight into these mechanisms.
Most stress theories, such as the Michigan model (Kahn, 1981) and the de-
mand/control model (Karasek & Theorell, 1990), minimally specify the
mechanisms mediating work demands and health outcomes.

Given that stress and disease are inter-related via a complex interplay of
psychological and physiological processes, associations between stress and
disease cannot simply be described in terms of cause and effect. It is assumed
that the dysfunctional activation under stress may trigger pathophysiological
processes, which in turn may cause changes in regulatory mechanisms, lead
to tissue damage, provoke the development of malignant processes, or suppress the immune system (see also Kamarck & Jennings, 1991; Krantz &
Manuck, 1984; Steptoe, 1989; Ursin & Olff, 1993).

An integrative framework for the regulation of energetic states should
be able to specify how the reactivity to work demands affects efficiency, well-
being, and long-term health risks. It should also be able to explain how it is
possible that these effects are so strongly modulated by psychological factors
(e.g., social support) and personality characteristics (e.g., temperament,
coping). Recent epidemiological as well as laboratory studies have demonstrated the impact of lack of social support on cardiovascular reactivity and
risk for coronary heart disease (e.g., Commerce, Manuck, & Jennings, 1990;
Orth-Gomér & Undén, 1990).

The construction of such a framework may not yet be possible; too little is
known about the relation between psychological processes and energetic
mechanisms. Instead of attempting to build an integrative, multidimensional theory, it may be more realistic to develop models of state regulation
and a taxonomy of states. On the basis of such a taxonomy, it is possible to
specify the negative effects that particular work demands may have for performance efficiency, well-being, and health. On the basis of this analysis, the
work environment may be (re)designed, and recommendations may be
given to enhance the work efficiency of employees while reducing the probability that stress reactions occur.

BIOBEHAVIORAL STATES

The reactivity pattern a particular task tends to evoke is not only determined
by the type and the amount of work but also by psychosocial factors (e.g.,
social support) and the characteristics of the employee (e.g., temperament

and coping strategies). The current energetic state, which is continuously adapted to the fluctuations in work demands, is also modulated by the work strategies of the employee, the progress made, the feedback received, and so on. Because the critical characteristics of a particular work environment are rather stable, work demands may have enduring effects on mind and body that appear to be specific for a particular type of work. If work demands evoke a typical and enduring pattern of reactions, one could speak of a biobehavioral (BB) state (see also Frankenhaeuser, 1986, 1989; Gaillard & Wientjes, 1994; Leonova, 1994), psychobiological states (Neiss, 1988), or adjustment modes (Hockey, 1997). A BB-state is characterized by a typical pattern of cognitive, emotional, and physiological, and physiological re-actions that are highly interactive and interdependent. The notion of BB-states differs from one-dimensional theories, in that physiological activation is no longer regarded as causal agent: Cognition and emotion and physio-logical reactions are assumed to be evoked (in parallel) by external events (for example, demands, threats, etc.).

Although this is not the place to give an overview of studies that have dealt with BB-states, a few examples are mentioned. Frankenhaeuser (1986, 1989) identified states based on different neuroendocrine response pat-terns. These responses were collected in a series of studies in which the con-trol that subjects had over their performance was manipulated. The "effort without distress" state involves effort and engagement and is often accom-panied by positive emotions. This state is dominated by sympathetic activity and the release of catecholamines (e.g., adrenaline) by the adrenal-medullary system. The "distress" state is characterized by passive coping and feelings of helplessness. This state is dominated by the release of cortisol by the adrenal-cortical system. In the third state, "effort with distress," the work demands require effort, but also evoke negative feelings (e.g., anxiety). The state is associated with sympathetic dominance and increased excretion of both catecholamines and cortisol. Dienstbier (1989) has made a similar dis-tinction between a "tough"-arousal pattern related to efficient performance and a pattern related to defensiveness. Physiological toughness is character-ized as an arousal pattern encompassing strong catecholamine responses to challenging and stressful situations, but also a fast recovery and low resting values. A "tough"-arousal pattern is associated with efficient performance, positive temperament, emotional stability, and immune system enhance-ment. This pattern is contrasted with a defensive pattern that is dominated by high cortisol responses, protracted recovery, and increased baselines in rest conditions. Earlier research of Ursin (Ursin & Olff, 1993) also found differences in coping style and mood. A tough-arousal pattern is associated with problem-oriented coping and positive emotions. A defensive pattern is associated with emotion-oriented coping and negative emotions (anxiety and depressions).

The model of Karasek and Theorell (1990) encompasses two factors (psychological demands and autonomy) that result in two reactivity dimensions (active-passive and contentment versus strain). According to this model, stress reactions occur when the work environment encompasses high psychological demands in combination with a low level of control and little social support. The symptoms vary from reduced work satisfaction and well-being to life-threatening cardiovascular diseases. In contrast, active jobs that are highly demanding but have control possibilities do not produce stress reactions, psychosomatic complaints, or increase health risks. Thus, a work environment that is exacting but stimulating and challenging may evoke efficient performance, active coping, enhanced activation, and positive emotions. In contrast, when employees have to work below their level of competence, this may evoke noncreative behavior and passive coping, which results in reduced activation accompanied by apathy and depression. Similarly, a hectic environment in which employees have limited control over their behavior and experience little social support evokes strain, anxiety, psychosomatic complaints, and dysfunctional physiological reactions.

Although these results originated from quite different disciplines and research areas, they have several aspects in common:

- The theoretical frameworks encompass reactivity patterns that consist of typical reactions at different levels: Behavioral, physiological, and emotional.

- These reactivity patterns are assumed to be more or less enduring.

- High levels of activation are not necessarily accompanied by negative emotions, psychosomatic complaints, or increased health risks.

- Two types of energy mobilization are distinguished: one results in an activated state that is functional and increases performance efficiency, and the second leads to a disorganized state that is associated with strain, negative emotions, and increased health risks.

- The patterns observed can be described in terms of two dimensions proposed by Thayer (1989) on the basis of a factor-analytic study of self-reports on mood: *activation,* ranging from a passive to an active state, and *affectivity,* ranging from an emotionally negative to a positive state. As is illustrated in Fig. 3.12.1, combining the two dimensions results in four states which could be labeled "enthusiastic" (active, positive), "relaxed" (passive, positive), "strain" (active, negative) and "depressive" (passive, negative).

The states "mental load" and "stress" described previously can be put in this framework. A high mental workload is regarded as an active/positive state, and stress as an active/negative state. These two states may be contrasted with the situation of a normal workload, in which the person is content and the task does not require extra energy mobilization. In addition to an active/negative stress state, a low active stress state may be distinguished

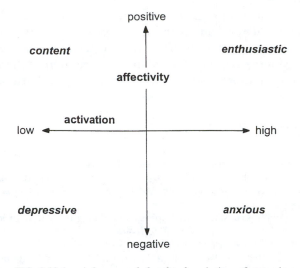

FIG. 3.12.1.   A framework for the description of states in
the work environment based on two dimensions: activation
and affectivity (see also Thayer, 1989).

that encompasses dissatisfaction and irritation, leading on the long term to
burnout and depression. In addition to the dimensions of activation and affectivity, one could think of other dimensions, such as the level of control,
aftereffects, and coping strategy (discussed earlier). In this framework, different types of fatigue can be distinguished depending on the particular states
that have produced the feelings of fatigue; thus, each of the four states defined
in terms of the two dimensions (see Fig. 3.12.1) has its own type of fatigue.
For example, after a heavy working day involving a high workload, feelings
of accomplishment may prevail when all went well; the resulting fatigue may
be experienced as positive, because it is accompanied with satisfaction. After
a hectic day full of frustration and agony, feelings of anger and restlessness
may be dominant, which may protract recovery and cause sleeping problems.

   The goal of this description is to offer a framework for explaining why
some types of activation improve performance while others reduce well-
being and endanger health. By using these dimensions, other finer-grained
states may be specified. A taxonomy of states that prevail in the work environment may be quite helpful to reveal the critical characteristics of persons
and work environment that promote efficient and healthy behavior. States
should be categorized not only on their *intensity* (high/low activation), but
also with regard to the *pattern* of reactions. Research efforts should concentrate on the emotional and physiological mechanisms that mediate between
cognitive processes and bodily reactions and underlie overreactivity, recov-

ery, and accumulation of stress effects over time and across different stressors. In this way, adaptive and maladaptive patterns may be distinguished. Adaptive patterns enhance performance efficiency and promote well-being. Maladaptive patterns are associated with the disruption of the homeostase and with inefficient functioning, which may have negative health effects.

REFERENCES

Cannon, W. B. (1932). *The wisdom of the body.* New York: Norton.

Commerce, T. W., Manuck, S. B., & Jennings, J. R. (1990). Social support reduces cardiovascular reactivity to psychological challenge: A laboratory model. *Psychosomatic Medicine, 52,* 42–58.

Dienstbier, R. A. (1989). Arousal and psychophysiological toughness: Implications for mental and physical health. *Psychological Review, 96,* 84–100.

Eysenck, M. W. (1982). *Attention and arousal.* Berlin: Springer-Verlag.

Frankenhaeuser, M. (1986). A psychobiological framework for research on human stress and coping. In M. H. Appley & R. Trumball (Eds.), *Dynamics of stress* (pp. 101–116). New York: Plenum.

Frankenhaeuser, M. (1989). A biopsychosocial approach to work life issues. *International Journal of Health Services, 19,* 747–758.

Gaillard, A. W. K. (1993). Comparing the concept of mental load and stress. *Ergonomics, 9,* 991–1005.

Gaillard, A. W. K., & Wientjes, C. J. E. (1994). Mental load and workstress as two types of energy mobilization. *Work and Stress, 8,* 141–152.

Gopher, D., & Donchin, E. (1986). Workload—An examination of the concept. In K. R. Boff, L. Kaufman, & J. P. Thomas (Eds.), *Handbook of perception and human performance* (pp. 41:1–41:49). New York: Wiley.

Hancock, P. A. (1986). Stress and adaptability. In G. R. J. Hockey, A. W. K. Gaillard, & M. G. H. Coles (Eds.), *Energetics and human information processing* (pp. 243–251). Dordrecht, Netherlands: M. Nijhoff.

Hancock, P. A. (1987). *Human factors psychology.* Amsterdam: North-Holland.

Hebb, D. O. (1955). Drives and the C.N. S. (conceptual nervous system). *Psychological Review, 62,* 243–254.

Hockey, G. R. J. (1997). Compensatory control in the regulation of human performance under stress and high workload: A cognitive-energetical framework. *Biological Psychology, 45,* 73–93.

Hockey, G. R. J., Coles, M. G. H., & Gaillard, A. W. K. (1986). Energetical issues in research on human information processing. In G. R. J. Hockey, A. W. K. Gaillard, & M. G. H. Coles (Eds.), *Energetics and human information processing* (pp,. 3–21). Dordrecht, Netherlands: M. Nijhoff.

Holmes, T. H., & Rahe, R. H. (1967). The social readjustment rating scale. *Journal of Psychosomatic Research, 11,* 213–218.

Johnson, J. V., Hall, E. M., & Theorell, T. (1989). Combined effects of job strain and social isolation on cardiovascular disease morbidity and mortality in a random sample of the Swedish working population. *Scandinavian Journal of Work and Environmental Health, 15,* 271–279.

Kahn, R. L. (1981). *Work and health.* New York: Wiley.

Kahneman, D. (1973). *Attention and effort.* Englewood Cliffs, NJ: Prentice Hall.

Kamarck, T., & Jennings, J. R. (1991). Behavioral factors in sudden cardiac death. *Psychological Bulletin, 109,* 42–75.

Karasek, R. A., & Theorell, T. (1990). *Healthy work.* New York: Basic Books.

Krantz, D. S., & Manuck, S. B. (1984). Acute psychophysiologic reactivity and risk of cardiovascular disease: A review and methodologic critique. *Psychological Bulletin, 96,* 435–464.

Lazarus, R. S., & Folkman, S. (1984). *Stress, appraisal, and coping.* New York: Springer.

Leonova, A. B. (1994). Industrial and organizational psychology in Russia: The concept of human functional states and applied stress research. In C. L. Cooper & I. T. Robertson (Eds.), *International review of industrial and organizational psychology* (pp, 173–212). Chichester, England: Wiley.

Neiss, R. (1988). Reconceptualizing arousal: Psychobiological states in motor performance. *Psychological Bulletin, 103,* 345–366.

Orth-Gomér, K., & Undén, A. L. (1990). Type A behaviour, social support, and coronary risk: Interaction and significance for mortality in cardiac patients. *Psychosomatic Medicine, 52,* 59–72.

Sanders, A. F. (1983). Toward a model of stress and human performance. *Acta Psychologica, 53,* 61–97.

Selye, H. (1956). *The stress of life.* New York: McGraw-Hill.

Steptoe, A. (1989). Psychophysiological interaction in behavioral medicine. In G. Turpin (Ed.), *Handbook of clinical psychophysiology* (pp. 213–239). New York: Wiley.

Thayer, R. E. (1989). *The biophysiology of mood and arousal.* Oxford: Oxford University Press.

Ursin, H., & Olff, M. (1993). The stress response. In C. Stanford, P. Salmon, & J. Gray (Eds.), *Stress: An integrated response* (pp. 3–22). New York: Academic Press.

Wickens, C. D. (1992). *Engineering psychology and human performance.* New York: Harper-Collins.

Author Index

Subject Index